The Elements of Life

The Elements

ALAN SHERMAN
SHARON J. SHERMAN

Middlesex County College
Edison, New Jersey

# of Life

An Approach to Chemistry for the Health Sciences

Prentice-Hall, Inc., Englewood Cliffs, N.J. 07632

Library of Congress Cataloging in Publication Data

Sherman, Alan.
    The elements of life.

    Includes index.
    1. Chemistry.   2. Biological chemistry.
I. Sherman, Sharon, joint author.   II. Title.
QD31.2.S484 1980   540′.2′461   79-24453
ISBN   0-13-266130-6

The Elements of Life
An Approach to Chemistry for the Health Sciences
Alan Sherman and Sharon J. Sherman

Printed in the United States of America

10 9 8 7 6 5 4 3 2 1

Some of the discussion in the text, including figures, is taken from
pages 120–126, 136–146, 224–229, 242–246, 265–285, 299–325,
Alan Sherman, Sharon Sherman, and Leonard Russikoff, *Basic
Concepts of Chemistry*. Copyright © 1976 by Houghton Mifflin
Company. Used by permission.

Interior and cover design by Mark A. Binn
Line illustrations by Danmark & Michaels, Inc.
Manufacturing buyer: Edmund Leone
Cover photograph © Howard Sochurek

Prentice-hall of Australia Pty. Limited, *Sydney*
Prentice-Hall of Canada, Ltd., *Toronto*
Prentice-Hall of India Private Limited, *New Delhi*
Prentice-Hall of Japan, Inc., *Tokyo*
Prentice-Hall of Southeast Asia Pte. Ltd., *Singapore*
Whitehall Books Limited, *Wellington, New Zealand*

To our friend Dorothy Birdsey,
whose warmth and kindness made
her a very special person

Contents

CHAPTER 6

The Three Faces of Matter

Solid, Liquid, and Gas 89

Some Things You Should Know After Reading This Chapter 90

CHAPTER 8

Nuclear Radiation

Cure All or Kill All   157

Some Things You Should Know After Reading This Chapter   158

CHAPTER 11

The Carbohydrate Story

How Sweet It Is! 259

Some Things You Should Know After Reading This Chapter 260

CHAPTER 12

A Look at Lipids

A Heavy Experience   287

Some Things You Should Know After Reading This Chapter   288

CHAPTER 15

Nucleic Acids

The Basis of Life   362

Some Things You Should Know
After Reading This Chapter   363

CHAPTER 16

Blood

The Flow of Life   389

Some Things You Should Know
After Reading This Chapter   390

CHAPTER 17

The Urinary System and Urine

The End Product 413

Some Things You Should Know After Reading This Chapter 414

APPENDIX B

The Metric System and Measurement

The Backbone of the Sciences 472

Some Things You Should Know After Reading This Supplement 473

Important Tables

Preface

The Elements of Life: An Approach to Chemistry for the Health Sciences has been written as a one-semester or two-quarter introductory text for health science students. Students who will benefit most from this text have generally taken few courses in science and mathematics, but they usually have some knowledge of algebra. Our simple basic approach allows students who may be wary of taking a chemistry course to gain confidence in their ability to achieve the needed competence in the subject.

The health science chemistry course is a requirement for all health science students at most colleges and universities. It is important that these students have a good understanding of chemistry and its relationship to life processes. We have arranged the style and format of our text with a view to helping students achieve this understanding.

Some of the outstanding features of the text are:

1. The use of *simple language*. The reader does not need to sit by a dictionary to comprehend the written material.
2. Clear presentation, usually in a *conversational style*.

3. Visual illustration through *numerous cartoons, figures, photographs, and tables*. The cartoons are used to explain difficult concepts and to add humor. Some photographs display the latest types of equipment used by health practitioners. Others portray the effects of various diseases in human beings. The figures reinforce the written material, and the tables summarize the information given in the text. The tables were chosen with great care, and will provide an excellent reference source for health science students long after they complete their chemistry course.

4. *Learning goals* listed at the beginning of each chapter. The student will learn from these what he or she is expected to know after completing the chapter.

5. A chapter *summary* at the end of each chapter. These summaries reinforce the learning goals.

6. Numerous *worked-out examples* in each chapter. Some of the worked-out examples show step by step how to solve a quantitative problem; the strategy as well as all the mathematics is shown in solution of the problem. Other worked-out examples supply sample test questions, with the complete solution following each question.

7. An abundant supply of *self-test exercises* at the end of each chapter. The exercises relate back to the learning goals.

8. An extremely useful set of *supplements*, including a glossary, an appendix on basic mathematics, an appendix on the metric system and measurement, and important tables.

We have also prepared a *laboratory manual* and an *instructor's guide* to accompany the text.

Unlike texts that tend to be too long and too technical for the intended audience, this one is directed to the needs and abilities of the health science student. The topics included reflect those that were suggested by faculty who teach health science students—for example, nursing faculty, dental hygiene faculty, and chemistry faculty—in their replies to a questionaire. The questionaire listed hundreds of objectives for the health science chemistry course that had been obtained from a review of current textbooks in the area. Faculty members were asked to check the ones they felt were most important for health science students and to add additional objectives if they wished.

We believe that our text presents the selected student with a basic chemistry course for the health sciences. We *do not* "short change" the student by omitting concepts in order to simplify the presentation; nor do we overwhelm the student with an overabundance of material. We attempt to give the student a fundamental knowledge of general, organic, and biological chemistry that will be useful to him or her as a health scientist (for example, nurse, dental hygienist, x-ray technologist).

We are always open to new ideas. The methods we use in the text work well with our students, but we welcome any and all constructive criticism from the experience of others with these materials, so that we can make future editions of this text even more effective and useful.

We would like to thank some of our friends and relatives whose efforts and encouragement helped us to complete this project. Our deep appreciation goes to Professors Dominic Macchia, John Murray, Barbara Drescher, Len Russikoff, Frank Spano, and Natalie Rener, all of Middlesex County College. Our thanks also to Pat Benson, Max Sapirstein, Gertrude Sapirstein, Pamela Kance, Richard Sherman, Fay Sherman, Robert Schulman, Robert Scott, and Michael Gregory.

We would also like to thank several people who contributed significantly to the development of this text. A very special thank you to Professor Mary S. Vennos of Essex Community College, Baltimore County, Maryland, and to Professor Henry Rick of Miami-Dade Community College, Miami, Florida, whose thorough review of the manuscript resulted in making this a better text. We thank you both again.

Our thanks also to all the people who answered our questionaire and to our students who class-tested the manuscript.

Finally, our thanks to the staff of Prentice Hall, who are true professionals in their work. Our sincere thanks to Fred Henry, our editor, whose constant interest in our project was a real help to us. We would also like to thank Diane Heyn for her help in preparing the manuscript. Last, a special thanks to Zita de Schauensee, who handled beautifully all aspects of production.

ALAN SHERMAN
SHARON J. SHERMAN
North Brunswick, New Jersey

Chemistry and the Health Sciences

A Synthesis of Knowledge

You should be able to:

1. State the relationship between chemistry and the health sciences.
2. State some of the contributions of the Egyptians to early science.
3. State some of the contributions of the Greeks to early science.
4. State some of the contributions of the alchemists to modern science and medicine.

Introduction

What am I doing here? Sure I want a career in the health sciences, but what does chemistry have to do with health science, and how much chemistry will I have to know once I start working? Besides, who wants to put up with a chemist for an instructor? They always have their heads in a test tube (Fig. 1-1).

These are some of the questions that many students ask upon entering a course in chemistry for the health sciences. Certainly these are legitimate questions and deserve to be answered. After all, you're probably in this course because it's a requirement for your curriculum and you should at least know why the faculty at your college or university feel that this course should be part of your program. Perhaps the best way to answer some of your questions and show the connection between chemistry and the health sciences is to look at some history.

The Greatest Story Ever Told

From our early beginnings, humankind has always felt a need to know. We have been constantly exploring the environment around us and proposing answers for the many natural but seemingly mysterious phenomena.

FIGURE 1-1
Professor Emeritus at the beach. (After a cartoon by Marchant in the August 1972 I^{2}R calendar. Used by permission of Instruments for Research and Industry.)

It was with this attitude that the early Egyptian and Greek civilizations came on the scene. The Egyptians, as long ago as perhaps 1000 B.C., were practicing a sort of mystical chemistry called *khemia*. It was an applied chemistry of sorts and involved the formation of embalming fluids and other potions. It also involved the mining and refining of metals and the extraction of pigments and juices from various plants. Some of these plant extracts had powerful healing and pain-killing properties. In Egypt it was the priests who practiced khemia and much of their time was spent in trying to understand the processes of life and death.

The Greeks, on the other hand, were more interested in the fundamental composition of things. They wanted to know if there was one thing in the universe from which all other things can be derived. By about 600 B.C. the Greek philosopher Thales thought he had the answer—water! By 546 B.C. the Greek philosopher Anaximenes thought he had the answer—air! By 450 B.C. the Greek philosopher Empedocles thought he had the answer*s*—earth, air, fire, and water! Today, we think we have the answers (Fig. 1-2), but we'll hold off on that for a little while.

The Chinese and Arabs were also practicing their own form of khemia. In fact, the Egyptian word "khemia" became the Arabic word *al-kimiya*, and today we use the word *alchemy* to define everything that happened in chemistry between A.D. 300 and A.D. 1600. The alchemists employed the techniques and the knowledge of both the Greeks and Egyptians.

One of the goals of the alchemists was to find the "elixir of life." This elixir would cure all diseases and prolong life indefinitely. Unfortunately, this

FIGURE 1-2

The answer. (After a cartoon by Marchant in the April 1973 I^2R calendar. Used by permission of Instruments for Research and Industry.)

FIGURE 1-3
The consequences.

goal was never realized. Just think of some of the consequences if it was (Fig. 1-3). But the work of the alchemists, especially the later ones (A.D. 500 to A.D. 1000), did prove important in setting the stage for modern chemistry and medicine. Many substances that chemists use today were discovered by the alchemists. Indeed, in the year 1000 the alchemist Avicenna published his "Book of the Remedy," and in 1330 the alchemist Bonus published his "Introduction to the Arts of Alchemy." Both of these texts proved valuable to later scientists and may have even helped Van Helmont, in 1620, lay the foundations of chemical physiology.

SUMMARY

As you may have gathered, it was from these ancient arts that the fields of chemistry and medicine are derived, and it is because of this heritage that they are so closely related. The discovery of the antibiotic sulfanilamide by Gelmo in 1908; of insulin by Banting, Best, and Macleod in 1922; and of penicillin by Fleming in 1928 are further proof of this. More recently, the discovery of the structure of DNA by Watson and Crick in 1953 is additional proof. The implications of the breaking of the genetic code are just now making their debut. Now it will be more important than ever for the chemist and his medical colleagues to work even more closely if the knowledge of gene manipulation is to be used wisely. The decades of the 1980's and 1990's will tell the story. The future history of humankind rests on what happens (Fig. 1-4).

We hope that this short discussion has answered some of your questions and has placed in proper perspective the course you are about to take.

FIGURE 1-4
The future of humankind.

EXERCISES (good for the body and the mind)

1. State the relationship between chemistry and the health sciences.

2. Make a table and list the contributions to chemistry and medicine by the Egyptians, Greeks, and alchemists.

3. Discuss the implications of gene manipulation as it concerns the future generations of man. Also discuss some of the ethical and social implications. (*Note:* You may want to look at some additional reference sources for your answer; for example, *Future Shock*, by Alvin Toffler, or various issues of *Scientific American*.)

CHAPTER 2

Health

A Fact of Matter

Some Things You Should Know After Reading This Chapter

You should be able to:

1. Describe the use of the scientific method.
2. Classify matter according to Figure 2-4.
3. Define the terms heterogeneous, homogeneous, mixture, solution, element, and compound.
4. Define the terms atom and molecule.
5. Write the chemical symbol of an element using either the periodic table or an alphabetical listing of the elements, and write the name of the element when given the symbol (for the more common elements).
6. Say what is meant by a relative system of atomic weights.
7. Define what is meant by a mole, and give an example.
8. Determine the formula weight of a compound, using the periodic table to look up atomic weights.
9. Determine the number of moles of a compound, given the number of grams of the compound and its formula weight.
10. Determine the number of moles of an element, given the number of grams of the element and its atomic weight.
11. Determine the number of grams of a compound when given the number of moles and the formula weight of the compound.
12. Determine the number of grams of an element when given the number of moles and the atomic weight of the element.

FIGURE 2-1
The Philosopher Scientist.

Introduction

In Chapter 1 we reviewed briefly the history of chemistry and some of its applications to the health sciences. Although the ancient Egyptian and Greek civilizations laid the foundations of modern science, the methods of scientists from the mid-1600's on differed greatly from the methods used by the "ancients." For example, the Greek philosophers never bothered to subject their ideas to experiment to see if they were true. In other words, they never bothered to test the many substances that they knew about to see if they were composed of only four elements; earth, air, fire, and water (Fig. 2-1). However, beginning with Robert Boyle (1627–1691) scientists did subject their ideas to experimental proof. A sort of method was developed, a *scientific method*.

The Scientific Method: Doctors and Nurses Use It, Too

What is this scientific method; what does it consist of? In simple terms, it consists of four phases:

1. Observation.

2. Hypothesizing.

3. Prediction.

4. Experimentation.

(Step 4 can lead back again to step 2, depending on the results of the experiment; Fig. 2-2.)

FIGURE 2-2
Professor Emeritus and the scientific method.

We have all used the scientific method at some time in our lives. In fact, just the other day, Professor Emeritus (remember him?) was examining an object he found in his backyard. It looked like a vase (an "observation"). He felt the object and said that it was brittle (a "hypothesis"). Based on this "hypothesis" he "predicted" that if he hit the vase with a hammer it would break. Figure 2-2d shows the results of Emeritus's "experiment."

So much for the good professor, but it is the use of the "scientific method" that makes possible the advances of modern science. You, too, will be using this method as a future health practitioner. Also, much of the chemistry you are about to learn has been developed by scientists who have used the scientific method.

Matter and Energy: That's All There Is!

Classification has always been associated with the scientist. It goes hand in hand with using the scientific method. More important, it allows us to take a great amount of information and arrange it in a convenient form (Fig. 2-3). For example, let's say that we want to classify everything in our universe. That's a pretty tall order! But if we want to answer the question of the Greeks— "What is the one substance of which all other substances are composed?"— devising a classification system might help. Let's try to keep it simple, and let's make use of our general knowledge of our environment.

FIGURE 2-3
Classification—it sure helps! (After a cartoon by Marchant in the January 1973 I^{2}R calendar. Used by permission of Instruments for Research and Industry.)

Just about all that exists in our universe falls into two broad categories: matter and energy. *Matter* is anything that occupies space and has mass. Some examples are clothes, furniture, water, and you. *Energy* is often defined as the ability to do work. It comes in various forms: heat, electrical, and chemical energy, to mention a few. We'll talk more about energy later in the course, but for now, let's turn our attention to the matter.

What's the Matter?

Our classification scheme at this point is quite simple, and based upon our common experience. But just to say that the universe is composed of matter and energy is not very helpful. A classification system should be simple, but not to the point where the categories are too general. There are many different types of matter, so we may want to devise a few more categories for our classification system. One method would be to say that matter exists in three states: solid, liquid, and gas. However, a more useful classification system is shown in Figure 2-4.

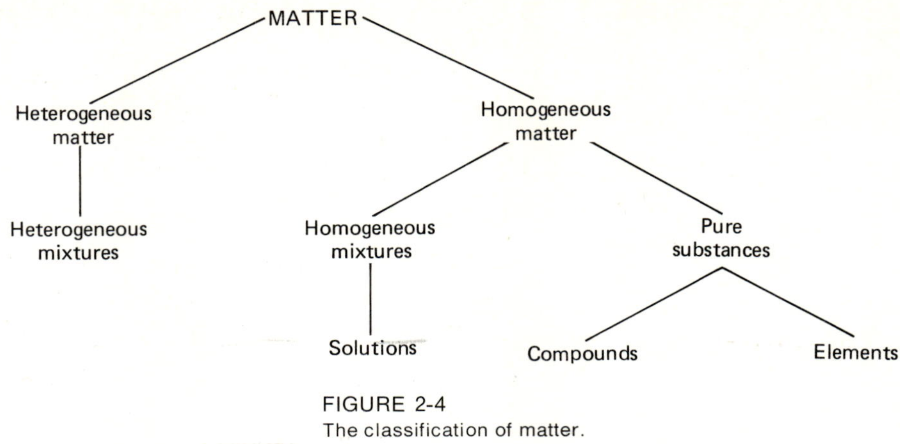

FIGURE 2-4
The classification of matter.

The Types of Matter Explained

Let's briefly discuss our classification system.

1. HETEROGENEOUS MATTER   *Heterogeneous* means having different parts with different properties.

    (a) *Heterogeneous mixtures*   This is the common type of mixture, such as salt and sand. It is a mixture because it consists of two or more substances "tossed" together. It is a heterogeneous mixture because each part of the mixture retains its own characteristic properties. In other words, the salt and sand each retains its own properties, and so the mixture has different parts with different properties.

2. HOMOGENEOUS MATTER   *Homogeneous* means having similar consistency throughout.

    (a) *Solutions*   These are homogeneous mixtures. For example, salt and water mix to form a solution. The saltwater solution is said to be homogeneous because there is uniformity throughout. All parts of the solution have the same properties.

    (b) *Elements*   These are the basic building blocks of all matter. This is in keeping with the definition formulated by the Greeks. Remember, they thought that there were only four elements; earth, air, fire and water. However, today we know of 105 elements (and earth, air, fire, and water are not even on the list, since they aren't really elements). What are some of the elements? Examples are gold, silver, oxygen, nitrogen, and calcium. Each element has its own specific properties and has been given a chemical symbol. The elements have been arranged in what is known as the *periodic table of the elements* (see the inside front cover of the book). There is also an alphabetical listing of the elements on the inside back cover of the book.

Elements may be classified into three major groups: metals, metalloids, and nonmetals. Examples of *metallic* elements are sodium (which has the symbol Na), calcium (Ca), iron (Fe), cobalt (Co), and silver (Ag). The reason that these elements are classified as metals is that they have certain properties, among which are luster (they are shiny), they conduct electricity well, and they conduct heat well.

Some examples of *nonmetals* are chlorine (which has the symbol Cl), oxygen (O), carbon (C), and iodine (I). Nonmetals don't shine, don't conduct electricity well, nor do they conduct heat well.

The *metalloids* fall halfway between the metals and nonmetals. Metalloids have some properties which are like those of metals and other properties which are like those of nonmetals. Some examples are arsenic (As), germanium (Ge), and silicon (Si).

(c) *Compounds* Compounds are substances composed of two or more elements, chemically combined in a definite proportion by mass or weight. Unlike mixtures, which can be physically tossed together with varying composition, compounds have a definite composition. Water, for example, is made from the elements hydrogen and oxygen in the ratio 11.1% hydrogen to 88.9% oxygen by weight. Regardless of the source of the water, it always has this hydrogen/oxygen ratio.

Compounds can be broken down into their elements by various chemical means. For example, you can use an electric current to decompose water into hydrogen and oxygen. Approximately 3 million compounds have been reported to date, and there are possibilities for millions more to be discovered. Remember that compounds are combinations of elements, and there are many ways to combine over 100 elements (Fig. 2-5).

FIGURE 2-5
Compounds are combinations of elements.

Elements may be the basic building blocks of matter, but what—if anything—makes up the elements? In other words, what would be the result of taking an element, a piece of gold for example, and cutting it in half, and in half again, ad infinitum? We would soon reach the point of having such a small piece of gold that it would be beyond our ability to cut it. It is at times like these when scientists must use their knowledge about how elements react to continue the experiment in their minds. Scientists have done just that and have agreed that if they could continue to cut a piece of gold in half, they would eventually reach a particle called the *atom* (in this case, an "atom" of gold). The atom is the smallest part of an element that retains the chemical properties of the element. One gold atom is so small that billions of them are required to make a tiny speck of gold that can be seen with a microscope. The atom, therefore, is the ultimate particle that constitutes the elements. Gold is composed of gold atoms, iron of iron atoms, and oxygen of oxygen atoms. Each element is composed of its own type of atoms.

At this point there are usually some skeptics in the room. They say, "How do you know that atoms exist, if no one has ever seen an atom?" The answer is by indirect evidence. We postulate the existence of atoms based upon how elements react to form compounds. Much of what we know about chemical reactions is easily explained by using the concept of the atom. (However, it's not quite true that no one has ever seen an atom. In a manner of speaking, atoms have been seen. In 1970, Albert Crewe and his staff at the University of Chicago's Enrico Fermi Institute took the first black-and-white pictures of single atoms using a special type of electron microscope. In late 1978, Crewe and his staff reported taking the first color time-lapse moving pictures of individual atoms with the aid of a high-resolution scanning electron microscope. In this case it was uranium atoms that were observed.)

FIGURE 2-6

Molecules are composed of *combinations* of atoms. (This figure is not intended to describe a chemical reaction. It only shows that molecules are composed of combinations of atoms.)

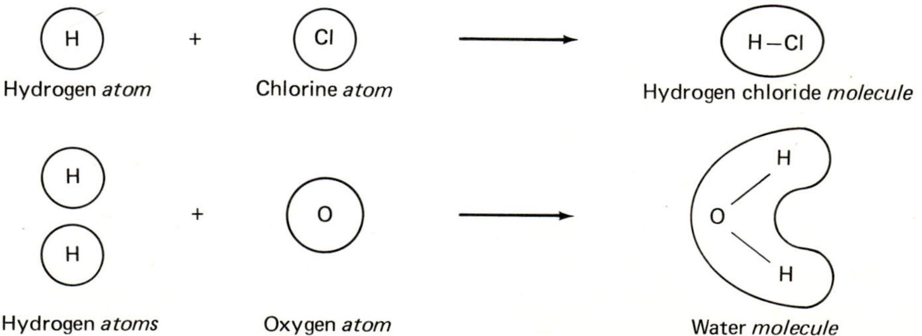

We have just seen what happens when we continually subdivide a sample of an element; we eventually reach an atom. What happens when we continually subdivide a compound? Consider the compound sugar. As we continually divide a sugar grain, we come to a point where further breakdown of the sugar will result in a loss of its physical and chemical properties. This ultimate particle of a compound is called a *molecule*. Molecules, too, are extremely small, but with the aid of an electron microscope, molecules of some of the larger and more complex molecules have been observed (Fig. 2-6).

Symbols and Formulas of Elements and Compounds

Scientists have developed a chemical shorthand to abbreviate the names of the elements. You don't have to write out the word "oxygen"; you can just use the symbol O. The names of most elements are derived from Latin or English, and their chemical symbols may be the first letter or first two letters of their names. The first letter in the symbol is always a capital. The symbol for hydrogen is H, the symbol for neon is Ne, and the symbol for cobalt is Co. (Note this: CO does not represent cobalt but the elements carbon and oxygen, which is a compound, carbon monoxide. We'll discuss compounds in a moment.) Some symbols come from the Latin names of the elements; iron is Fe (from the Latin *ferrum*), lead is Pb (from the Latin *plumbum*).

Since a *chemical compound* is composed of two or more elements, we combine symbols of the particular elements to represent the *chemical formula* of the compound. For example, hydrogen chloride contains one atom of hydrogen and one atom of chlorine. The chemical formula for hydrogen chloride is HCl. Water is another example: the water molecule contains two atoms of hydrogen and one atom of oxygen, so we write H_2O. The number 2 in the formula is called a *subscript*. It indicates the number of atoms of a particular element present in the molecule. A molecule of ethyl alcohol contains two atoms of carbon, one of oxygen, and six of hydrogen: C_2H_6O.

Atomic Weight

If it were possible to assemble one atom of each of the 105 known elements and compare their weights, what would we find? We would find that the hydrogen atom was lightest, that the helium atom weighed four times as much as the hydrogen atom, the oxygen 16 times as much, and the gold atom 197 times as much. Based on this information we could establish a relative atomic weight scale. We could let hydrogen be one weight unit, and let everything else be some multiple of this unit. The atomic weights of all 105 elements are shown below the symbol for each element in the periodic table (inside the front cover). A method similar to the one just discussed was used by scientists to determine the atomic weights of the elements. (Actually, the atomic weight scale is based on an isotope of the element carbon. This isotope, called carbon-12, is assigned a weight of 12 units (sometimes called 12 atomic mass units, or 12 amu, for short.) If you don't know what an isotope is, see Chapter 3, p. 31.)

Gram Atomic Weight and the Mole

It's all well and good to talk about atoms and relative atomic weights, but scientists work with large amounts of materials, not a few atoms or molecules. What we need is a way to relate the numbers of atoms of an element (or molecules of a compound) to a weight, let's say in grams. Fortunately, there is a way to do this; it's by defining a quantity called the *mole*.

What is a mole? It is simply a number of items, like a dozen (12 items) or a gross (144 items). How many items are there in a mole? A huge number,

$$602{,}300{,}000{,}000{,}000{,}000{,}000{,}000$$

items, which can be more conveniently written as 6.023×10^{23}. You can see why we normally wouldn't use the mole when we talk about purchasing eggs, apples, or oranges (Fig. 2-7). But it is a very convenient number for the chemist.

FIGURE 2-7
A mole of apples.

Why? Because by definition, *the atomic weight of an element in grams is 1 mole of that element.* In other words, 1 gram of hydrogen is 1 mole of H atoms (remember that hydrogen has an atomic weight of 1). Sixteen grams of oxygen is 1 mole of 0 atoms (remember that 0 has an atomic weight of 16), and 238 g of uranium is 1 mole of U atoms (remember that U has an atomic weight of 238).

The mole concept is very important in chemistry. It gives the chemist a convenient way to describe a large number of atoms or molecules and to relate this number of atoms or molecules to a weight (usually in grams). We will now learn how to convert moles of atoms of an element to its weight in grams, and vice versa. This is like converting inches to feet and feet to inches. Let's try some examples.

EXAMPLE 2-1   Do the following conversions:

    (a) 48 inches = ? feet

    (b) 3.0 feet = ? inches

SOLUTION   In this text all math problems will be solved by the *factor-unit method.* This method involves setting up the problem in the following manner.

$$\text{What I want to know} = (\text{quantity given})(\text{factor unit})$$

The idea is that when you multiply the "quantity given" by the "factor unit," the proper units cancel to give you the desired quantity. Let's see how this works.

    (a) 48 inches = ? feet. In this problem the factor unit is 12 inches = 1 foot, which can be written in two ways:

$$\frac{12 \text{ inches}}{1 \text{ foot}} \quad \text{or} \quad \frac{1 \text{ foot}}{12 \text{ inches}}$$

Our job is to choose the proper factor unit so that our answer comes out in feet.

$$\text{What I want to know} = (\text{quantity given})(\text{factor unit})$$

$$? \text{ feet} = (48 \text{ inches})\left(\frac{1 \text{ foot}}{12 \text{ inches}}\right)$$

$$= 4.0 \text{ feet}$$

Notice how the term "inches" cancels to give "feet." If we had chosen the other factor unit, the terms wouldn't have canceled. Let's do it so you can see for yourself.

$$? \text{ feet} = (48 \text{ inches})\left(\frac{12 \text{ inches}}{1 \text{ foot}}\right)$$

$$= 576 \frac{\text{inches}^2}{\text{foot}}$$

The units don't cancel and we don't get the answer feet!

(b) 3.0 feet = ? inches

$$\text{What I want to know} = (\text{quantity given})(\text{factor unit})$$

$$? \text{ inches} = (3.0 \text{ feet})\left(\frac{12 \text{ inches}}{1 \text{ foot}}\right)$$

$$= 36 \text{ inches}$$

Notice that in this problem we use the factor unit $\dfrac{12 \text{ inches}}{1 \text{ foot}}$, in order for the proper terms to cancel.

Now let's apply our knowledge of moles, atomic weight, and the factor-unit method to do some chemical arithmetic.

EXAMPLE 2-2   Determine the moles of atoms in each weight of the following elements.

    (a) 36 g of C   (b) 64 g of O
    (c) 6.4 g of O   (d) 23.8 g of U

SOLUTION   (a) The atomic weight of C is 12. This gives us the factor unit $\dfrac{12 \text{ g}}{1 \text{ mole}}$ or $\dfrac{1 \text{ mole}}{12 \text{ g}}$.

$$\text{What I want to know} = (\text{quantity given})(\text{factor unit})$$

$$? \text{ moles} = (36 \text{ g})\left(\frac{1 \text{ mole}}{12 \text{ g}}\right)$$

$$= 3.0 \text{ moles}$$

(b) The atomic weight of O $= 16$. This gives us the factor unit $\dfrac{16 \text{ g}}{1 \text{ mole}}$ or $\dfrac{1 \text{ mole}}{16 \text{ g}}$.

$$\text{What I want to know} = (\text{quantity given})(\text{factor unit})$$

$$? \text{ moles} = (64 \text{ g})\left(\frac{1 \text{ mole}}{16 \text{ g}}\right)$$

$$= 4.0 \text{ moles}$$

(c) The atomic weight of O is 16. The same factor unit applies here as in part (b).

$$\text{What I want to know} = (\text{quantity given})(\text{factor unit})$$

$$? \text{ moles} = (6.4 \text{ g})\left(\frac{1 \text{ mole}}{16 \text{ g}}\right)$$

$$= 0.40 \text{ mole}$$

(d) The atomic weight of U is 238. This gives us the factor unit $\dfrac{238 \text{ g}}{1 \text{ mole}}$ or $\dfrac{1 \text{ mole}}{238 \text{ g}}$.

$$\text{What I want to know} = (\text{quantity given})(\text{factor unit})$$

$$? \text{ moles} = (23.8 \text{ g})\left(\frac{1 \text{ mole}}{238 \text{ g}}\right)$$

$$= 0.100 \text{ mole}$$

EXAMPLE 2-3   Determine the grams of each element in

    (a) 3.0 moles of He   (b) 0.10 mole of C
    (c) 2.0 moles of K   (d) 0.0100 mole of U

SOLUTION   (a) The atomic weight of He is 4. This gives us the factor unit $\dfrac{4 \text{ g}}{1 \text{ mole}}$ or $\dfrac{1 \text{ mole}}{4 \text{ g}}$.

$$\text{What I want to know} = (\text{quantity given})(\text{factor unit})$$

$$? \text{ g} = (3.0 \text{ moles})\left(\frac{4 \text{ g}}{1 \text{ mole}}\right)$$

$$= 12 \text{ g}$$

(b) The atomic weight of C is 12. This gives us the factor unit $\dfrac{12 \text{ g}}{1 \text{ mole}}$ or $\dfrac{1 \text{ mole}}{12 \text{ g}}$.

$$\text{What I want to know} = (\text{quantity given})(\text{factor unit})$$

$$? \text{ g} = (0.10 \text{ mole})\left(\frac{12 \text{ g}}{1 \text{ mole}}\right)$$

$$= 1.2 \text{ g}$$

(c) The atomic weight of K is 39. This gives us the factor unit $\dfrac{39 \text{ g}}{1 \text{ mole}}$ or $\dfrac{1 \text{ mole}}{39 \text{ g}}$.

What I want to know = (quantity given)(factor unit)

$$? \text{ g} = (2.0 \text{ moles})\left(\frac{39 \text{ g}}{1 \text{ mole}}\right)$$

$$= 78 \text{ g}$$

(d) The atomic weight of U is 238. This gives us the factor unit $\dfrac{238 \text{ g}}{1 \text{ mole}}$ or $\dfrac{1 \text{ mole}}{238 \text{ g}}$.

What I want to know = (quantity given)(factor unit)

$$? \text{ g} = (0.0100 \text{ mole})\left(\frac{238 \text{ g}}{1 \text{ mole}}\right)$$

$$= 2.38 \text{ g}$$

The Formula Weight of a Compound

The formula of water is H_2O, and that for table salt is NaCl. What is the formula weight of these compounds? The formula weight is simply the *sum* of the atomic weights of the elements that compose the compound.

For H_2O, the atomic weight of H = 1 and O = 16.

$$\text{Formula weight} = (2 \times \text{at. wt. of H}) + (1 \times \text{at. wt. of O})$$

$$= (2 \times 1) + (1 \times 16) = 2 + 16 = 18$$

atoms   atomic
per   weight
molecule   of hydrogen

For NaCl, the atomic weight of Na = 23 and Cl = 35.5.

$$\text{Formula weight} = (1 \times 23) + (1 \times 35.5) = 23 + 35.5 = 58.5$$

EXAMPLE 2-4   Determine the formula weights of the following compounds:

(a) Fe_2O_3 (b) CO_2
(c) Na_3PO_4 (d) $Cu_3(PO_4)_2$

SOLUTION

(a) The atomic weight of Fe = 56 and O = 16.

$$\text{Formula weight} = (2 \times 56) + (3 \times 16)$$
$$= 112 + 48$$
$$= 160$$

(b) The atomic weight of C = 12 and O = 16.

$$\text{Formula weight} = (1 \times 12) + (2 \times 16)$$
$$= 12 + 32$$
$$= 44$$

(c) The atomic weight of Na = 23, P = 31, and O = 16.

$$\text{Formula weight} = (3 \times 23) + (1 \times 31) + (4 \times 16)$$
$$= 69 + 31 + 64$$
$$= 164$$

(d) The atomic weight of Cu = 64, P = 31, and O = 16.

$$\text{Formula weight} = (3 \times 64) + (2 \times 31) + (8 \times 16)$$
$$= 192 + 62 + 128$$
$$= 382$$

Compounds and Moles

As you may have guessed, the formula weight of a compound in grams is the weight of 1 mole of molecules. Therefore, 18 g of H_2O (remember that the formula weight of water is 18) is 1 mole of water molecules. And 44 g of CO_2 (remember that the formula weight of carbon dioxide is 44) is 1 mole of carbon dioxide molecules. Let's apply this knowledge to the following problems.

EXAMPLE 2-5 Determine the moles of molecules in each weight of the following compounds.

(a) 500 g of $CaCO_3$
(b) 13.2 g of $(NH_4)_2SO_4$

SOLUTION

(a) First determine the formula weight of $CaCO_3$.
Atomic weight of Ca = 40, C = 12, and O = 16.

$$\text{Formula weight} = (1 \times 40) + (1 \times 12) + (3 \times 16)$$
$$= 40 + 12 + 48$$
$$= 100$$

This gives us the factor unit $\dfrac{100 \text{ g}}{1 \text{ mole}}$ or $\dfrac{1 \text{ mole}}{100 \text{ g}}$.

$$\text{What I want to know} = (\text{quantity given})(\text{factor unit})$$

$$? \text{ moles} = (500 \text{ g})\left(\frac{1 \text{ mole}}{100 \text{ g}}\right)$$

$$= 5 \text{ moles}$$

(b) First determine the formula weight of $(NH_4)_2SO_4$.
Atomic weight of $N = 14$, $H = 1$, $S = 32$, and $O = 16$.

$$\text{Formula weight} = (2 \times 14) + (8 \times 1) + (1 \times 32) + (4 \times 16)$$

$$= \quad 28 \quad + \quad 8 \quad + \quad 32 \quad + \quad 64$$

$$= 132$$

This gives us the factor unit $\dfrac{132 \text{ g}}{1 \text{ mole}}$ or $\dfrac{1 \text{ mole}}{132 \text{ g}}$.

$$\text{What I want to know} = (\text{quantity given})(\text{factor unit})$$

$$? \text{ moles} = (13.2 \text{ g})\left(\frac{1 \text{ mole}}{132 \text{ g}}\right)$$

$$= 0.100 \text{ mole}$$

EXAMPLE 2-6   Determine the grams of each compound in

    (a) 2.5 moles of $Ca(NO_3)_2$
    (b) 0.050 mole of MgO

SOLUTION   (a) First determine the formula weight of $Ca(NO_3)_2$.
Atomic weight of $Ca = 40$, $N = 14$, and $O = 16$.

$$\text{Formula weight} = (1 \times 40) + (2 \times 14) + (6 \times 16)$$

$$= \quad 40 \quad + \quad 28 \quad + \quad 96$$

$$= 164$$

This gives us the factor unit $\dfrac{164 \text{ g}}{1 \text{ mole}}$ or $\dfrac{1 \text{ mole}}{164 \text{ g}}$.

$$\text{What I want to know} = (\text{quantity given})(\text{factor unit})$$

$$? \text{ g} = (2.5 \text{ moles})\left(\frac{164 \text{ g}}{1 \text{ mole}}\right)$$

$$= 410 \text{ g}$$

(b) First determine the formula weight of MgO.
Atomic weight of Mg $= 24$ and O $= 16$.

$$\text{Formula weight} = (1 \times 24) + (1 \times 16)$$

$$= \quad 24 \quad + \quad 16$$

$$= 40$$

This gives us the factor unit $\dfrac{40 \text{ g}}{1 \text{ mole}}$ or $\dfrac{1 \text{ mole}}{40 \text{ g}}$.

$$\text{What I want to know} = (\text{quantity given})(\text{factor unit})$$

$$? \text{ g} = (0.050 \text{ mole})\left(\frac{40 \text{ g}}{1 \text{ mole}}\right)$$

$$= 2.0 \text{ g}$$

SUMMARY

In this chapter we discussed the scientific method. We talked about how scientists use this method to try and understand our universe (Fig. 2-8). We also studied the different kinds of matter and talked about atoms, molecules, chemical symbols, and moles. What we have tried to do is give you a basic understanding of important chemical principles. We will use these principles in subsequent chapters. Before you proceed, however, try to answer the questions and problems at the end of this chapter.

FIGURE 2-8

The Scientist. (After a cartoon by Marchant in the May 1964 I^2R calendar. Used by permission of Instruments for Research and Industry.)

1. Give an example of how a physician might use the scientific method.

2. State whether each of the following is an element, compound, or mixture.
    (a) milk   (b) glass   (c) gold   (d) calcium   (e) wood
    (f) sugar   (g) paper   (h) salt   (i) iron   (j) sand

3. Choose the word atom or molecule for each of the following statements.
    (a) The smallest part of an element that can enter into a chemical reaction is called a(n) _________.
    (b) The smallest part of a compound that can enter into a chemical reaction is called a(n) _________.

4. Look up the symbol and write the name of the first 20 elements in the periodic table.

5. If the periodic table was set up so that calcium had an atomic weight of 1, what would be the weight of (a) bromine?   (b) neon?

6. If you had 1 mole of dollar bills and divided it equally among the 4 billion people in the world, how long would it take you to spend your share, if you spent $1 million a day?

7. Determine the formula weights of the following compounds. (You may round off the atomic weights in the periodic table to whole numbers.)
    (a) Br_2 (b) OF_2 (c) H_3PO_4
    (d) K_2SO_4 (e) $(NH_4)_3PO_4$ (f) $(NH_4)_2SO_4$

8. Determine the number of grams of each of the following.
    (a) 2.0 moles of Na   (b) 0.20 mole of Na
    (c) 0.50 mole of NO_2 (d) 4.0 moles of H_2

9. Determine the number of moles of each of the following.
    (a) 78 g of K   (b) 7.8 g of K
    (c) 640 g of SO_2 (d) 150 g of $CaCO_3$

10. Cholesterol is a compound suspected of causing hardening of the arteries. The formula for cholesterol is $C_{27}H_{46}O$. Determine the formula weight of cholesterol. How many moles is 3.86 g?

11. Vitamin C may be a cure for colds or perhaps a preventative. The formula for vitamin C is $C_6H_8O_6$. How many moles is 8.8 g of vitamin C?

CHAPTER 3

Inside the Atom

It's What's Inside That Counts

Some Things You Should Know After Reading This Chapter

You should be able to:

1. State that atoms are composed of protons, electrons, and neutrons.
2. State that the protons and neutrons are in the center of the atom called the nucleus, and that the electrons are outside the nucleus.
3. Explain the concept of energy levels.
4. Reproduce the following table from memory:

PARTICLE	CHARGE	APPROX. MASS (amu)
Proton	+1	1
Electron	−1	Negligible
Neutron	0	1

5. Define the term isotope and give examples.
6. Use the standard notation for describing isotopes, for example $^{238}_{92}U$.
7. Determine the number of protons, electrons, and neutrons in an isotope of an element given the standard notation.
8. Determine the number of electrons in the outermost energy level of an A-group element.
9. Distinguish between a period and a group in the periodic table.
10. State that elements in the same group have similar chemical properties.

> Matter matter everywhere,
> But what are we to think?
> What it's all composed of,
> Is essentially the link.

In Chapter 2 we talked about atoms. We said that elements were composed of atoms, each element being composed of its own type of atoms. The element helium is composed of helium atoms, and the element lithium is composed of lithium atoms. But what, if anything, are atoms composed of? Up until the mid-1800's most scientists thought that atoms were indivisible (Fig. 3-1). They thought that atoms were solid through and through. But in the 1850's, experiments performed by various scientists proved that the atom was indeed divisible.

These experiments showed that at least two subatomic particles composed atoms—particles that had a negative charge, and particles that had a positive charge. The particles with a positive charge were called *protons*, and those with a negative charge were called *electrons*. Although the atoms of various elements were composed of protons and electrons, *the atoms themselves*

FIGURE 3-1
Things are not always what they seem.

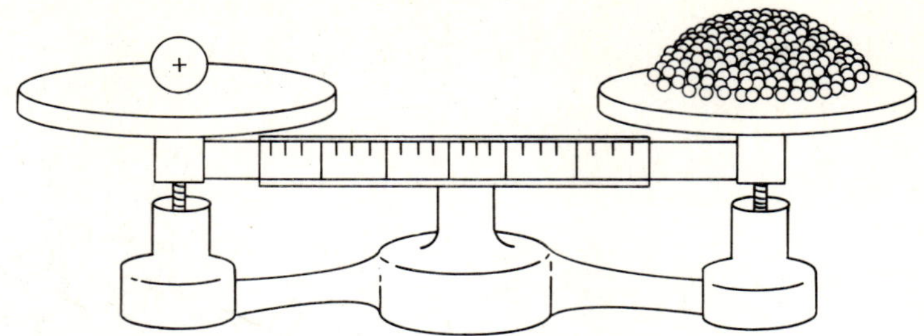

FIGURE 3-2
It takes about 2,000 electrons to equal the mass of a proton.

were neutral because they had the same number of protons and electrons. It was also determined that protons had a relative mass of one unit, called 1 *atomic mass unit* (1 amu), whereas the mass of the electron was negligible in comparison (Fig. 3-2).

By 1930, a third major subatomic particle was discovered—the *neutron*. The neutron, as its name implies, has no charge, but has a mass of one unit (like the proton). (Just think what this means. It is the number of protons and neutrons that give an atom its weight.) A summary of the three major subatomic particles is given in Table 3-1.

You may be wondering how the protons, electrons, and neutrons are arranged in an atom. And how do the protons, electrons, and neutrons that compose an atom of helium differ from those that compose an atom of uranium or an atom of gold? Let's try to answer these questions in the next section.

The Atomic Model: It's Almost Like a Volkswagen

No one has ever seen an atom. Therefore, the best that scientists can do is to develop a model of what they think the atom looks like. They base their model on the available evidence of how matter reacts. The model of the atom has changed significantly over the years. It has become more and more refined as more and more knowledge about matter has been obtained. There have been no major changes in the model of the atom for some years now. However, there have been some minor refinements. The scientist, of course, wants to

TABLE 3-1   A summary of the three major subatomic particles

PARTICLE	CHARGE	APPROXIMATE MASS (amu)
Proton	+1	1
Electron	−1	Negligible
Neutron	0	1

FIGURE 3-3
A look at dolls over the years shows that they have become more humanlike.

develop a model of the atom that closely resembles what the atom is like. This is almost analogous to the way doll making has progressed over the years. The first dolls were crude images of people, representing only their forms. As time passed, and tools and technology advanced, the model of the doll became more and more humanlike. Today, dolls can do many of the things that people do (Fig. 3-3).

The model of the atom that will be most useful to us in this course can be described as follows. Think of an atom as a ball. In the center of the atom (ball) are the protons and neutrons. This is the *nucleus* of the atom (Fig. 3-4). The electrons are situated outside the nucleus and revolve around it, very much like the planets revolve about the sun. We'll say more about this in a moment.*

FIGURE 3-4
A model of an atom. The protons and neutrons are in the nucleus of the atom, while the electrons orbit outside the nucleus.

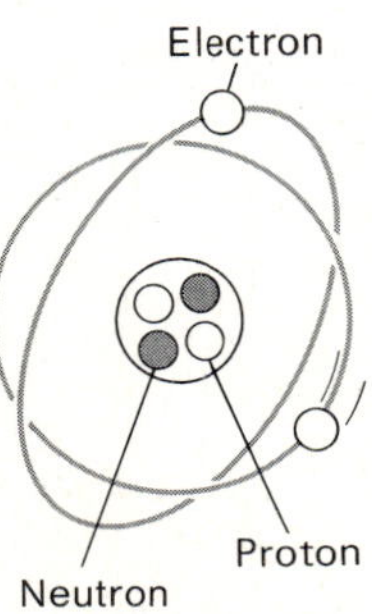

* A more in-depth discussion of atomic theory can be found in texts such as Theodore L. Brown and Eugene H. LeMay, *Chemistry—The Central Science* (Englewood Cliffs, N.J.: Prentice-Hall, Inc., 1977), Chapter 6.

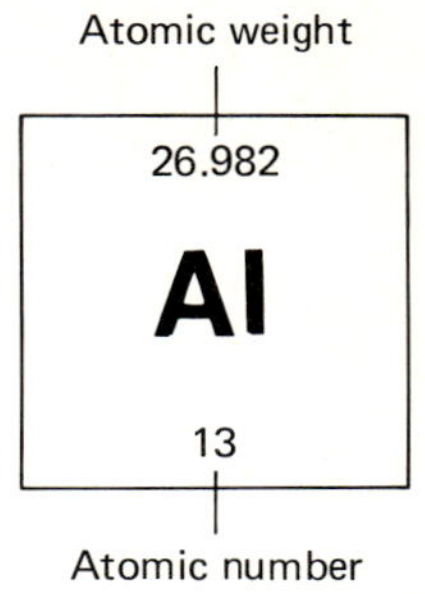

FIGURE 3-5
The atomic weight of an element appears above its symbol in the periodic table (of this text).
The atomic number of an element appears below its symbol in the periodic table (of this text).

What makes the atoms of lithium different from that of beryllium or boron? It is simply the numbers of protons, electrons, and neutrons that compose the atoms of these elements. To find out just how many protons and electrons are in an atom of a specific element, all you need to do is consult the periodic table (inside front cover). The number below the symbol for each element is called the *atomic number* (Fig. 3-5). It tells you how many protons and electrons are in an atom of that element. For example, hydrogen, atomic number 1, has one proton and one electron, and helium, atomic number 2, has two protons and two electrons.

EXAMPLE 3-1   Determine the number of protons and electrons in an atom of the following elements.

        (a) nitrogen (N)   (b) neon (Ne)
        (c) calcium (Ca)   (d) uranium (U)
        (e) silver (Ag)   (f) gold (Au)

SOLUTION   Use the periodic table on the inside front cover to obtain the atomic number of each element. The atomic number tells you the number of protons and electrons in an atom of the element.

    (a) Nitrogen is atomic number 7; therefore, an atom of N has 7 protons and 7 electrons.
    (b) Neon is atomic number 10; therefore, an atom of Ne has 10 protons and 10 electrons.
    (c) Calcium is atomic number 20; therefore, an atom of Ca has 20 protons and 20 electrons.
    (d) Uranium is atomic number 92; therefore, an atom of U has 92 protons and 92 electrons.

(e) Silver is atomic number 47; therefore, an atom of Ag has 47 protons and 47 electrons.

(f) Gold is atomic number 79; therefore, an atom of Au has 79 protons and 79 electrons.

But how about the number of neutrons in an atom of an element? Before we discuss this we must first learn about isotopes.

Isotopes: It's All in the Family

Imagine that it is possible to put billions of hydrogen atoms into a hat and pick 1 million of them at random (Fig. 3-6). Would they all be the same? In other words, would all the hydrogen atoms have the same number of protons, electrons, and neutrons? You might think so. In fact, all 1 million atoms would have one proton and one electron. The difference would be in the number of neutrons. Most of the 1 million hydrogen atoms would have zero neutrons, some would have one neutron, and still others would have two neutrons. Scientists have determined that out of 1 million hydrogen atoms,

999,985 would have 1 proton, 1 electron, 0 neutrons
14 would have 1 proton, 1 electron, 1 neutron
1 would have 1 proton, 1 electron, 2 neutrons

FIGURE 3-6
If you could put billions of hydrogen atoms into a hat and pick 1 million of them at random, would they all be the same?

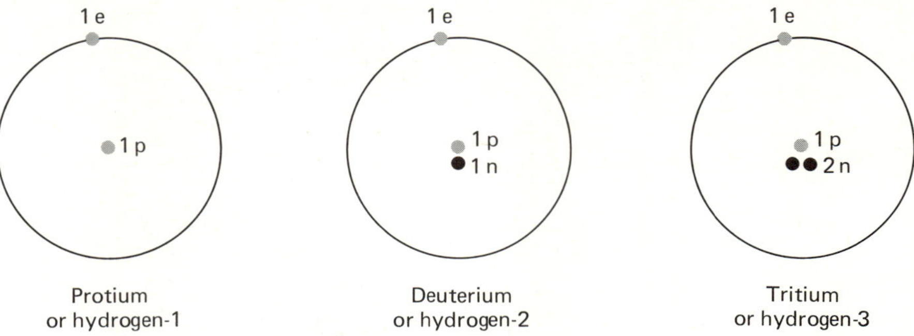

FIGURE 3-7
The three isotopes of hydrogen.

These three types of hydrogen atoms are collectively referred to as *isotopes* of hydrogen (Fig 3-7). The first type, and most common type is called hydrogen-1 (since the mass of this atom is 1). (Remember that the mass of an atom is the sum of its protons and neutrons.) The second type of hydrogen atoms is called hydrogen-2 (since the mass of this atom is 2). The third type of hydrogen atom is called hydrogen-3 (since the mass of this atom is 3). All three forms of hydrogen exist in nature in the relative abundances that we previously mentioned.

The atoms of most elements exist in more than one isotopic form. For example, lithium comes in two isotopic forms, lithium-6 and lithium-7. Scientists have determined the relative abundances of isotopes for all the known elements. Table 3-2 lists these relative abundances for the first 15 elements.

Because the atoms of isotopes have different numbers of neutrons, their atomic masses are different. Remember, it is the sum of the protons and neutrons that give an atom its weight. A notation has been developed to represent the isotopes of different elements. The notation is as follows.

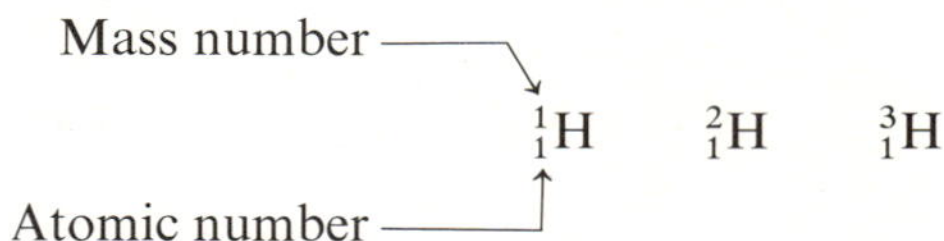

The superscript is the mass number of the isotope, and the subscript is the atomic number. Using this notation it is easy to determine the number of protons, electrons, and neutrons in an isotope of an element. To do this simply look at the atomic number for the protons and electrons. The number of neutrons is determined by subtracting the atomic number from the mass number.

EXAMPLE 3-2   Determine the number of protons, electrons, and neutrons in the isotopes of the following elements.

(a) ^3_1H (b) $^{238}_{92}U$ (c) $^{56}_{26}Fe$ (d) $^{12}_6C$

TABLE 3-2   Naturally occurring isotopes of the first 15 elements

NAME	SYMBOL	ATOMIC NUMBER	MASS NUMBER	PERCENTAGE NATURAL ABUNDANCE
Hydrogen-1	^1_1H	1	1	99.985
Hydrogen-2	^2_1H	1	2	0.015
Hydrogen-3	^3_1H	1	3	Negligible
Helium-3	^3_2He	2	3	0.00013
Helium-4	^4_2He	2	4	99.99987
Lithium-6	^6_3Li	3	6	7.42
Lithium-7	^7_3Li	3	7	92.58
Beryllium-9	^9_4Be	4	9	100
Boron-10	$^{10}_5\text{B}$	5	10	19.6
Boron-11	$^{11}_5\text{B}$	5	11	80.4
Carbon-12	$^{12}_6\text{C}$	6	12	98.89
Carbon-13	$^{13}_6\text{C}$	6	13	1.11
Nitrogen-14	$^{14}_7\text{N}$	7	14	99.63
Nitrogen-15	$^{15}_7\text{N}$	7	15	0.37
Oxygen-16	$^{16}_8\text{O}$	8	16	99.759
Oxygen-17	$^{17}_8\text{O}$	8	17	0.037
Oxygen-18	$^{18}_8\text{O}$	8	18	0.204
Fluorine-19	$^{19}_9\text{F}$	9	19	100
Neon-20	$^{20}_{10}\text{Ne}$	10	20	90.92
Neon-21	$^{21}_{10}\text{Ne}$	10	21	0.257
Neon-22	$^{22}_{10}\text{Ne}$	10	22	8.82
Sodium-23	$^{23}_{11}\text{Na}$	11	23	100
Magnesium-24	$^{24}_{12}\text{Mg}$	12	24	78.70
Magnesium-25	$^{25}_{12}\text{Mg}$	12	25	10.13
Magnesium-26	$^{26}_{12}\text{Mg}$	12	26	11.17
Aluminum-27	$^{27}_{13}\text{Al}$	13	27	100
Silicon-28	$^{28}_{14}\text{Si}$	14	28	92.21
Silicon-29	$^{29}_{14}\text{Si}$	14	29	4.70
Silicon-30	$^{30}_{14}\text{Si}$	14	30	3.09
Phosphorus-31	$^{31}_{15}\text{P}$	15	31	100

SOLUTION   The number of protons and electrons is obtained from the atomic number of the element. The number of neutrons is the difference between the mass number and atomic number.

(a) ^3_1H has 1 proton, 1 electron, and 2 neutrons.

(b) $^{238}_{92}\text{U}$ has 92 protons, 92 electrons, and 146 neutrons.

(c) $^{56}_{26}\text{Fe}$ has 26 protons, 26 electrons, and 30 neutrons.

(d) $^{12}_{6}\text{C}$ has 6 protons, 6 electrons, and 6 neutrons.

Please note that the atomic weight in the periodic table is the average of all the isotopes of a particular element, taking into account the percent abundance of each. We'll have a lot more to say about isotopes in Chapter 8 when we discuss their medical applications.

Electron Configuration: It's on the Level

Why do atoms of elements combine to form molecules of compounds? And why do elements in the same vertical column of the periodic table have similar chemical properties? These two questions are of great interest to the chemist. The answers to these questions lie in something called the *electron configuration* of the elements. For you see, it is the electrons that determine how the atoms

FIGURE 3-8
The names of the energy levels in atoms.

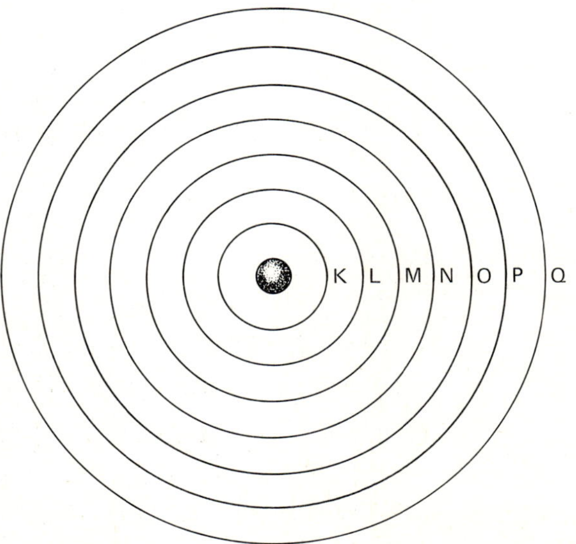

ENERGY LEVEL	MAXIMUM NUMBER OF ELECTRONS
K	2
L	8
M	18
N	32
O	50
P	72
Q	98

of a particular element will react. More precisely, it is how the electrons *are
arranged* outside the nucleus of the atom that determines the chemical properties of that element.

Remember that a short while ago we asked you to think of an atom as a
ball, with the protons and neutrons in the nucleus and electrons revolving
around the nucleus like planets revolving about the sun. We'd like you to
continue to think of an atom like that even more so, in that electrons, like
planets, can have different orbits about the nucleus of the atom. We call these
different orbits *energy levels*. These energy levels have letter names. The level
closest to the nucleus is called the K energy level. Then comes the L, M, N,
O, P, and Q levels (Fig. 3-8). Each energy level can hold only a certain number
of electrons. Table 3-3 lists the *maximum* number of electrons that each energy
level can hold. To determine the electron configuration of an element you
should know that the lowest energy level (the K level) fills first, followed by
the next lowest energy level (the L level), and so on. Following these rules,
see if you can write the electron configuration of the first 10 elements (see
Example 3-3).

EXAMPLE 3-3   Write the electron configuration of elements 1 through 10.

SOLUTION

(a) Hydrogen has one electron; therefore, K has one electron.

(b) Helium has two electrons; therefore, K has two electrons.

(c) Lithium has three electrons; therefore, K has two electrons and L
has one.

(d) Beryllium has four electrons; therefore, K has two electrons and L
has two.

(e) Boron has five electrons; therefore, K has two electrons and L has three.

(f) Carbon has six electrons; therefore, K has two electrons and L has four.

(g) Nitrogen has seven electrons; therefore, K has two electrons and L has five.

(h) Oxygen has eight electrons; therefore, K has two electrons and L has six.

(i) Fluorine has nine electrons; therefore, K has two electrons and L has seven.

(j) Neon has 10 electrons; therefore, K has two electrons and L has eight.

Table 3-4 gives the electron configuration of the first 18 elements.

TABLE 3-4   Electron configurations of the first 18 elements

ELEMENT	ATOMIC NUMBER	ELECTRON CONFIGURATION		
		K	L	M
H	1	1		
He	2	2		
Li	3	2	1	
Be	4	2	2	
B	5	2	3	
C	6	2	4	
N	7	2	5	
O	8	2	6	
F	9	2	7	
Ne	10	2	8	
Na	11	2	8	1
Mg	12	2	8	2
Al	13	2	8	3
Si	14	2	8	4
P	15	2	8	5
S	16	2	8	6
Cl	17	2	8	7
Ar	18	2	8	8

Predicting the electron configuration of an element at this point may seem quite simple. All you do is place electrons in the lowest energy level until it becomes filled, and then proceed to the next lowest energy level, and so on, until you have placed all your electrons. This works fine up to element 18. But then we run into some problems. What does this mean? Well, try predicting the electron configuration for potassium, element 19. You would expect

K to have two, L to have eight, and M to have nine electrons

After all, the M level can hold up to 18 electrons. However, chemists know that this *is not* the electron configuration of potassium. Instead, the electron configuration of potassium is

K has two, L has eight, M has eight, and N has one electron

How come? Why hasn't the M level continued to fill as expected? We can account for these strange results by introducing you to the concept of energy *sublevels*. We could tell you that the sublevels of some energy levels overlap (have less energy than) the sublevels of other energy levels. However, it's not important for us to go into that in this course. Instead, let's say that once an energy level has achieved eight electrons it would rather not alter that configuration. There seems to be something very stable about having eight electrons in an energy level. We'll talk about this in Chapter 4 when we discuss the *octet rule*. Anyway, it is because of this stability associated with eight electrons that the element potassium has the electron configuration we've just described.

The electron configuration of calcium (element 20) again shows the N level filling before the M level is filled to capacity. The 20 electrons of calcium are arranged in the following way:

K has two, L has eight, M has eight, and N has two electrons

However, starting with scandium (element 21) the M energy level begins to fill again, and this continues until we reach zinc (element 30), when the M level is filled to capacity. The electrons of zinc are arranged in the following configuration:

K has two, L has eight, M has 18, and N has two electrons

Beginning with gallium (element 31) the N level begins to fill again. Table 3-5 lists the electron configuration of elements 19 to 36.

A similar process is seen for elements 37 to 54 (see Table 3-6). How does knowing all this help us to predict electron configuration? Well, it might not help you at all, but on the other hand you might be able to see a pattern. Look at the periodic table (inside front cover). You will see that each column of

ELEMENT	ATOMIC NUMBER	ELECTRON CONFIGURATION			
		K	*L*	*M*	*N*
K	19	2	8	8	1
Ca	20	2	8	8	2
Sc	21	2	8	9	2
Ti	22	2	8	10	2
V	23	2	8	11	2
Cr	24	2	8	13	1
Mn	25	2	8	13	2
Fe	26	2	8	14	2
Co	27	2	8	15	2
Ni	28	2	8	16	2
Cu	29	2	8	18	1
Zn	30	2	8	18	2
Ga	31	2	8	18	3
Ge	32	2	8	18	4
As	33	2	8	18	5
Se	34	2	8	18	6
Br	35	2	8	18	7
Kr	36	2	8	18	8

elements has a roman numeral, followed by the letter A or B. These columns are called *groups* or *families* of elements. There are two kinds of groups, the A-group elements and the B-group elements. The A-group elements can be thought of as always filling the outermost energy level, whereas the B-group elements fill inner energy levels. For example, it is the next-to-outermost energy level that is filling for elements 21 to 30. Look back at Tables 3-5 and 3-6 to see what we mean. However, predicting electron configuration is not the all-important question.   To find out what is, read on.

Periodicity and Electron Configuration: Play It Again, Sam!

The important thing about electron configuration is that elements in the same group of the periodic table have the same number of electrons in their outermost energy levels. For elements in the *A-groups*, the group number tells you how many electrons are in the outermost energy level. For example, elements in group IA have one electron in their outermost energy level, elements

TABLE 3-6   Electron configurations of elements 37 to 54

ELEMENT	ATOMIC NUMBER	ELECTRON CONFIGURATION				
		K	*L*	*M*	*N*	*O*
Rb	37	2	8	18	8	1
Sr	38	2	8	18	8	2
Y	39	2	8	18	9	2
Zr	40	2	8	18	10	2
Nb	41	2	8	18	12	1
Mo	42	2	8	18	13	1
Tc	43	2	8	18	14	1
Ru	44	2	8	18	15	1
Rh	45	2	8	18	16	1
Pd	46	2	8	18	18	
Ag	47	2	8	18	18	1
Cd	48	2	8	18	18	2
In	49	2	8	18	18	3
Sn	50	2	8	18	18	4
Sb	51	2	8	18	18	5
Te	52	2	8	18	18	6
I	53	2	8	18	18	7
Xe	54	2	8	18	18	8

in group VIA have six electrons in their outermost energy level (Fig. 3-9). (*Note:* Although elements in the *same B-group* have the same number of electrons in their outermost energy level, the number of the group *does not* tell you how many electrons that is.) It is because elements in the same group have the same number of electrons in their outermost energy level that elements in the same group have similar chemical properties.

There are some very practical consequences of all this. One good example involves the strontium-90 scare of the early 1960's. During the 1950's and early 1960's the United States and Soviet Union were testing nuclear weapons in the atmosphere. One of the consequences of these tests was the production of strontium-90, a radioactive isotope of strontium. The strontium-90 was part of the radioactive fallout and eventually settled to the earth. Once on the soil, the strontium-90 would get incorporated into various plant life, such as grass, upon which cows grazed. The cows would pick up the strontium-90 and pass it along to us in their milk.

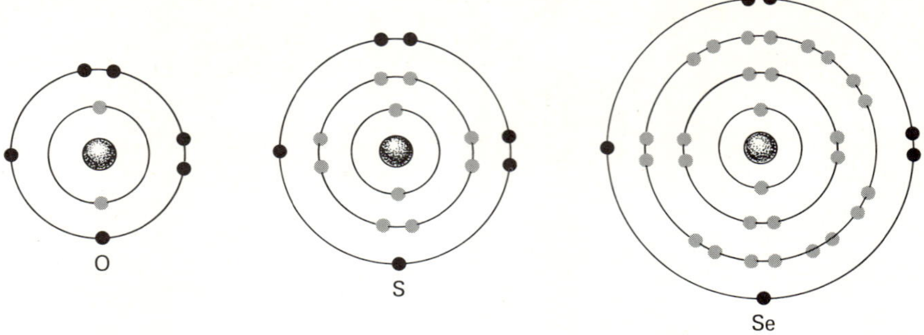

FIGURE 3-9
Elements in group VIA all have *six* electrons in their outermost energy level.

What was the problem? Strontium and calcium are in the same group of the periodic table. That means that they are chemically similar. Our bodies can't distinguish between calcium and strontium. Therefore, if we drink milk containing strontium-90, our bodies will act as if it were calcium. The result is that radioactive strontium-90 is incorporated into our bones and teeth. Fortunately, in 1963, the United States, the Soviet Union, and Great Britain signed a nuclear test ban treaty to halt atmospheric testing.

The Periodic Table: A Brief Summary

Throughout this chapter and the previous one we've talked about the periodic table. Let's briefly summarize some of the things we've discussed as well as some of the terminology we used when we mentioned the periodic table.

We've seen that the table is divided into vertical columns called *groups* or *families* of elements. The A-group elements are sometimes called the *representative elements*. The B-group elements are sometimes called the *transition metals*. The periodic table also has seven horizontal rows called *periods*. Period 1 consists of the elements hydrogen and helium. Period 2 consists of the elements lithium through neon. Periods 6 and 7 are so long that part of these periods are written below the rest of the table. (Elements 50 to 71 and 90 to 103 are unique in that it is the second-to-outermost energy level that is filling.)

The arrangement of the periodic table is such that the metals appear on the left side of the table, and the nonmetals on the right side of the table. The metalloids intervene between the metals and nonmetals (see the inside front cover of the book).

Each element on the table is represented by its symbol. Above the symbol is the atomic weight of the element (actually, it is the average of the masses of all the isotopes of that element). Below the symbol is the atomic number of the element. The atomic number tells you the number of protons and also the number of electrons in an atom of that element.

Elements in the same group have similar chemical properties because they have the same number of electrons in their outermost energy levels.

All of this information will be of great use to us as we continue our study of chemistry.

EXERCISES

1. Name the three major subatomic particles that compose atoms.

2. Where does one find the protons and neutrons in an atom?

3. (a) Name the seven energy levels found in atoms.
   (b) An easy way to find the maximum electron number for an energy level is to use the formula

$$\text{Maximum number of electrons} = 2(n)^2$$

   where n is the number of the energy level (for example, K is level 1, L is level 2, etc.). Use the $2(n)^2$ rule to calculate the maximum number of electrons that the P energy level can hold.

4. Fill in the following table.

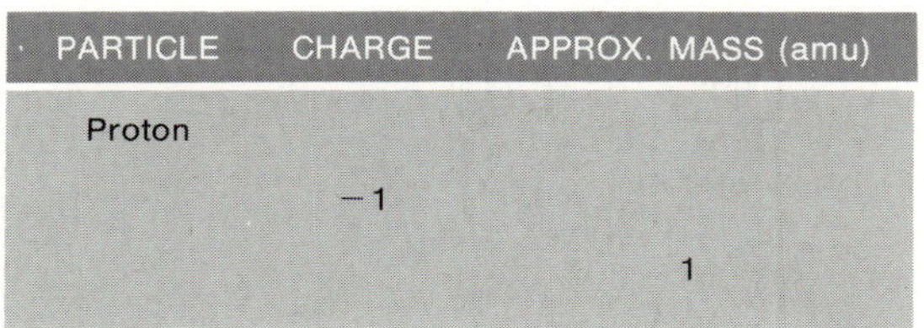

5. Determine the number of protons, electrons, and neutrons in each of the following isotopes.
   (a) $^{235}_{92}U$ (b) $^{119}_{48}Cd$ (c) $^{90}_{38}Sr$ (d) $^{160}_{66}Dy$

6. Write the electron configuration for the following elements.
   (a) $_{12}Mg$ (b) $_{8}O$ (c) $_{18}Ar$

7. How many electrons are in the outermost energy level of the following elements?
   (a) Ca   (b) Cs   (c) I   (d) Kr   (e) Sn

8. Elements 4, 12, 20, 38, 56, and 88 are in the same ________ (fill in the word group or period).

9. Element 111 would have properties similar to what other elements?

10. Given the following information, name the element.
    (a) K has two, L has eight, M has 18, N has eight, and O has two electrons.
    (b) K has two, L has eight, M has eight, and N has one electron.

11. Given the following information, use the standard notation for describing the isotope.
    (a) 12 protons, 12 electrons, 12 neutrons
    (b) 35 protons, 35 electrons, 46 neutrons
    (c) 97 protons, 97 electrons, 249 neutrons
    (d) 4 protons, 4 electrons, 5 neutrons

CHAPTER 4

The Molecular Connection

An Rx for Lonely Atoms

Some Things You Should Know After Reading This Chapter

You should be able to:

1. Write Lewis dot structures for the A-group elements.
2. Define and give examples of ionic and covalent bonds.
3. Write bonding structures for various covalent compounds.
4. Define single covalent bond, double covalent bond, and triple covalent bond.
5. Use the concept of electronegativity to decide whether a compound is bonded ionically or covalently.
6. Define and give an example of a polar covalent bond.
7. Define and give an example of a nonpolar covalent bond.
8. Write the formula of a chemical compound given its name.
9. Write the name of a chemical compound given its formula.
10. Decide whether a molecule is polar or nonpolar when given its three-dimensional shape.

FIGURE 4-1
Looking for elements with Professor Emeritus.

Introduction

In the previous chapters we discussed the elements that compose our world and our universe. Yet, if you were to look for these elements in the earth, you would not find many of them in their pure elemental form (Fig. 4-1). Of course, you would find some of them, such as gold and platinum. But most of the elements would be found as parts of chemical compounds. For example, you would not find pure sodium metal occurring naturally on the earth, but you would find plenty of sodium comprising the compound sodium chloride (table salt). How come? Why are most of the elements found in compounds and not as pure elements? The answer must have something to do with the stability of elements in compounds. In other words, elements must be more stable in the form of compounds then as the pure elements. This is a good thing for us! After all, we biological species are composed of compounds.

In this chapter we are going to look at the various ways that elements get together to form compounds. It is something that we like to call "The Molecular Connection."

Lewis Dot Structures, or I Keep Seeing Dots Before My Eyes

In Chapter 3 we learned that elements in the same column of the periodic table have similar chemical properties because they have *the same number of electrons in the outermost energy levels*. Chemists have discovered that it is these outermost electrons that play a major role in determining how the atoms of these

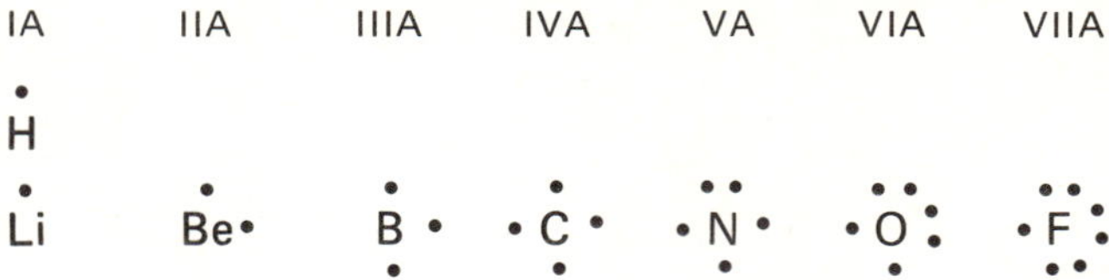

FIGURE 4-2
Electron dot diagrams of some A-group elements.

elements bond to form molecules of compounds. In this chapter we are going to describe the bonding of some compounds. It will therefore be a great help to us if we have a way of depicting this bonding, in other words, a standard notation. Fortunately, such a notation exists. It was developed by a famous physical chemist, G. N. Lewis, back in 1916.

The Lewis dot notation shows the symbol of the element and the number of outermost electrons that an atom of the element contains (Fig. 4-2). The rules for writing the Lewis dot notation for elements are as follows:

RULE 1.   Write the symbol for the element.

RULE 2.   Use a dot to represent each outer electron, beginning on top and going clockwise.

RULE 3.   Don't put two dots next to each other until you have distributed the first four dots.

The nice thing about the Lewis dot notation is that all elements in the same group of the periodic table have the same notation. This is because they all have the same number of outermost electrons. For example, the notation for a sodium atom (group IA) is the same as that for a lithium atom (group IA).

$$\text{Na} \qquad \text{Li}$$

And the notation for an oxygen atom (group VIA) is the same as that for an atom of selenium (group VIA).

$$\text{O} \qquad \text{Se}$$

EXAMPLE 4-1   Write the Lewis dot notation for the following elements.

(a) gallium   (b) cesium   (c) iodine   (d) antimony

SOLUTION   Look up the symbol of the element. Then find its position in the periodic table. The group number tells you the number of electrons in the outermost energy level.

(a) Gallium is in group IIIA: $\overset{\cdot}{Ga}\cdot$

(b) Cesium is in group IA: $\overset{\cdot}{Cs}$

(c) Iodine is in group VIIA: $\cdot\overset{\cdot\cdot}{\underset{\cdot\cdot}{I}}{\cdot}$

(d) Antimony is in group VA: $\cdot\overset{\cdot\cdot}{\underset{\cdot}{Sb}}\cdot$

Now let's see how we use the Lewis dot notation to show how the atoms of elements bond to form molecules of compounds.

Covalent Bonding: This Is the Way We Share Our Dots!

Scientists have long known that elements in group VIIIA of the periodic table (the *noble gases*) tend to be unreactive. In other words, they tend not to react with other elements to form compounds. The interesting thing about these group VIIIA elements is that their atoms all have *eight* electrons in their outermost energy level. We say that they have an *octet* of electrons. (An exception to this statement is helium, whose atoms have two electrons in the outermost energy level.) It appears that when atoms have an octet of electrons in their outermost energy level (or in the case of helium, a filled K energy level) they become very stable. The question is: "How can atoms which don't have eight electrons in their outermost energy level obtain an octet?" The answer is that they can attain eight electrons in their outermost energy level by sharing electrons with other atoms which also want an octet. In fact, chemists have found that this is one of the main reasons why elements bond to form compounds—to get eight electrons in their outermost energy level. This observation has come to be known as the *octet rule*. Let's see how this is done in *covalent bonding*. Remember that *covalent bonding* is a sharing of outermost electrons between two atoms, so that each can attain an octet of electrons. (An exception to this rule is hydrogen, which needs only two electrons in its outermost energy level. In this manner it will have an electron configuration similar to the noble gas helium.)

The Diatomic Elements: Some Nice Examples of Covalent Bonding

A number of elements are known to exist naturally in a diatomic form. This means that there are two atoms to a molecule. The elements hydrogen, oxygen, nitrogen, chlorine, bromine, iodine, and fluorine all exist naturally as diatomic elements. This means that they exist as H_2, O_2, N_2, Cl_2, Br_2, I_2, and F_2 (Fig. 4-3). You can remember these diatomic elements by simply remembering the name HONClBrIF (pronounced "honkelbrif"). Why do these elements exist as diatomics? Because by doing so they can form *covalent bonds* and obtain an octet of electrons. (In the case of hydrogen, it would be a duet of electrons.) Let's see what we mean by this (Fig. 4-4). Notice in Figure

FIGURE 4-3
A diatomic molecule may be pictured in this manner.

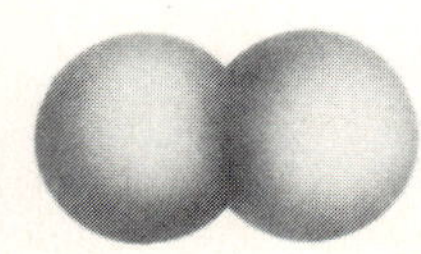

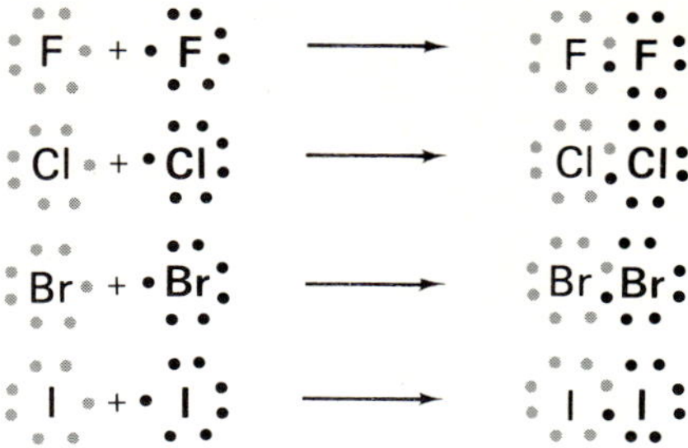

FIGURE 4-4
Covalent bonding of the group VIIA diatomic elements.

4-4 that each fluorine atom has seven electrons in its outermost energy level. But by sharing the "lone" electron, each fluorine atom can have eight electrons in its outermost energy level. The same is true for a molecule of Cl_2, Br_2, and I_2.

In hydrogen the situation is similar, except that each H atom has one electron, and by the process of sharing, each H atom can have two electrons (Fig. 4-5). (Remember what we said about hydrogen wanting only two electrons in its outermost energy level.)

The type of covalent bond we have discussed so far is called a *single bond*. This is because each atom has donated one electron to form the bond.

In a molecule of oxygen, we see that the bond is slightly different from the previous examples (Fig. 4-6). In order for each oxygen atom to have an octet of electrons, it is necessary that each oxygen atom donate two electrons to form the bond. This type of covalent bond is called a *double bond*. A double bond occurs when each atom donates two electrons to form the bond. Notice that this is the only structure that will give each oxygen atom an octet. (We should point out that although the structure in Figure 4-6 is not entirely accurate for the oxygen molecule, it does serve our needs and gets across the general idea.) There are many compounds that have double bonds. We shall look at a few of them in the forthcoming pages.

FIGURE 4-5
Covalent bonding in diatomic hydrogen.

H • + • H ⟶ H : H

FIGURE 4-6
Covalent bonding in diatomic oxygen.

: O • + • O : ⟶ : O :: O :

$$\cdot \overset{\cdot\cdot}{\underset{\cdot}{N}} \cdot + \cdot \overset{\cdot\cdot}{\underset{\cdot}{N}} \cdot \longrightarrow \quad N ::: N$$

FIGURE 4-7
Covalent bonding in diatomic nitrogen.

In the molecule nitrogen, we see another type of covalent bond (Fig. 4-7). Notice that in this molecule each nitrogen atom donates three electrons to form the bond. This type of covalent bond is called a *triple bond*. A triple bond occurs when each atom donates three electrons to form the bond. Again, this is the only structure that we can show such that each nitrogen atom has an octet of electrons. There are many compounds that have triple bonds. We shall look at a few in a moment.

Other Molecules Having Covalent Bonds

Up to this point we've looked at covalent bonds formed between diatomic elements. Let's now turn our attention to covalent bonds formed between the atoms of different elements.

One of the most important compounds to biological organisms is water (H_2O). The water molecule bonds in the following manner:

$$\overset{\cdot\cdot}{\underset{\cdot\cdot}{:O:}} H$$
$$H$$

Notice, that by sharing electrons in the manner shown, the oxygen atom has eight electrons, and each hydrogen atom has two electrons. (Remember that the hydrogen atom only needs two electrons.)

The compound chloroform ($CHCl_3$), aside from its use in spy stories, was used by the medical profession as an anesthetic. The bonding picture for chloroform looks like this:

$$\overset{\cdot\cdot}{:Cl:}$$
$$H : C : \overset{\cdot\cdot}{Cl} :$$
$$\underset{\cdot\cdot}{:Cl:}$$

As you probably know, we breathe in oxygen gas (O_2), whose bonding we described earlier, and we exhale carbon dioxide (CO_2). This process is known as *respiration*. We'll have more to say about respiration later in this text. However, the bonding structure for carbon dioxide is as follows:

$$:\overset{\cdot\cdot}{O}: :C: :\overset{\cdot\cdot}{O}:$$

Notice that there are two double bonds in this compound.

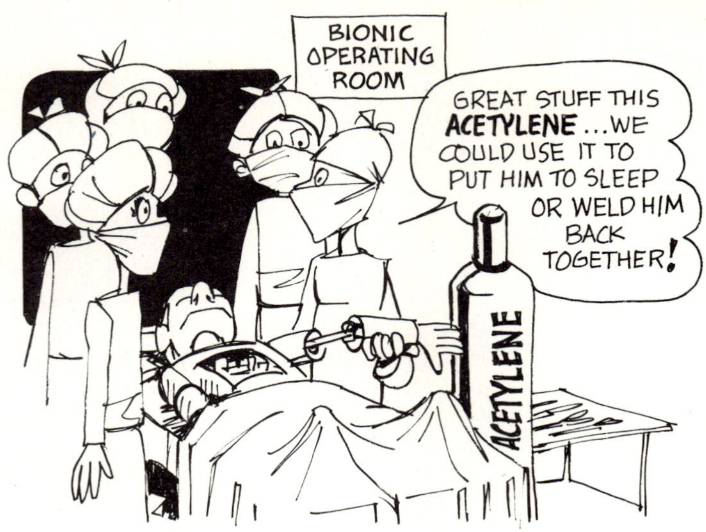

FIGURE 4-8
Acetylene, a very useful compound.

The compound acetylene (C_2H_2), which is used today for welding (for example, as the fuel in an oxyacetylene torch), was at one time used in hospitals as a surgical anesthetic (Fig. 4-8). The bonding in a molecule of acetylene looks like this:

$$H \!:\! C \!:\!:\! C \!:\! H$$

Notice that this is the only way that each carbon atom can have an octet of electrons and each hydrogen atom a duet.

Now see if you can write the bonding structures for some covalent compounds in the example below.

EXAMPLE 4-2   Write electron dot pictures for the following covalent compounds.

(a) H_2Se (Both hydrogens are bonded to the selenium.)
(b) CH_2O (This is the compound formaldehyde. The oxygen and each hydrogen is bonded to the carbon.)
(c) NH_3 (This is ammonia gas. Each hydrogen is bonded to the nitrogen.)
(d) HCN (This is hydrogen cyanide. The hydrogen and nitrogen are both bonded to the carbon.)

SOLUTION

(a) H : Se :
         H

(b)   H
     H : C :: O :

(c) H : N : H
         H

(d) H : C :: N :

Before we leave the topic of covalent bonding, we should mention the use of the *dash*. The dash (—) is used to simplify the writing of electron dot pictures. The dash represents a bond [in other words a shared pair of electrons (:)]. Let's see how this works.

EXAMPLE 4-3   Write the bonding structures in Example 4-2 using the dash notation.

SOLUTION

(a) H—S̈e:
          |
          H

(b)   H
           |
     H—C=Ö:

(c) H–N̈–H
          |
          H

(d) H–C≡N

Ionic Bonding: This Is the Way We Transfer Our Dots

Another way that atoms can attain an octet is by transferring electrons among each other. Consider the compound sodium chloride, table salt (NaCl). The electron dot notation for a sodium atom is Na and the electron dot notation for a chlorine atom is ·Cl: . A sharing of electrons might help the chlorine atom attain an octet, but it certainly wouldn't help the sodium atom. Therefore, this compound cannot bond covalently. However, if the sodium atom transfers its lone electron to the chlorine atom, both atoms can have an octet (Fig. 4-9). Do you see why this is so? For the chlorine atom it's easy to see. But how does the sodium atom gain an octet by giving an electron away? Remember, the

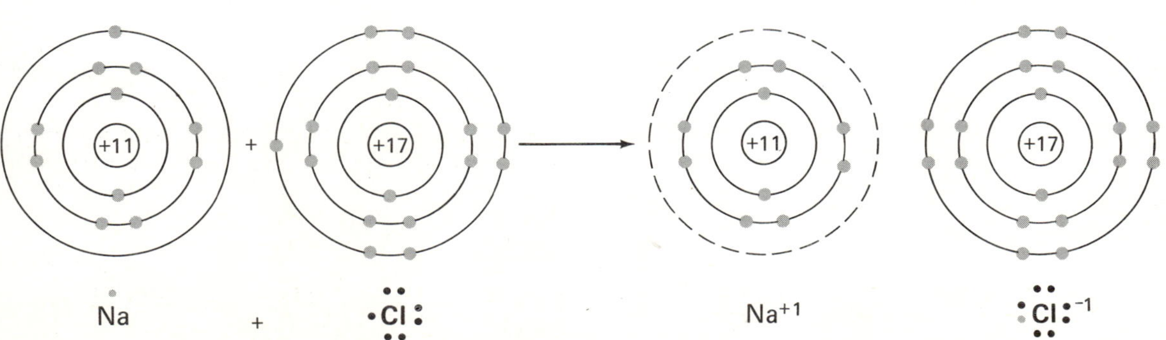

FIGURE 4-9
Ionic Bonding in sodium chloride.

$$Mg \cdot \; + \; \cdot \overset{\displaystyle ..}{\underset{\displaystyle ..}{O}} : \; \longrightarrow \; Mg^{+2} \quad : \overset{\displaystyle ..}{\underset{\displaystyle ..}{O}} :^{-2}$$

$$Mg \cdot \; + \; \cdot \overset{\displaystyle ..}{\underset{\displaystyle ..}{Cl}} : \; + \; \cdot \overset{\displaystyle ..}{\underset{\displaystyle ..}{Cl}} : \; \longrightarrow \; Mg^{+2} \quad : \overset{\displaystyle ..}{\underset{\displaystyle ..}{Cl}} :^{-1}$$

$$: \overset{\displaystyle ..}{\underset{\displaystyle ..}{Cl}} :^{-1}$$

FIGURE 4-10

Ionic bonding in magnesium oxide (MgO), and magnesium chloride ($MgCl_2$).

electron configuration of a sodium atom is

K has two electrons, L has eight electrons,   and M has one electron

If the sodium atom gives away its M electron, the L level becomes the outermost level, and this level has eight electrons.

But what holds the sodium and chlorine atoms together to form the bond? If you remember, the sodium atom has given up one of its electrons. It's no longer a neutral sodium atom. It now contains 11 protons but only 10 electrons. We say that the sodium atom has become a *positively charged sodium ion* (Na^{+1}). The chlorine, on the other hand, has gained an electron. It now has 17 protons and 18 electrons. We say that the chlorine atom has become a *negatively charged chloride ion* (Cl^{-1}).

The word "ion" is used because it indicates that we are talking about atoms that have *either gained or lost electrons*, and therefore have a *positive or negative charge*. By the way, notice that we said chlor*ide* ion, not chlor*ine* ion. We'll explain why the negative ion has an *ide* ending when we discuss the topic of chemical naming, later in this chapter. The important thing is that because the sodium ion and chloride ion are oppositely charged, they are attracted to each other. It is this positive-negative attraction that is the *ionic bond*. Many compounds bond ionically. Some examples are given in Figure 4-10.

Some Exceptions to the Octet Rule

We have now looked at two ways that atoms bond to form molecules. We have also seen that the major force behind this bonding is the attaining of an octet of electrons in the outermost energy level. However, it is only fair to point out that there are exceptions to this octet phenomenon. There are molecules that exist in which an atom has 10 electrons in its outermost energy level (Fig. 4-11). And other molecules that exist where an atom has six electrons in its outermost energy level (Fig. 4-12). But for most compounds, we find that the octet rule leads to the proper bonding structure.

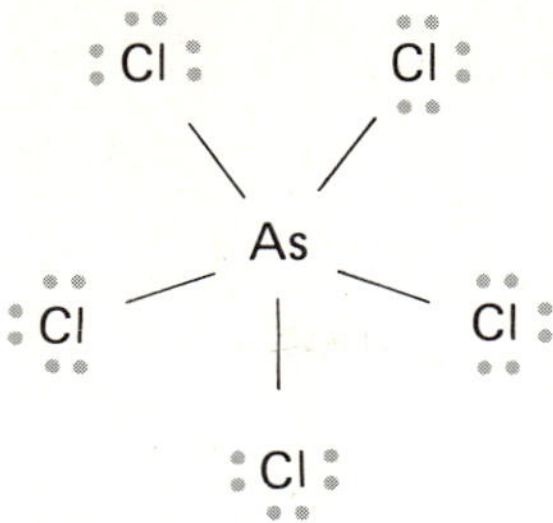

FIGURE 4-11
Covalent bonding in arsenic pentachloride.

FIGURE 4-12
Covalent bonding in boron trifluoride.

Electronegativity: A Way to Diagnose Ionic or Covalent Bonds

At this point you may be asking, "How can I tell whether the atoms of two elements bond ionically or covalently?" The answer is simple! We use the concept of *electronegativity*, which is the attraction that an atom has for the electrons it is sharing with another atom. The electronegativities of the elements are shown in Table 4-1. The concept of electronegativity was devised by Linus Pauling, a two-time Nobel-prize-winning scientist (probably better known to all you future health practitioners as the "vitamin C man"). Pauling assigned fluorine an electronegativity value of 4.0 because it attracts electrons more strongly than any other element. The other elements have various values, depending on their ability to attract electrons.

You can use the electronegativity table to help you decide on whether a bond is ionic or covalent. All you need do is calculate the difference in electronegativity values between the two elements forming the bond. Once you obtain this difference you can check Table 4-2. This table will help you convert your electronegativity difference into a percentage covalent and percentage ionic character of the bond. If the electronegativity difference is between zero and 1.6, the percentage covalent character is greater than the percentage ionic. The bond may therefore be considered to be covalent. If the electronegativity difference is greater than 1.6, the percentage ionic character is greater than the percentage covalent. The bond may therefore be considered to be ionic. Let's try some examples to see how this works.

TABLE 4-1   Periodic table of electronegativities

Period	Group IA	IIA	IIIB	IVB	VB	VIB	VIIB	VIII			IB	IIB	IIIA	IVA	VA	VIA	VIIA	VIIIA
1	2.1 H 1																	He 2
2	1.0 Li 3	1.5 Be 4				Transition elements							2.0 B 5	2.5 C 6	3.0 N 7	3.5 O 8	4.0 F 9	Ne 10
3	0.9 Na 11	1.2 Mg 12											1.5 Al 13	1.8 Si 14	2.1 P 15	2.5 S 16	3.0 Cl 17	Ar 18
4	0.8 K 19	1.0 Ca 20	1.3 Sc 21	1.5 Ti 22	1.6 V 23	1.6 Cr 24	1.5 Mn 25	1.8 Fe 26	1.8 Co 27	1.8 Ni 28	1.9 Cu 29	1.6 Zn 30	1.6 Ga 31	1.8 Ge 32	2.0 As 33	2.4 Se 34	2.8 Br 35	Kr 36
5	0.8 Rb 37	1.0 Sr 38	1.2 Y 39	1.4 Zr 40	1.6 Nb 41	1.8 Mo 42	1.9 Tc 43	2.2 Ru 44	2.2 Rh 45	2.2 Pd 46	1.9 Ag 47	1.7 Cd 48	1.7 In 49	1.8 Sn 50	1.9 Sb 51	2.1 Te 52	2.5 I 53	Xe 54
6	0.7 Cs 55	0.9 Ba 56	La-Lu 57–71	1.3 Hf 72	1.5 Ta 73	1.7 W 74	1.9 Re 75	2.2 Os 76	2.2 Ir 77	2.2 Pt 78	2.4 Au 79	1.9 Hg 80	1.8 Tl 81	1.8 Pb 82	1.9 Bi 83	2.0 Po 84	2.2 At 85	Rn 86
7	0.7 Fr 87	0.9 Ra 88	Ac-Lr 89–103	[Rf] 104	[Ha] 105													

1.1 La 57	1.1 Ce 58	1.1 Pr 59	1.1 Nd 60	1.1 Pm 61	1.1 Sm 62	1.1 Eu 63	1.1 Gd 64	1.1 Tb 65	1.1 Dy 66	1.1 Ho 67	1.1 Er 68	1.1 Tm 69	1.1 Yb 70	1.2 Lu 71
1.1 Ac 89	1.3 Th 90	1.5 Pa 91	1.7 U 92	1.3 Np 93	1.3 Pu 94	1.3 Am 95	1.3 Cm 96	1.3 Bk 97	1.3 Cf 98	1.3 Es 99	1.3 Fm 100	1.3 Md 101	1.3 No 102	Lr 103

Key: box shows 2.5 ← Electronegativity, C, 6 ← Atomic number

Source: Alan Sherman, Sharon Sherman, and Leonard Russikoff, *Basic Concepts of Chemistry*, Houghton Mifflin Company, Boston, 1976. By permission of the publisher.

EXAMPLE 4-4   Determine whether the bonds in the following compounds are ionic or covalent.

(a) HBr   (b) KCl   (c) O_2 (d) CO_2 (e) NaF

SOLUTION   (a) We use Table 4-1 to find the electronegativity values of hydrogen and bromine.

H is 2.1 and Br is 2.8.

The electronegativity difference is $2.8 - 2.1 = 0.7$.

Table 4-2 tells us that an electronegativity difference of 0.7 is a bond that is 12% ionic and 88% covalent. The bond is therefore considered to be covalent.

TABLE 4-2 The relationship between electronegativity difference and the ionic percentage and covalent percentage of a chemical bond

DIFFERENCE IN ELECTRONEGATIVITY	IONIC PERCENTAGE	COVALENT PERCENTAGE
0.0	0.0	100
0.1	0.5	99.5
0.2	1.0	99.0
0.3	2.0	98.0
0.4	4.0	96.0
0.5	6.0	94.0
0.6	9.0	91.0
0.7	12.0	88.0
0.8	15.0	85.0
0.9	19.0	81.0
1.0	22.0	78.0
1.1	26.0	74.0
1.2	30.0	70.0
1.3	34.0	66.0
1.4	39.0	61.0
1.5	43.0	57.0
1.6	47.0	53.0
1.7	51.0	49.0
1.8	55.0	45.0
1.9	59.0	41.0
2.0	63.0	37.0
2.1	67.0	33.0
2.2	70.0	30.0
2.3	74.0	26.0
2.4	76.0	24.0
2.5	79.0	21.0
2.6	82.0	18.0
2.7	84.0	16.0
2.8	86.0	14.0
2.9	88.0	12.0
3.0	89.0	11.0
3.1	91.0	9.0
3.2	92.0	8.0

Source: Alan Sherman, Sharon Sherman, and Leonard Russikoff, *Basic Concepts of Chemistry*, Houghton Mifflin Company, Boston, 1976. By permission of the publisher.

(b) The electronegativity value of K is 0.8.
The electronegativity value of Cl is 3.0.
The electronegativity difference is $3.0 - 0.8 = 2.2$.
Table 4-2 tells us that an electronegativity difference of 2.2 is a bond that is 70% ionic and 30% covalent. The bond is therefore considered to be ionic.

(c) The electronegativity of oxygen is 3.5; however, the bond in question is between two O atoms.
Therefore, the electronegativity difference is zero.
Table 4-2 tells us that this bond has 0% ionic character and 100% covalent character.

(d) The electronegativity of C is 2.5.
The electronegativity of O is 3.5.
The electronegativity difference is $3.5 - 2.5 = 1.0$.
Table 4-2 tells us that an electronegativity difference of 1.0 is a bond that is 22% ionic and 78% covalent. The bond is therefore considered to be covalent.

(e) The electronegativity of Na is 0.9.
The electronegativity of F is 4.0.
The electronegativity difference is $4.0 - 0.9 = 3.1$.
Table 4-2 tells us that an electronegativity difference of 3.1 is a bond that is 91% ionic and 9% covalent. The bond is therefore considered to be ionic.

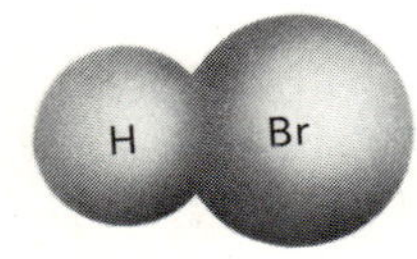

FIGURE 4-13
The electronegativity difference between H and Br gives rise to a polar covalent bond in the compound hydrogen bromide (HBr).

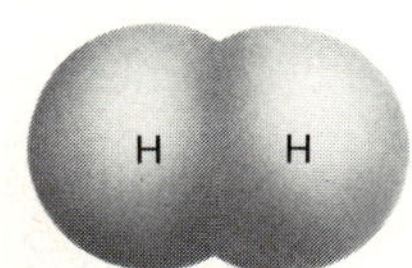

FIGURE 4-14
The diatomic molecule hydrogen (H₂) has no electronegativity difference. This gives rise to a nonpolar covalent bond.

The Polar and Nonpolar Covalent Bond

Although a *covalent bond* is defined as a sharing of electrons between two atoms, the sharing is usually not on an equal basis. This is because the electronegativity of the two elements will probably not be the same (Fig. 4-13). Only if there is no electronegativity difference between the two atoms forming the covalent bond will there be equal sharing, for example in the diatomic elements (Fig. 4-14). Such a bond is said to be a *nonpolar covalent bond*. Whenever there is an electronegativity difference between two atoms forming a covalent bond, we say that the bond is a *polar covalent bond*. The greater the difference, the more polar the bond. (Of course, an electronegativity difference of greater than 1.6 means the bond is ionic.) The fact that there are polar and nonpolar covalent bonds is extremely important to life on this planet. A case in point is the water molecule, whose bonds are polar covalent. It is because of this that water has its unique properties, for example, its unusually high boiling point. If it weren't for polar covalent bonding, water might have turned out to be a gas, at the temperatures that normally exist on our planet, and life, as we know it, would have never evolved.

EXAMPLE 4-5   List the following covalent compounds in order of increasing polarity of their bonds: Br_2, SO_2, H_2O, and CCl_4.

SOLUTION   Use Table 4-1 to determine the electronegativity difference between the elements in each compound. Then rank the compounds in order of least difference to most difference.

Br_2: The electronegativity difference between two bromines is zero.
SO_2: The electronegativity difference between S and O is 1.0.
H_2O: The electronegativity difference between H and O is 1.4.
CCl_4: The electronegativity difference between C and Cl is 0.5.
Therefore, the ranking would be:

$$Br_2 \qquad CCl_4 \qquad SO_2 \qquad H_2O$$

Least   Most
polar   polar

How to Write a Chemical Formula

We learned from our discussion on bonding that elements transfer or share electrons in order to obtain an octet of electrons. The number of electrons transferred or shared determines how the atoms of one element will combine with the atoms of another element. It is this observation that gives rise to the concept of *combining capacity* of elements. The combining capacity of an element is sometimes called its *oxidation number*. The oxidation numbers of some elements are listed in Table 4-3. (Actually, we should say that the *oxidation numbers of some ions* are listed in Table 4-3, since we are showing the charges of the *ions* when they enter into chemical combinations.) Also shown in Table 4-3 are some complex ions (also called *polyatomic ions*). These polyatomic ions, for example nitrate ion $(NO_3)^{-1}$, and sulfate ion $(SO_4)^{-2}$, are covalently bonded, and are found in nature as parts of numerous compounds. For example, if you take vitamins containing iron, you're not really consuming the element iron, but the compound iron(II) sulfate, $FeSO_4$. Notice that sulfate group, $(SO_4)^{-2}$, as part of the compound. And if you're suffering from indigestion and consume an antacid, you might be taking some calcium carbonate, $CaCO_3$. Notice that carbonate ion, $(CO_3)^{-2}$, as part of the compound.

By using the table of oxidation numbers we can easily learn to write chemical formulas by following a few simple rules.

RULE 1.   Write the symbol of each ion, along with its charge. For example, if we want to write the formula of calcium chloride, we first write

$$Ca^{+2}Cl^{-1}$$

	+1		+2		+3
Hydrogen	H^{+1}	Calcium	Ca^{+2}	Iron(III)	Fe^{+3}
Lithium	Li^{+1}	Magnesium	Mg^{+2}	Aluminium	Al^{+3}
Sodium	Na^{+1}	Barium	Ba^{+2}		
Potassium	K^{+1}	Zinc	Zn^{+2}		
Mercury(I)	Hg^{+1}	Mercury(II)	Hg^{+2}		
Copper(I)	Cu^{+1}	Tin(II)	Sn^{+2}		
Ammonium	$(NH_4)^{+1}$	Iron(II)	Fe^{+2}		
Silver	Ag^{+1}	Lead(II)	Pb^{+2}		
		Copper(II)	Cu^{+2}		

	−1		−2		−3
Fluoride	F^{-1}	Oxide	O^{-2}	Nitride	N^{-3}
Chloride	Cl^{-1}	Sulfide	S^{-2}	Phosphate	$(PO_4)^{-3}$
Hydroxide	$(OH)^{-1}$	Sulfite	$(SO_3)^{-2}$	Arsenate	$(AsO_4)^{-3}$
Nitrite	$(NO_2)^{-1}$	Sulfate	$(SO_4)^{-2}$		
Nitrate	$(NO_3)^{-1}$	Carbonate	$(CO_3)^{-2}$		
Acetate	$(C_2H_3O_2)^{-1}$	Chromate	$(CrO_4)^{-2}$		

RULE 2.   Because chemical compounds are electrically neutral, we must choose proper subscripts to balance the positive and negative charge of each ion. A simple way to do this is to criss-cross the numbers.

$$Ca_1^{+2}Cl_2^{-1}$$

Notice that the subscript numbers are written without charge, in other words, without a plus or minus sign.

RULE 3.   Now rewrite the formula in a more professional-looking manner. To do this, don't show the oxidation numbers, just the subscript numbers. Also, the number "1" is not shown. Therefore,

$$Ca_1^{+2}Cl_2^{-1} \quad \text{becomes} \quad CaCl_2$$

RULE 4.   Be sure that the subscript numbers are written in least common denominator form. This means that they must be reduced to lowest terms. For example, if you were asked to write the formula for aluminum nitride, you would do the following:

$$Al^{+3}N^{-3} \qquad \text{(Rule 1)}$$
$$Al_3^{+3}N_3^{-3} \qquad \text{(Rule 2)}$$
$$Al_3N_3 \qquad \text{(Rule 3)}$$

But since the subscripts are divisible by *three*, we simplify the formula to

$$AlN \qquad \text{(Rule 4)}$$

EXAMPLE 4-6   Write the formulas for the following compounds.

(a) aluminum oxide   (b) sodium chloride
(c) potassium sulfate   (d) ammonium phosphate
(e) barium arsenate   (f) ammonium sulfite

SOLUTION   Use Table 4-3 to find the oxidation numbers of the ions in each compound, then follow the rules we've just learned.

(a) $Al^{+3}O^{-2}$ $Al_2^{+3}O_3^{-2}$ which becomes Al_2O_3

(b) $Na^{+1}Cl^{-1}$ $Na_1^{+1}Cl_1^{-1}$ which becomes $NaCl$

(c) $K^{+1}(SO_4)^{-2}$ $K_2^{+1}(SO_4)_1^{-2}$ which becomes K_2SO_4

Notice that we handle the polyatomic ion by simply keeping parentheses around it. The parentheses were dropped in the final step because the subscript outside the parentheses was "1."

(d) $(NH_4)^{+1}(PO_4)^{-3}$ $(NH_4)_3^{+1}(PO_4)_1^{-3}$ which becomes $(NH_4)_3PO_4$

Notice that the ammonium ion must have the parentheses since the subscript outside the parentheses is "3." However, the parentheses around the phosphate group is unnecessary since the subscript outside the parentheses is "1."

(e) $Ba^{+2}(AsO_4)^{-3}$ $Ba_3^{+2}(AsO_4)_2^{-3}$ which becomes $Ba_3(AsO_4)_2$

(f) $(NH_4)^{+1}(SO_3)^{-2}$ $(NH_4)_2^{+1}(SO_3)_1^{-2}$ which becomes $(NH_4)_2SO_3$

What would you do if we asked you to write the formula of iron oxide? The first thing you would have to find out is which iron oxide we mean: iron(II) oxide or iron(III) oxide. Each is a distinct chemical compound with its own properties. *The roman numeral indicates the oxidation number of the iron in the compound.* The formula of iron(II) oxide would be determined as follows:

$$Fe^{+2}O^{-2} \qquad Fe_2^{+2}O_2^{-2} \qquad \text{which becomes } FeO$$

The formula of iron(III) oxide would be determined as follows:

$$Fe^{+3}O^{-2} \qquad Fe_2^{+3}O_3^{-2} \qquad \text{which becomes } Fe_2O_3$$

EXAMPLE 4-7   Write the formulas for the following compounds.

(a) copper(I) carbonate   (b) tin(II) phosphide
(c) iron(III) acetate   (d) lead(II) phosphate

SOLUTION   (a) $Cu^{+1}(CO_3)^{-2}$ $Cu_2^{+1}(CO_3)_1^{-2}$ which becomes
            Cu_2CO_3

            (b) $Sn^{+2}P^{-3}$ $Sn_3^{+2}P_2^{-3}$ which becomes Sn_3P_2

            (c) $Fe^{+3}(C_2H_3O_2)^{-1}$ $Fe_1^{+3}(C_2H_3O_2)_3^{-1}$ which becomes
            $Fe(C_2H_3O_2)_3$

            (d) $Pb^{+2}(PO_4)^{-3}$ $Pb_3^{+2}(PO_4)_2^{-3}$ which becomes
            $Pb_3(PO_4)_2$

Whenever you are dealing with chemistry, you need to be able to write chemical formulas. Writing them is much easier and faster if you have the information in Table 4-3 right at your fingertips. So spend some time mulling over this table and getting acquainted with these oxidation numbers. You can predict the oxidation numbers of most A-group elements; just keep in mind the following:

Group IA is $+1$.
Group IIA is $+2$.
Group IIIA is $+3$.
Group IVA is ±4.
Group VA is usually ±3 or $+5$.
Group VIA is usually -2.
Group VIIA is usually -1.

However, it's a good idea to memorize the charges of the polyatomic ions and transition metals in Table 4-3, if you possibly can.

How to Write a Chemical Name

Naming chemical compounds is a vast, but systematic, area of chemistry. Chemists from all over the world get together regularly and agree on rules for naming compounds (Fig. 4-15). This is done at meetings if the IUPAC (International Union of Pure and Applied Chemistry). Some of the more important rules are:

RULE 1.   The positive ion is always written first: for example, NaCl, not ClNa. (If the compound contains a metal and a nonmetal, the name of the metal comes first.)

FIGURE 4-15
Rules on naming compounds are decided at the IUPAC convention(?) (After a cartoon by Marchant in the April, 1964 I²R calendar. Used by permission of Instruments for Research and Industry.)

RULE 2. The name of the compound consists of the names of both *ions*. (Once the elements join together to form compounds, they are not the same elements any more, but ions.) For example, NaCl is sodium chlor*ide*, not sodium chlor*ine*. Names of compounds made up of only two elements always end in *-ide*.

RULE 3. Compounds made up of elements with variable oxidation numbers must include the roman numeral as part of the compound name. For example, $Fe(NO_3)_3$ is called iron(III) nitrate, not just iron nitrate.

RULE 4. For compounds made up of oxide ions, we sometimes use a Greek prefix in the compound name (Table 4-4). For example: CO is carbon *mon*oxide, CO_2 is carbon *di*oxide, and SO_3 is sulfur *tri*oxide.

TABLE 4-4   Greek prefixes

mono = 1	di = 2	tri = 3	tetra = 4	penta = 5
hexa = 6	hepta = 7	octa = 8	nona = 9	deca = 10

EXAMPLE 4-8   Name the following compounds.

(a) Fe_2O_3 (b) CuO (c) Cu_2O (d) Ag_2SO_4
(e) $Sr(NO_3)_2$ (f) Cs_2S (g) $Hg(NO_3)_2$ (h) $Zn(C_2H_3O_2)_2$
(i) SO_2 (j) UF_6

SOLUTION

(a) iron(III) oxide
(b) copper(II) oxide
(c) copper(I) oxide
(d) silver sulfate
(e) strontium nitrate
(f) cesium sulfide
(g) mercury(II) nitrate
(h) zinc acetate
(i) sulfur dioxide
(j) uranium hexafluoride or uranium(VI) fluoride

FIGURE 4-17
A three-dimensional model of HCl.

HCl

The 3-D Characteristics of Molecules: Some Are Polar; Some Are Not

Up to this point, we have neglected the fact that molecules are three-dimensional. This fact, however, is very important to the understanding of how some molecules behave. Let's look at the three-dimensional shape of some molecules to see what properties appear.

If we could enlarge a single molecule of any diatomic element (for instance, H_2, O_2, or N_2), it would seem to be linear (in a line) (Fig. 4-16). This would also be true for covalently bonded compounds consisting of two *unlike* atoms, such as HCl (Fig. 4-17).

But if we could enlarge a single molecule of water, we would see a bent molecule (Fig. 4-18). Experiments have shown that the angle between the hydrogen atoms is about $105°$.

FIGURE 4-18
A three-dimensional model of H_2O.

H_2O

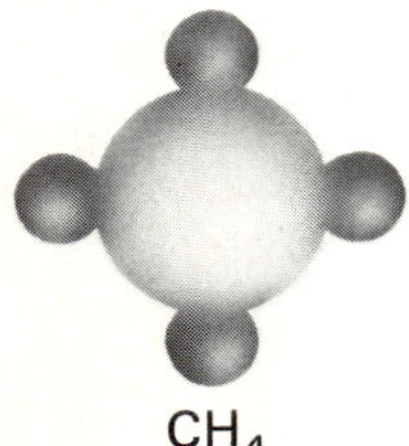

FIGURE 4-19
A three-dimensional model of CH_4.

CH_4

FIGURE 4-16
The three-dimensional models of some diatomic elements.

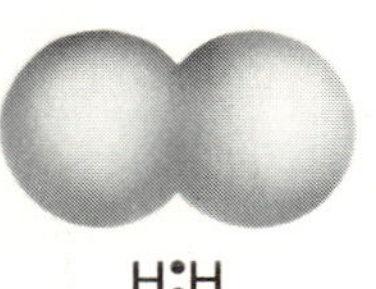

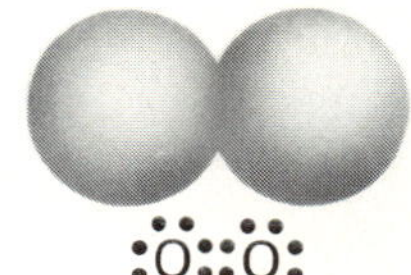

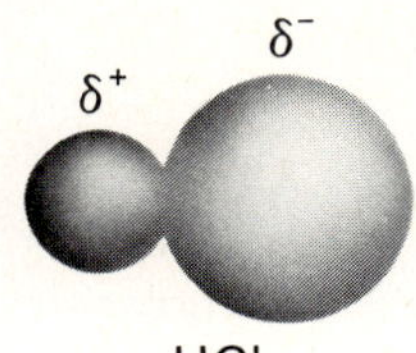

HCl

FIGURE 4-20
The polar molecule
HCl, showing the
positive center on the
hydrogen and the
negative center on the
chlorine.

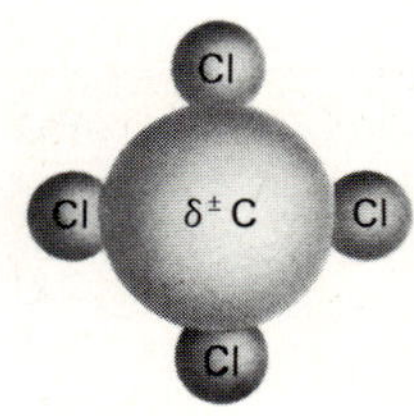

FIGURE 4-21
The nonpolar molecule
carbon tetrachloride
(CCl_4), showing that
the centers of positive
and negative charge
coincide.

If we look at a molecule of methane (CH_4), we would see a pyramid-shaped symmetrical molecule (Fig. 4-19). The angle between neighboring hydrogen atoms in methane has been shown to be 109.5°.

Molecules of other compounds exist in many other shapes. But remember, when you are discussing shapes of molecules, that these shapes are always three-dimensional.

The fact that molecules are three-dimensional, and the fact that in covalent compounds the sharing of electrons is not always equal, gives some molecules the property of being *polar* and other molecules the property of being *nonpolar*. Let's see what this means.

In a molecule of HCl (Fig. 4-20) the hydrogen represents the positive center of the molecule. (We say that the hydrogen appears to have a partial positive charge, δ^+.) The chloride ion represents the negative center of the molecule. (We say that the chloride ion appears to have a partial negative charge, δ^-.) This is because the electronegativity of Cl is greater than that of H. Because the two centers of charge do not coincide, the molecule is said to be *polar*.

In the compound CCl_4 (Fig. 4-21) there are four carbon–chlorine bonds, each of which is a polar covalent bond. This because the electronegativity of C and Cl are different. Because of this electronegativity difference, the carbon appears to have a partial positive charge, and each chlorine atom a partial negative charge. But the molecule is *nonpolar*! This is because the *center of positive charge* (on the carbon atom) coincides with the *center of negative charges* from each of the chloride ions (also on the carbon atom). This is due to the symmetrical shape of CCl_4, which is pyramidal.

EXAMPLE 4-9   Determine (a) if the bonds in each of the following molecules are polar or nonpolar, and (b) if the molecule is polar or nonpolar.

    (a) Br_2 (shape is linear)
    (b) CH_4 (shape is pyramidal)
    (c) CO_2 (shape is linear, with each oxygen bonded to the carbon)
    (d) H_2O (shape is bent as in Fig. 4-18)

SOLUTION   (a) Bromine is a diatomic element; therefore, the Br—Br bond is nonpolar and the molecule is nonpolar.

    (b) The bonding in CH_4 is similar to the bonding in CCl_4. Each bond in methane (CH_4) is a polar covalent bond. But the molecule is nonpolar since the centers of positive and negative charges coincide.

(c) Each C$=$O bond in CO_2 is polar, but the molecule is nonpolar. This is because the centers of positive and negative charges coincide.

$$O = \delta\pm = O$$

(d) Each O—H bond in H_2O is polar. The molecule is also polar since the centers of positive and negative charges do not coincide.

$$H \underset{\delta+}{\overset{\delta-}{\diagup\diagdown}} H$$

SUMMARY

In this chapter we learned how the chemical elements bond to form compounds. We saw that there are two basic types of bonds: ionic and covalent. And we learned that by using the concept of electronegativity we could tell whether a bond in a compound was ionic or covalent. We found out that there are three types of covalent bonds: single, double, and triple bonds. We also found out that these covalent bonds may be polar or nonpolar.

Our study of bonding led us to the idea of combining capacity of elements, which we called oxidation numbers. Using oxidation numbers we learned how to write the formulas of chemical compounds. We also learned how to name these compounds from their formulas.

Finally, we looked at the three-dimensional characteristics of some covalently bonded molecules. We found that there are two types: polar molecules and nonpolar molecules. In future chapters we will see how polar and nonpolar molecules play an important role in living organisms.

EXERCISES

1. Write the Lewis dot notation for the following elements.
   (a) barium   (b) fluorine   (c) rubidium   (d) tin

2. Write the Lewis dot notation for
   (a) chloride ion   (b) sulfide ion   (c) phosphide ion
   (*Hint:* Note that we asked for the *ions*, Cl^{-1}, S^{-2}, and P^{-3}, not the elements!)

3. With the exception of helium, all the other group VIIIA elements have ________ electrons in their outermost energy level.

4. Write electron dot diagrams for the following covalent compounds.
   (a) Br_2 (b) HBr   (c) C_2H_4 (d) PH_3 (e) C_2Cl_2

5. Name the diatomic elements from memory.

6. Stability, associated with eight electrons in the outermost energy level, has come to be known as the _________ rule. An exception to this rule is the element hydrogen, which needs only _________ electrons in its outermost energy level.

7. (a) When each atom donates two electrons to form the bond, it is called a _________ covalent bond.
   (b) When each atom donates three electrons to form the bond, it is called a _________ covalent bond.
   (c) When each atom donates one electron to form the bond, it is called a _________ covalent bond.

8. In the formation of the compound magnesium chloride ($MgCl_2$), the magnesium ion has given up _________ electrons. And each chloride ion has gained _________ electron.

9. Determine whether the bonds in the following compounds are ionic or covalent.
   (a) HI   (b) $SrCl_2$ (c) CO   (d) MgF_2 (e) AsH_3

10. List the following compounds in order of increasing polarity of their bonds: H_2Se, H_2O, H_2S, H_2Te.

11. Define the following terms.
    (a) ionic bond   (b) covalent bond
    (c) polar covalent bond   (d) nonpolar covalent bond
    (e) polar molecule   (f) nonpolar molecule

12. Write the formulas for the following compounds.
    (a) sodium arsenate   (b) potassium sulfite
    (c) barium chloride   (d) iron(II) nitrate
    (e) copper(II) sulfate   (f) aluminum sulfide
    (g) rubidium oxide   (h) cobalt(II) nitrite
    (i) carbon tetrachloride   (j) sulfur dioxide
    (k) hydrogen acetate   (l) phosphorus pentachloride
        (also known as
         acetic acid)

13. Write the names of the following compounds.
    (a) AgBr   (b) Cu_2S (c) $Fe(NO_3)_2$ (d) $Fe(NO_3)_3$
    (e) $Ca(OH)_2$ (f) $Al_2(SO_3)_3$ (g) Zn_3P_2 (h) $CuCrO_4$
    (i) $Hg_3(PO_4)_2$ (j) $CoCl_2$ (k) OsO_4 (l) Cu_3N

14. Determine the oxidation number of the underlined element in each compound.
    (a) $\underline{Ni}Cl_2$ (b) $\underline{Os}O_4$ (c) $\underline{V}_2O_5$ (d) $\underline{U}F_6$

15. Determine whether the bond in each compound is polar or nonpolar, and then whether the molecule is polar or nonpolar.

MOLECULE	SHAPE

(a) NH_3

$$\begin{array}{ccc} & N & \\ H & | & H \\ & H & \end{array}$$

(b) H_2S

$$\begin{array}{ccc} & S & \\ H & & H \end{array}$$

(c) N_2 $N\equiv N$

(d)   HCN $H-C\equiv N$

16. Write the formulas for the following compounds.
    (a) sodium sulfite   (b) potassium arsenate
    (c) barium nitrate   (d) iron(II) chloride
    (e) copper(II) sulfide   (f) aluminum sulfate
    (g) rubidium nitrite   (h) cobalt(II) oxide
    (i) carbon dioxide   (j) sulfur trioxide
    (k) uranium(VI) oxide   (l) phosphorus trichloride

17. Write the names of the following compounds.
    (a) Ag_2S (b) $CuBr$ (c) $Hg(NO_3)_2$ (d) $Fe(NO_2)_3$
    (e) $CaSO_3$ (f) $Al(OH)_3$ (g) $ZnCrO_4$ (h) Cu_3P_2
    (i) HgS (j) $Co_3(PO_4)_2$ (k) Cu_3P (l) V_2O_5

CHAPTER 5

Chemical Reactions and Chemical Equations

That's What's Happening

Some Things You Should Know After Reading This Chapter

You should be able to:

1. Write a formula equation from a word equation.
2. Balance a formula equation.
3. Name the four major types of inorganic reactions: combination, decomposition, single replacement, and double replacement.
4. Give an example of each type of reaction mentioned in objective 3.
5. Use a solubility table to predict if a substance is water-insoluble.
6. Use the activity series to predict the products of a single replacement reaction.
7. Recognize an oxidation-reduction reaction.
8. Define oxidation as the loss of electrons by a substance undergoing a chemical reaction.
9. Define reduction as the gain of electrons by a substance undergoing a chemical reaction.
10. Define reducing agent as a substance that causes another substance to be reduced, and oxidizing agent as a substance that causes another to be oxidized.

Within our own bodies and without, chemical reactions are taking place all the time (Fig. 5-1). As we breathe oxygen gas (O_2) into our lungs at this moment, some previously inhaled oxygen gas is returned to the atmosphere as carbon dioxide (CO_2). This is a result of a series of chemical reactions known as *respiration.* In our stomachs, small intestines, and large intestines, food that we've consumed over the past hours is being broken down into basic nutrients by a series of chemical reactions known as *digestion.* Numerous other chemical reactions are taking place in our bodies all the time. There are reactions that build new cells and reactions in which old cells are broken down. There are reactions that store energy, and there are reactions that produce energy so that we may move around.

Outside our bodies countless other reactions are taking place. Gasoline is reacting with oxygen, allowing automobiles to run. Plants are absorbing carbon dioxide from the atmosphere and producing oxygen gas in a series of reactions known as *photosynthesis.* There are metals, such as iron, reacting with oxygen in the atmosphere to produce rust. There are also chemists in numerous laboratories around the world reacting various elements and compounds for the purpose of producing the many products we use in our daily lives. Some of the reactions carried out by chemists are for the purpose of producing pharmaceutical compounds, for use by health practitioners in their never-ending fight against disease and illness.

As you can see, chemical reactions are *what's happening*! Even the production of sunlight is due to a chemical reaction. And although some chemical reactions occur instantaneously while others occur over a long period of time, all these reactions have a profound effect on our daily lives. As future health practitioners you will want to study the chemical reactions that are important to living organisms.

In this chapter we intend to look at some very basic types of chemical reactions. Most of these reactions are of the inorganic type. However, later in this text we will look at more complex reactions, specifically those which are involved with living organisms. In order for us to study chemical reactions we must learn how to represent these reactions using chemical equations. The first part of the chapter will be devoted to this task.

Formula Equations from Word Equations: the Chemist's Shorthand

Just about any chemical reaction can be put down in words. For example:

$$\text{Hydrogen gas} + \text{oxygen gas} \xrightarrow[\text{spark}]{\text{electric}} \text{water}$$

The "+" sign means *and,* and the "→" means *yields.* The words *electric spark* over the *yields* sign indicates the necessary conditions for the reaction to occur. This word equation tells us that water can be produced from its

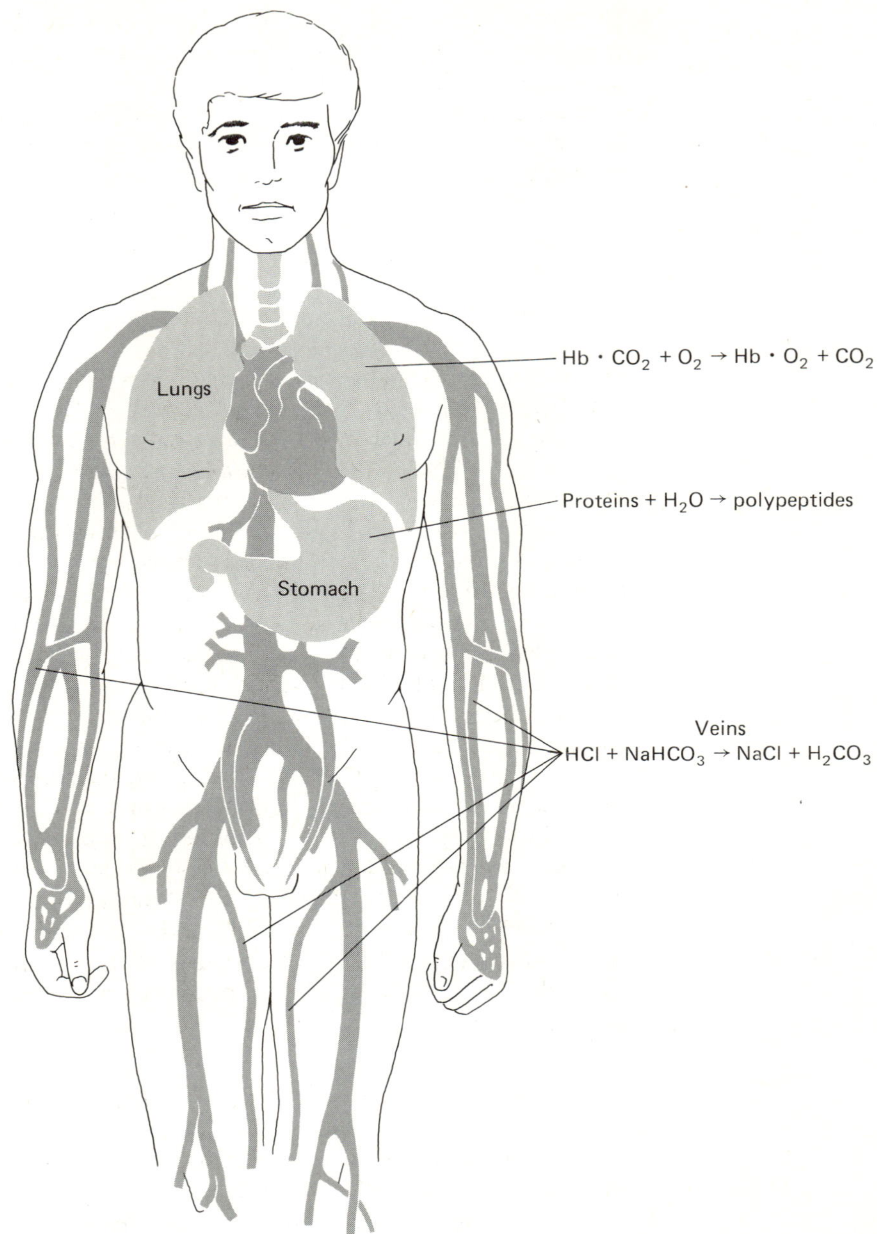

FIGURE 5-1
Chemical reactions are taking place in our bodies all the time.

elements hydrogen and oxygen. In this equation, the hydrogen and oxygen gas are called the *reactants*. This means that they are the starting substances. The water is called the *product*. This means that it is the substance produced. However, word equations tell us nothing about the chemistry of the reaction. We can remedy this situation by writing the formula of each substance in place of the words. The information we learned in Chapter 4 will help us with that.

$$\text{Hydrogen gas} + \text{oxygen gas} \xrightarrow{\text{electric spark}} \text{water}$$

$$H_2 + O_2 \xrightarrow{\text{electric spark}} H_2O$$

We have written the symbols in place of the words. However, we must perform one additional task, and that is to balance the equation. What does this mean? It means that we must be sure that the same number of atoms of each element appear on both sides of the equation—the reactant side and the product side. This is because chemists have found that atoms are not lost or gained in a chemical reaction. This is what has come to be known as the *law of conservation of mass*. (We should point out that atoms are not lost or gained in *common* types of chemical reactions. These are all types of chemical reactions *except* those involving *nuclear fission* or *fusion*. In nuclear fission and fusion, atoms can be lost or gained by being turned into energy. However, we'll save further discussion of this topic for Chapter 8.)

In the equation for the formation of water that we've just written, you will see that there are two hydrogen atoms and two oxygen atoms on the left side of the equation, and two hydrogen atoms and one oxygen atom on the right side of the equation. This equation is unbalanced. There is one less oxygen atom on the right side of the equation as opposed to the left. The way we balance a chemical equation is to use *coefficients*, that is, numbers that we place in front of the formulas of elements and compounds (Fig. 5-2).

We choose the numbers by trial and error, until we have the same number of atoms of each element on both sides of the equation. By the way, we try using the *smallest whole numbers* possible. In our example for water, we do the following:

$$2H_2 + 1O_2 \xrightarrow{\text{electric spark}} 2H_2O \qquad \text{(balanced)}$$

This equation is now balanced. There are four hydrogen atoms and two oxygen atoms on the left side of the equation, and four hydrogen atoms and two oxygen atoms on the right side of the equation. The balanced equation tells us that two molecules of hydrogen gas plus one molecule of oxygen gas produce two molecules of water (Fig. 5-3). The previous equation may also be written in the following manner:

$$2H_2 + O_2 \xrightarrow{\text{electric spark}} 2H_2O$$

FIGURE 5-2
Coefficients are used to balance chemical equations.

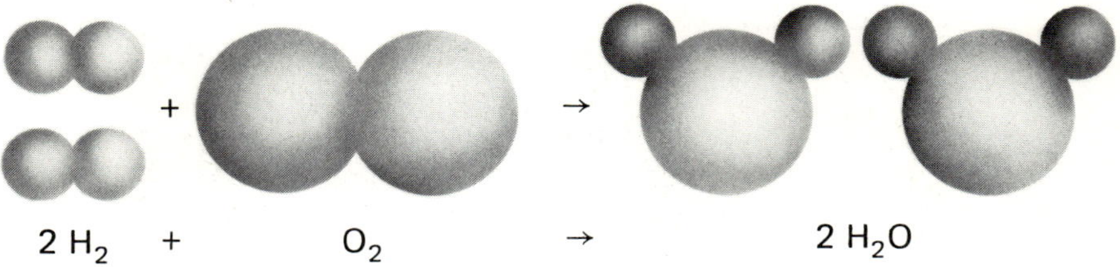

FIGURE 5-3
Two molecules of hydrogen gas plus one molecule of oxygen gas produce two molecules
of water.

The coefficient in front of the oxygen doesn't have to be shown if it is the number 1.

When you attempt to balance a chemical equation remember that:

1. The formulas of the reactants and products must be written properly and *never* changed.

2. The coefficients must be chosen so that the same number of atoms of each element appear on both sides of the equation.

EXAMPLE 5-1   Write a balanced chemical equation for

$$\text{Sodium chloride} \xrightarrow{\text{electricity}} \text{sodium metal} + \text{chlorine gas}$$

SOLUTION   Write the formulas for the reactants and products.

$$NaCl \xrightarrow{\text{electricity}} Na + Cl_2$$

There is one chlorine atom on the left side of the equation, but two chlorine atoms on the right side. We can balance the chlorines by placing a 2 in front of the NaCl.

$$2NaCl \xrightarrow{\text{electricity}} Na + Cl_2$$

Now the chlorines are balanced but the sodiums are not. We can balance the sodiums by placing a 2 in front of the Na atom on the right. Now we have

$$2NaCl \xrightarrow{\text{electricity}} 2Na + Cl_2 \quad \text{(balanced)}$$

A very important thing to remember in balancing a chemical equation is *never change or use subscripts.* For example, you may *not* balance the sodium chloride equation by doing the following:

STEP 1. $NaCl \xrightarrow{\text{electricity}} Na + Cl_2$

STEP 2. $NaCl_2 \xrightarrow{\text{electricity}} Na + Cl_2 \quad \text{(balanced)}$

What's wrong in step 2? The answer is that the formula for sodium chloride is NaCl, *not* $NaCl_2$. Also, never insert a coefficient between the ions in a molecule. Coefficients must be placed in front of the symbol for the molecule. Remember that we choose coefficients by trial and error. Do one element at a time; then after you think the equation is balanced, recheck your results.

EXAMPLE 5-2   Write a balanced equation for

$$\text{Zinc} + \text{phosphoric acid} \longrightarrow \text{zinc phosphate} + \text{hydrogen gas}$$

SOLUTION   Write the formulas for the reactants and products.

$$Zn + H_3PO_4 \longrightarrow Zn_3(PO_4)_2 + H_2$$

There is one zinc on the left side of the equation and three on the right. Therefore, place a coefficient of 3 in front of the zinc atom on the left side of the equation.

$$3Zn + H_3PO_4 \longrightarrow Zn_3(PO_4)_2 + H_2$$

The zinc atoms are balanced. However, there is one phosphate group on the left side of the equation and two on the right. Therefore, place a coefficient of 2 in front of the phosphate group on the left side of the equation.

$$3Zn + 2H_3PO_4 \longrightarrow Zn_3(PO_4)_2 + H_2$$

The zinc atoms and phosphate groups are balanced. However, there are six hydrogen atoms on the left side of the equation and only two on the right.

Therefore, place a coefficient of 3 in front of the hydrogen on the right side of the equation.

$$3Zn + 2H_3PO_4 \longrightarrow Zn_3(PO_4)_2 + 3H_2 \quad \text{(balanced)}$$

The equation is now balanced. There are the *same number of atoms of each element on both sides of the yield sign.*

What follows are some word equations. See if you can balance them yourself before looking at the solutions that follow.

EXAMPLE 5-3   Write a balanced equation for each of the following chemical reactions.

(a) potassium oxide + water $\longrightarrow$ potassium hydroxide

(b) hydrogen gas + fluorine gas $\longrightarrow$ hydrogen fluoride

(c) silver nitrate + barium chloride $\longrightarrow$ silver chloride + barium nitrate

(d) calcium hydroxide + nitric acid $\longrightarrow$ calcium nitrate + water

(e) aluminum + sulfuric acid $\longrightarrow$ aluminum sulfate + hydrogen gas

SOLUTION   Write the formulas for the reactants and products. Then balance the equation.

(a) $K_2O + H_2O \longrightarrow KOH$
$K_2O + H_2O \longrightarrow 2KOH \quad \text{(balanced)}$

(b) $H_2 + F_2 \longrightarrow HF$
$H_2 + F_2 \longrightarrow 2HF \quad \text{(balanced)}$

(c) $AgNO_3 + BaCl_2 \longrightarrow AgCl + Ba(NO_3)_2$
$2AgNO_3 + BaCl_2 \longrightarrow 2AgCl + Ba(NO_3)_2 \quad \text{(balanced)}$

(d) $Ca(OH)_2 + HNO_3 \longrightarrow Ca(NO_3)_2 + H_2O$
$Ca(OH)_2 + 2HNO_3 \longrightarrow Ca(NO_3)_2 + 2H_2O \quad \text{(balanced)}$

(e) $Al + H_2SO_4 \longrightarrow Al_2(SO_4)_3 + H_2$
$2Al + 3H_2SO_4 \longrightarrow Al_2(SO_4)_3 + 3H_2 \quad \text{(balanced)}$

Some Major Types of Chemical Reactions

As future health practitioners you will study hundreds of different chemical reactions that take place both in and out of the body. However, the hundred or so you may study are but a small part of the tens of thousands of chemical reactions that chemists have investigated. Because there are so many chemical reactions known, chemists have tried to organize these reactions into a small number of groups or types. In the remainder of this chapter we will study four major types of chemical reactions that are important to inorganic chemistry.

In Chapters 9 to 17 we will study some additional types of reactions important to organic chemistry and biochemistry. Remember, however, that all three classes of reactions (inorganic, organic, and biochemical) are important to living organisms.

The four types of inorganic reactions we are going to look at are:

1.   Combination reactions.

2.   Decomposition reactions.

3.   Single-replacement reactions.

4.   Double-replacement reactions.

Combination Reactions

Combination reactions are those in which two or more substances combine to form a more complex substance. For example:

$$A + B \longrightarrow AB$$

A common type of combination reaction involves two elements forming a compound (Fig. 5-4). For example:

$$\text{Sulfur} + \text{oxygen gas} \longrightarrow \text{sulfur dioxide:}$$
$$S + O_2 \longrightarrow SO_2 \quad \text{(balanced)}$$

Sulfur dioxide is one of the air pollutants that is responsible for causing respiratory diseases like chronic bronchitis and pulmonary emphysema (Fig. 5-5).

FIGURE 5-4
A combination reaction.

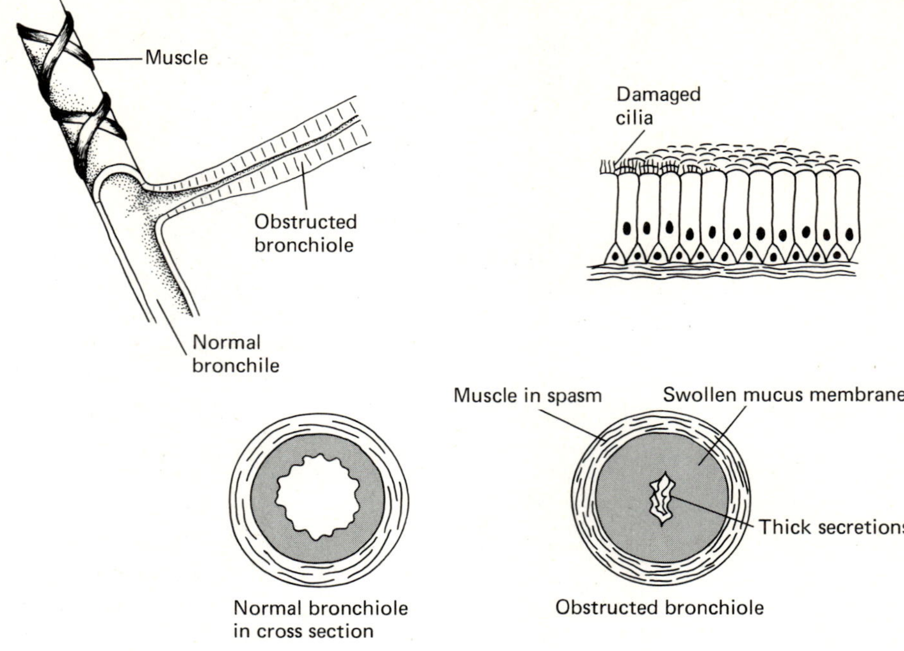

FIGURE 5-5
Phenomena of typical bronchitis.

A combination reaction may also involve two simple compounds forming a more complex compound. For example:

Water + carbon dioxide $\longrightarrow$ carbonic acid

$$H_2O + CO_2 \longrightarrow H_2CO_3 \qquad \text{(balanced)}$$

This reaction is very important in cellular respiration, and accounts for the way that carbon dioxide is transported through the blood,

Decomposition Reactions

A *decomposition reaction* is the reverse of a combination reaction (Fig. 5-6). It involves the breakdown of a complex substance into simpler substances. For example:

$$AB \longrightarrow A + B$$

A common type of decomposition reaction involves a compound decomposing into its elements. For example:

Mercury(II) oxide $\xrightarrow{\text{heat}}$ mercury metal + oxygen gas

$$2HgO \xrightarrow{\text{heat}} 2Hg + O_2 \qquad \text{(balanced)}$$

FIGURE 5-6
A decomposition reaction.

Mercury(II) oxide has been used as a topical antiseptic, but it has also been used for diluting pigments for painting on porcelain. However, care must be taken in the use of mercury(II) oxide since it is as poisonous as the element mercury, into which it can be made to decompose.

A decomposition reaction may also involve the breakdown of a more complex substance into simpler substances. For example:

$$\text{Sulfurous acid} \xrightarrow{\text{heat}} \text{water} + \text{sulfur dioxide}$$

$$H_2SO_3 \xrightarrow{\text{heat}} H_2O + SO_2 \qquad \text{(balanced)}$$

We've already mentioned the health hazards of sulfur dioxide.

Single-Replacement Reactions

Single-replacement reactions are those in which an uncombined element replaces another element from its compound (Fig. 5-7). For example:

$$A + BC \longrightarrow AC + B$$

A common type of single-replacement reaction involves metals and acids. For example:

$$\text{Aluminum} + \text{hydrochloric acid} \longrightarrow \text{aluminum chloride} + \text{hydrogen gas}$$

$$2Al + 6HCl \longrightarrow 2AlCl_3 + 3H_2$$

$$\text{(balanced)}$$

This reaction shows you why it's not a good idea to store hydrochloric acid in an aluminum container—the container will be turned into the powdery substance, aluminum chloride.

FIGURE 5-7
A single-replacement reaction.

Another type of single-replacement reaction involves an uncombined metal replacing another metal which is already in a compound. For example:

Zinc + copper(II) sulfate $\longrightarrow$ zinc sulfate + copper

$$Zn + CuSO_4 \longrightarrow ZnSO_4 + Cu \quad \text{(balanced)}$$

Copper sulfate solutions have long been used topically as fungicides, and zinc sulfate has been used medically as an astringent.

Double-Replacement Reactions

Double-replacement reactions are those in which two compounds exchange ions with each other. For example:

$$A^+B^- + C^+D^- \longrightarrow A^+D^- + C^+B^-$$

A common type of double-replacement reaction is one that involves an acid and a base (Fig. 5-8). For example:

Hydrochloric acid + magnesium hydroxide $\rightarrow$ magnesium chloride + water

$$2HCl + Mg(OH)_2 \rightarrow MgCl_2 + 2H_2O$$

$$\text{(balanced)}$$

FIGURE 5-8
A double-replacement reaction.

Hydrochloric acid is produced in the stomach for the process of digestion, and is sometimes called stomach acid. Magnesium hydroxide is commonly called *milk of magnesia*. Its major uses are pharmaceutical. In low dosages of a few hundred milligrams, magnesium hydroxide acts as an antacid. In large dosages of 2 to 4 grams (g), magnesium hydroxide acts as a laxative (Fig. 5-9). Magnesium hydroxide neutralizes the excess stomach acid by reacting with the hydrochloric acid and producing the salt, magnesium chloride, and water.

Another type of double-replacement reaction involves two salts which are soluble in water. (*Soluble* means that the salt can be dissolved in the water.) The salts are usually dissolved in water separately, and then combined. A reaction occurs if one of the products (which is also a salt) is insoluble in water (Fig. 5-10). (*Insoluble* means that the salt is not able to dissolve in the water.) An example of such a reaction is

Barium chloride + silver nitrate $\rightarrow$ barium nitrate + silver chloride

$$BaCl_2(aq) + 2AgNO_3(aq) \rightarrow Ba(NO_3)_2(aq) + 2AgCl(s)$$

(balanced)

The symbol (*aq*) means aqueous; in other words, the salt is soluble in water. The symbol (*s*) means solid, and indicates that the substance is insoluble in

FIGURE 5-9
In small doses magnesium hydroxide is an antacid; in large doses, a laxative.

water. Double-replacement reactions of this type are important to us as biological species because these reactions represent one way that soluble ions can be incorporated into our bodies and then complexed as insoluble compounds. For example, one can imagine soluble compounds containing calcium ions and phosphate ions being consumed in our diets, and then eventually

FIGURE 5-10
The reaction between aqueous barium chloride and silver nitrate produces the soluble salt barium nitrate and the insoluble salt silver chloride.

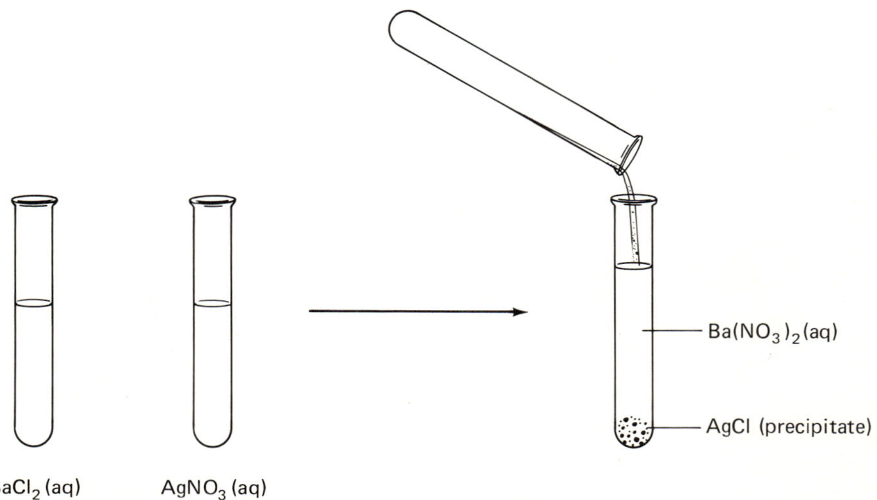

being combined in our bones in the form of calcium phosphate (a relatively insoluble substance). In order to determine whether a substance is soluble or insoluble in water, you may consult a solubility table. Such a table may be found on page 496.

EXAMPLE 5-4   Predict the products of the following reactions, then write the correct formulas and balance the equation.

COMBINATION REACTIONS

(a) $K + Cl_2 \longrightarrow$

(b) $SO_3 + H_2O \longrightarrow$

(c) $H_2 + I_2 \longrightarrow$

DECOMPOSITION REACTIONS

(d) $NH_3 \longrightarrow$

(e) $MgCO_3 \longrightarrow$

(f) $H_2O \longrightarrow$

SINGLE-REPLACEMENT REACTIONS

(g) zinc + hydrochloric acid $\longrightarrow$

(h) magnesium + copper(II) nitrate $\longrightarrow$

(i) magnesium + phosphoric acid $\longrightarrow$

DOUBLE-REPLACEMENT REACTIONS

(j) phosphoric acid + calcium hydroxide $\longrightarrow$

(k) $AgNO_3 + BaCl_2 \longrightarrow$

(l) $H_2SO_4 + NaOH \longrightarrow$

SOLUTION   COMBINATION REACTIONS

(a) $K + Cl_2 \longrightarrow KCl$ (It's KCl because in compounds
                    K is $+1$ and Cl is -1.)
$2K + Cl_2 \longrightarrow 2KCl$ (balanced)

(b) $SO_3 + H_2O \longrightarrow H_2SO_4$ (balanced)

(c) $H_2 + I_2 \longrightarrow HI$
$H_2 + I_2 \longrightarrow 2HI$ (balanced)

(d) $NH_3 \longrightarrow N_2 + H_2$ (Remember, nitrogen and hydrogen are diatomic elements in their free states.)

$2NH_3 \longrightarrow N_2 + 3H_2$ (balanced)

(e) $MgCO_3 \longrightarrow MgO + CO_2$ (balanced)

(f) $H_2O \longrightarrow H_2 + O_2$
$2H_2O \longrightarrow 2H_2 + O_2$ (balanced)

SINGLE-REPLACEMENT REACTIONS

(g) $Zn + HCl \longrightarrow ZnCl_2 + H_2$
$Zn + 2HCl \longrightarrow ZnCl_2 + H_2$ (balanced)

(h) $Mg + Cu(NO_3)_2 \longrightarrow Mg(NO_3)_2 + Cu$ (balanced)

(i) $Mg + H_3PO_4 \longrightarrow Mg_3(PO_4)_2 + H_2$
$3Mg + 2H_3PO_4 \longrightarrow Mg_3(PO_4)_2 + 3H_2$ (balanced)

DOUBLE-REPLACEMENT REACTIONS

(j) $H_3PO_4 + Ca(OH)_2 \longrightarrow Ca_3(PO_4)_2 + H_2O$
$2H_3PO_4 + 3Ca(OH)_2 \longrightarrow Ca_3(PO_4)_2 + 6H_2O$ (balanced)

(k) $AgNO_3 + BaCl_2 \longrightarrow AgCl + Ba(NO_3)_2$
$2AgNO_3 + BaCl_2 \longrightarrow 2AgCl + Ba(NO_3)_2$ (balanced)

(l) $H_2SO_4 + NaOH \longrightarrow Na_2SO_4 + H_2O$
$H_2SO_4 + 2NaOH \longrightarrow Na_2SO_4 + 2H_2O$ (balanced)

Oxidation-Reduction Reactions: Another Way to Classify Reactions

We've just finished learning about four types of inorganic reactions: combination, decomposition, single replacement, and double replacement. Now here we are again with a fifth type—the *oxidation–reduction reaction*. Actually, however, this is not really a fifth type of reaction, because some of the reactions we've already studied fall into this category of reaction. Let's just say that this is simply another way to classify the reactions we've just studied. In other words, reactions may be classified as oxidation–reduction reactions (called *redox reactions*, for short), or nonredox reactions. You'll see what we mean in a moment, but first let's understand what the terms oxidation and reduction mean.

Oxidation and Reduction Defined

In the past, oxidation was defined as the combination of a substance with oxygen. For example, glucose in our cells reacts with oxygen to produce carbon dioxide and water.

$$C_6H_{12}O_6 + 6O_2 \longrightarrow 6CO_2 + 6H_2O \quad \text{(balanced)}$$

Another example is the rusting of an iron pipe.

$$4Fe + 3O_2 \longrightarrow 2Fe_2O_3 \quad \text{(balanced)}$$

However, even though this is a useful definition, the concept of oxidation has been expanded to include reactions where oxygen is not involved. *Oxidation* is presently defined as

> the loss of electrons by a substance undergoing a chemical reaction or, in other words, the change in oxidation number of a substance to a more positive oxidation number.

For example, in the reaction of iron plus oxygen, the oxidation number of iron changes from 0 to +3, because each iron atom gives up three electrons in the reaction. (Remember, the oxidation number of an element in its uncombined state is zero.)

$$4Fe^0 + 3O_2 \longrightarrow 2Fe_2^{+3}O_3$$

By our definition this means that the iron is oxidized. But what happens to the oxygen? We say that the oxygen has been reduced. *Reduction* is defined as

> the gain of electrons by a substance undergoing a chemical reaction or, in other words, the change in oxidation number of a substance to a more negative oxidation number.

In the reaction of iron plus oxygen, the oxidation number of the oxygen changes from 0 to -2, because each oxygen atom gains two electrons from the iron.

$$4Fe + 3O_2^{\,0} \longrightarrow 2Fe_2O_3^{\,-2}$$

Notice that in this reaction one substance (the iron) gets oxidized, while the other substance (oxygen) gets reduced. The processes of oxidation and reduction always occur together. In other words, you can't have oxidation in a reaction unless there is also reduction. And you can't have reduction in a reaction unless there is also oxidation.

EXAMPLE 5-5 For each reaction, see if you can determine which substance gets oxidized and which substance gets reduced.

(a) $2Na + Cl_2 \longrightarrow 2NaCl$
(b) $Zn + 2HCl \longrightarrow ZnCl_2 + H_2$
(c) $C + O_2 \longrightarrow CO_2$
(d) $2HgO \longrightarrow 2Hg + O_2$

SOLUTION   Use your knowledge of oxidation numbers from Chapter 4. Also remember that the oxidation number of an element in its uncombined state is zero.

(a) $2Na^0 + Cl_2{}^0 \longrightarrow 2Na^{+1}Cl^{-1}$

The oxidation number of the sodium changes from 0 to $+1$. This means that the sodium is oxidized. The oxidation number of the chlorine changes from 0 to -1. This means that the chlorine is reduced.

(b) $Zn^0 + 2H^{+1}Cl^{-1} \longrightarrow Zn^{+2}Cl_2{}^{-1} + H_2{}^0$

The oxidation number of the zinc changes from 0 to $+2$. This means that the zinc is oxidized. The oxidation number of the hydrogen changes from $+1$ to 0. This means that the hydrogen is reduced. The oxidation number of the chlorine does not change. It is neither oxidized or reduced.

(c) $C^0 + O_2{}^0 \longrightarrow C^{+4}O_2{}^{-2}$

The oxidation number of the carbon changes from 0 to $+4$. This means that the carbon is oxidized. The oxidation number of the oxygen changes from 0 to -2. This means that the oxygen is reduced.

(d) $2Hg^{+2}O^{-2} \longrightarrow 2Hg^0 + O_2{}^0$

The oxidation number of the oxygen changes from -2 to 0. This means that the oxide ion is oxidized. The oxidation number of the mercury changes from $+2$ to 0. This means that the mercury(II) ion is reduced.

Oxidizing Agents and Reducing Agents: They Need No Commission

The purpose here is not to make things more difficult for you by adding new terminology. But it is important for you to become familiar with some words you will be seeing frequently. The words are oxidizing agent and reducing agent. Simply defined,

an oxidizing agent is a substance that causes something else to be oxidized, and

a reducing agent is a substance that causes something else to be reduced.

As things work out, the oxidizing agent is usually the substance being reduced. And the reducing agent is usually the substance being oxidized. Let's take, for example, the reaction involving iron and oxygen to form iron(III) oxide.

$$4Fe + 3O_2 \longrightarrow 2Fe_2O_3$$

The iron is the substance being oxidized, and the oxygen is the substance being reduced. Therefore, the iron is the reducing agent, and the oxygen is the oxidizing agent.

$$4Fe^0 \; + \; 3O_2{}^0 \; \longrightarrow \; 2Fe_2{}^{+3}O_3{}^{-2}$$

$$\begin{array}{cc} \text{Oxidized} & \text{Reduced} \\ \text{(reducing} & \text{(oxidizing} \\ \text{agent)} & \text{agent)} \end{array}$$

EXAMPLE 5-6   Using the reactions in Example 5-5, state which substance is the oxidizing agent, and which substance is the reducing agent.

SOLUTION   Just remember that the substance oxidized is the reducing agent, and the substance reduced is the oxidizing agent.

(a) $2Na^0 \; + \; Cl_2{}^0 \; \longrightarrow \; 2Na^{+1}Cl^{-1}$

$$\begin{array}{cc} \text{Oxidized} & \text{Reduced} \\ \text{(reducing} & \text{(oxidizing} \\ \text{agent)} & \text{agent)} \end{array}$$

(b) $Zn^0 \; + 2H^{+1}Cl^{-1} \; \longrightarrow \; Zn^{+2}Cl_2{}^{-1} + H_2{}^0$

$$\begin{array}{cc} \text{Oxidized} & \text{Reduced} \\ \text{(reducing} & \text{(oxidizing} \\ \text{agent)} & \text{agent)} \end{array}$$

Actually, the compound HCl is considered to be the oxidizing agent.

(c) $C^0 \; + \; O_2{}^0 \; \longrightarrow \; C^{+4}O_2{}^{-2}$

$$\begin{array}{cc} \text{Oxidized} & \text{Reduced} \\ \text{(reducing} & \text{(oxidizing} \\ \text{agent)} & \text{agent)} \end{array}$$

(d) $2Hg^{+2}O^{-2} \; \longrightarrow \; 2Hg^0 + O_2{}^0$

The mercury(II) ion is reduced. The oxide ion is oxidized. The compound, mercury(II) oxide, acts as both the oxidizing agent and reducing agent.

The Activity Series

After many experiments, chemists have been able to arrange a list of elements according to their reactivities (Table 5-1). In this table, called the *activity series*, the most active (or readily reactive) elements appear at the top. As we move from the top to the bottom of the table, the reactivity of the elements decreases. Each element in the list can replace any element below it in a compound; that is, by a single-replacement reaction, it can take the other element's place. For example,

$$2Al + 3Zn(NO_3)_2 \; \longrightarrow \; 2Al(NO_3)_3 + 3Zn$$

Aluminum replaces zinc in the compound zinc nitrate, since aluminum is higher than zinc in the activity series.

The activity series helps us determine whether a reaction can take place. It is especially helpful when we are dealing with single-replacement reactions.

EXAMPLE 5-7   Predict whether the following reactions can occur.

(a) $Cu + 2HCl \xrightarrow{\text{heat}} CuCl_2 + H_2$

(b) $H_2 + CuO \xrightarrow{\text{heat}} H_2O + Cu$

SOLUTION   (a) When we look at Table 5-1, we see that copper is below hydrogen in the activity series. Therefore, copper cannot replace hydrogen in the compound HCl.

$$Cu + HCl \longrightarrow \text{no reaction}$$

(b) Looking at Table 5-1, we see that hydrogen is above copper in the activity series. Therefore, this reaction can occur, since hydrogen can replace copper in the compound CuO.

$$H_2 + CuO \longrightarrow H_2O + Cu$$

TABLE 5-1   The activity series

METALS	NONMETALS
Lithium	Fluorine
Potassium	Chlorine
Calcium	Bromine
Sodium	Iodine
Magnesium	
Aluminum	
Zinc	
Chromium	
Iron	
Nickel	
Tin	
Lead	
Hydrogen[a]	
Copper	
Mercury	
Silver	
Platinum	
Gold	

[a] Hydrogen is in italic type because the activities of the other elements were calculated relative to hydrogen.

SUMMARY

In this chapter we learned how to write a formula equation from a chemical word equation. We also learned how to balance chemical equations using a trial-and-error method. We saw that in order to balance a chemical equation, we must use coefficients, which we place in front of the formulas of each substance. We also discovered that there are four types of inorganic reactions: combination reactions, decomposition reactions, single-replacement reactions, and double-replacement reactions. We looked at examples of each type of reaction. We also learned about redox reactions, and how we could tell whether a given substance is being oxidized or reduced. We also learned the meaning of oxidizing agent and reducing agent. Finally, we looked at something called the activity series, and saw how this series could help us predict the products of a single-replacement reaction.

EXERCISES

1. Write a balanced chemical equation from the following word equations.
    (a) potassium + water $\longrightarrow$ potassium hydroxide + hydrogen gas
    (b) magnesium + copper(II)nitrate $\longrightarrow$ magnesium nitrate + copper
    (c) sulfuric acid + iron(III) hydroxide $\longrightarrow$ iron(III) sulfate + water
    (d) sodium + bromine $\longrightarrow$ sodium bromide
    (e) aluminum + mercury(II) acetate $\longrightarrow$ aluminum acetate + mercury

2. Identify each of the reactions in Exercise 1, as a combination reaction, decomposition reaction, single-replacement reaction, or double-replacement reaction.

3. For each of the combination reactions that follow, predict the product(s) and balance the equation.
    (a) $Mg + O_2 \xrightarrow{\text{heat}}$
    (b) $H_2 + I_2 \xrightarrow{\text{heat}}$
    (c) $SO_3 + H_2O \longrightarrow$
    (d) $N_2 + H_2 \xrightarrow[\text{high pressure}]{\text{heat}}$

4. For each of the decomposition reactions that follow, predict the product(s) and balance the equation.
    (a) $H_2O \xrightarrow{\text{electrolysis}}$
    (b) $HgO \xrightarrow{\text{heat}}$
    (c) $MgCO_3 \xrightarrow{\text{heat}}$
    (d) $NaCl \xrightarrow{\text{electrolysis}}$

5. For each of the single-replacement reactions that follow, predict the product(s) and balance the equation. Don't forget to check the activity series to make sure that the reaction occurs.

(a) $Mg + HCl \longrightarrow$
(b) $Cu + H_2SO_4 \longrightarrow$
(c) $Zn + Ni(NO_3)_2 \longrightarrow$
(d) $Mg + H_3PO_4 \longrightarrow$

6. For each of the double-replacement reactions that follow, predict the products and balance the equation. Don't forget to check the solubility table to make sure that the reaction occurs.
   (a) $HBr + Mg(OH)_2 \longrightarrow$
   (b) sulfuric acid + iron(III) hydroxide $\longrightarrow$
   (c) $NaCl\ (aq) + Ba(NO_3)_2\ (aq) \longrightarrow$
   (d) barium chloride (aq) + sodium sulfate $(aq) \longrightarrow$

7. For each of the redox reactions that follow, determine which substance gets oxidized and which substance gets reduced.
   (a) $Mg + H_2SO_4 \longrightarrow MgSO_4 + H_2$
   (b) $2Na + 2H_2O \longrightarrow 2NaOH + H_2$
   (c) $2Hg + O_2 \longrightarrow 2HgO$
   (d) $2Al + Fe_2O_3 \longrightarrow Al_2O_3 + 2Fe$

8. For each of the redox reactions in Exercise 7, determine which substance is the oxidizing agent, and which substance is the reducing agent.

9. The redox reactions in Exercise 7 may also be classified as combination reactions, decomposition reactions, single-replacement reactions, or double-replacement reactions. Determine, for each example, the type of reaction it is.

10. For each of the reactions that follow, write a balanced chemical equation. Be sure to check the activity series or solubility table if necessary.
    (a) potassium + water $\longrightarrow$ potassium hydroxide + hydrogen gas
    (b) $H_2 + Br_2 \longrightarrow$
    (c) lithium iodide $\xrightarrow{\text{electrolysis}}$
    (d) $Zn + Cu(NO_3)_2 \longrightarrow$
    (e) $CuCl_2(aq) + Na_3PO_4(aq) \longrightarrow$
    (f) $Ag + ZnCl_2 \longrightarrow$
    (g) sulfurous acid (H_2SO_3) + oxygen $\longrightarrow$ sulfuric acid
    (h) $SrCl_2(aq) + KNO_3(aq) \longrightarrow$

CHAPTER 6

The Three Faces of Matter

Solid, Liquid, and Gas

Some Things You Should Know After Reading This Chapter

You should be able to:

1. Name the three states of matter: solid, liquid, and gas.
2. Compare and contrast the properties of solids, liquids, and gases.
3. Describe what is meant by crystalline and amorphous solids.
4. Describe the differences among atomic, molecular, and ionic crystals.
5. Define what is meant by a crystal lattice.
6. Describe the forces that hold crystals together.
7. Describe the process of evaporation.
8. Explain what happens when a solid changes to a liquid, and a liquid changes to a gas.
9. Discuss the major points of the kinetic theory of gases.
10. Change °C to °K and °K to °C.
11. Change atmospheres to torr and torr to atmospheres.
12. Use the combined gas law to solve gas law problems.
13. Explain Dalton's law of partial pressures.
14. Explain Henry's law.
15. Discuss the process of respiration in terms of how oxygen moves from the lungs to the cells of the body, and how carbon dioxide moves from the cells of the body to the lungs.
16. Explain the concept behind oxygen hyperbaric therapy, and give some of its uses.

Of course, the title of this chapter should be *The Three Phases of Matter: Solid, Liquid, and Gas.* But in all seriousness, *The Three Faces of Matter...* seems to indicate so much more. This is because the *three phases of matter*, or *states* as they are called, appear to differ so much from one another—even for the same chemical compound (Fig. 6-1). But like an actor who wears different faces for different roles, yet is the same person underneath the makeup, so are the three states of matter. A comparison of the three states of matter can be seen in Table 6-1.

Each of the three states of matter is important to living organisms. Our bodies are composed of solids, liquids, and gases. Solids compose our bones,

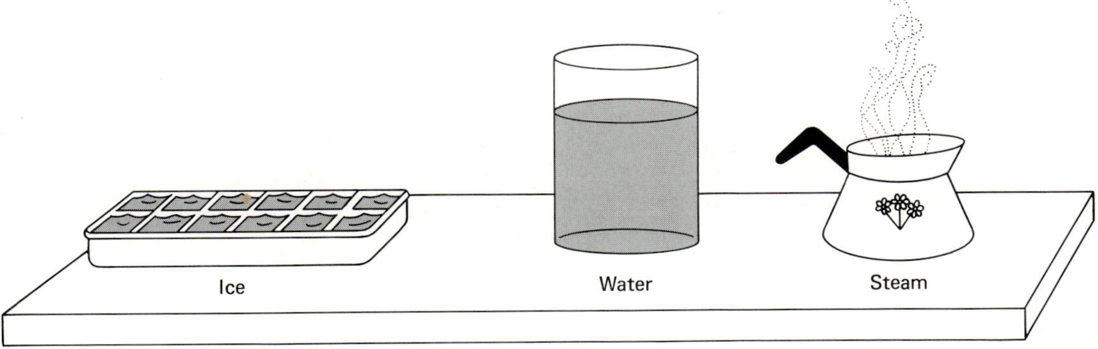

FIGURE 6-1
The three states of H_2O: solid ice, liquid water, and gaseous steam.

TABLE 6-1   Properties of solids, liquids, and gases

PROPERTY	SOLIDS	LIQUIDS	GASES
Volume	Have a definite volume	Have a definite volume	Have no definite volume (gases expand to fill their containers)
Shape	Have a definite shape (most solids have a well-defined crystalline arrangement)	Have no definite shape (liquids assume the shape of their containers)	Have no definite shape
Compressibility	Cannot be compressed (they are practically incompressible)	Tend to be incompressible	Are easily compressed

skin, and teeth. Liquids compose the fluids in our cells and blood. And gases are used in our bodies in the process of respiration. And let's not forget that the foods we eat are also composed of solids, liquids, and gases. For these reasons and more, it is important that we study the three states of matter.

The Solid State

The major characteristics of *solids* are their fixed shape and fixed volume. Most solids are of the crystalline variety. This means that the particles which compose the solid are arranged in a definite geometric pattern (Fig. 6-2). Some of these geometric patterns are shown in Table 6-2. Some solids, such as glass and paraffin, have no definite geometric pattern. In other words, they are not crystalline. These solids are said to be *amorphous*. Although amorphous solids are interesting to study, we'll confine our discussion to crystalline solids.

Solids can be composed of three types of particles: atoms, molecules, and ions. Solids such as carbon, silicon, copper, and gold are composed of atoms. Solids such as dry ice (CO_2), iodine, and water (ice) are composed of molecules. Solids such as sodium chloride, calcium chloride, and sodium bromide are composed of ions. These three types of solids have very different properties. Table 6-3 sums up some of the differences.

FIGURE 6-2
A crystal of sodium chloride. (American Museum of Natural History)

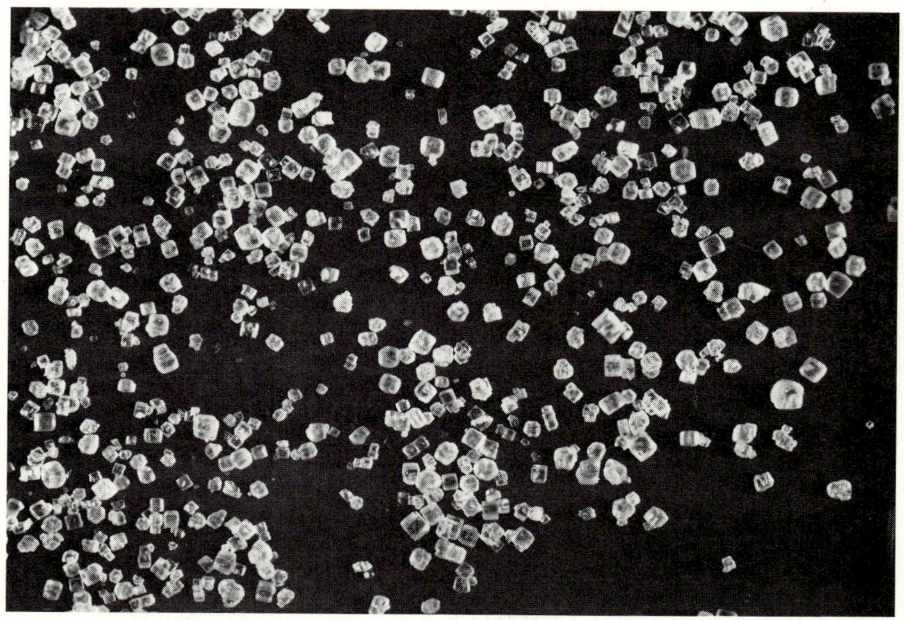

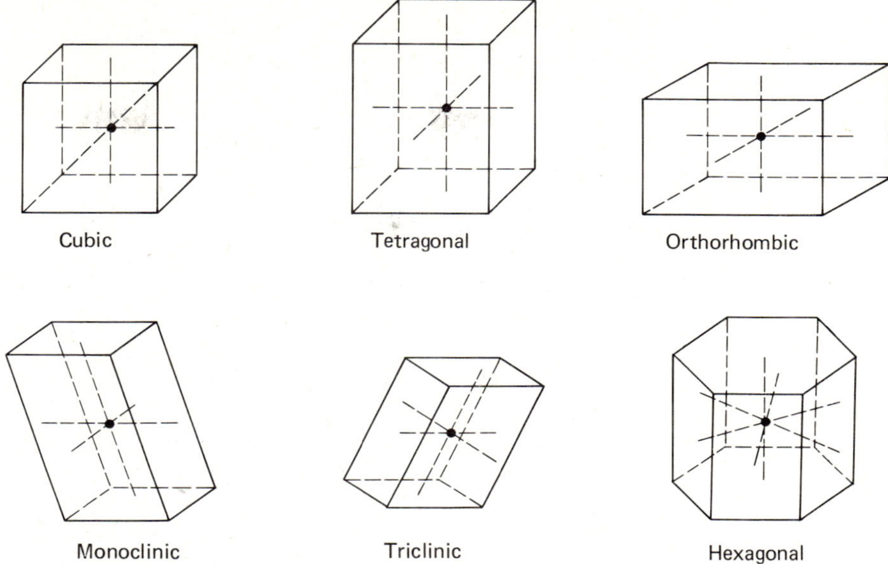

TABLE 6-3   Some properties of crystalline solids

TYPE	MELTING POINT	HARDNESS	CONDUCTIVITY	EXAMPLE
Ionic	High	Hard and brittle	Nonconductors of electricity	Sodium chloride, NaCl
Molecular	Low	Soft	Nonconductors of electricity	Dry ice, CO_2
Atomic				
Metallic	High	Hard	Good conductors of electricity	Copper, Cu
Nonmetallic	High	Hard and brittle	Nonconductors of electricity	Diamond, C

All About Crystals

A crystalline solid is obviously composed of crystals. But what is a crystal? The dictionary says that a *crystal* is *a solid body having a characteristic internal structure and enclosed by symmetrically arranged plane surfaces, intersecting at definite and characteristic angles.* In simpler terms this means that a crystal is a symmetrical object. The symmetrical structure formed by the

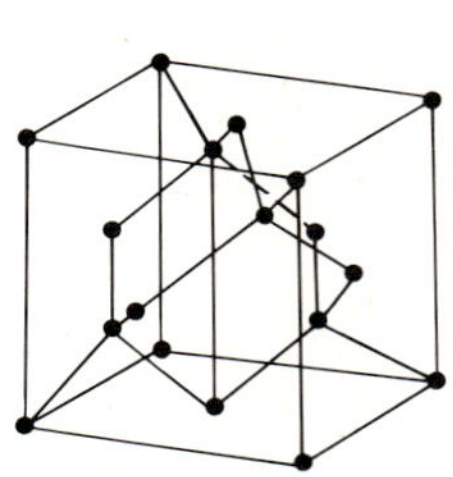
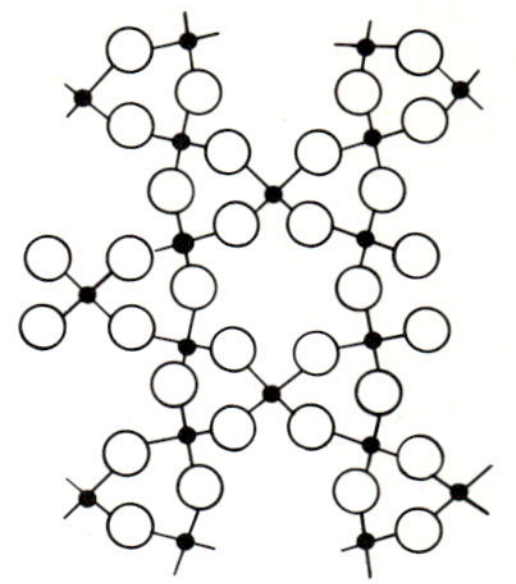
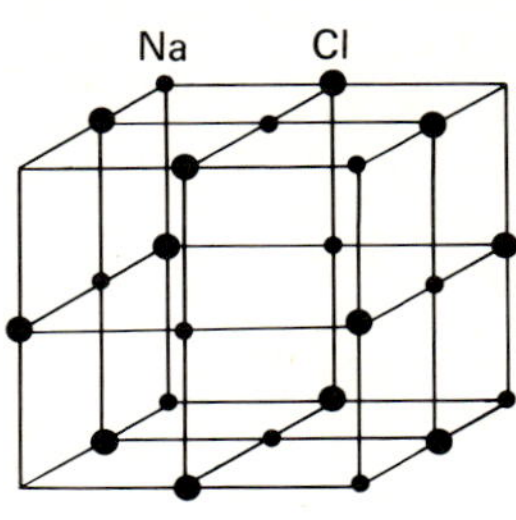

Diamond

Atomic solid
(atoms occupy the
lattice positions)

Silicon dioxide

Molecular solid
(molecules occupy
the lattice positions)

Sodium chloride

Ionic solid
(ions occupy the
lattice positions)

FIGURE 6-3
Some examples of different types of crystalline solids.

particles in a crystal is called the *crystal lattice*. The crystal lattice is the repeating unit within the crystal that gives the crystal its pattern. Actually, the crystal lattice is repeated over and over again in the crystal. One repeating unit, however, is called a *unit cell*. As we said before, there are three types of particles that compose solids: atoms, molecules, and ions. It is also these three particles that occupy the lattice positions in a crystal (Fig. 6-3).

Atomic, Molecular, and Ionic Crystals:
A Look at What Holds Them Together

The properties of a crystalline solid are partly the result of the forces that bind together the particles that make up the crystal. Let's look briefly into the bonding forces of the three basic types of crystals (ionic, molecular, and atomic).

In an *ionic* crystal, it is the attractive forces between oppositely charged ions that hold the crystal together. Since this kind of bonding force is very strong, ionic solids have very high melting points. For example, table salt (NaCl) has a melting point of 800°C. It takes a lot of heat energy to force those ions apart.

Molecular crystals, on the other hand, are usually held together by weak electrical forces. Only molecular crystals whose molecules are polar have binding forces that are fairly strong. The forces binding most molecular crystals together are so weak that these crystals frequently have much lower melting points than ionic crystals. For example, ice melts at 0°C.

Solids composed of *atomic* crystals have strong binding forces. Table 6-4 shows how strong. Look at the melting points of some of these substances. When the atoms are nonmetallic, they are held together by covalent bonds. An example is diamond, made up of carbon atoms that bond covalently. One of the reasons that diamonds are a girl's best friend is that it takes a temperature of 3500°C to break those bonds.

SUBSTANCE	MELTING POINT (°C)
Iron	1535
Copper	1083
Diamond	3500
Silicon	1410

When atomic crystals are composed of metallic atoms, it is the attraction of opposite charges that makes the bonds so strong. Scientists think of the metallic lattice as being composed of positive ions surrounded by a cloud of electrons. The electrons are given off by the metallic atoms, but are considered to be part of the entire crystal—electrons at large, you might say. Since the electrons are free to wander throughout the crystal, they are able to conduct electricity. This is the reason that metals are good conductors of electricity.

The Liquid State

The major characteristics of *liquids* are their fixed volume and indefinite shape. This means that the particles of a liquid are not held together as rigidly as a solid. In other words, the particles of a liquid may move about each other, so that they can occupy the shape of their container. The fact that the particles of a liquid can move about gives rise to an interesting property of liquids called evaporation.

FIGURE 6-4
Liquid molecules move about and collide with one another. In these collisions some liquid molecules transfer energy to other liquid molecules.

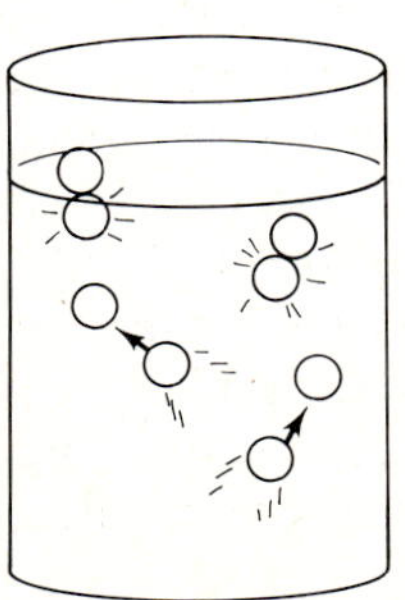

Evaporation: The Case of the Disappearing Liquid

When a liquid *evaporates* it seems to disappear. But actually, the molecules of liquid have turned into molecules of gas. If you leave a glass of water on a table for a few days, you will notice that the water level in the glass drops due to the process of evaporation. But what causes the evaporation to take place? Surely the temperature of the liquid as well as the exposed surface area of the liquid should affect the speed of evaporation. But there's more to the process than that and it has to do with the fact that *the molecules of a liquid are in constant motion about each other*. This means that the molecules of liquid have a specific amount of energy, and as these molecules move about and collide with each other, some liquid molecules transfer some of their energy to other liquid molecules (Fig. 6-4). If a molecule on its way up to the surface of the liquid collides with a molecule already on the surface, the surface molecule may get enough energy from the collision to escape from the liquid. After this process has continued for a period of time, we notice that there is less liquid in the container. We say that the liquid has evaporated.

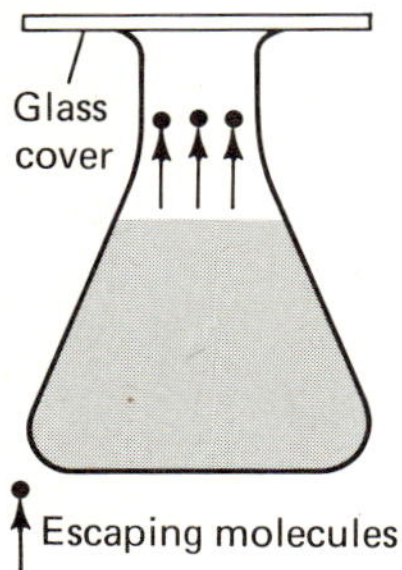

FIGURE 6-5
Liquid molecules escaping from the surface of the liquid.

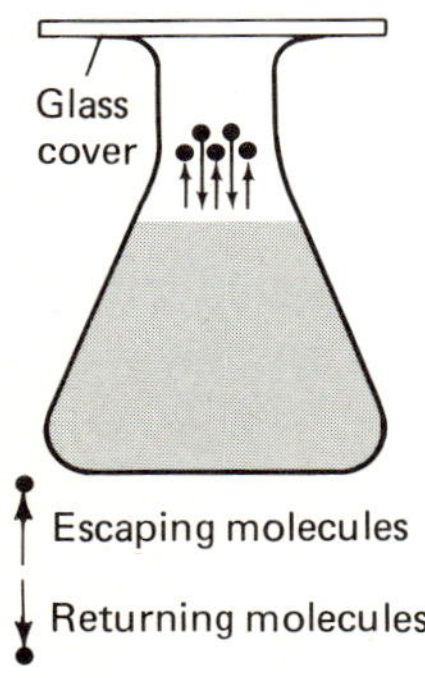

FIGURE 6-6
A dynamic equilibrium.

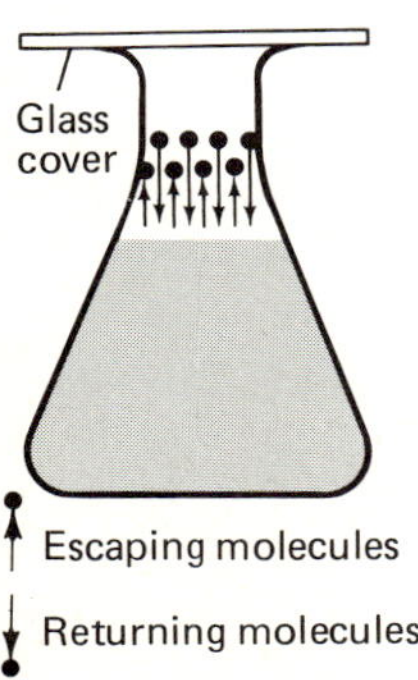

FIGURE 6-7
Liquids have a higher equilibrium vapor pressure at a higher temperature.

As a liquid evaporates, its temperature decreases. This is because the more energetic molecules (the warmer ones) are leaving. Think how important this is to us! When we exercise or perform hard work, heat is brought to the skin, where it is transferred to water molecules. These molecules of water evaporate and our skin feels cooler. In this manner our body temperature can be maintained at 37°C (98.6°F). This is extremely critical for good health because at temperatures a few degrees above 37°C we can suffer from heat prostration and even death as a result of the malfunction of our central nervous system. This critical temperature is about 41°C (106°F). (If you don't know how to convert from °C to °F, and vice versa, see Appendix B.)

Equilibrium Vapor Pressure

If we put a cover on a container that is partly filled with water, evaporation begins, but it seems to stop after a time. How can we explain this? Picture the situation shown in Figure 6-5. Water molecules begin to escape from the surface of the liquid. These molecules occupy the air space in the partly filled container. As this process continues, more and more escaping molecules occupy the same space. This increases the chance that they will collide with one another, and that some of them will be bumped back into the water. At first many molecules escape, and few return; but after a while, as the concentration of the molecules in the vapor state increases, the number of molecules returning to the liquid also increases. Eventually we reach a point at which the number of molecules escaping from the water equals the number of molecules returning to it. The net result is that no additional liquid seems to evaporate from the container. But, in reality, a *dynamic equilibrium* is in progress. This means that the two opposing processes are proceeding at equal rates (Fig. 6-6).

The pressure of the molecules that have escaped in our partly filled container at the equilibrium point is called the *equilibrium vapor pressure* of the liquid. We can define the equilibrium vapor pressure of any substance as the pressure exerted by a vapor when it is in equilibrium with its liquid at any given temperature. For every liquid at a given temperature there is a characteristic vapor pressure at equilibrium.

Let's go back to our covered container that is partly filled with water. Remember that the water and the water vapor are at equilibrium. What happens to the equilibrium vapor pressure if we raise the temperature of the water? The water molecules should gain energy as their temperature is increased. As a result, a greater number of molecules should have enough energy to escape from the surface of the liquid and become vapor. In fact, that's just what happens. The equilibrium that had existed is now upset, but it eventually reestablishes itself at the new temperature. The overall effect is a higher equilibrium vapor pressure at the higher temperature (Fig. 6-7).

Comparing equilibrium vapor pressures of different liquids at the same temperature tells us how *volatile* a liquid is (that is, how readily it vaporizes). Liquids that have high equilibrium vapor pressures tend to evaporate easily. Some examples are gasoline, alcohol, and ether. Liquids that have low equilibrium vapor pressures tend to evaporate less easily. Examples of these are motor oil, water, and glycerine.

The Phenomenon of Boiling

Using our information about the equilibrium vapor pressures of liquids, let's now try to explain the phenomenon of boiling.

If we heat a beaker of water over a flame, bubbles form on the bottom of the beaker. What are these bubbles? Is it steam being formed? No, these first bubbles are merely dissolved air (oxygen, nitrogen, and so forth) being driven out of the water by the heat. As the temperature rises, the molecules become more active. Soon other bubbles form and begin to rise. These bubbles are water vapor (steam). The reason they form on the bottom of the beaker is that the water is hottest there. These bubbles disappear as they rise and reach the cooler part of the water. After a while the bubbles stop disappearing and rise all the way to the surface. At this point we say that the water is *boiling*. Actually, the vapor pressure exerted by the bubbles of steam has become equal to the atmospheric pressure (Fig. 6-8).

The *boiling point* of a liquid is the temperature at which the equilibrium vapor pressure of the liquid equals the atmospheric pressure. Knowing this,

FIGURE 6-8

Boiling water: the vapor pressure exerted by the bubbles of steam has become equal to the atmospheric pressure.

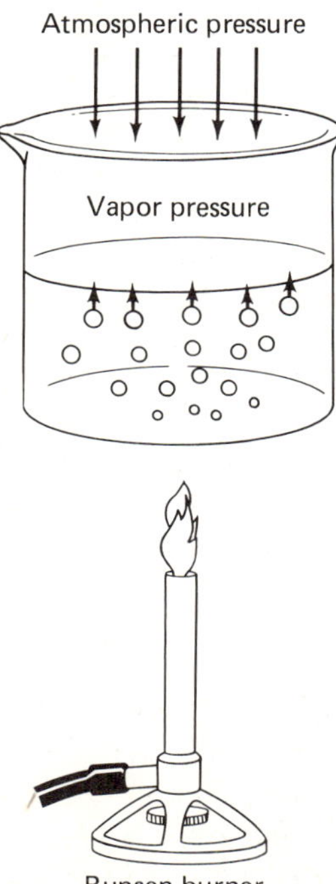

PRESSURE (torr)	BOILING POINT (°C)
18	20
355	80
526	90
760	100
906	105
2026	130

we can say that increasing the atmospheric pressure on the liquid should increase the boiling point. By the same token, decreasing the atmospheric pressure should decrease the boiling point. Table 6-5 shows that this is actually true. So remember that when we discuss the normal boiling point of a liquid, we are talking about its boiling point at what is called *normal pressure*—that is, 1 atmosphere (atm).

FIGURE 6-9
How water at 1 atm pressure changes from a solid to a liquid to a gas. Below 0°C water is in the solid state—ice. In this diagram we begin with ice at −10°C. Energy (heat) is supplied so that the temperature of the ice rises. When it reaches 0°C, the ice starts to melt into water. As more energy is supplied, more ice melts into water. This water remains at 0°C until all the ice melts. After all the ice has melted, the energy supplied starts to warm the water. This continues until the water temperature reaches 100°C. At this point the water starts to vaporize into steam. The temperature of the water remains constant at 100°C as the liquid is turned into gas.

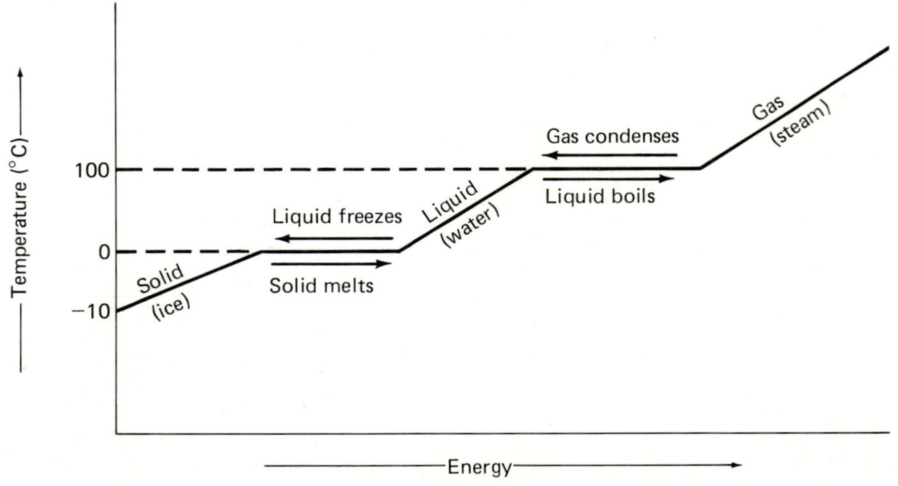

If we measure the temperature of the liquid at its boiling point, we find that it remains constant. (The boiling liquid is of course the same temperature as its vapor.) For example, if you heat a container of water, you find that it boils at 100°C (at 1 atm pressure) and remains at this temperature *regardless of how much heat you give to the water.* The reason that the temperature of the water remains constant during boiling is that it takes energy to change the liquid water into steam. The heat energy you are supplying is used to separate the molecules of water, so that the water changes from the arrangement of a liquid to the arrangement of a gas (Fig. 6-9).

The Gaseous State

The major characteristics of *gases* are their indefinite shape and volume. This means that the particles of a gas behave as if they are independent of each other. In other words, the molecules of a gas move about a container until they fill it. Gases have been extensively studied since the 1500's and scientists have discovered a number of their properties. We can summarize these as follows:

1. Nearly all gases are composed of molecules. (The so-called noble gases are composed of atoms, not molecules. Remember, atoms of noble gases don't combine readily with other atoms.)

2. The forces of attraction between gas molecules increase as the molecules move closer together. Under conditions of normal atmospheric pressure and room temperature, the distances between the gas molecules are large compared to the size of the molecules.

3. Gas molecules are always in motion. They often collide with other gas molecules or with their container. After collisions occur, the molecules do not stick together; and they do not lose any energy as a result of the collisions.

4. Gas molecules move faster when the temperature rises and slower when the temperature falls.

5. All gas molecules (heavy as well as light) have the same average kinetic energy—that is, energy of motion—at the same temperature.

These five statements are known as the *kinetic theory of gases.* Keep these five statements in mind as we continue our study of gases.

The *P, V, T* of Gases

The kinetic theory can be used to explain the properties of gases. However, a more quantitative explanation can be obtained from a series of gas law equations. The first gas law was proposed by Robert Boyle in 1660, and related the pressure of a gas with its volume (at a constant temperature, although Boyle forgot to say so). Another gas law was discovered in 1780 by the French physicist Jacques Charles, and related the temperature of a gas with its volume (at constant pressure). In mathematical terms, *Boyle's law* can be stated as

$$\frac{P_i}{P_f} = \frac{V_f}{V_i} \qquad (6\text{-}1)$$

The P stands for pressure and the V for volume. The subscript i indicates initial conditions, and the subscript f indicates final conditions. The formula says that the volume of a gas varies inversely with pressure (at a constant temperature).

By the way, pressure is measured in units called *torr*, in honor of Evangelista *Torricelli*, the man who invented the barometer, a device used to measure air pressure.

Torricelli filled a glass tube, which was sealed on one end, with mercury. He then turned the tube upside down and placed it in a bowl, also filled with mercury (Fig. 6-10). Torricelli noticed that most of the mercury remained in the upright tube. He concluded that the surrounding air exerted pressure on the surface of the mercury in the bowl, which in turn supported the column of mercury in the tube.

At sea level the height of mercury supported in the tube is 760 mm. This height is now expressed in pressure units of torr. In other words, normal atmospheric pressure at sea level is 760 torr. For larger pressures the unit *atmosphere* is used.

$$1 \text{ atm} = 760 \text{ torr}$$

FIGURE 6-10
A simple barometer.

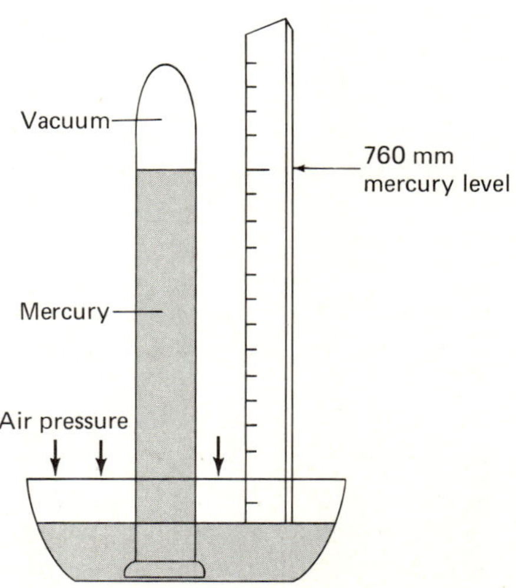

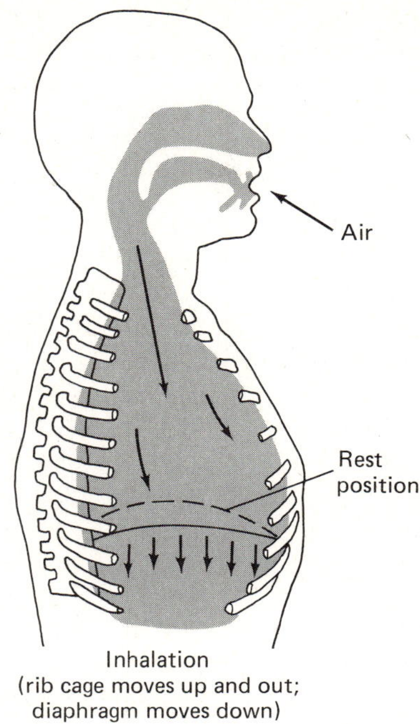

FIGURE 6-11
The process of breathing: inhalation.

Boyle's law allows us to understand the process of breathing. Our lungs are elastic, which means that they can expand and contract. They do so within an airtight chamber called the *thoracic cavity*. A muscle called the *diaphragm* is part of the thoracic cavity. The contraction of the diaphragm causes an *increase in the volume* of the thoracic cavity and a corresponding *increase in the volume of the lungs*. This causes a *pressure decrease in the lungs*. The pressure of the atmosphere is now greater than the pressure within the lungs, and this pressure difference causes air to rush into the lungs (Fig. 6-11). When the diaphragm returns to its normal position, the volume of the thoracic cavity decreases, and so does the volume of the lungs. This causes the air pressure in the lungs to be greater than atmospheric, and air flows out of the lungs (Fig. 6-12).

Now that we've discussed the importance of Boyle's law, let's see if we can solve a gas law problem involving equation 6-1.

EXAMPLE 6-1   An adult moves about $5\overline{0}0$ ml of air in and out of his or her lungs in one inhalation–exhalation cycle (this is called the *tidal volume*). Let's say that this occurs at $76\overline{0}$ torr pressure. If the pressure on the lungs should be increased to 2280 torr, what would be the volume of air in the lungs?

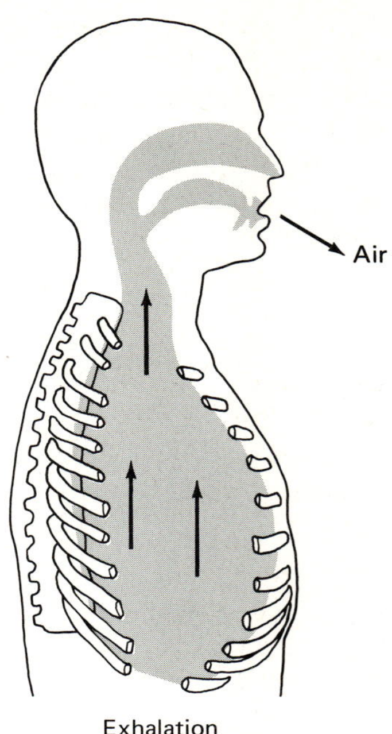

Exhalation
(resting position)

FIGURE 6-12
The process of breathing: exhalation.

SOLUTION   Let's organize the data.

$$P_i = 76\overline{0} \text{ torr} \qquad P_f = 2280 \text{ torr}$$

$$V_i = 50\overline{0} \text{ ml} \qquad V_f = ?$$

Now we'll write Boyle's law (equation 6-1), and solve it for V_f.

$$\frac{P_i}{P_f} = \frac{V_f}{V_i}$$

$$V_f = \frac{P_i V_i}{P_f}$$

$$= \frac{(76\overline{0} \text{ torr})(50\overline{0} \text{ ml})}{2280 \text{ torr}} = 167 \text{ ml}$$

Charles' Law and Kelvin Temperature

In mathematical terms, *Charles' law* can be stated as

$$\frac{V_i}{V_f} = \frac{T_i}{T_f}$$

The V stands for volume, and the capital T stands for temperature in degrees Kelvin. The *Kelvin temperature scale*, also called the *absolute temperature scale*, was named after Lord Kelvin, the scientist who proposed it. This temperature scale is related to the activity of gas molecules. Scientists theorize that at a temperature of $0°K$ ($-273°C$), the molecular motion of gas molecules would cease. The value of $0°K$ is sometimes called *absolute zero*. The relationship between Celsius ($°C$) and Kelvin ($°K$) temperature is

$$°K = °C + 273 \qquad (6\text{-}3)$$

By using the Kelvin temperature in the Charles' law equation, we can express the relationship between temperature and volume as a direct proportion. In other words, the volume of a gas varies directly with its Kelvin temperature, at constant pressure.

EXAMPLE 6-2   Figure 6-13 illustrates the situation discussed in this example. Let's say that we have $5\overline{0}0$ ml of a gas in a balloon at a temperature of $27°C$. What would be the volume of the balloon if the temperature of the gas is increased to $327°C$?

SOLUTION   Let's organize the data.

$$V_i = 5\overline{0}0 \text{ ml} \qquad V_f = ? \text{ ml}$$
$$t_i = 27°C \qquad t_f = 327°C$$

FIGURE 6-13
A balloon attached to a can has a temperature of 27°C and a volume of 500 ml. The can, with the balloon attached, is placed in a hot oil bath which has a temperature of 327°C. The air in the can and in the balloon soon reaches this temperature and the balloon expands, increasing its volume.

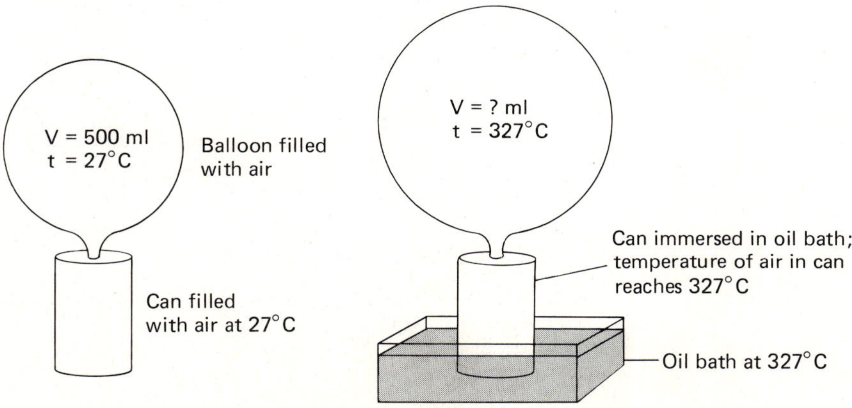

Use equation 6-3 to change °C to °K:

$$T_i = 27°C + 273 = 3\overline{00}°K$$

$$T_f = 327°C + 273 = 6\overline{00}°K$$

Now we'll write Charles' law (equation 6-2), and solve for V_f.

$$\frac{V_i}{V_f} = \frac{T_i}{T_f}$$

$$V_f = \frac{V_i T_f}{T_i}$$

$$= \frac{(5\overline{00}\ ml)(6\overline{00}°K)}{3\overline{00}°K} = 1\overline{000}\ ml$$

We can combine Boyle's law and Charles' law into a general law called the *combined gas law*:

$$\frac{P_i V_i}{T_i} = \frac{P_f V_f}{T_f} \tag{6-4}$$

This equation relates the initial pressure, volume, and Kelvin temperature of a gas with its final pressure, volume, and Kelvin temperature. If we are given any five quantities in equation 6-4, we can solve for the remaining quantity. Also, if one of the quantities in equation 6-4 should be a constant, the equation reduces to a *four-letter* equation. For example, if you are given a problem with T constant, equation 6-4 becomes

$$P_i V_i = P_f V_f$$

which is simply another form of Boyle's law. The following examples will show you how to use equation 6-4 to solve all types of P, V, T problems.

EXAMPLE 6-3 A gas has a volume of $1\overline{00}$ ml at a pressure of 3.00 atm. What will be the volume if the pressure is decreased to 0.500 atm? Assume that the temperature is constant.

SOLUTION Let's organize the data.

$$P_i = 3.00\ atm \qquad P_f = 0.500\ atm$$

$$V_i = 1\overline{00}\ ml \qquad V_f = ?\ ml$$

Now we'll use equation 6-4.

$$\frac{P_i V_i}{T_i} = \frac{P_f V_f}{T_f}$$

However, T is constant, so equation 6-4 becomes

$$P_i V_i = P_f V_f$$

Now we'll solve for V_f and plug in the values.

$$V_f = \frac{P_i V_i}{P_f}$$

$$= \frac{(3.00 \text{ atm})(10\overline{0} \text{ ml})}{0.500 \text{ atm}} = 60\overline{0} \text{ ml}$$

EXAMPLE 6-4   A gas is enclosed in a flexible container. It has a volume of $80\overline{0}$ ml at a pressure of $76\overline{0}$ torr and a temperature of 127°C. What temperature would be necessary to increase the volume to $160\overline{0}$ ml at constant pressure?

SOLUTION   Let's organize the data.

$$P_i = 76\overline{0} \text{ torr} \qquad P_f = 76\overline{0} \text{ torr}$$

$$V_i = 80\overline{0} \text{ ml} \qquad V_f = 160\overline{0} \text{ ml}$$

$$t_i = 127°\text{C} \qquad t_f = ?°\text{C}$$

Now we'll use equation 6-4.

$$\frac{P_i V_i}{T_i} = \frac{P_f V_f}{T_f}$$

However, P is constant, so equation 6-4 becomes

$$\frac{V_i}{T_i} = \frac{V_f}{T_f}$$

Before we substitute the values into the equation, we must first change t_i (127°C), into °K, using equation 6-3.

$$T_i = 127°\text{C} + 273 = 40\overline{0}°\text{K}$$

$$T_f = \frac{T_i V_f}{V_i}$$

$$= \frac{(40\overline{0}°\text{K})(160\overline{0} \text{ ml})}{80\overline{0} \text{ ml}} = 800°\text{K}$$

Now we'll change °K back to °C using equation 6-3.

$$°\text{K} = °\text{C} + 273$$

$$°\text{C} = °\text{K} - 273$$

Therefore,

$$t_f = 80\overline{0}°\text{K} - 273 = 527°\text{C}$$

EXAMPLE 6-5   If your lungs can hold $5\overline{0}0$ ml of air at $2\overline{0}°C$ and 1.0-atm pressure, what would happen if you took a deep breath, held it, then dove into the ocean where the pressure was 1.5 atm and the temperature was $1\overline{0}°C$? In other words, what would be the volume of air in your lungs under these conditions?

SOLUTION   Let's organize the data.

$$P_i = 1.0 \text{ atm} \qquad P_f = 1.5 \text{ atm}$$

$$V_i = 5\overline{0}0 \text{ ml} \qquad V_f = ? \text{ ml}$$

$$t_i = 2\overline{0}°C \qquad t_f = 1\overline{0}°C$$

Now we'll use equation 6-4:

$$\frac{P_i V_i}{T_i} = \frac{P_f V_f}{T_f}$$

Before we solve for V_f and substitute the values into the equation, we must first change t_i and t_f into °K.

$$T_i = 2\overline{0}°C + 273 = 293°K$$

$$T_f = 1\overline{0}°C + 273 = 283°K$$

$$V_f = \frac{P_i V_i T_f}{P_f T_i}$$

$$= \frac{(1.0 \text{ atm})(5\overline{0}0 \text{ ml})(283°K)}{(1.5 \text{ atm})(293°K)} = 320 \text{ ml}$$

Dalton's Law of Partial Pressures

It was around the year 1800 when John Dalton discovered an important property of gases, which came to be known as *Dalton's law of partial pressures*. This law states that the gases in a mixture do not affect each other unless they react chemically. In other words, the pressure exerted by each gas in a mixture remains the same and is not affected by other gases in the mixture. This means that the pressures of gases in a mixture are additive. Therefore, if you place oxygen gas in a container so that it has a pressure of 300 torr, and then add to it nitrogen gas with a pressure of 200 torr, the total pressure in the container will be 500 torr, the sum of the individual pressures. The individual pressure of each gas in the mixture is called the *partial pressure* of the gas (Fig. 6-14).

EXAMPLE 6-6   Our atmosphere is a mixture of gases. There are, however, three major components in this mixture: nitrogen gas, 78.1% by volume; oxygen gas, 20.9% by volume; and argon gas, 0.9% by volume. If normal atmospheric pressure is $76\overline{0}$ torr, what is the partial pressure of each gas?

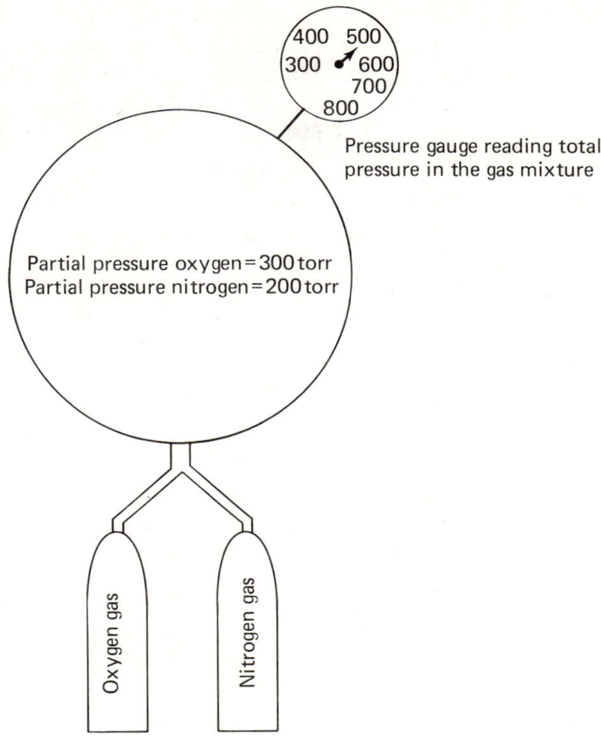

FIGURE 6-14
The sum of the partial pressures of each gas equals the total pressure of the gas mixture.

SOLUTION   Dalton's law tells us that the partial pressure of each gas should be proportional to its percentage in the gas mixture.

$$\text{Partial pressure of N}_2 = (0.781)(76\overline{0}\text{ torr}) = 594\text{ torr}$$

$$\text{Partial pressure of O}_2 = (0.209)(76\overline{0}\text{ torr}) = 159\text{ torr}$$

$$\text{Partial pressure of Ar} = (0.009)(76\overline{0}\text{ torr}) = 7\text{ torr}$$

Notice that the sum of the partial pressures add up to the total pressure of the gas mixture, $76\overline{0}$ torr.

We can use Dalton's law, and the fact that gases flow from areas of high pressure to areas of low pressure, to explain the movement of oxygen and carbon dioxide in our own bodies (Table 6-6). If you review Table 6-6, you will see that the partial pressure of oxygen in the air we inhale (*inspired air*) is about 158 torr, whereas it is only 100 torr in the *alveoli*, which are the small sacs at the end of the lungs. It is in the alveoli where the transfer of oxygen and carbon dioxide takes place (Fig. 6-15). Because of this difference in pressure,

GAS	INSPIRED AIR	ALVEOLI	ARTERIAL BLOOD AS IT PASSES THROUGH THE ALVEOLI	ARTERIAL BLOOD AFTER IT LEAVES ALVEOLI	TISSUES IN THE BODY	VENOUS BLOOD	EXPIRED AIR
Oxygen, O_2	158	100	40	100	35	40	116
Carbon dioxide, CO_2	0.3	40	40	40	50	46	28

oxygen, from inspired air, flows to the alveoli. Blood, which is circulating through the capillaries of the alveoli, has a partial pressure of oxygen of about 40 torr, so oxygen moves into the blood. The blood is now called *arterial blood*, and has a built up oxygen supply of about 100 torr.

FIGURE 6-15
It is in the alveoli that the transfer of oxygen and carbon dioxide takes place.

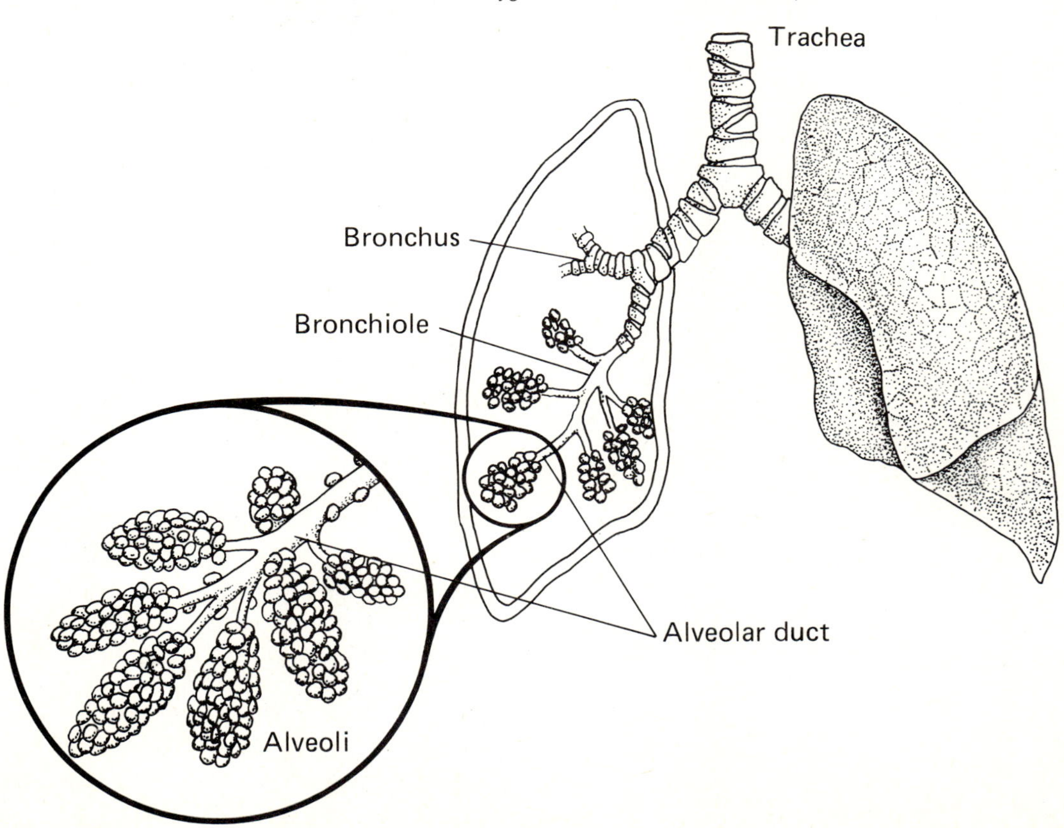

The oxygenated blood, which has passed through the alveoli, now continues on its way to the tissues of the body. The partial pressure of oxygen in these tissues is about 30 to 35 torr. This causes the oxygen to move quickly into the cells of these tissues (Fig. 6-16). The pressure of oxygen gas in the blood now drops to about 40 torr, at which point it becomes *venous blood* and is ready to return to the lungs to repeat the cycle.

At the same time that oxygen is moving from the lungs to the cells, the same forces are moving carbon dioxide from the cells to the lungs. The carbon dioxide pressure in an active cell is about 50 torr, whereas the carbon dioxide pressure in arterial blood is about 40 torr. This pressure difference causes the carbon dioxide to move out of the cells and into the blood. The blood (now called venous blood) has a carbon dioxide pressure of 46 torr by the time it reaches the alveoli. The carbon dioxide is now transferred to the alveoli, which only has a carbon dioxide pressure of 40 torr. From here the carbon dioxide

FIGURE 6-16

Flow of oxygen gas through the body. The oxygen moves from areas of high partial pressure to low partial pressure.

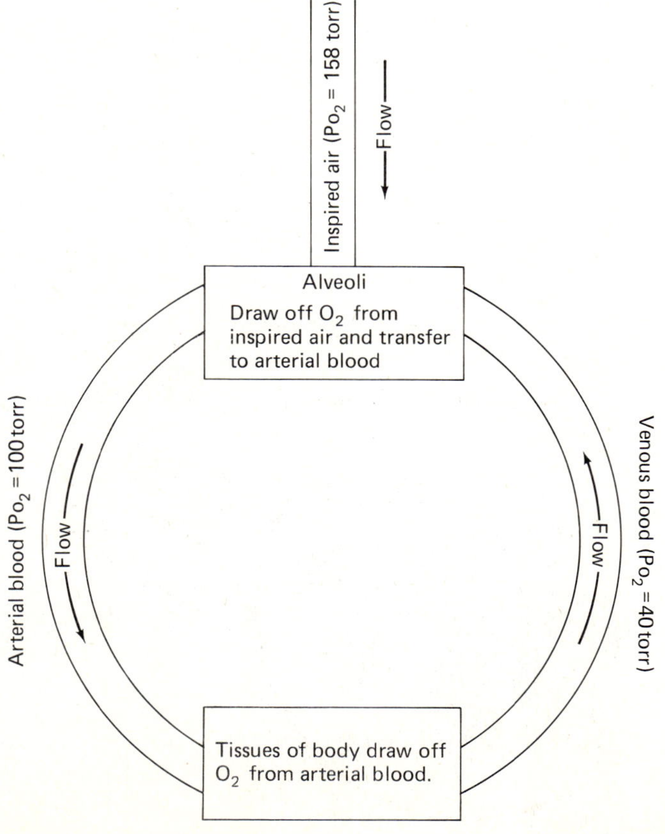

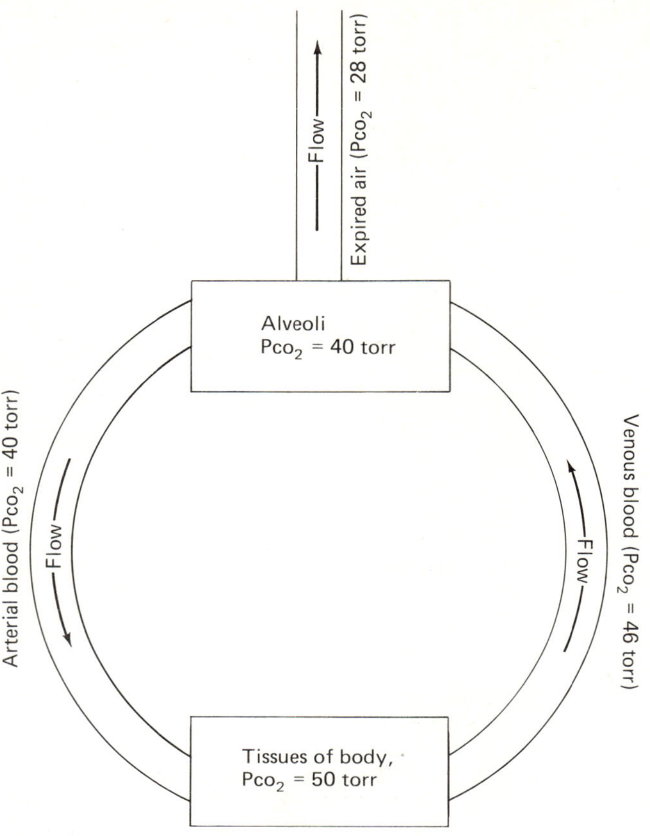

FIGURE 6-17

Flow of carbon dioxide gas through the body. The carbon dioxide moves from areas of high partial pressure to low partial pressure.

moves out of the lungs with the *expired* (exhaled) air, which has a partial pressure of carbon dioxide of about 28 torr (Fig. 6-17).

Henry's Law: Just One More Gas Law

We've just been talking about transportation of gases through the blood. Blood is, of course, a liquid. In 1801, a doctor by the name of William Henry discovered that the solubility of a gas in a liquid is directly proportional to the pressure of the gas at the surface of the liquid. In other words, the greater the pressure of the gas above the liquid, the greater is the amount of gas that can be dissolved in that liquid (Fig. 6-18). It is this principle that gave rise, in the 1960's, to *oxygen hyperbaric therapy*. This therapy involves the use of a high-pressure chamber where oxygen can be administered to an individual at a pressure of 2 to 3 atm (Fig. 6-19). This increased pressure of oxygen, over a

FIGURE 6-18
Henry's law says that the greater the pressure of the gas above the liquid, the greater is the amount of gas that can be dissolved in that liquid. Professor Emeritus has discovered that the reverse of this statement is also true.

FIGURE 6-19
A hyperbaric chamber. (Vacudyne Altair)

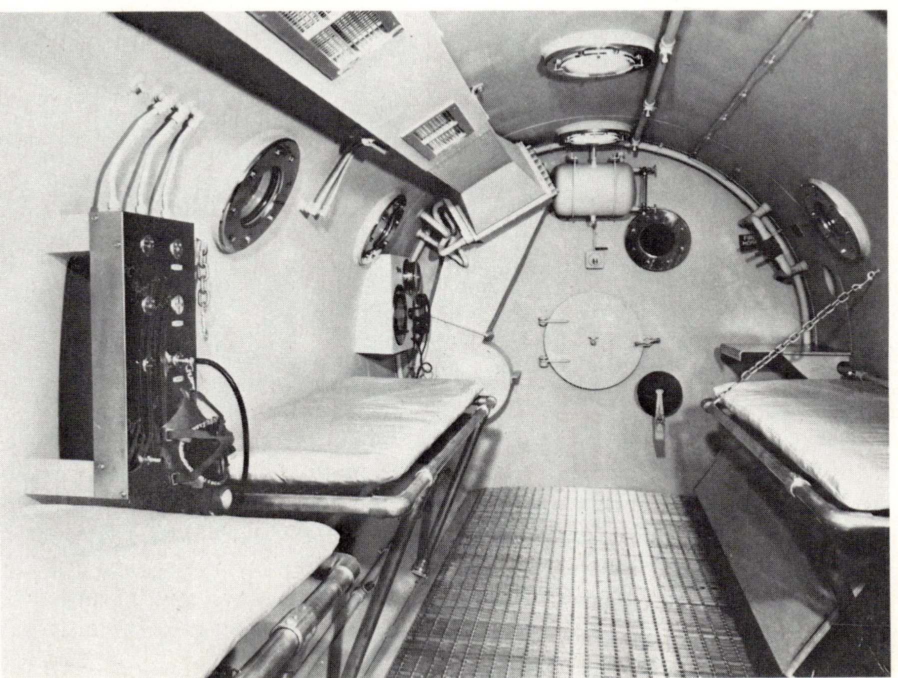

period of 5 hours, allows nearly 20 times more oxygen to be dissolved in the patient's blood. This is a great advantage to people recovering from heart attacks, strokes, gas gangrene, and various other illnesses. Babies born with hyaline membrane conditions of the lung can often be saved by treatment in a hyperbaric chamber. Although oxygen hyperbaric therapy is a great lifesaving technique, it must be monitored very carefully because high concentrations of oxygen over a prolonged period of time can cause the alveoli in the lungs to collapse.

SUMMARY

In this chapter we discussed the three phases of matter: solids, liquids, and gases. We looked at crystalline solids, and learned about the crystal lattice. We discovered that there are three types of crystalline solids: ionic, molecular, and atomic, and that each type has its distinct properties.

We looked at the liquid state and the process of evaporation. We learned that the process of evaporation allows us to maintain normal body temperature when we exercise. We also discussed the equilibrium vapor pressure of liquids and the phenomenon of boiling.

Our discussion of gases began with a review of the kinetic theory of gases. We then learned about Boyle's law, Charles' law, and the combined gas law. We discussed the gas variables: pressure, volume, and temperature, and how Torricelli measured gas pressure with his barometer. We also discussed the implications of Boyle's law as regards respiration. Our study of Dalton's law of partial pressures allowed us to understand the movement of oxygen and carbon dioxide through the body. And our study of Henry's law enabled us to see how oxygen hyperbaric therapy can force more oxygen into the blood than is normally possible.

EXERCISES

1. Define the following terms.
   (a) crystal lattice
   (b) equilibrium vapor pressure
   (c) normal boiling point
   (d) torr

2. Compare and contrast the three types of crystalline solids.

3. Describe the major characteristics of solids, liquids, and gases.

4. An unknown solid has a low melting point. It is soft and does not conduct electricity. What type of solid is it (ionic, molecular, or atomic)?

5. How will the boiling point of a liquid be affected by
   (a) an increase in atmospheric pressure?
   (b) a decrease in atmospheric pressure?

6. How will the melting point of a substance be affected by
   (a) an increase in atmospheric pressure?
   (b) a decrease in atmospheric pressure

7. Explain how the evaporation of moisture from our skin helps us to maintain our body temperature.

8. Scientists believe that the molecules of a liquid are constantly in motion. What simple experiment could you perform to show that this is true?

9. Explain why oxygen flows from the alveoli into the blood, and not the reverse.

10. If the pressure of a gas is increased at constant temperature, what happens to its volume? Does it increase or decrease?

11. If the temperature of a gas is increased at constant pressure, what happens to its volume? Does it increase or decrease?

12. When you open a bottle of soda pop, carbon dioxide seems to escape from the liquid. This is an example of what gas law?

13. A device called an *iron lung* was used in the past to help children who were afflicted with infantile paralysis. The iron lung was used to help the patient breathe. What gas law is the principle behind the iron lung?

14. Change the following pressures in *torr* to pressures in *atmosphere*.
    (a) $38\overline{0}$ torr   (b) 152 torr   (c) $190\overline{0}$ torr   (d) 4560 torr

15. Change the following pressures in *atmosphere* to pressures in *torr*.
    (a) 1.50 atm   (b) 0.10 atm   (c) 0.75 atm   (d) 3.50 atm

16. Change the following temperatures in °C to temperatures in °K.
    (a) 25°C   (b) 37°C   (c) -25°C   (d) -273°C

17. Change the following temperatures in °K to temperatures in °C.
    (a) 373°K   (b) 300°K   (c) 0°K   (d) 100°K

18. In a hospital, oxygen is sometimes brought into a patient's room in a large cylinder. This cylinder may contain about 423 liters of oxygen at a pressure of 136 atm. What would be the volume of this oxygen at a pressure of 1 atm? Assume the temperature of the gas remains constant.

19. A balloon is filled with $20\overline{0}$ ml of gas at a pressure of $76\overline{0}$ torr and a temperature of 25°C. What would the volume be if the temperature is increased to $5\overline{0}$°C and the pressure remains at $76\overline{0}$ torr?

20. A child's lungs can hold about $2\overline{0}0$ ml of air. Suppose that a child took a deep breath ($V = 2\overline{0}0$ ml, $t = 2\overline{0}°C$, $P = 76\overline{0}$ torr), then dove into an ocean where the temperature was 5°C and the pressure was $10\overline{0}0$ torr. What would be the volume of air in his lungs?

21. A mixture of carbon dioxide (5% by volume) and oxygen (95% by volume) is sometimes used as a respiratory stimulant. If a container of this gas mixture has a total pressure of $76\overline{0}$ torr, what is the partial pressure of each gas in the mixture?

CHAPTER 7

Solutions of Acids, Bases, and Salts

A Stirring Experience

Some Things You Should Know After Reading This Chapter

You should be able to:

1. Define the terms solution, solute, and solvent.
2. Discuss the differences among a true solution, a colloidal dispersion, and a suspension.
3. Calculate the percent by weight-volume of a solution given the necessary information.
4. Calculate the molarity of a solution given the necessary information.
5. State the four classes of inorganic compounds; acids, bases, salts, and oxides.
6. Define the terms acid, base, and salt.
7. State some of the important characteristics of acids and bases.
8. Discuss the concept of ionization.
9. State what is meant by a strong electrolyte and weak electrolyte.
10. Write an equation to show the ionization of water.
11. State that pure water has a hydrogen ion concentration of 10^{-7} mole per liter and a hydroxide ion concentration of 10^{-7} mole per liter.
12. Define pH as the negative logarithm of the hydrogen ion concentration in moles per liter.
13. Calculate the pH of a solution given its hydrogen ion concentration.
14. State the relationship between hydrogen ion concentration and hydroxide ion concentration in aqueous solutions.
15. Calculate the pH of a solution given its hydroxide ion concentration.
16. Define what is meant by a buffer solution.
17. Explain how the carbonic acid/bicarbonate buffer system maintains the pH of the blood fairly constant.
18. Define the terms metabolic acidosis and alkalosis and tell how they occur.
19. Define the terms respiratory acidosis and alkalosis and tell how they occur.
20. State what is meant by hyperventilation and hypoventilation.
21. Describe the processes of osmosis and diffusion.
22. Define what is meant by osmotic pressure.
23. Define the terms hypertonic, hypotonic, and isotonic solutions.
24. Describe the conditions of crenation and hemolysis and tell why they occur.
25. Discuss the process of dialysis and hemodialysis.

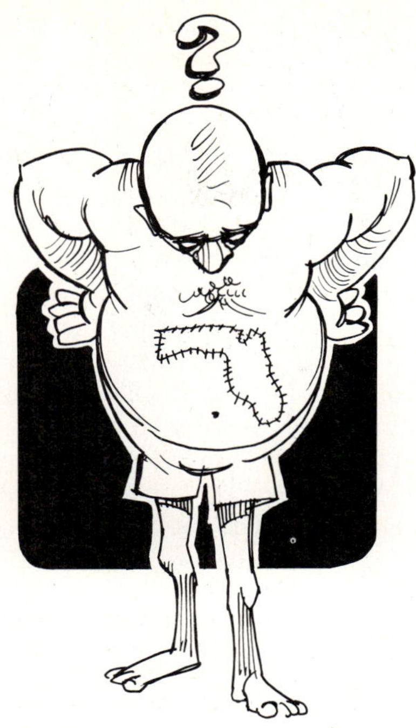

FIGURE 7-1
Uncle Charley's stomach, or is it the map of Florida?

Introduction: The Strange Case of Uncle Charley*

Uncle Charley is an individual whom you as a health practitioner would love. After all, in his 65 years he's had just about every ailment one human being could have. You might think that Uncle Charley would be a bitter, angry man, but he is not. Throughout it all he has kept his sense of humor. In fact, Uncle Charley is a 24-hour comedy show.

Uncle Charley's problems began in 1932, when he was stricken with diabetes mellitus at the age of 20. Luckily, just 10 years before, in 1922, three researchers, Banting, Best, and Macleod, discovered that insulin from pigs could be injected into humans to control some forms of diabetes. Uncle Charley was one of the lucky people whose diabetes was controlled (at least to some degree). In general, life went well for Uncle Charley until about 1960. It was then when his medical problems really started to surface, perhaps in part due to his diabetes. In rapid succession he suffered a major heart attack (a myocardial infarction), gallbladder stones (requiring an operation for their

---

* Uncle Charley, who is also known as Charles Skolnick, is one of this world's special people. Although he really is our uncle, he's also called that by everyone who knows him. Uncle Charley, we dedicate this chapter to you.

removal), glaucoma (which almost cost him his sight), and an enlarged prostate (which required surgery). But the most serious thing that happened to Uncle Charley was the one day he tried to do two things at the same time; take an aspirin for his headache and test his urine for sugar with Clinitest tablets. These tablets, we should point out, are highly caustic, and can cause severe skin burns and death if taken internally. Now, if things had been done properly, Uncle Charley would have swallowed the two aspirin tablets and placed the Clinitest tablets into a cup that contained a sample of his urine. Unfortunately, that's not what happened. Uncle Charley swallowed the Clinitest tablets and placed the aspirin in the cup of urine. Realizing what he had done, Uncle Charley ran to the refrigerator and drank a 16-ounce bottle of lemon juice. This action saved his life, but unfortunately damage had already been done to his esophagus. An emergency 6-hour operation and 6-months in the hospital were required to repair the damage. Today, Uncle Charley is well and living in Miami Beach. And he's still telling jokes, like the one about his stomach. "You know," he says, "I have the only stomach that looks like the map of Florida, and I owe it all to my operations" (Fig. 7-1).

Why did we tell you about Uncle Charley? Because all his problems had something to do with solution chemistry. And the last problem we discussed had to do with solutions of acids and bases. Many of the reactions that take place in our bodies occur in solution. Some of these solutions contain acids, bases, and salts. In this chapter we'll see just how important solutions are to our body chemistry.

Solutions Defined

In Chapter 2 we defined a solution as a homogeneous mixture. A solution can contain two substances, or three, or more. The most common types of solutions are made by dissolving a solid in a liquid—for example, salt in water. However, you can make solutions by mixing any of the three states of matter. Table 7-1 shows six possible types of solutions and gives an example of each.

Since solutions are always composed of at least two substances, we need to be able to identify the role that each substance plays. The *solute* is the substance that is being dissolved. The *solvent* is the substance that is doing the dissolving. For example, in a salt-and-water solution, salt is the solute and water is the solvent. When we deal with solutions that are composed of the same states of matter—such as liquid-liquid solutions—it's hard to establish which substance is the solute and which is the solvent. A rule to go by is that the substance present in the larger amount is the solvent. If we have a solution containing 10 ml of ethyl alcohol and 90 ml of water, the water is the solvent and the ethyl alcohol is the solute.

EXAMPLE 7-1   Which is the solute and which is the solvent in the following solutions?

(a) sugar and water
(b) hydrogen chloride and water

TABLE 7-1   Different types of solutions

	SOLID	LIQUID	GAS
Solid	Copper metal dissolved in silver metal (for example, coins)	—	—
Liquid	Salt dissolved in water	Ethyl alcohol dissolved in water	—
Gas	Hydrogen dissolved in platinum metal	Carbon dioxide dissolved in water (soda water)	Oxygen gas dissolved in nitrogen gas

(c) 70 ml of ethyl alcohol and 30 ml of water
(d) 80 ml of nitrogen and 20 ml of hydrogen
(e) soda water (which contains carbon dioxide gas)

SOLUTION

(a) Sugar is the solute; water is the solvent.
(b) Hydrogen chloride is the solute; water is the solvent.
(c) Water is the solute; ethyl alcohol is the solvent.
(d) Hydrogen is the solute; nitrogen is the solvent.
(e) Carbon dioxide is the solute; water is the solvent.

True Solutions Versus Colloidal Dispersions and Suspensions

A *true solution* is one where the solute particles have dissolved to the point of ions, atoms, or molecules into the solvent. In a true solution you can't visually distinguish the solute and solvent. Also, the solute particles will not settle out of the solvent after a period of time.

In a *colloidal dispersion*, the solute particles don't dissolve to the point that they do in a true solution. Instead, the solute particles form groups of ions, atoms, or molecules. Of course, the solute particles in a colloidal dispersion are evenly dispersed through the solvent. Because of this, we can think of a colloidal dispersion as a solution of sorts. But unlike a true solution, a colloidal dispersion will appear cloudy when a beam of light shines through it (Fig. 7-2). This is called the *Tyndall effect*, named after a famous nineteenth-century British physicist.

FIGURE 7-2
A colloidal dispersion will appear cloudy when a beam of light shines through it. This is called the Tyndall effect. The light beam appears to become visible when it passes through the colloidal dispersion. A true solution allows the light to pass through.

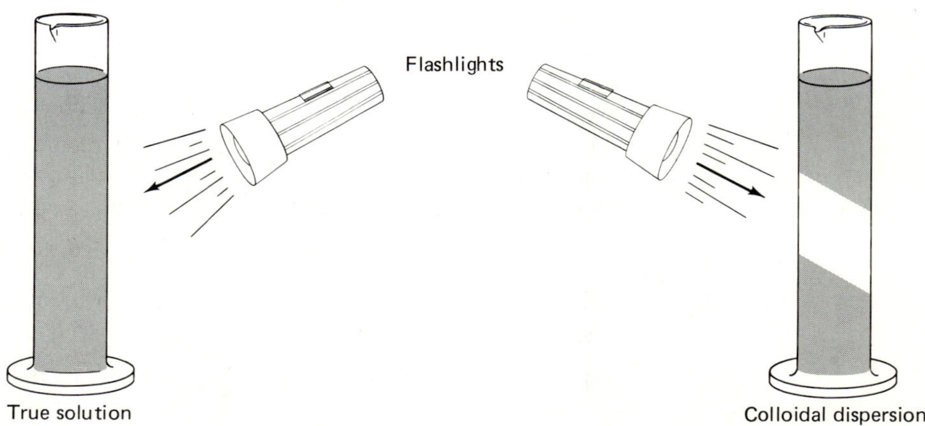

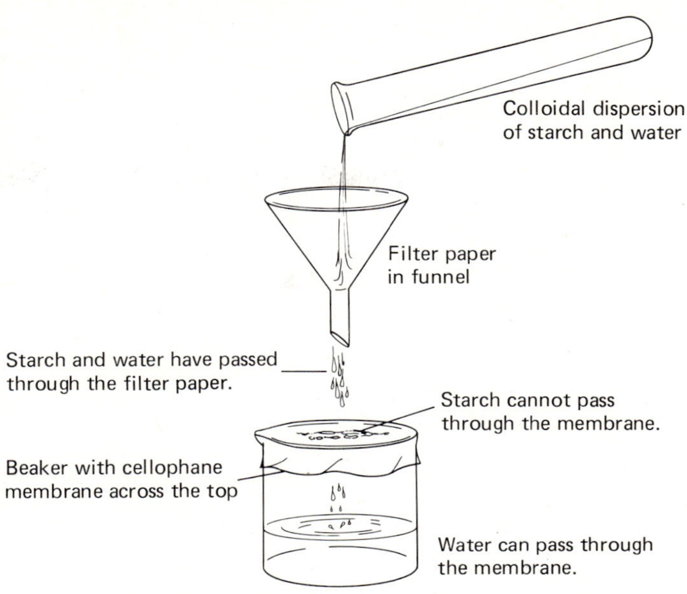

FIGURE 7-3
Colloidal particles will pass through most filter papers, but not through semipermeable membranes.

Some examples of colloidal dispersions are fog, airborne dust, egg white, and homogenized milk (the colloidal dispersion is between the cream and the skim milk fraction). The solute particles in a colloidal dispersion will, like the solute particles in a true solution, pass through most paper filters. But colloidal particles will not pass through semipermeable membranes, such as cellophane or cell walls (Fig. 7-3). This is important in the process of digestion. The membranes that line the small and large intestine allow the particles of a true solution to pass into the blood and lymphatic systems while keeping colloid-size particles out.

Suspensions are not solutions in any sense of the word; they're really just mixtures. In a suspension the suspended particles settle out and can be easily separated from the solvent by filtering. Many common pharmaceutical agents are suspensions and must be shaken well before they are used. Milk of magnesia is an excellent example. Table 7-2 summarizes the major characteristics of true solutions, colloidal dispersions, and suspensions.

EXAMPLE 7-2   Determine whether each of the following are true solutions, colloidal dispersions, or suspensions.

(a) The solute and solvent are shaken well, but the solute settles upon standing.

TABLE 7-2   Some major characteristics of true solutions, colloids, and suspensions

	TRUE SOLUTIONS	COLLOIDS	SUSPENSIONS
Particles that compose them	Ions, atoms, or molecules	Groups of ions, atoms, or molecules	Large groups of insoluble particles
Size of particles	Less than 1 nm	1–100 nm	Greater than 100 nm
Separation of solute and solvent by filtering	Will not effect separation	Will not effect separation	Will effect separation
Separation of solute and solvent by a semipermeable membrane	Will not effect separation	Will effect separation	Will effect separation

(b) The solute and solvent are shaken well, then filtered. Nothing remains on the filter paper. The solution is then poured through a semi-permeable membrane. The solute and solvent separate.

SOLUTION

(a) This must be a suspension, because the solute settles upon standing.

(b) This must be a colloidal dispersion, because the solute and solvent are separated by a semipermeable membrane.

Concentrations of Solutions

A particular solution can be made in various concentrations simply by adding more solute to a given amount of solvent. However, we should also mention that you can only vary the concentration of a solution up to a point, and that is the *point of saturation*. The point of saturation occurs when no more solute dissolves in the solution. In this chapter we will look at two ways that health practitioners measure the concentrations of solutions: percent by weight-volume, and molarity.

Percent by Weight-Volume

Every solution has a solute and a solvent. In a percent by weight-volume solution, the solute is usually measured in grams, and the solvent in milliliters. The formula for calculating the percent by weight-volume of a solution is

$$\% \text{ by weight-volume} = \frac{\text{grams of solute}}{\text{ml of solution}} \times 100 \qquad (7\text{-}1)$$

For example, if we dissolve 1.8 g of NaCl in enough water to make 200 ml of solution, we would have a 0.9% NaCl solution.

$$\% \text{ NaCl by weight-volume} = \frac{1.8 \text{ g NaCl}}{200 \text{ ml soln.}} \times 100 = 0.9\%$$

A 0.9% weight-volume sodium chloride solution is sometimes called *physiological saline solution.* We'll have a lot more to say about physiological saline solution later in this chapter. In the meantime see if you can solve the percent by weight-volume problems that follow.

EXAMPLE 7-3   A solution is made by dissolving 5 g of glucose in enough water to make 250 ml of solution. What is the percent weight-volume of the solution?

SOLUTION   We use equation 7-1.

$$\% \text{ glucose by weight-volume} = \frac{5 \text{ g glucose}}{250 \text{ ml soln.}} \times 100 = 2\%$$

EXAMPLE 7-4   How many grams of sucrose must you mix with 200 ml of water solution to prepare a 5% by weight-volume sucrose solution?

SOLUTION   A 5% weight-volume sucrose solution means that there are 5 g of sucrose per 100 ml of solution, or

$$\frac{5 \text{ g sucrose}}{100 \text{ ml soln.}}$$

Therefore we can use the factor-unit method to solve for the number of grams of sucrose in 200 ml of solution.

$$? \text{ g of sucrose} = (200 \text{ ml soln.})\left(\frac{5 \text{ g sucrose}}{100 \text{ ml soln.}}\right) = 10 \text{ g sucrose}$$

Molarity

Another frequently used unit of concentration is molarity. We may define *molarity* as *the number of moles of solute per liter of solution,* or in terms of a formula,

$$\text{Molarity} = \frac{\text{moles of solute}}{\text{liter of solution}}$$

or in an abbreviated form as

$$M = \frac{\text{moles}}{\text{liter}} \qquad (7\text{-}2)$$

Molarity is an extremely useful concentration unit because it is based on the mole. Therefore, the molarity of a solution actually expresses the number of molecules of solute per liter of solution. This is important information for a chemist who is carrying out chemical reactions in solutions.

If you wanted to prepare a 2 molar sodium chloride solution (which you would label 2 M NaCl), you would need to obtain 2 moles of NaCl for every liter of solution you prepared. A mole of NaCl weighs 58.5 g; therefore, you would need 117 g of NaCl for each liter of solution. Figure 7-4 shows how you would prepare 1 liter of 2 M NaCl solution. Sometimes the concentrations of certain pharmaceutical agents are given in units of molarity. See if you can use this concentration unit in the examples that follow.

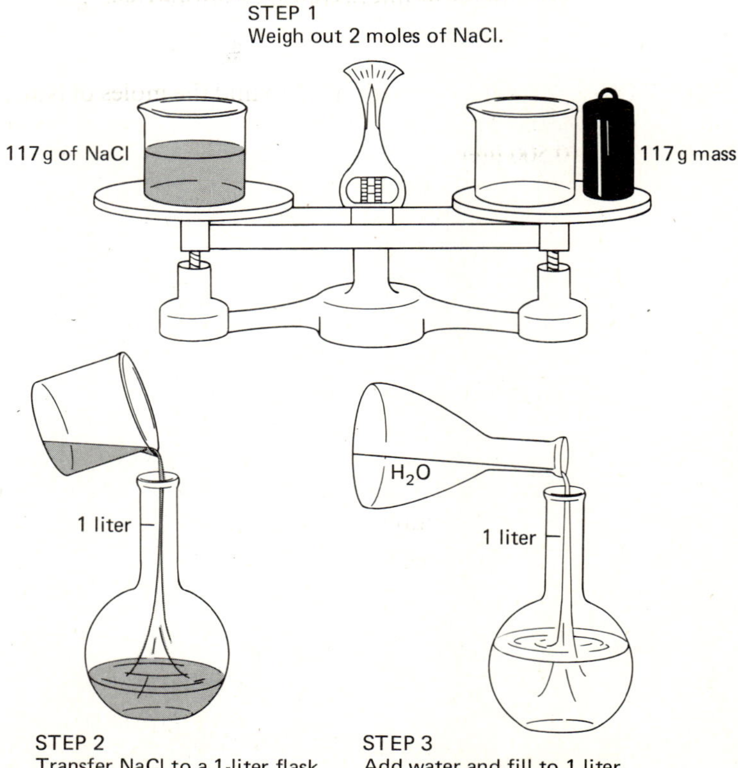

FIGURE 7-4
How to prepare 1 liter of 2 M NaCl.

EXAMPLE 7-5   The antineoplastic (anticancer) agent 5-fluorouracil (abbreviated 5-FU) has a molecular weight of 130 and comes in a strength of 50.0 mg/ml. This means that there are 50.0 mg of 5-FU for every 1 ml of solution, or 50.0 g of 5-FU for every liter of solution. What is the concentration of this material in terms of molarity?

SOLUTION   Molarity is defined as moles of solute per liter of solution. We know that there are 50.0 g of solute per liter of solution. So in order to use equation 7-2 we must change the 50.0 g of 5-FU into moles.

$$? \text{ moles of 5-FU} = (50.0 \text{ g})\left(\frac{1 \text{ mole}}{130 \text{ g}}\right) = 0.38 \text{ mole}$$

$$? M = \frac{\text{moles}}{\text{liter}} = \frac{0.38 \text{ mole}}{1.0 \text{ liter}} = 0.38 \ M$$

EXAMPLE 7-6   Lactated Ringer's solution is used intravenously to maintain the electrolytic balance in patients who otherwise might go into shock. One of the ingredients in lactated Ringer's is sodium chloride, whose concentration is about 0.13 M. How many grams of sodium chloride (NaCl) would be in $50\overline{0}$ ml of this solution?

SOLUTION   We will use equation 7-2 to find the moles of NaCl in $50\overline{0}$ ml of 0.13 M solution. Then we will convert the moles of NaCl to grams of NaCl. Note: $50\overline{0}$ ml = $0.50\overline{0}$ liter.

$$M = \frac{\text{moles}}{\text{liter}}; \qquad \text{therefore, moles} = (M)(\text{liter})$$

$$? \text{ moles} = (0.13 \ M)(0.50\overline{0} \text{ liter}) = 0.065 \text{ mole}$$

$$? \text{ grams of NaCl} = (0.065 \text{ mole})\left(\frac{58.5 \text{ g}}{1 \text{ mole}}\right) = 3.80 \text{ g}$$

EXAMPLE 7-7   A popular cough syrup and expectorant is elixir terpin hydrate with codeine. The concentration of codeine in this medicine is 0.0070 M. How many milliliters of cough syrup would you have to consume to ingest $1\overline{0}$ mg of codeine? The molecular weight of codeine is $30\overline{0}$.

SOLUTION   First, $1\overline{0}$ mg of codeine is 0.010 g of codeine. Now let's change this to moles of codeine.

$$? \text{ moles} = (0.010 \text{ g})\left(\frac{1 \text{ mole}}{30\overline{0} \text{ g}}\right) = 3.3 \times 10^{-5} \text{ mole}$$

Now we can use equation 7-2 to solve for the liters of cough syrup that contain 3.3×10^{-5} mole of codeine. Note that the cough syrup is 0.0070 M with respect to the codeine. This is the same as 7.0×10^{-3} M.

$$M = \frac{\text{moles}}{\text{liter}} \qquad \text{therefore, liter} = \frac{\text{moles}}{M}$$

$$? \text{ liter} = \frac{3.3 \times 10^{-5} \text{ mole}}{7.0 \times 10^{-3} \, M} = 4.8 \times 10^{-3} \text{ liter or 4.8 ml}$$

In other words, you would have to drink 4.8 ml of cough syrup to ingest $1\overline{0}$ mg of codeine.

Classes of Inorganic Compounds:
Acids, Bases, Salts, and Oxides

If we attempted to classify the thousands and thousands of inorganic compounds, we would find that they fall into four broad categories:

1. ACIDS: Substances that release hydrogen ions in solution and counteract bases.

2. BASES: Substances that release hydroxide ion in solution and counteract acids.

3. SALTS: Substances composed of the positive ions of a base and the negative ions of an acid.

4. OXIDES: Substances composed of any element combined with oxygen.

Although all four classes of compounds are important, we'll concentrate our efforts on acids, bases, and salts. It is these three classes of compounds that play a major role in our bodily functions.

Acids: What Have They Got in Common?

The acids listed in Table 7-3 have many common characteristics.

1. EACH HAS A SOUR TASTE. (The word "acid" comes from the Latin *acidus*, meaning sour.)

2. EACH CAN CHANGE THE COLOR OF CERTAIN DYES. For example, blue litmus dye or litmus paper turns red in the presence of an acid. These dyes are called *indicators* because they indicate whether a substance is an acid or a base (Fig. 7-5).

3. EACH CAN REACT WITH CERTAIN METALS, SUCH AS ZINC AND MAGNESIUM, TO PRODUCE HYDROGEN GAS.

$$Zn + 2HCl \longrightarrow ZnCl_2 + H_2(g)$$

$$Zn + H_2SO_4 \longrightarrow ZnSO_4 + H_2(g)$$

NAME OF ACID	FORMULA
Hydrochloric acid	HCl
Nitric acid	HNO_3
Sulfuric acid	H_2SO_4
Phosphoric acid	H_3PO_4
Acetic acid	$HC_2H_3O_2$
Boric acid	H_3BO_3

4.   EACH CAN NEUTRALIZE BASES.   (We'll hold our discussion of this until later.)

But the most important characteristic of acids is the one we used in the definition: *They dissolve in water to produce hydrogen ions* (H^{+1}). These ions are responsible for the other common properties of acids.

Let's take a closer look at how an acid dissolves in water. In Chapter 4 we said that water is a polar molecule, with the oxygen end negatively charged and the hydrogen end positively charged. Suppose that we take hydrogen

FIGURE 7-5
Hydrochloric acid changes blue litmus dye to red.

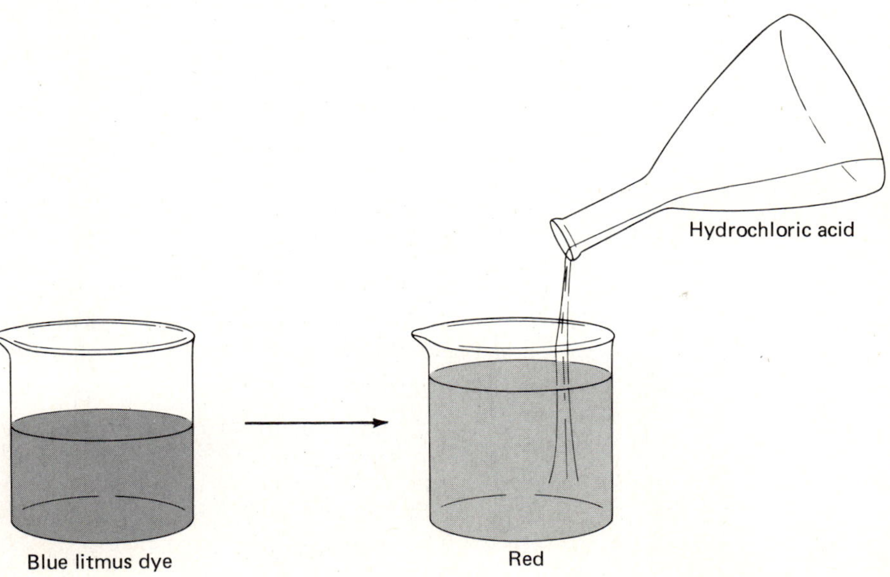

$$H^{\oplus}\!\!-\!Cl^{\ominus} + H\!-\!\overset{\ominus}{\underset{\oplus}{O}}\!-\!H \longrightarrow \left[H\!-\!\overset{H}{\underset{}{O}}\!-\!H\right]^{\oplus} + Cl^{\ominus}$$

Hydrogen chloride gas reacts with water. The covalently bonded hydrogen chloride molecule is ionized by the water into a hydrogen ion and chloride ion. The positive hydrogen ion is attracted to the negative end of the water molecule forming a hydronium ion.

chloride gas (HCl) and bubble it into a beaker of water. What happens? The gas seems to dissolve in the water. But actually much more is happening. The HCl gas is reacting with the water (Fig. 7-6). The covalently bonded hydrogen chloride molecules are being broken apart (ionized) by the water into hydrogen ions (H^{+1}) and chloride ions (Cl^{-1}). The hydrogen ions are attracted to the negative ends of the water molecules, forming ions of a new type. The new kind of ion, $(H_3O)^{+1}$, is called a *hydronium ion*. The reaction can be written

$$HCl + H_2O \longrightarrow (H_3O)^{+1} + Cl^{-1}$$

If we examine the other acids in Table 7-3, we find that similar reactions occur.

$$H_2SO_4 + 2H_2O \longrightarrow 2(H_3O)^{+1} + (SO_4)^{-2}$$
$$H_3PO_4 + 3H_2O \longrightarrow 3(H_3O)^{+1} + (PO_4)^{-3}$$

Some Acids: Their Discovery and Uses

The ancient Greeks knew how to ferment grapes to make wine. They also knew that if they let the fermentation process continue for too long, they got vinegar (which means sour wine). Vinegar was the strongest acid known to the Greeks. Centuries later an Arab alchemist named Geber distilled vinegar and got the substance responsible for vinegar's acidic properties. Today we call this substance acetic acid, and we use it as a solvent for manufacturing rubber, plastics, acetate fibers, drugs, and photographic chemicals.

We get acids such as acetic acid and citric acid from living things, and we therefore call them *organic acids*. For many centuries, organic acids were the only acids alchemists knew. However, in the 1200's, another alchemist, also called Geber, found a method of preparing acids from minerals. Two of the *mineral acids* he prepared were *sulfuric acid* and *nitric acid*. Since these were much stronger than the organic acids, alchemists could dissolve many substances that they thought were inert. This meant that new chemical reactions were possible. Mineral acids were essential to the great advances of modern chemistry and medicine. Let's look at how the acids in Table 7-3 are used.

Hydrochloric acid (HCl), also known commercially as *muriatic acid*, is used in its more concentrated forms to clean brick, cement, and metals. Although hydrochloric acid is considered to be a mineral acid, it is also produced in the stomach, where it is used for the digestion of proteins. In this capacity, hydrochloric acid has also been called *stomach acid.*

Nitric acid (HNO_3), another mineral acid, is very reactive. If nitric acid gets on your skin, it will turn it yellow. This is because of a reaction between the protein in your skin and the nitric acid. Because it has the ability to coagulate protein, nitric acid is also used to test for the presence of albumin in urine. However, the major uses of nitric acid are in the production of fertilizers, dyes, plastics, and explosives.

Sulfuric acid (H_2SO_4) is used primarily in the production of fertilizers, explosives, dyestuffs, other acids, paper, and glue. In its concentrated form, sulfuric acid is highly corrosive to our skin, and inhalation of its vapors can cause serious damage to our lungs. In its diluted form, sulfuric acid has been used as an appetite stimulant and in the control of serious diarrhea.

Phosphoric acid (H_3PO_4) is a weak mineral acid. Its uses range from flavoring soft drinks to producing fertilizers. The false teeth of millions of people are held in place by dental cements containing phosphoric acid.

Acetic acid is found in vinegar, as we have seen. According to the pure-food laws in the United States, vinegar must contain no less than 4% acetic acid. Concentrated acetic acid is strong enough to be used in making synthetic fibers called *acetates.* It is also used in manufacturing various plastics. Remember, though, that acetic acid—unlike the other acids we have just discussed—is an organic acid.

Boric acid (H_3BO_3) is a weak acid that has been used frequently and safely by people as an antiseptic, germicide, and eye wash. However, misuse of this acid has led to boric acid poisoning in certain individuals, and even to death. This has occurred from boric acid absorption through the skin when it was applied to certain types of wounds. Ingestion of less than 5 g of boric acid in infants and from 5 to 20 g in adults has also caused death.

Bases: What Have They Got in Common?

The bases listed in Table 7-4 have many common characteristics.

1. Each has a bitter taste.

2. Each can change the color of certain dyes. For example, red litmus turns blue in the presence of base (Fig. 7-7).

3. Each has a slimy or soapy feeling.

4. Each can neutralize acids.

But the most important characteristic of bases is the one we used in the definition: *They dissolve in water to produce hydroxide ions,* $(OH)^{-1}$. It is these hydroxide ions that are responsible for the other common properties of bases.

NAME OF BASE	FORMULA
Lithium hydroxide	$LiOH$
Sodium hydroxide	$NaOH$
Potassium hydroxide	KOH
Calcium hydroxide	$Ca(OH)_2$
Magnesium hydroxide	$Mg(OH)_2$
Iron(III) hydroxide	$Fe(OH)_3$
Ammonium hydroxide	NH_4OH

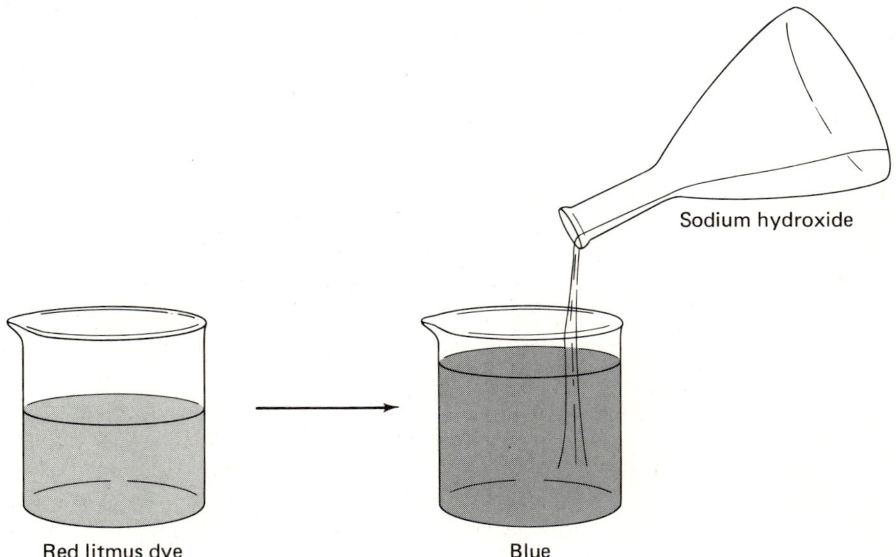

FIGURE 7-7
Sodium hydroxide changes red litmus dye to blue.

How do bases dissolve in water to produce hydroxide ions? Take, for example, solid sodium hydroxide ($NaOH$). Sodium hydroxide is an ionic compound made up of sodium ions and hydroxide ions, Na^{+1} and $(OH)^{-1}$. The ions are held together by the attraction of a plus charge to a minus charge. When you add sodium hydroxide to water, the negative end of the water molecule (the oxygen) attracts the sodium ions, and the positive end (the hydrogen) attracts the hydroxide ions. This attraction breaks the bonds between the sodium ions and the hydroxide ions (Fig. 7-8). There are now free

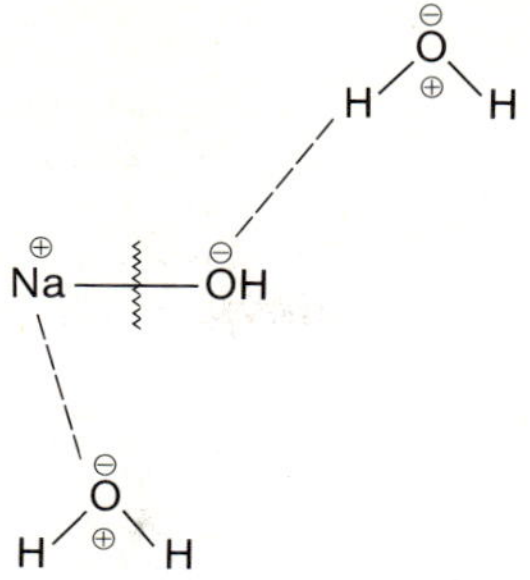

FIGURE 7-8

When you add sodium hydroxide to water, the negative end of the water molecule attracts the sodium ions, and the positive end of the water molecule attracts the hydroxide ions.

sodium ions and free hydroxide ions in solution. The reaction can be represented by the following equation.

$$NaOH \xrightarrow{\text{water}} Na^{+1} + (OH)^{-1}$$

Some Bases and Their Uses

Like acids, bases play an important role in our daily lives and are just as dangerous to our skin and clothes. They are used in our homes, in industrial processes, and in manufacturing various pharmaceutical products.

For example, *sodium hydroxide*, $NaOH$, is a base used in drain cleaners, or what we sometimes call lye. It is also used to make soaps and cellophane.

Calcium hydroxide, $Ca(OH)_2$, which in solution is also known as *lime-water*, is sometimes used as an antacid. However, its major uses are in the construction industry, where it is used in mortar, plaster, cement, and paving materials.

Ammonium hydroxide, NH_4OH, is the household cleaner we know as ammonia water, and it has an intense, almost suffocating odor. Ammonium hydroxide is the chief ingredient of smelling salts. It is excellent for removing stains, and it is also used in manufacturing textiles such as rayon, and in making plastics and fertilizers (Fig. 7-9).

Magnesium hydroxide, $Mg(OH)_2$, commonly called *milk of magnesia*, has many uses as a medicine. In small doses of a few hundred milligrams, the magnesium hydroxide acts as an antacid. In large doses of 2 to 4 g, it acts as a laxative.

Salts: What Have They Got in Common?

We have learned that acids and bases react with each other in what are called neutralization reactions (Chap. 5). The products of these reactions are salts and water.

FIGURE 7-9
Ammonium hydroxide is the chief ingredient of smelling salts.

$$HCl + NaOH \longrightarrow NaCl + H_2O$$

Sodium
chloride

$$HCl + NH_4OH \longrightarrow NH_4Cl + H_2O$$

Ammonium
chloride

$$H_2SO_4 + 2NH_4OH \longrightarrow (NH_4)_2SO_4 + 2H_2O$$

Ammonium
sulfate

If we examine these reactions closely, we can see what takes place. The hydrogen ions from the acid combine with the hydroxide ions from the base to produce water. At the same time, the positive ions of the base combine with the negative ions of the acid to produce the compound that we call a *salt*. Most salts are composed of a metal ion combined with a nonmetal (or polyatomic) ion that has a negative oxidation number. Examples of these salts are sodium chloride ($NaCl$), which is everyday table salt, silver bromide ($AgBr$), potassium sulfate (K_2SO_4), and iron(III) phosphate ($FePO_4$). Some salts are composed of a polyatomic ion with a positive oxidation number combined with a nonmetal (or polyatomic) ion that has a negative oxidation number. Examples of these salts are ammonium chloride (NH_4Cl) and ammonium nitrate (NH_4NO_3).

TABLE 7-5   Some salts and their uses

FORMULA	CHEMICAL NAME	MEDICAL USE
$AlCl_3$	Aluminum chloride	Deodorant
$AlPO_4$	Aluminum phosphate	Antacid
NH_4Cl	Ammonium chloride	Expectorant
$BaSO_4$	Barium sulfate	X-ray work
$CaCO_3$	Calcium carbonate	Antacid
$CaCl_2$	Calcium chloride	Electrolyte replacement
$FeSO_4$	Iron(II) sulfate	As an iron supplement
$MgCl_2$	Magnesium chloride	Laxative
$MgSO_4$	Magnesium sulfate (as the heptahydrate, $MgSO_4 \cdot 7H_2O$, this substance is known as epsom salts)	Laxative
$HgCl_2$	Mercury(II) chloride	Topical antiseptic
$NaCl$	Sodium chloride	Used in physiological saline solution
NaF	Sodium fluoride	Prevents dental decay
SnF_2	Tin(II) fluoride	Prevents dental decay

Some Salts and Their Uses

Salts have a tremendous number of uses in the body. For example, calcium phosphate is the main constituent of our bones and teeth. Iron salts are necessary for the production of hemoglobin. Sodium and potassium salts help maintain the acid–base balance in our bodies. And the salt sodium iodide is necessary for the proper functioning of our thyroid gland. Salts are also responsible for the proper functioning of our muscles, including the heart, and they also maintain the fluid balance in our cells. A summary of some salts and their uses can be found in Table 7-5.

Ionization and the Concept of Electrolytes

What happens when things dissolve? For years chemists have tried to find out what happens to one substance as it dissolves in another. Chemists of the 1800's were also puzzled because some solutions would conduct electric current while others wouldn't.

Michael Faraday was one scientist who tried to explain this phenomenon. Faraday classified two kinds of solutions in the following way.

1. *Electrolytic solutions* are those solutions that conduct electric current.

2. *Nonelectrolytic solutions* are those solutions that do not conduct electric current.

Faraday said that electrolytic solutions—such as sodium chloride (NaCl) in water—have charged particles in them. He called these particles *ions* (from a Greek word that could be translated as "wanderer"), and said that ions wandered through the solution carrying the electric current (Fig. 7-10). In Faraday's time, questions arose about the nature of these ions. For example, why are there ions in a salt-and-water solution and no ions in a sugar-and-water solution? It was left to Arrhenius to come up with the answer.

In 1884, in his Ph.D. thesis, Arrhenius advanced his ideas about ions. He suggested that Faraday's ions were really simple atoms (or groups of atoms) carrying a positive or negative charge. He said that when some substances dissolve in solution, they break up into ions. This process is called *ionization*.

$$NaCl(s) + H_2O \longrightarrow Na^{+1}(aq) + Cl^{-1}(aq)$$

Solutions that contain ions are electrolytic, since they have charged particles that carry electric current (Fig. 7-11).

FIGURE 7-10
An electrolytic solution conducts electric current.

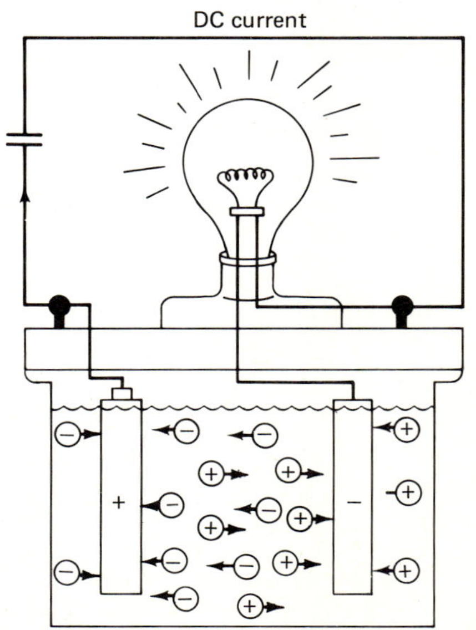

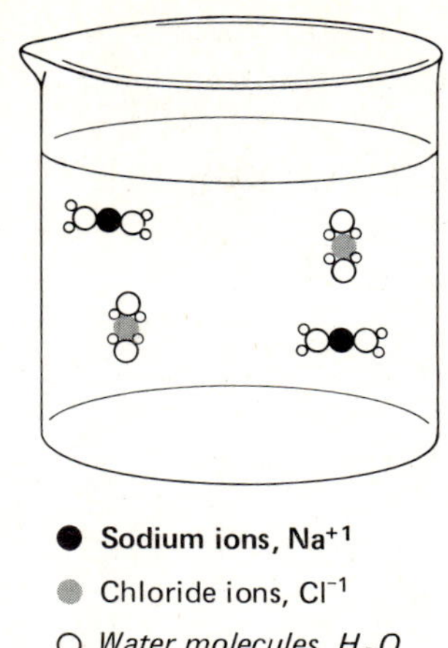

FIGURE 7-11
An electrolytic solution contains ions.

On the other hand, there are some substances that dissolve in solution and do *not ionize*; they simply break up into their neutral molecules and become surrounded by the molecules of solvent.

$$C_6H_{12}O_6(s) + H_2O \longrightarrow C_6H_{12}O_6(aq)$$

Glucose

These solutions are nonelectrolytic, since they have *no* charged particles to carry electric current (Fig. 7-12).

FIGURE 7-12
Nonelectrolytic solutions have no charged particles or ions.

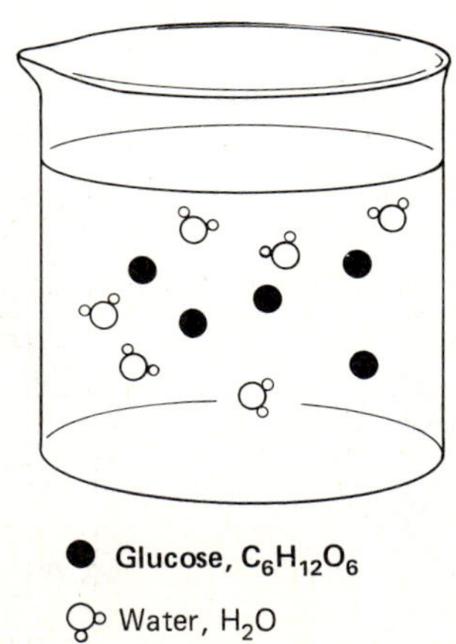

TABLE 7-6   Some strong electrolytes

ACIDS	BASES	SALTS
Hydrochloric acid, HCl	Sodium hydroxide, NaOH	Sodium chloride, NaCl
Nitric acid, HNO_3	Potassium hydroxide, KOH	Potassium chloride, KCl

Strong and Weak Electrolytes

If we study solutions of electrolytes very carefully we will discover that we can further subdivide them into two categories: strong electrolytes and weak electrolytes. The *strong electrolytes* are those which essentially ionize completely. For example, 1 mole of sodium chloride in 1 liter of solution contains essentially 1 mole of sodium ions (Na^{+1}) and 1 mole of chloride ions (Cl^{-1}). There are no sodium chloride molecules! Strong electrolytes conduct current well and light the bulb in Figure 7-10 very brightly. Strong electrolytes can be compounds of acids, bases, or salts. Therefore, an acid that ionizes completely in solution is called a *strong acid*, and a base that ionizes completely in solution is called a *strong base*. A list of some strong electrolytes can be found in Table 7-6.

Weak electrolytes are those that don't ionize completely. For example, if you were to analyze a solution containing 1 mole of $HC_2H_3O_2$ in 1 liter of solution (this would be a 1 *M* acetic acid solution), you would find that most of the acetic acid stays in the form of molecules and only about 0.4% ionizes to form hydrogen ions and acetate ions. We can represent the dissociation of a weak electrolyte in the following manner.

$$HC_2H_3O_2 \xrightleftharpoons{\text{water}} H^{+1} + (C_2H_3O_2)^{-1}$$

The *double arrow* indicates that in a given solution of acetic acid an equilibrium exists among the acetic acid molecules, the hydrogen ions, and the acetate ions. Weak electrolytes can also be compounds of acids, bases, or salts (Table 7-7). An acid that does not ionize completely is called a *weak acid*, and a base that does not ionize completely is called a *weak base* (Fig. 7-13). We'll have more to say about the importance of electrolytes in the body later in this chapter.

TABLE 7-7   Some weak electrolytes

ACIDS	BASES	SALTS
Acetic acid, $HC_2H_3O_2$	Ammonium hydroxide, NH_4OH	Barium fluoride, BaF_2
Carbonic acid, H_2CO_3	Dimethylamine, $(CH_3)_2NH$	Silver sulfate, Ag_2SO_4

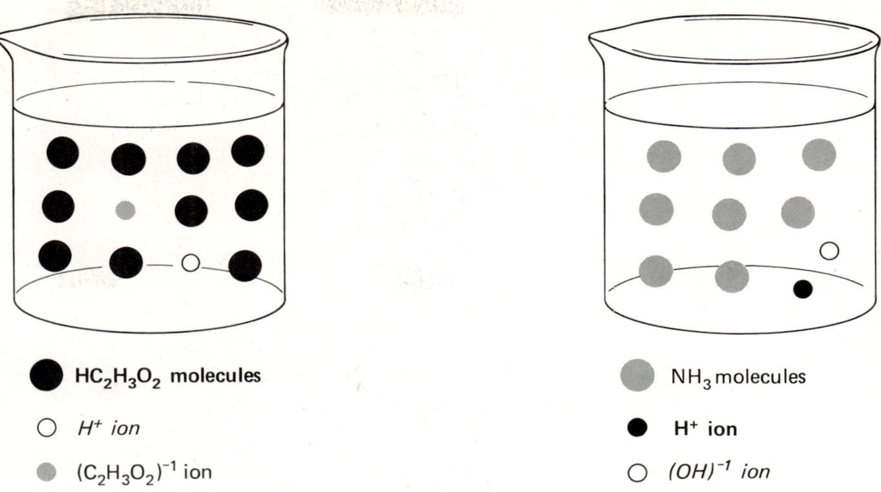

FIGURE 7-13

An acid that does not ionize completely is called a weak acid, and a base that does not ionize completely is called a weak base. Acetic acid and ammonia water (also known as ammonium hydroxide) are examples.

The Ionization of Water

We've just finished discussing the ionization of weak acids and weak bases, but did you know that water also has a tendency to ionize? We can represent the ionization of water as follows.

$$H\!-\!OH + H_2O \;\rightleftharpoons\; (H_3O)^{+1} \;+\; (OH)^{-1} \qquad (7\text{-}3)$$

$$\text{Hydronium} \qquad \text{Hydroxide}$$
$$\text{ion} \qquad\quad \text{ion}$$

We purposely wrote the formula of the first water molecule as H—OH so that you could see how the ionization occurs. A hydronium ion and hydroxide ion are formed. However, we can represent this ionization in a simpler way as follows:

$$H_2O \;\rightleftharpoons\; H^{+1} + (OH)^{-1} \qquad (7\text{-}4)$$

This shows that a hydrogen ion and hydroxide ion are formed. Even though equation 7-3 is more technically correct, equation 7-4 is easier to use and gets the point across. However, we must point out that the ionization of water occurs only to a slight extent. In fact, if you had exactly 1 liter of water, which is 55.6 moles of water, only 0.0000001 mole (which can be written as 10^{-7} mole) of water would actually dissociate into hydrogen and hydroxide ions. In other

FIGURE 7-14
Pure water is neutral.

words, there would be 10^{-7} mole of hydrogen ions and 10^{-7} mole of hydroxide ions present per liter of water. Because there are equal amounts of hydrogen ions and hydroxide ions, pure water is said to be neutral (Fig. 7-14).

But what happens if you add hydrochloric acid to pure water? Adding hydrochloric acid to pure water increases the hydrogen ion content. (Remember, we've just learned hydrochloric acid is a strong acid, which means that it dissociates into H^{+1} ions and Cl^{-1} ions.) This, of course, makes the solution of water and hydrochloric acid acidic. Just how acidic depends on the concentration of the hydrogen ions. It is just this idea that gives rise to the concept of pH.

The pH Scale

The concept of *pH* is a means of measuring the acid-base strength of solutions. The formal definition is that pH is equal to the negative logarithm of the hydrogen ion concentration (in moles per liter). Symbolically, this is

$$pH = -\log[H^{+1}]$$

The brackets, [ ], indicate moles per liter. The pH scale runs from zero to 14 (Fig. 7-15). A solution whose pH is equal to 7 is neutral. Why? Remember that in our discussion of the ionization of water, the concentration of hydrogen ion was 10^{-7} mole per liter. Therefore, the pH of water, which is neutral because it has equal amounts of hydrogen ion and hydroxide ion, is calculated as follows.

$$pH = -\log[H^{+1}] = -\log[10^{-7}] = 7$$

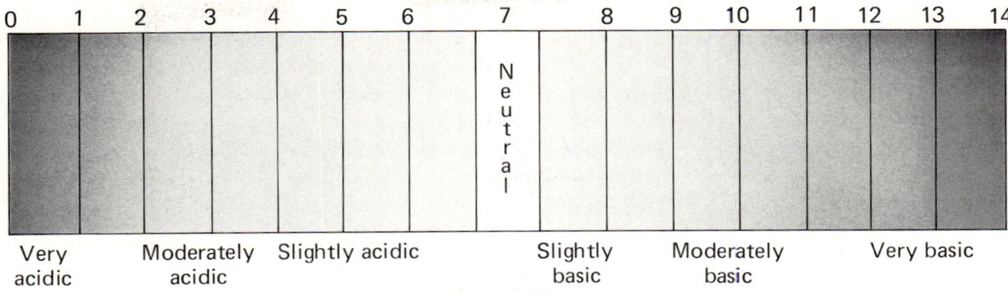

FIGURE 7-15
The pH scale.

(The *logarithm* of an exponential number is simply that exponent. A more detailed discussion of logarithms can be found in the appendix of this text.)

A solution that has a pH between zero and 2 is considered to be very acidic. A solution that has a pH between 2 and 5 is moderately acidic, and one with a pH between 5 and 7 is slightly acidic. A solution whose pH is 7 is neutral. A solution whose pH is between 7 and 9 is slightly basic, one with a pH between 9 and 11 is moderately basic, and one with a pH between 11 and 14 is very basic. Table 7-8 lists the pH values of some common substances.

You can determine the pH of a solution by many methods. A simple method is to place a few drops of a chemical dye or a piece of paper tape that contains the dye in the solution. The dye or paper tape turns color according to the pH of the solution. There is also a device called a pH meter, which directly reads the pH of a solution.

TABLE 7-8 The pH value of some common substances

SUBSTANCE	pH
Lemons	2.3
Vinegar	2.8
Soft drinks	3.0
Oranges	3.5
Tomatoes	4.2
Bananas	4.5
Rainwater	6.2
Milk	6.5
Pure water	7.0
Seawater	8.5
Blood	7.4
Urine	5.5–7.0
Saliva	6.5–7.5
Gastric juices	1.0–3.0

EXAMPLE 7-8 Determine the pH of the following solutions, then use Fig. 7-15 to state how acidic or basic the solution is.

(a) a solution whose $[H^{+1}] = 10^{-3}$ M

(b) a solution whose $[H^{+1}] = 10^{-12}$ M

(c) a solution whose $[H^{+1}] = 10^{0}$ M

SOLUTION $pH = -\log[H^{+1}]$; therefore,

(a) $pH = -\log[10^{-3}] = 3$, which is moderately acidic.

(b) $pH = -\log[10^{-12}] = 12$, which is very basic.

(c) $pH = -\log[10^{0}] = 0$, which is very acidic.

The Relationship Between Hydrogen Ions and Hydroxide Ions

There is an interesting relationship between the hydrogen ion concentration and the hydroxide ion concentration in aqueous (water) solutions. The product of the $[H^{+1}]$ and $[OH^{-1}]$ is always equal to 10^{-14}. In other words,

$$[H^{+1}][OH^{-1}] = 10^{-14}$$

Remember that neutral water had an H^{+1} concentration of 10^{-7} M, and an $(OH)^{-1}$ concentration of 10^{-7} M. The product of these two numbers would be 10^{-14}:

$$[H^{+1}][OH^{-1}] = [10^{-7}][10^{-7}] = 10^{-14}$$

The value of 10^{-14} is called the *dissociation constant* of water. It is usually given the symbol K_w. Therefore,

$$[H^{+1}][OH^{-1}] = K_w = 10^{-14}$$

If a solution has a hydrogen ion concentration of 10^{-5} M, its hydroxide ion concentration has to be 10^{-9} M, because

$$[10^{-5}][OH^{-1}] = 10^{-14}; \quad \text{therefore, } [OH^{-1}] = \frac{10^{-14}}{[10^{-5}]} = [10^{-9}]$$

A summary of the relationships among hydrogen ion concentration, hydroxide ion concentration, and pH can be found in Table 7-9.

EXAMPLE 7-9 Determine the pH of the following solutions.

(a) a 0.001 M HCl solution
(b) a 0.01 M HNO_3 solution

$[H^{+1}]$	pH	$[(OH)^{-1}]$
$10^0 = 1$	0	10^{-14}
$10^{-1} = 0.1$	1	10^{-13}
$10^{-2} = 0.01$	2	10^{-12}
$10^{-3} = 0.001$	3	10^{-11}
$10^{-4} = 0.0001$	4	10^{-10}
$10^{-5} = 0.00001$	5	10^{-9}
$10^{-6} = 0.000001$	6	10^{-8}
$10^{-7} = 0.0000001$	7	10^{-7}
$10^{-8} = 0.00000001$	8	10^{-6}
$10^{-9} = 0.000000001$	9	10^{-5}
$10^{-10} = 0.0000000001$	10	10^{-4}
$10^{-11} = 0.00000000001$	11	10^{-3}
$10^{-12} = 0.000000000001$	12	10^{-2}
$10^{-13} = 0.0000000000001$	13	10^{-1}
$10^{-14} = 0.00000000000001$	14	10^0

(c) a solution whose $[OH^{-1}] = 10^{-3}\ M$
(d) a solution whose $[OH^{-1}] = 10^{-11}\ M$
(e) a 0.0001 M NaOH solution

SOLUTION

(a) A 0.001 M HCl solution releases 0.001 M hydrogen ions, which is $10^{-3}\ M$ hydrogen ions. Therefore, the

$$pH = -\log[H^+] = -\log[10^{-3}] = 3$$

(b) A 0.01 M HNO$_3$ solution releases 0.01 M hydrogen ions, which is $10^{-2}\ M$ hydrogen ions. Therefore, the

$$pH = -\log[H^{+1}] = -\log[10^{-2}] = 2$$

(c) A solution whose $[OH^{-1}]$ is $10^{-3}\ M$ has a $[H^{+1}] = 10^{-11}\ M$ (see Table 7-9); therefore, the

$$pH = -\log[H^{+1}] = -\log[10^{-11}] = 11$$

(d) A solution whose $[OH^{-1}]$ is $10^{-11}\ M$ has a $[H^{+1}] = 10^{-3}\ M$ (see Table 7-9); therefore, the

$$pH = -\log[H^{+1}] = -\log[10^{-3}] = 3$$

(e) A 0.0001 M NaOH solution releases 0.0001 M hydroxide ions, which is $10^{-4}\,M$ hydroxide ions. A solution whose $[OH^{-1}]$ is $10^{-4}\,M$ has a $[H^{+1}] = 10^{-10}\,M$ (see Table 7-9); therefore, the

$$pH = -\log[H^{+1}] = -\log[10^{-10}] = 10$$

PART 2

Solution Chemistry in the Body

Now that we've learned the basics about solution chemistry, let's see how all this applies to us. By the way, if you've been reading this chapter straight through up till now, this is a good time to put the book down and have a drink!

Buffer Solutions: Maintaining the Normal pH of the Blood

"Don't take aspirin, take buffered aspirin," says a popular commercial, "It won't upset your stomach." What does this mean? Why does buffered aspirin offer more protection against stomach upset than ordinary aspirin? To understand this we must explain what is meant by a buffer.

To begin with, you should recall that if you added an acid to neutral water, the resulting solution would be acidic. Depending on the strength of the acid added, the pH of the resulting solution could change drastically. The same would be true if you added a base. The resulting solution could become quite basic. However, if the water contained a buffer, the pH would not change drastically upon the addition of an acid or a base. How do you make a buffer solution? You can use:

1. A weak acid and its salt.

2. A weak base and its salt.

Table 7-10 lists some buffer systems. Let's look at a simple example to show you how a buffer system works. Perhaps the best example we could choose is the carbonic acid/bicarbonate buffer system, which maintains the pH of our blood at a fairly constant level.

Carbonic acid in our blood is formed by the reaction of carbon dioxide and water, which are already present in the blood.

$$H_2O + CO_2 \rightleftharpoons H_2CO_3 \qquad (7\text{-}5)$$

$$\text{Carbonic}$$
$$\text{acid}$$

As equation 7-5 indicates, an equilibrium exists among the water, carbon dioxide, and carbonic acid. The carbonic acid further ionizes to form an equilibrium mixture with hydrogen ion and bicarbonate ion.

TABLE 7-10   Some buffer systems

WEAK ACIDS AND THEIR SALTS
Acetic acid, $HC_2H_3O_2$, and sodium acetate, $NaC_2H_3O_2$
Carbonic acid, H_2CO_3, and sodium bicarbonate, $NaHCO_3$

WEAK BASES AND THEIR SALTS
Ammonium hydroxide, NH_4OH, and ammonium chloride, NH_4Cl
Dimethylamine, $(CH_3)_2NH$, and dimethylammonium chloride, $(CH_3)_2NH_2Cl$

$$H_2CO_3 \rightleftharpoons H^{+1} + (HCO_3)^{-1} \qquad (7\text{-}6)$$

$$\text{Bicarbonate}$$
$$\text{ion}$$

In essence we can represent this entire system in the blood as follows:

$$H_2O + CO_2 \rightleftharpoons H_2CO_3 \rightleftharpoons H^{+1} + (HCO_3)^{-1} \qquad (7\text{-}7)$$

What equation 7-7 says is that the blood contains water, carbon dioxide, carbonic acid, hydrogen ions, and bicarbonate ions in equilibrium with each other. The fact that these substances are in equilibrium with each other is very important. Why? Well, if the concentration of any one of these substances changes, the equilibrium shifts in a direction to make up for this change. For example, if too many hydrogen ions should form in the blood, the equilibrium shifts toward the left-side of the equation, to get rid of these extra hydrogen ions. Too much carbon dioxide would shift the equilibrium to the right side of equation 7-7. Keeping this in mind, let's look at what could happen to the pH of our blood.

The normal pH of blood is 7.4. However, what would happen if as a result of the ingestion of a basic substance, or an internal disorder such as kidney disease, the hydroxide ion concentration of the blood increased (a condition known as *metabolic alkalosis*)? If the blood had *no* buffer system, its pH would increase greatly. This could prove fatal to an individual. However, because the blood has the carbonic acid/bicarbonate buffer system, the following reaction occurs:

$$H_2CO_3 + (OH)^{-1} \rightleftharpoons H_2O + (HCO_3)^{-1}$$

The carbonic acid reacts with the hydroxide ions, producing water and bicarbonate ions, thus maintaining the pH at about 7.4. The excess bicarbonate is eventually eliminated via the kidneys and urine.

What would happen if the hydrogen ion concentration of the blood were to increase suddenly as a result of the ingestion of an acidic substance, kidney failure, or diabetes mellitus (a condition known as *metabolic acidosis*)? The excess hydrogen ions would react with the bicarbonate ions, producing carbonic acid. The excess carbonic acid would then decompose to form water and carbon dioxide, which would eventually be eliminated from the blood. The reactions can be summarized as follows.

$$(HCO_3)^{-1} + H^{+1} \rightleftharpoons H_2CO_3 \rightleftharpoons H_2O + CO_2$$

Again the pH would remain almost constant.

Respiratory Acidosis and Alkalosis

The pH of the blood can also be affected by a change in the level of carbon dioxide, due to rapid or slow breathing. Individuals who have emphysema, congestive heart failure, or some other lung ailment can suffer *hypoventilation*, which means a below-normal breathing rate. This condition causes an increase of carbon dioxide in the blood. The increase in carbon dioxide causes the formation of more carbonic acid, which further decomposes to produce hydrogen ions (equation 7-7). The increased hydrogen ion concentration causes a decrease in pH, which makes the blood more acidic. This condition is known as *respiratory acidosis*. When the pH of the blood drops from its normal value of 7.4 to about 7.2, the buffer system, which we previously discussed under metabolic acidosis, takes over to bring about a normal pH value.

A condition known as *hyperventilation* results from an above-normal breathing rate. (The word *hyper* means above normal.) An individual who is suffering from fear or trauma can hyperventilate. In this condition the individual exhales more carbon dioxide than is normal. If this occurs, the equilibrium represented in equation 7-7 shifts to the left. This causes a decrease in the hydrogen ion concentration of the blood, with a resultant increase in pH. The blood becomes too alkaline (too basic). This condition is called *respiratory alkalosis*. When the pH of the blood reaches a value of 7.5, the buffer system we discussed under metabolic alkalosis goes to work to restore the normal pH.

The Transport of Electrolytes in the Body: Diffusion and Osmosis

We've already discussed the definition of electrolytes earlier in this chapter, but we didn't discuss the importance of electrolytes in maintaining the normal body processes. One of the first indicators of body malfunction is a change in the level of various electrolytes. Table 7-11 shows the normal value of electrolytes in body fluid, both intracellular (in the cells) and extracellular (outside the cells). Table 7-12 shows what happens to the concentration of these

TABLE 7-11   Some normal values of electrolytes in body fluids

ION	INTRACELLULAR (mg/liter)	EXTRACELLULAR (mg/liter)
Na^{+1}	230	3,335
K^{+1}	6,240	117
Mg^{+2}	420	24
Cl^{-1}	71	3,905
$(HCO_3)^{-1}$	488	1,830
$(HPO_4)^{-1}$	13,440	384

TABLE 7-12   Changes in the concentrations of electrolytes due to a body malfunction

CONDITION	ELECTROLYTE LEVEL				
	Na^{+1}	K^{+1}	Cl^{-1}	$(HCO_3)^{-1}$	H^{+1}
Burns	Down	Down	Down	—	—
Dehydration	Down	Down	Down	—	—
Diabetes	Down	Down	—	Down	Up
Diarrhea	Down	Down	—	Down	Up
Diuretic (increased urine flow)	Down	Down	Down	—	Down
Enema	Down	—	—	—	—
Head injury (hemorrhage)	Up	—	Up	—	—
Surgery	Down	Down	—	—	—
Sweating	Down	—	Down	—	—
Vomiting	Down	Down	Down	Down	Down

electrolytes when a body malfunction occurs. When the electrolyte balance of the body is upset to a great degree, as a result of serious injury or illness, the physician uses intravenous solutions of various electrolytes to restore the normal balance. Some of these i.v. solutions, as they are called, are listed in Table 7-13. The question you should now be asking is, "How do the electrolytes in the i.v. solutions get to the place where they are needed?" To answer this question we must look at the processes of diffusion and osmosis.

NAME OF SOLUTION	ELECTROLYTES IN SOLUTION	USE OF SOLUTION
Sodium chloride (0.9%)	Na^{+1}, Cl^{-1}	Replace fluid loss
Ammonium chloride (21.4%)	$(NH_4)^{+1}$, Cl^{-1}	Acidifying agent
Sodium bicarbonate (7.5%)	Na^{+1}, $(HCO_3)^{-1}$	Alkalinizing agent
Dextran 75 (6% injection in sodium chloride 0.9%)	Na^{+1}, Cl^{-1}	Replacement solution
Ringer's	Na^{+1}, K^{+1}, Ca^{+2}, Cl^{-1}	Replacement solution
Dextrose (5%) (maintenance solution)	Glucose, Na^{+1}, K^{+1}, Cl^{-1}, lactate, $(HPO_4)^{-2}$	Maintain fluid and electrolyte balance

The Processes of Diffusion and Osmosis

If you take a few drops of a dye and place it in a glass of water, what happens? Very quickly the dye disperses itself through the water. Given enough time, the dye will distribute itself evenly throughout the water (Fig. 7-16). This process is known as *diffusion*. It occurs because the dye moves from an area of high dye concentration to an area of low dye concentration. In other words, there is a *concentration gradient* between the place in the water where the dye is dropped and the rest of the water. The diffusion process stops when the dye is evenly distributed throughout the water, so that the concentration gradient no longer exists. It is this process of diffusion which moves most of the substances in our body about. For example, the process of respiration is based on the diffusion of gases from areas of high concentration to areas of low concentration. And the process of diffusion is the system by which electrolytes in the body are carried to where they are needed.

FIGURE 7-16

A dye will diffuse through a solution from areas of high dye concentration to areas of low dye concentration. Eventually, the dye will distribute itself evenly throughout the solution.

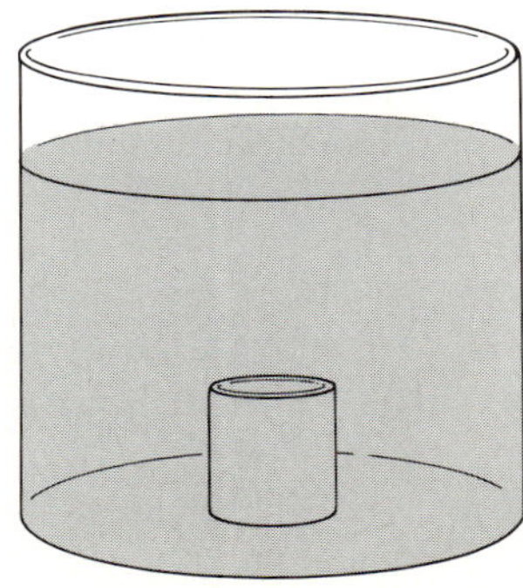

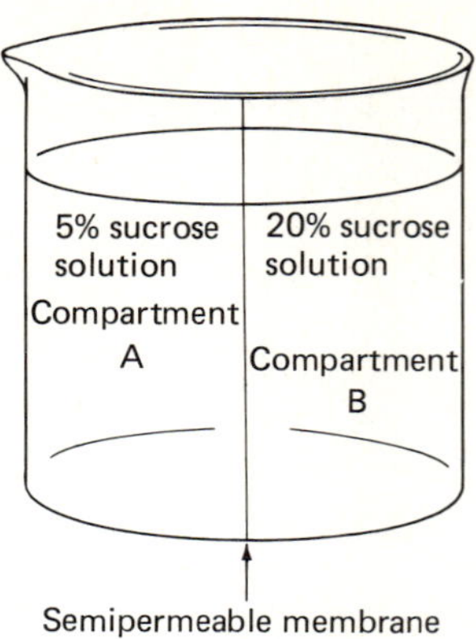

FIGURE 7-17
A beaker with two compartments separated by a semipermeable membrane.

A special type of diffusion process called *osmosis* involves the *passage of water through a semipermeable membrane, like cellophane or cell walls.* In this process, water moves from an area of *low solute concentration* to an area of *high solute concentration.* The best way to describe the process of osmosis is through an example. A commonly used example involves a beaker having two compartments, separated by a semipermeable membrane (Fig. 7-17). One compartment, which we'll call compartment A, is filled with a 5% sucrose solution. The other compartment, which we'll call compartment B, is filled with an equal volume of a 20% sucrose solution. Soon after the solutions are placed in the beaker, we notice a shift in the water levels of the compartments. Water moves from the compartment containing the 5% sucrose solution to the

FIGURE 7-18
This is what happens to our two solutions after osmosis has proceeded for a while. Water has left compartment A, increasing the concentration of that sucrose solution. The water has gone into compartment B, diluting that sucrose solution.

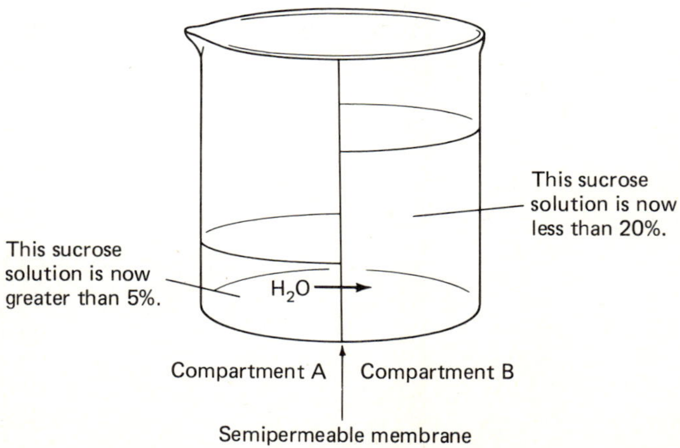

compartment containing the 20% sucrose solution, in an attempt to equalize the sucrose concentrations in both compartments (Fig. 7-18). This movement of water is the process of osmosis. Of course, you may be thinking, "Why doesn't the sucrose in the 20% solution move through the membrane to equalize the concentration of the two solutions?" This is surely a possibility. However, if this did occur, the process would be diffusion, not osmosis. Remember, *osmosis is the passage of water through a semipermeable membrane.* In most instances, however, the membrane tends to retard the movement of the solute particles, while allowing the solvent particles to pass freely.

Osmotic Pressure

Consider the two sucrose solutions we've just been talking about. After the process of osmosis has proceeded for a while, it finally stops. This can occur because the number of water molecules passing from compartment A to compartment B becomes equal to the number of water molecules passing from compartment B to compartment A (Fig. 7-19). In other words, a dynamic equilibrium is in progress. But, why does this occur? The answer lies in looking at the heights of the water in both compartments. The height of the water in compartment B is greater than that in A (Fig. 7-20). At first water molecules moved from compartment A to B in an attempt to equalize the concentration of the two sucrose solutions. But eventually the height of the water in compartment B caused a pressure to be exerted on the water molecules in that compartment, pushing them back into compartment A. When these two opposing forces became equal, the process of osmosis stopped. The pressure exerted by the water in compartment B at this point is called the *osmotic pressure.* In general, the osmotic pressure can be defined as *the amount of pressure that must be applied to prevent the flow of water through a membrane.* This concept is extremely important in the transport of body fluids.

FIGURE 7-19

The process of osmosis stops when the number of water molecules moving from compartment A to compartment B equals the number of water molecules moving from compartment B to compartment A.

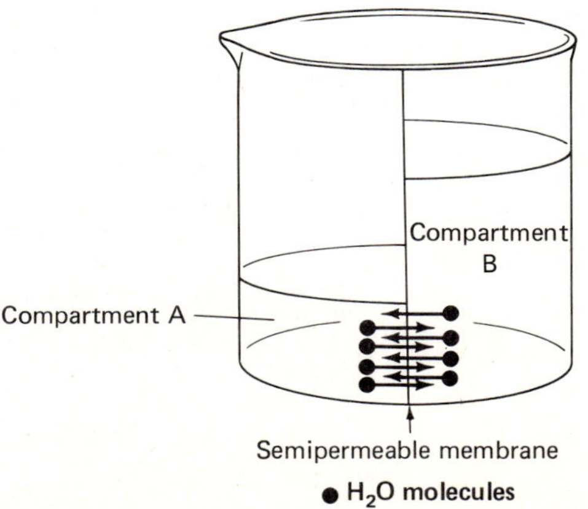

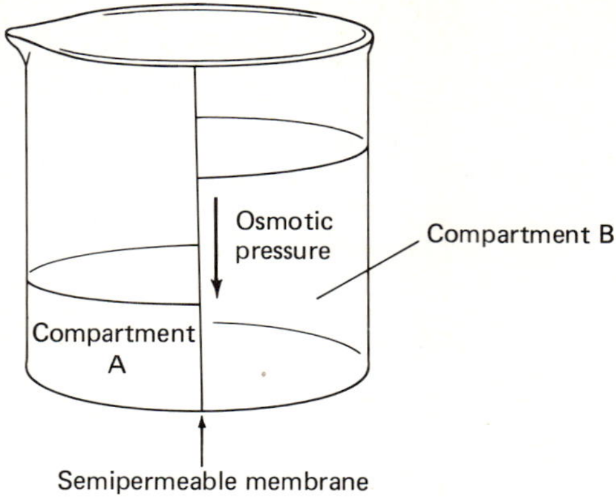

FIGURE 7-20
Osmotic pressure is due to the height of the water column in compartment B.

Isotonic, Hypotonic, and Hypertonic Solutions

We've already discussed the use of i.v. solutions to restore the electrolyte balance in the body of those who have suffered from injury or illness. However, the concentration of electrolytes in these i.v. solutions are very important in terms of what they will do to the cells of the body. "What does this mean?" you ask. Well, remember that a cell has fluid and electrolytes within it (intracellular fluid). There is also fluid and electrolytes outside the cell (extracellular fluid). If the osmotic pressure of the intracellular fluid is the same as that of the extracellular fluid, the cell will maintain its normal volume (Fig. 7-21). If the osmotic pressure of the extracellular fluid is lower than that of the intracellular fluid, then water will flow into the cell. If this sounds confusing, just remember that a lower osmotic pressure in the extracellular fluid means that the electrolyte concentration inside the cell is greater than that outside the cell; therefore, water will move into the cell in an attempt to equalize the electrolyte concentra-

FIGURE 7-21
If the osmotic pressure of the intracellular fluid is the same as that of the extracellular fluid, the cell will maintain its normal volume.

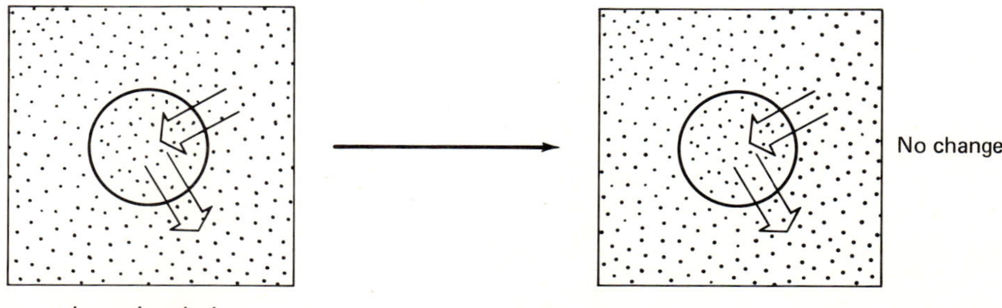

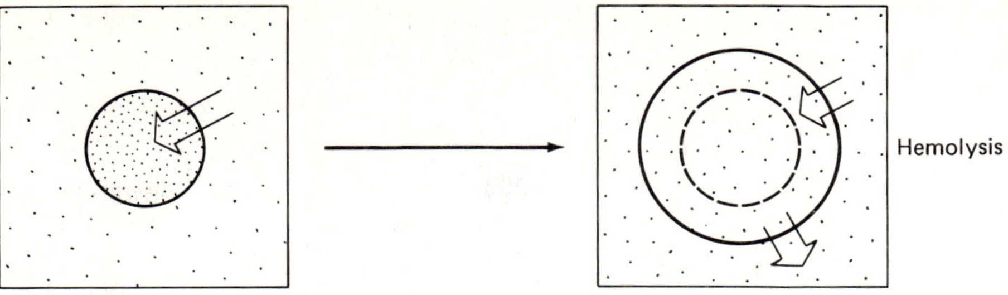

FIGURE 7-22
If the osmotic pressure of the extracellular fluid is lower than that of the intracellular fluid,
water will flow into the cell, causing a condition known as hemolysis.

tion (Fig. 7-22). The cell will expand due to the water moving into it, and it may burst. This condition is known as *hemolysis*.

If the osmotic pressure of the extracellular fluid is greater than that of the intracellular fluid, then water will flow out of the cell. This will cause the cell to shrink, a condition known as *crenation* (Fig. 7-23). Intravenous solutions can affect cells in these three ways. An i.v. solution whose osmotic pressure is the same as that of the cells is called *isotonic*. Two commonly used isotonic solutions are 0.9% sodium chloride and 5% glucose. An i.v. solution whose osmotic pressure is lower than that of the cells is called *hypotonic*. Hypotonic solutions can cause hemolysis. And an i.v. solution whose osmotic pressure is greater than that of the cells is called *hypertonic*. Hypertonic solutions can cause crenation.

EXAMPLE 7-10   For each solution given, state whether it is isotonic, hypotonic, or hypertonic. Then state the effect of the solution on the cells.

(a) 5% glucose
(b) 10% glucose
(c) 0.5% sodium chloride

FIGURE 7-23
If the osmotic pressure of the extracellular fluid is greater than that of the intracellular
fluid, water will flow out of the cell, causing a condition known as crenation.

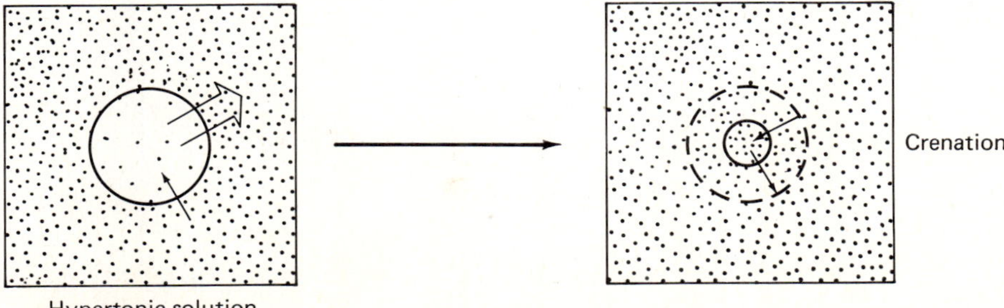

SOLUTION

(a) A 5% glucose solution is isotonic to the cells; therefore, there is no effect on the cells.

(b) A 10% glucose solution is hypertonic to the cells; therefore, it will cause crenation.

(c) A 0.5% sodium chloride solution is hypotonic to the cells; therefore, it will cause hemolysis.

The Process of Dialysis

Dialysis is a means by which particles of a solution can be separated from the particles of a colloid through the use of a membrane. A dialyzing membrane allows the ions of electrolytes and small molecules to pass through, while retaining colloids and large molecules. Consider a solution of sodium chloride and starch sealed in a cellophane bag. The bag is immersed in a beaker of pure water (Fig. 7-24). The bag acts as a dialyzing membrane. Because the concentration of sodium chloride in the cellophane bag is greater than that in

FIGURE 7-24
The process of dialysis. The dialyzing membrane allows the ions of electrolytes and small molecules to pass through, while retaining colloids and large molecules. The sodium chloride passes through the dialyzing membrane because of the concentration gradient that exists between the inside and outside of the membrane. The starch cannot pass through the membrane.

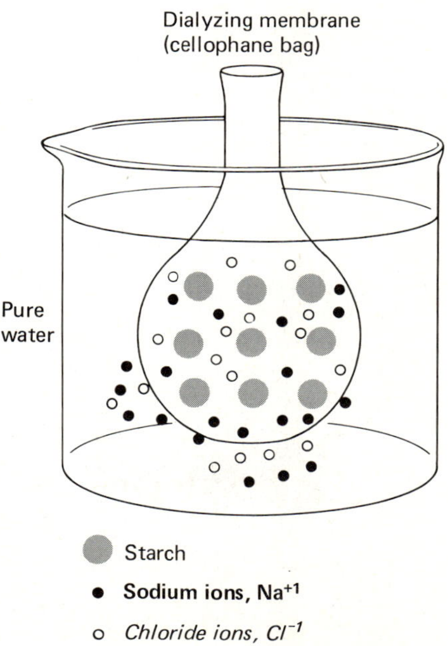

the beaker of water, salt passes out of the bag and into the beaker. The starch, a colloidal material, can't pass through the membrane. Eventually, the salt concentration in the bag and beaker become equal and the dialysis stops. However, if we keep the water in the beaker running, the salt concentration in the beaker never accumulates and so salt keeps passing out of the bag until it is completely depleted. Throughout all of this, the starch has remained in the cellophane bag.

Hemodialysis is the method used to remove waste materials from an individual whose kidneys are not functioning. The same technique we've just described is used. The patient's blood passes through a dialyzing membrane. The dialyzing membrane is submerged in a solution of distilled water containing various electrolytes called a *dialysate* (Fig. 7-25). Because the concentration of

FIGURE 7-25
A modern unit for hemodialysis. (From "The Hospital People," published by the Blue Cross Association)

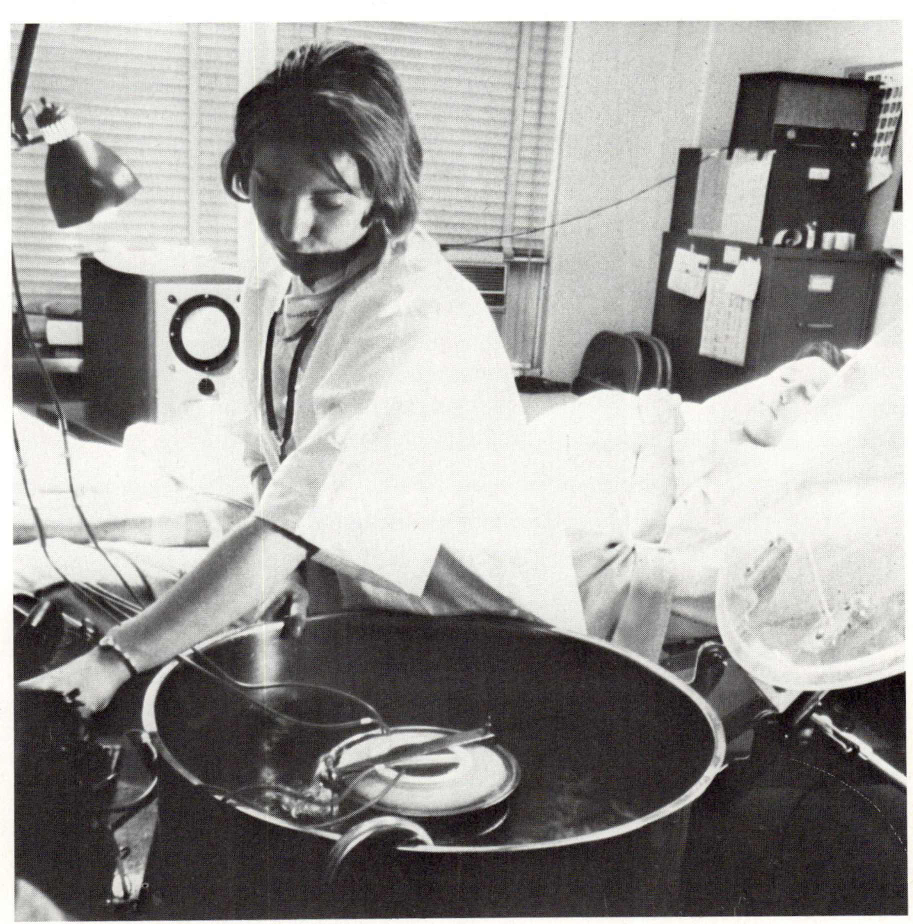

wastes in the blood is high, the electrolytes pass through the membrane into the dialysate. At the same time, needed electrolytes pass from the dialysate into the blood. No red blood cells are lost in the process because they cannot pass through the membrane. Hemodialysis can thus be used for patients whose kidneys are malfunctioning, to restore the normal electrolyte balance in the blood and remove toxic wastes.

SUMMARY

In this chapter we learned about solutions of acids, bases, and salts. We also talked about colloidal dispersions and suspensions. We discussed how concentrations of solutions are measured using percent weight-volume and molarity, and we talked about ionization and the concept of strong and weak electrolytes. We also learned about pH and how the pH of the blood is maintained fairly constant by buffers. We discussed the conditions of metabolic acidosis and alkalosis, as well as the conditions of respiratory acidosis and alkalosis. We read about the transport of electrolytes in the body and discussed the processes of diffusion and osmosis. We ended the chapter with a discussion of dialysis. All and all, we tried to show you the importance of solution chemistry in the body. Many of the topics we discussed here will be important to you as a future health practitioner.

EXERCISES

1. Which is the solute and which is the solvent in the following solutions?
   (a) potassium chloride in water
   (b) 7% weight-volume sodium chloride solution
   (c) 5 liters of oxygen gas and 3 liters of nitrogen gas
   (d) 3 M HCl solution

2. Determine whether each of the following are true solutions, colloidal dispersions, or suspensions.
   (a) A bottle of milk of magnesia from the pharmacy. The directions on the bottle say, "Shake well before using."
   (b) A solute and solvent are combined and shaken well, then filtered. Nothing remains on the filter paper. The solution is then poured through a semipermeable membrane. The solute and solvent do not separate.

3. What is the percent weight-volume of each of the following solutions?
   (a) 25.0 g of sodium chloride dissolved in $20\overline{0}$ ml of water solution.
   (b) 9 g of alcohol dissolved in $3\overline{0}$ ml of water solution.
   (c) 0.8 g of glucose dissolved in $4\overline{0}$ ml of water solution.

4. How many grams of potassium chloride must you mix with $5\overline{0}0$ ml of water to prepare a 15% ~~by weight-volume~~ potassium chloride solution? (Assume that the final volume of the solution is 500 ml.)

5. Determine the molarity of the following solutions.
   (a) $18\overline{0}$ g of glucose ($C_6H_{12}O_6$) in $5\overline{0}0$ ml of solution.
   (b) 5.85 g of NaCl in $25\overline{0}$ ml of solution
   (c) 72 g of HCl in 2.0 liters of solution

6. How many grams of calcium hydroxide, $Ca(OH)_2$, are necessary to prepare $6\overline{0}0$ ml of a 2.0 M solution?

7. How many milliliters of a 0.10 M glucose solution are necessary to obtain 36 g of glucose?

8. A popular pharmaceutical agent used in the control of seizures is phenobarbital. The concentration of phenobarbital in its elixir is 0.017 M. The typical dosage of this elixir is 5.0 ml. How many milligrams of phenobarbital are there in one dose? (*Hint:* The molecular weight of phenobarbital is 232.)

9. Define the following terms.
   (a) acid   (b) base   (c) salt   (d) oxide

10. List three properties of all acids.

11. List three properties of all bases.

12. Complete the equation: $HCl + H_2O \longrightarrow$

13. Complete the equation and balance the following neutralization reactions.
   (a) $HCl + NaOH \longrightarrow$ (b) $H_2SO_4 + KOH \longrightarrow$
   (c) $HC_2H_3O_2 + Ca(OH)_2 \longrightarrow$ (d) $H_3PO_4 + Mg(OH)_2 \longrightarrow$

14. Define electrolytic and nonelectrolytic solutions. Explain what makes them different from each other in terms of the particles that compose them.

15. What is the difference between a strong and weak electrolyte? Give some examples of each.

16. Write the reaction for the ionization of water.

17. Given the hydrogen ion concentration of each of the following aqueous solutions, determine the hydroxide ion concentration.
   (a) $[H^{+1}] = [10^{-6}]$ (b) $[H^{+1}] = [10^{-9}]$ (c) $[H^{+1}] = [10^{-1}]$

18. Determine the pH of each solution in Exercise 17.

19. Determine the pH of each solution.
   (a) 1 M HCl   (b) 0.01 M HCl
   (c) 0.01 M NaOH   (d) 0.0001 M LiOH

20. Given the following reaction at equilibrium:

$$H_2O + CO_2 \;\rightleftharpoons\; H_2CO_3 \;\rightleftharpoons\; H^{+1} + (HCO_3)^{-1}$$

decide which way the equilibrium shifts if
(a) excess H^{+1} is added
(b) excess CO_2 is added
(c) bicarbonate ion, $(HCO_3)^{-1}$, is removed

21. Describe the conditions of metabolic acidosis and metabolic alkalosis.

22. Hypoventilation can lead to a condition known as ________. (Fill in the terms *respiratory acidosis* or *alkalosis*.)

23. Hyperventilation can lead to a condition known as ________. (Fill in the terms *respiratory acidosis* or *alkalosis*.)

24. Define and give an example of
(a) diffusion　　(b) osmosis　　(c) osmotic pressure

25. For each solution given, state whether it is isotonic, hypotonic, or hypertonic.
(a) 5% glucose solution
(b) 10% glucose solution
(c) 0.5% sodium chloride solution
(d) 2% sodium chloride solution

26. State what would happen to the cells in the body if each of the solutions in Exercise 25 was administered i.v. State your answer in terms of no effect, crenation, or hemolysis.

27. Explain how dialysis can be used to save the life of a person who has consumed an electrolytic poison.

CHAPTER 8

Nuclear Radiation

Cure All or Kill All

Some Things You Should Know After Reading This Chapter

You should be able to:

1. Explain what is meant by radioactivity.
2. Name the three types of nuclear radiation: alpha rays, beta rays, and gamma rays.
3. State the charge and mass of an alpha particle.
4. State the charge and mass of a beta particle.
5. Complete a nuclear equation, given the starting isotope and the particle emitted.
6. Explain the term half-life.
7. Calculate the number of grams of isotope remaining after a given time period, when told its half-life.
8. Explain the terms chain reaction, nuclear fission, and nuclear fusion.
9. Explain the two ways that radiation can damage or destroy cells.
10. Explain what may happen when radiation reacts with water in cells.
11. Explain and give an example of a stage one radiation effect.
12. Explain and give an example of a stage two radiation effect.
13. Name the types of cells most sensitive to radiation.
14. Name four devices used to detect radiation and explain what each one is used for.
15. Define each of the following units of radiation: curie, roentgen, rad, and rem.
16. Explain what is meant by LD_{50}^{30}.
17. Explain the relationship between rem and rad.
18. Explain how the following isotopes are used in medical diagnosis and therapy: iodine-131, chromium-151, and cobalt-59 or cobalt-60.
19. Explain the three methods by which radiation can be delivered to a malignant area: teletherapy, brachytherapy, and radiopharmaceutical therapy.

Nuclear radiation has, over the past few decades, become a major weapon of the physician in the fight to treat diseases that were at one time untreatable (Fig. 8-1). Like invisible bullets, radiation can penetrate the skin and shrink tumors lodged deeply in the brain. And radioactively tagged isotopes can pinpoint abnormalities deep within a vital organ, allowing the abnormalities to make themselves known. "A panacea," you say. Unfortunately no, because radiation can kill as well as cure. Normal cells in the body can be destroyed or changed into cancerous cells by radiation. And the increased use of radioactive materials in medicine has only been paralleled by the increased use of radioactive materials in general, for example, in defense, power generation, and industrial applications. This has added additional sources of nuclear radiation to our environment, thus increasing the chance of exposure to all of us. We can only hope that the benefits derived from the use of nuclear radiation far outweigh the hazards.

In this chapter we are going to learn about nuclear radiation and look at some specific medical applications.

FIGURE 8-1
Nuclear radiation—a powerful weapon against previously untreatable diseases.

The Basics

History

Although the explosion of the atomic bomb over Hiroshima in 1945 is recognized as the beginning of the nuclear age, the study of nuclear chemistry actually began much earlier, in 1895, when a German physicist, Wilhelm Konrad Röntgen, accidentally discovered x rays. He found that x rays have a great penetrating power, so great that they can pass through walls.

In 1896, a French physicist by the name of Antoine Becquerel placed some salt that contained uranium on a photographic plate that was wrapped in black paper. When he developed the photographic plate, he found to his surprise that he had the image of the pile of salt. The uranium salt had taken a picture of itself! Becquerel concluded that the uranium was giving off some type of penetrating rays, and that these rays must be very strong to be able to pass through the black paper and expose the photographic plate (Fig. 8-2). However, when he placed a thick barrier of lead between the salt and the photographic plate, the lead absorbed the rays. Becquerel realized that these penetrating rays were probably Röntgen's x rays.

At the same time there was in Paris a young Polish chemist, Marie Curie, who, with her husband Pierre, was working in the laboratories at Sorbonne. The Curies became interested in Becquerel's problem. It was, in fact, Marie Curie who defined the ability of a substance to produce penetrating rays as

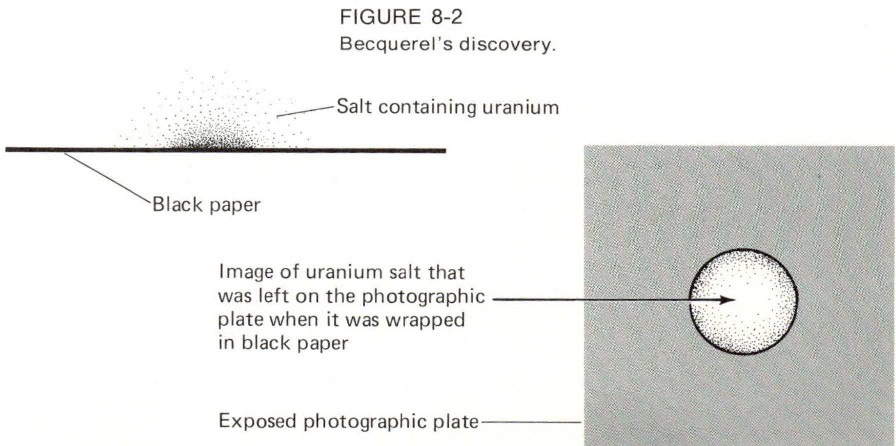

FIGURE 8-2
Becquerel's discovery.

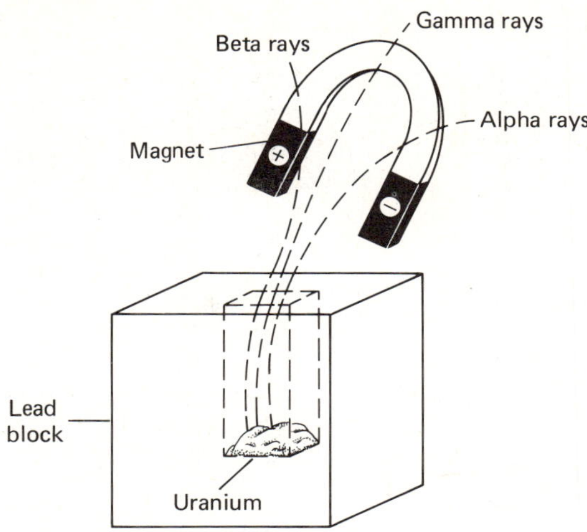

FIGURE 8-3
When the radiation produced by uranium passes through a magnetic field, it breaks up into three types.

radioactivity. The Curies found that a radioactive substance seems to keep on and on, year after year, emitting these powerful penetrating rays.

In 1898 they discovered that the element thorium is radioactive, and so is polonium. And so, most of all, is radium.

Although early experimenters first thought that radioactive materials produced only x rays, they soon discovered that the situation was more complex. For example, when they allowed the radiation produced by uranium to pass through a magnetic field, they could detect *three* types of radiation.

The English physicist Lord Rutherford called these three types of radiation *alpha rays*, *beta rays*, and *gamma rays*, from the first three letters of the Greek alphabet, α, β, and γ (Fig. 8-3).

Gamma Rays (γ rays)

Many experiments in passing radioactive materials through a magnetic field led to the conclusion that—although alpha and beta rays are deflected to one side or the other—*gamma rays* travel straight through and are not deflected. For this reason the early experimenters assumed that gamma rays carried no charge. Gamma rays seemed to have properties like those of x rays, but they were more energetic. In fact, gamma rays have a very high energy content and strong penetrating ability. Today we know that because of this penetrating ability a high concentration of gamma rays can harm human genes. The *nucleus* of an atom gives off gamma rays only under certain conditions, usually when the nucleus loses energy. And the atom, of course, has to be of the radioactive variety.

Beta Rays (β rays)

When Becquerel passed *beta* rays through a magnetic field, he found that they behaved like cathode rays—that is, like rays that come from a negative electrode. Therefore, Becquerel concluded that these rays were composed of speeding electrons. However, beta particles, like gamma and alpha particles, come from the *nucleus* of the atom, and electrons don't exist in the nucleus.

How, then, are beta particles formed? Scientists believe that they are formed when neutrons, during a nuclear reaction, are transformed into protons.

$$\text{Neutron} \longrightarrow \text{proton} + \beta \text{ particle}$$

A beta particle, like an electron, has a zero mass number and a charge of -1. We can represent this in the following way:

$$_{-1}^{0}e = \beta \text{ particle}$$

Beta particles have only slight penetrating power, compared to gamma rays. They can be stopped by a thin sheet (a few centimeters thick) of almost any metal.

Alpha Rays (α rays)

When scientists of the early 1900's (Becquerel, James Chadwick, and Werner Heisenberg) investigated *alpha* particles in a magnetic field, they found that these particles were deflected in the opposite direction from beta rays. Therefore, they assumed that alpha particles must be positively charged. Further experiments showed that these particles had a mass number of 4 and a charge of $+2$. Heisenberg suggested that we picture an alpha particle as a helium-4 atom with its two electrons removed; in other words, we can imagine it as composed of two protons and two neutrons. We can represent an alpha particle in this way:

$$_{2}^{4}\text{He} = \alpha \text{ particle}$$

Alpha particles have virtually no penetrating ability. They can be stopped by a thin sheet of paper.

Nuclear Transformation

Now that we are familiar with the basic particles emitted by radioactive materials, let's examine some natural nuclear reactions—that is, nuclear reactions that occur spontaneously in nature, as opposed to nuclear reactions that people set up in laboratories. Table 8-1 will help us get the picture. We

TYPE OF PARTICLE EMITTED	MASS NUMBER OF PARTICLE EMITTED	CHARGE OF PARTICLE EMITTED	WHAT HAPPENS TO ELEMENT'S MASS NUMBER?	ATOMIC NUMBER?
Alpha	4	$+2$	Decreases by 4	Decreases by 2
Beta	0	-1	No change	Increases by 1
Gamma	0	0	No change	No change

can use this table to help us determine what happens to the element that is doing the emitting, once the radioactive particle has been released. (And, incidentally, an element that is emitting such particles is said to be undergoing *radioactive decay*, or *nuclear decay*.)

Isotopes of some elements produce radiation naturally. An example is radium-226, $^{226}_{88}\text{Ra}$. When it is undergoing nuclear decay, radium emits alpha particles and actually becomes another element! We call this *nuclear transformation*, and we can represent it by the following equation:

$$^{226}_{88}\text{Ra} \longrightarrow \quad ^{4}_{2}\text{He} \quad + \; ?$$

$$\alpha \text{ Particle}$$

What is the unknown element? Subtraction tells us that it must be the element whose atomic number is 86, and whose mass number is 222. We look at the periodic table and find that element 86 is radon (Rn). So we can write the reaction in this way:

$$^{226}_{88}\text{Ra} \longrightarrow {}^{4}_{2}\text{He} + {}^{222}_{86}\text{Rn}$$

In the balanced nuclear equation, the sum of the superscripts, and also the sum of the subscripts, is the same on both sides of the arrow.

Take another example. The isotope thorium-234 produces radiation by emitting beta particles. We can write the nuclear reaction as

$$^{234}_{90}\text{Th} \longrightarrow \quad ^{0}_{-1}e \quad + \; ?$$

$$\beta \text{ Particle}$$

What is the other element that is being formed? Well, Table 8-1 tells us that the formation of a beta particle increases an element's atomic number by 1. Therefore, the unknown element must be the element whose atomic number is 91. The periodic table tells us that element 91 is protactinium. So we can write our equation as

$$^{234}_{90}\text{Th} \longrightarrow {}^{0}_{-1}e + {}^{234}_{91}\text{Pa}$$

Once again the sums of both superscripts and subscripts are the same on both sides of the arrow.

Half-Life

Just now we have been discussing some nuclear reactions, and we looked at what actually happens to a single atom of an element when it undergoes decay. We might now ask two questions:

1. Why are some elements radioactive? In other words, why do they undergo nuclear decay?

2. Can we predict when a particular atom of a radioactive substance will decay?

Although nobody really knows the exact answer to our first question, many scientists do feel that certain combinations of protons and neutrons make the nucleus of an atom unstable, and therefore make it tend to be radioactive.

The answer to our second question is no. No one can tell when a particular atom of a radioactive substance will decay. However, when we are dealing with large numbers of radioactive atoms (which is usually the case), we can predict how long it will take for a certain number or percentage (say half) of the atoms to decay. For each radioactive isotope, we can predict a certain *half-life*. *The half-life is the time it takes for one-half of the atoms originally present in a sample to decay.* The half-life of a radioactive isotope may be short or long. For example, the half-life of $^{238}_{92}U$ is 4.5 billion years, while the half-life of $^{257}_{103}Lr$ is 8 seconds.

EXAMPLE 8-1   The half-life of strontium-90 is 28 years. If we have 100 g of strontium-90 today, how many grams of strontium-90 will be left in our sample in 84 years?

SOLUTION   Since the half-life of strontium-90 is 28 years, this means that half of our strontium-90 will decay every 28 years. After 84 years, our sample will contain 12.5 g of strontium-90.

TIME (years)	STRONTIUM-90 REMAINING (g)
Now	100
28	50
56	25
84	12.5

While we're on the subject of strontium-90, we should say that it has a chemical similarity to calcium (notice that Ca and Sr are in the same chemical family), which causes it to replace the calcium in bone, and stay there, undergoing its radioactive decay. When nations test atomic bombs, strontium-90 gets into the atmosphere and begins to float around the world. Gradually, it falls down—on fields, for example. Since cows eat the grass from the fields, the strontium-90 gets into milk and other dairy products; it also gets into vegetables. People consume these farm products—we have, after all, no choice!—and this radioactive isotope, strontium-90, accumulates in our bones, side by side with the calcium. It then bombards the nearby tissues and organs with beta rays. That 28-year half-life means that it stays around for a long time. And if a person gets a really large dose, it can cause bone cancer. That's why thoughtful people were happy in the summer of 1963 when Russia, the United States, and Great Britain (three of the five nations with nuclear programs) signed the treaty banning atmospheric nuclear tests. Unfortunately, China and France didn't sign, and they've been testing. In late spring of 1974, India exploded a nuclear device underground, and became the sixth nation to join the nuclear club. It, too, has not signed any treaty.

Humanly Produced Radiation

The year is 1919. The British physicist Lord Rutherford, who first theorized the existence of the atomic nucleus, bombards nitrogen-14 with alpha rays. The results are astounding. Rutherford finds that he has produced oxygen-17, a nonradioactive isotope of oxygen.

$$\,^4_2\text{He} \;+\; \,^{14}_7\text{N} \;\longrightarrow\; \,^{17}_8\text{O} + \;\,^1_1\text{H}$$

$$\alpha \text{ Particle} \qquad\qquad\qquad\qquad \text{A proton}$$

The dreams of the alchemists have come true! Rutherford, by having triggered the first humanly produced nuclear reaction, has "transmuted" an element (Fig. 8-4).

FIGURE 8-4
When bombarded with alpha particles, nitrogen-14 becomes oxygen-17.

$$\,^4_2\text{He} \;+\; \,^{14}_7\text{N} \;\longrightarrow\; \,^{17}_8\text{O} \;+\; \,^1_1\text{H}$$

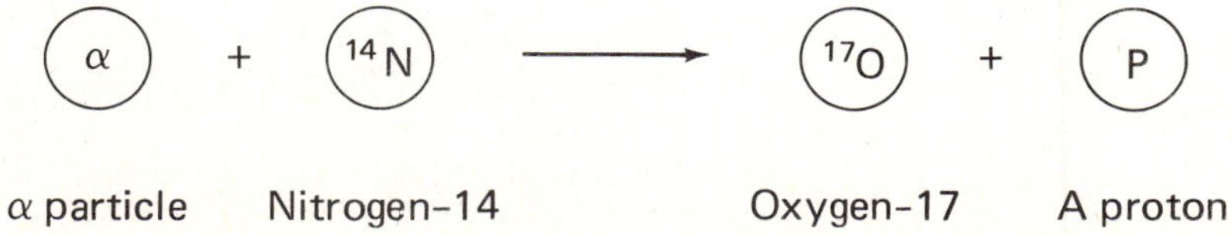

The year is 1934. The French physicists Irène Joliot-Curie and Frédéric Joliot-Curie (daughter and son-in-law of Marie and Pierre Curie) bombard boron-10 with alpha rays and obtain nitrogen-13, the first artificially produced radioactive isotope.

$$^{10}_{5}B + \ ^{4}_{2}He \longrightarrow \ ^{13}_{7}N + \ ^{1}_{0}n$$

$$\alpha \text{ Particle} \qquad\qquad\qquad \text{A neutron}$$

The Chain Reaction and Nuclear Fission

You can see that by the 1930's scientists were unraveling the atom's deepest secrets and were about to discover the power of the atom. They were, in fact, on the verge of producing a nuclear chain reaction, which could turn very small quantities of mass into huge quantities of energy.

In 1934, the Italian physicist Enrico Fermi (who won the Nobel prize in 1938) bombarded uranium (element 92) with neutrons in an attempt to produce element 93 (neptunium). But, instead, Fermi found himself with an isotope of barium (element 56), a mystifying outcome.

In 1938, the German physicist chemist Otto Hahn, who was also working on atomic power (many German scientists were, in this pre-World War II era), proposed an explanation. He said that uranium atoms split into different atoms when they were bombarded by neutrons. (This splitting process is called nuclear *fission*.) Furthermore, when these uranium atoms split, they produced *more* neutrons, which in turn were able to split still more uranium atoms. This is called a *chain reaction*. You can liken it to placing dominos in a triangular pattern. When you tip the first domino, it causes two others to tip over, and so on down the line (Fig. 8-5).

In September 1939 Europe exploded into World War II, and all the nations looked to their scientists for new and more terrible weapons. Many scientists all over the world who were working in the field of radioactivity knew that nuclear fission could be turned into a decisive weapon of war. It

FIGURE 8-5
The domino chain reaction.

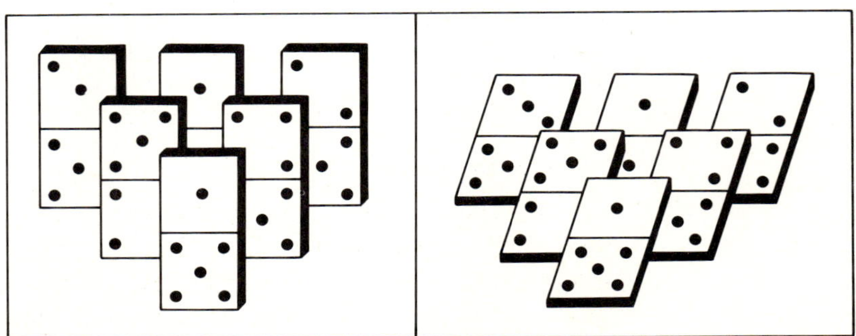

was only a matter of time before someone perfected the technique. Would the Germans be first? Were they racing toward that goal? Everyone knew that their scientists were second to none.

The United States, although it was not yet at war, recognized the danger and launched a research program to produce an atomic (fission) bomb. Under Enrico Fermi's leadership, the Manhattan Project (a code name) was started; it operated out of a windy, makeshift laboratory underneath the bleachers at Stagg Field at the University of Chicago. On December 2, 1942, Enrico Fermi and his team achieved the first sustained nuclear chain reaction. By July 1945, by incredible hard work and perseverance, the little band of scientists had managed to scrape together enough uranium-235 to make a fission bomb. To test it the United States had to fire it, so the first explosion of an A-bomb took place in a deserted area near Alamogordo, New Mexico. It created sickening devastation, and so, with feelings of foreboding that may be imagined, they set about making two more bombs at once. The major reaction produced by the exploding A-bomb may be represented as

$$^{235}_{92}U + {}^{1}_{0}n \longrightarrow {}^{141}_{56}Ba + {}^{92}_{36}Kr + 3{}^{1}_{0}n + energy$$

By August 1945 the other two bombs were ready. Yet such was the crisis— with America believing that they would lose 1 million American soldiers if they attempted to storm the Japanese mainland—that the United States went ahead and exploded them: the first at Hiroshima, and the second, 4 days later, at Nagasaki, Japan. Many Japanese people who lived in or near those two cities developed cancers as a result of their exposure to the radiation. The effects of this radiation on the second generation of Japanese from these two areas of Japan are just now being seen.

The Atom and Electric Power

Even though the first use of energy from nuclear fission was for making war, scientists could readily see that the atom was a tremendous storehouse of energy that could be used for peaceful purposes. Starting in the 1950's, several nations began to use atomic power for supplying electricity, and the number of nuclear-powered electric plants throughout the world has grown to more than 150. This enormous increase was a result of the development of nuclear reactors, which use the heat energy produced by controlled fission to generate steam. The steam in turn powers a generator, which produces electricity (Fig. 8-6).

Nuclear Fusion

Hydrogen is the most abundant element in the universe. It is hydrogen that is responsible for the tremendous quantities of energy produced by the sun and other stars. However, the sun and stars get their energy through the process of nuclear fusion: the fusing or combining of two hydrogen atoms to

FIGURE 8-6
A nuclear power plant. (Public Service Electric and Gas Company, Newark, New Jersey)

form a helium atom. This fusing process creates huge amounts of heat—on the order of 2 million degrees Celsius—and there are no waste products left over. However, there are some catches. How do you contain such fantastic heat? What vessel or tools can handle it? And also, in order to duplicate the sun's fusing process on earth, you would need a great amount of energy to start with. So far, the only device that scientists have developed from their knowledge of atomic fusion is the *hydrogen bomb* (Fig. 8-7). This is a bomb now possessed by the United States, Russia, and China. Because it uses atomic fusion, rather than fission, the hydrogen bomb is a lot more powerful than the old-fashioned atomic bomb (in fact, in the hydrogen bomb, an atomic bomb acts as the initiator or detonator).

So it's clear that atomic fusion can lead to our doom, if used unwisely. However, once we learn how to control the process, and tame it for peaceful

FIGURE 8-7
The explosion of an H-bomb, an example of nuclear fusion. (U.S. Department of Energy)

uses, fusion may give us an unlimited power supply, with no radioactive waste products. In this way atomic fusion can increase the human stay on earth, by providing enough energy for all human needs, millions of years after conventional fuels have been used up.

PART 2

Radiation and Its Medical Uses

Radiation and Cellular Damage

When alpha, beta, and gamma rays strike the complex molecules that compose the cells of our bodies, they can actually cause parts of these complex molecules to break away and ionize. Or the radiation can interact with these complex molecules, causing the formation of free radicals. [*Free radicals* are highly reactive *uncharged* species. They can be atoms or groups of atoms having an odd (unpaired) number of electrons. For example, a single chlorine atom, $:\ddot{Cl}\cdot$, would be a free radical.] These free radicals and ions can interact with other complex molecules in the body, causing their destruction (Fig. 8-8). If the molecules that are destroyed are necessary for important biological processes, the whole organism can suffer and even die. For example, if the

FIGURE 8-8
The effect of radiation on living cells. (a) A normal cell contains 23 pairs of chromosomes.
(b) Ionizing radiation causes the chromosomes in this cell to multiply abnormally.
(c) Radiation prevents this cell from dividing normally. The cell contains more than 700 chromosomes, and has grown to more than 10 times normal size. (Courtesy of Dr. C. K. Yu, National Tsing Hua University, Taiwan, China)

(a)	(b)	(c)

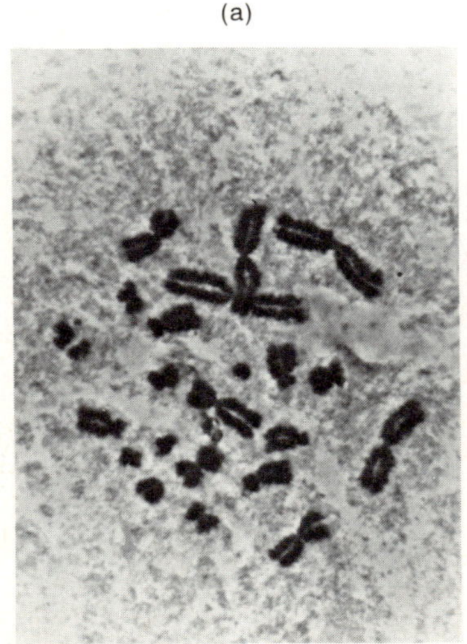 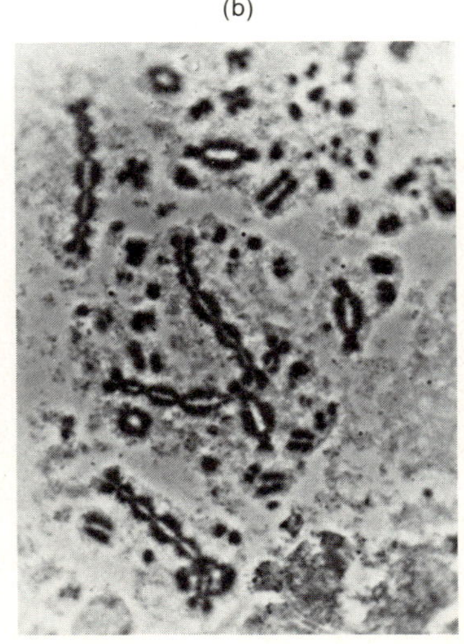 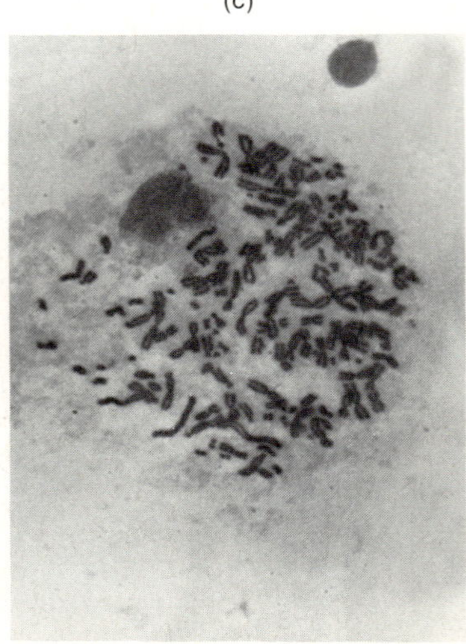

molecule attacked by the radiation is DNA, the cells involved will not be able
to reproduce themselves.

   Of course, the radiation could simply damage the cells and not destroy
them. But if this should happen, there is a possibility that the damaged cells
will not reproduce themselves in a normal fashion, which means they could
become cancerous.

   Radiation can also react with one of the major compounds in our bodies—
water! Cells contain about 80% water, and ionizing radiation can cause the
water to go through a series of free-radical reactions, forming a toxic substance,
hydrogen peroxide. This could be fatal to the cells. The reactions would look
something like this:

$$H \cdot \ddot{O} \colon + \text{radiation} \longrightarrow \quad \colon \ddot{O} \cdot H \quad + \quad \dot{H}$$

Hydroxide   Hydrogen
free radical   free radical

$$2 \colon \ddot{O} \cdot H \longrightarrow \quad \colon \ddot{O} \colon \ddot{O} \colon$$

Hydrogen
peroxide

   The effects of radiation in cells can be divided into two stages. *Stage 1
effects* are the actual changes in the chemical makeup of the cells. *Stage 2
effects* are the cells' inability to carry on their normal function, which results

FIGURE 8-9
Photographs of a normal thyroid (a), an enlarged thyroid (b), and a cancerous thyroid (c) taken
by a linear photo scanner using I-131. (Energy Research and Development Administration,
Oak Ridge, Tennessee)

(a)   (b)   (c)

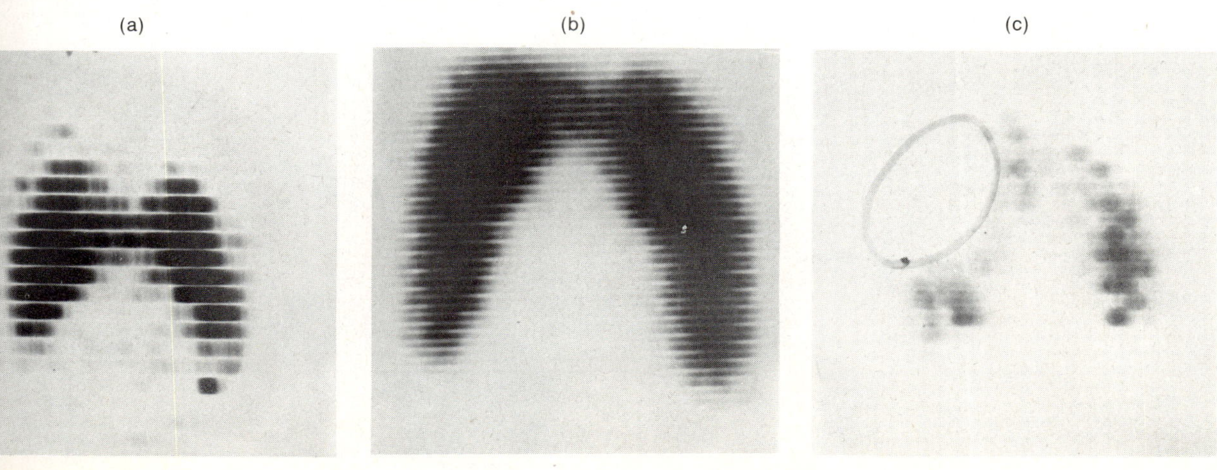

in a body malfunction. For example, when iodine-131 is administered to an individual, it will localize in the thyroid gland. The radiation from the iodine-131 will destroy some cells in the thyroid gland (a stage 1 effect). The thyroid gland will not be able to produce as much thyroxine (a stage 2 effect). For an individual with a hyperactive thyroid, this represents an excellent therapeutic method (Fig. 8-9).

The cells most sensitive to radiation are those undergoing rapid growth, for example, cells of the bone marrow and reproductive organs. Also, infants and children are very susceptible to radiation because a large number of cells in their bodies are undergoing rapid growth. By the way, cancer cells grow very rapidly, compared to normal cells. It is for this reason that radiation is very effective against these types of malignancies. Radiation directed at an area containing fast-growing cancer cells will have a greater devastating effect on these cells as compared to normal cells.

The Detection of Ionizing Radiation

There are various instruments and devices used for the detection of radiation. The best-known device is a Geiger–Müller counter. This device consists of two parts, a detecting tube and a counter (Fig. 8-10). The heart of the system is

FIGURE 8-10
A Geiger-Müller counter. (Dosimeter Corporation of America)

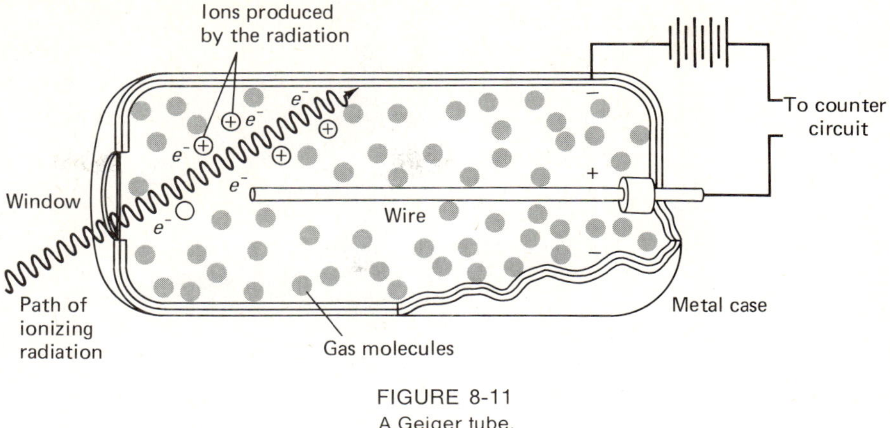

FIGURE 8-11
A Geiger tube.

the detecting tube, which consists of a pair of electrodes surrounded by an ionizable gas. As radiation enters the tube it ionizes the gas. The ions produced travel toward the electrodes, between which there is a high voltage. The ions cause pulses of current at the electrodes, which are picked up and recorded on the counter (Fig. 8-11).

The Geiger tube is most sensitive to beta radiation. Gamma radiation can pass right through the tube without being counted, and some alpha radiation can't make it through the window of the tube. The Geiger–Müller counter indicates the counts per minute of radiation entering the tube, but it doesn't tell you the energy of the radiation.

A scintillation counter is a device that not only counts radioactivity, but also enables the operator to determine the energy of the radiation. The principle of operation involves the radiation reacting with a crystal containing sodium iodide and thallium iodide, which produces a series of flashes of varying intensity. The intensity of the flashes is proportional to the energy of the radiation (Fig. 8-12).

Film badges are small portable devices that are worn by people such as x-ray technicians and nurses, who may be exposed to radiation. The badge contains a piece of photographic film which is removed monthly and developed. The darker the film badge, the greater is the degree of exposure (Fig. 8-13).

A newer device called a dosimeter is quickly replacing the film badge. One type of dosimeter works on the property of thermoluminescence, and is called a TLD for short. This TLD consists of a penlike device and a reading unit (Fig. 8-14). The penlike device, which is worn by the individual, contains a crystal such as lithium fluoride, which absorbs radiation. When the lithium fluoride crystal absorbs the radiation, its structure changes slightly. To determine the amount of radiation that the crystal has absorbed, the penlike device is placed in its reading unit, where it is heated quickly. This causes the lithium fluoride crystal to return to its original state. As it does, it gives off visible light. The visible light is proportional to the radiation absorbed by the lithium fluoride crystal.

FIGURE 8-13
A film badge.
(American Cancer
Society)

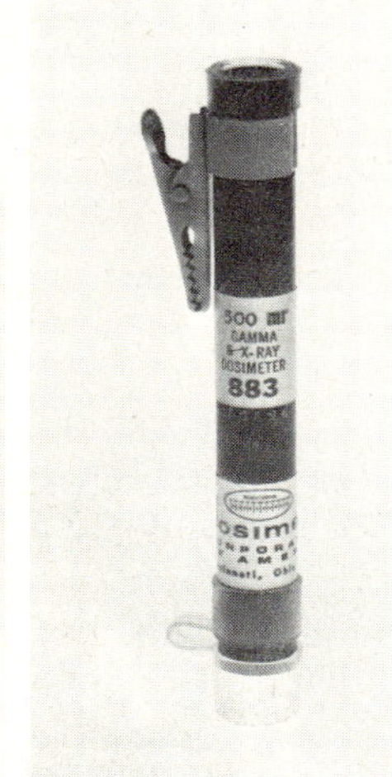

FIGURE 8-14
A thermoluminescence
dosimeter consists of a
penlike device and
a reading unit.
(Dosimeter Corporation
of America)

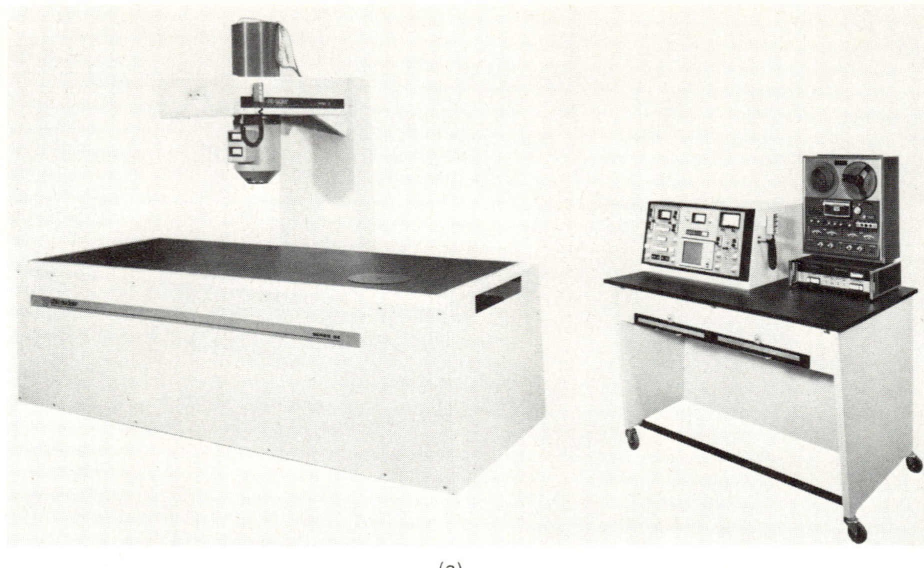

(a)

FIGURE 8-12
Some scintillation detectors: (a) scanner, (b) wide field gamma-ray camera, (c) mobile
gamma-ray camera. (Ohio-Nuclear)

Now that we've looked at some devices used to detect radiation, let's learn about the *units* used to express the concentration or doses of radiation.

Units Used for Expressing Dosages of Radiation

There are several units used for expressing the level of ionizing radiation (Table 8-2). Some of these units simply express the activity of the radiation source, while others relate the effect of a specific type of radiation to its effect on living tissue. Let's see how this works.

The *curie*, abbreviated Ci, is the unit used to describe the activity of the radiating source. One curie is equal to 3.7×10^{10} disintegrations per second, which happens to be the disintegration rate of 1 g of radium.

EXAMPLE 8-2　A gold-198 isotope rated at 0.50 Ci is used in the treatment of widespread abdominal cancers. What is the activity of this gold in disintegrations per second (dps)?

SOLUTION　We know that 1 Ci equals 3.7×10^{10} dps; therefore,

$$? \text{ dps} = (0.50 \text{ Ci})\left(\frac{3 \times 10^{10} \text{ dps}}{1 \text{ Ci}}\right) = 1.5 \times 10^{10} \text{ dps}$$

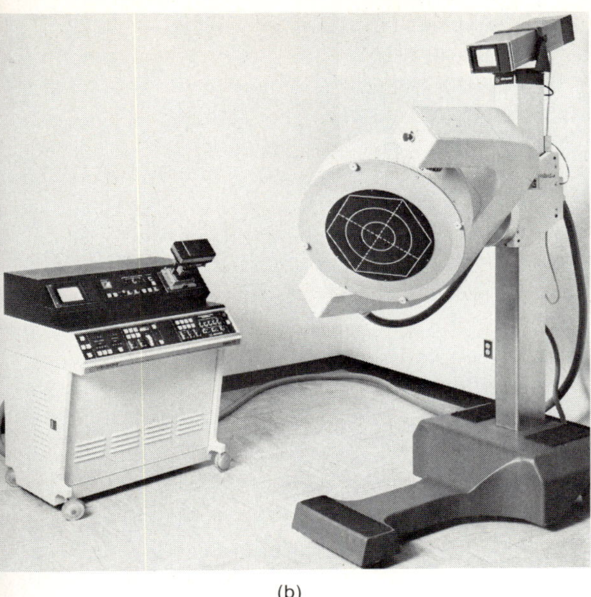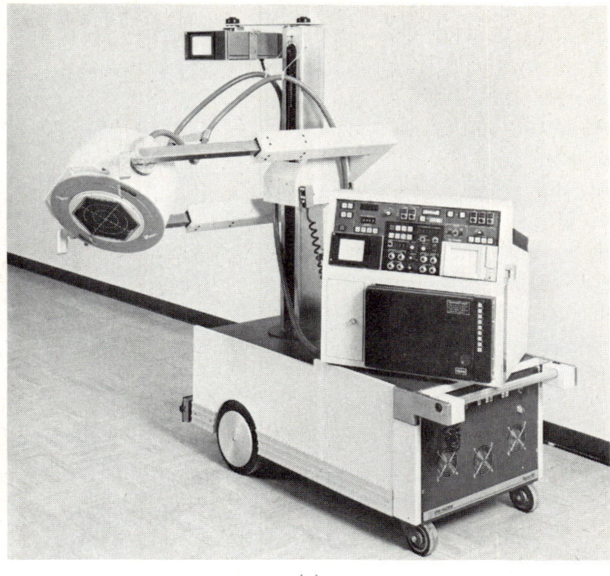

(b) (c)

TABLE 8-2   Unit used for expressing the level of ionizing radiation

UNIT	WHAT IT MEASURES	EXPLANATION
curie (Ci)	The activity of the radiation source.	One curie equals 3.7×10^{10} disintegrations per second.
roentgen (r)	Measures the ionizing ability of x rays and gamma rays.	One roentgen of radiation produces 2×10^9 ion pairs in 1 cm³ of air.
rad	*Rad*iation *a*bsorbed *d*ose (i.e., the amount of energy absorbed by living tissue).	One rad is equal to the 1 g of irradiated tissue absorbing 100 ergs of energy.
rem	*R*oentgen *e*quivalent *m*an (i.e., a *weighted* absorbed dose of radiation).	One rem equals 1 rad for beta and gamma radiation. One rem equals 0.1 rad for alpha radiation.

The *roentgen*, abbreviated r, measures the ionizing ability of x rays and gamma rays. One roentgen is the amount of radioactivity that produces 2×10^9 ion pairs in 1 cubic centimeter of air.

The *rad* (*r*adiation *a*bsorbed *d*ose) measures the amount of energy absorbed by living tissue, regardless of the type of radiation. (By the way, the energy absorbed is measured in units called ergs.* When 1 g of irradiated

* An erg is a unit of work in the metric system.

TABLE 8-3 Lethal dose values for
various organisms

ORGANISM	$LD_{50}^{30}(r)$
Guinea pig	200
Rabbit	300
Dog	325
Human being	400
Monkey	450
Rat	850
Bacteria	5,000–13,000
Viruses	100,000–200,000

tissue absorbs 100 ergs of energy, that is 1 rad.) In general, the exposure to radiation in roentgen units is numerically equal to the absorbed dose in rads.

EXAMPLE 8-3 Most authorities on radiation feel that a person exposed to 600 roentgens of radiation over his entire body will in all likelihood die. What is this value in rads?

SOLUTION In general, 1 roentgen equals 1 rad; therefore, 600 roentgens equals 600 rads.

The toxicity of radiation is sometimes expressed in terms of a *lethal dose*, which is the amount of radiation that will kill 50% of those exposed to it within 30 days. The symbol for this 30-day 50% kill lethal dose is LD_{50}^{30}. The units of radiation are usually expressed in roentgens. Table 8-3 is a summary of lethal dose values for various organisms.

Different types of radiation have different effects on body tissue. In other words, 1 rad of alpha radiation will have a different effect on the body than 1 rad of beta radiation. To account for this difference, scientists have developed a unit called the *rem* (roentgen *equivalent man*). The rem is calculated by multiplying the number of rads by a weighting factor. The weighting factor for alpha radiation is 10, whereas that for beta and gamma radiation is 1.

$$\text{Rem} = \text{rad} \times \text{weighting factor}$$

When a dose of radiation is expressed in rems, it doesn't matter what type of radiation it is since it has already been weighted (Fig. 8-15). The rem is a useful unit for expressing amounts of radiation that individuals receive each year (Table 8-4).

FIGURE 8-15
Sometimes it doesn't matter what the units are when measuring radiation.

TABLE 8-4   Annual per capita radiation doses in the United States

SOURCE OF RADIATION	DOSE PER YEAR (mrem)
Environmental	134
Medical	74
Occupational	1
Miscellaneous (e.g., television, etc.)	3

EXAMPLE 8-4   A recommended dose limit for the general public is 0.5 rem/year.

(a) If you received 0.5 rem of radiation during the year, and it was all in the form of alpha rays, what is your dose in rads?

(b) If you received 0.5 rem of radiation during the year, and it was all in the form of beta rays, what is your dose in rads?

SOLUTION   We know that: rem = rad × weighting factor. Therefore,

$$\text{Rad} = \frac{\text{rem}}{\text{weighting factor}}$$

(a) The weighting factor for alpha radiation is 10. Therefore,

$$? \, \text{rad} = \frac{0.5 \, \text{rem}}{10} = 0.05 \, \text{rad}$$

(b) The weighting factor for beta radiation is 1. Therefore,

$$? \text{ rad} = \frac{0.5 \text{ rem}}{1} = 0.5 \text{ rad}$$

Radioisotopes: Their Uses in Medical Diagnosis and Therapy

Radiation is used to both detect and treat abnormalities in the body. We've already discussed the use of iodine-131 therapy for hyperthyroidism. Iodine-131 can also be used in a diagnostic procedure to monitor the function of the thyroid. The rate at which the thyroid takes up the iodine-131 can be monitored with a scanning device, to see if it is functioning properly.

What makes radioisotopes so useful in diagnostic procedures is that the body treats the tagged isotope in the same way that it treats the nonradioactive element. Therefore, the tagged isotope goes right to the area of the body where you want it to go. For example, iodine, whether it's radioactive or not, goes right to the thyroid, where it is incorporated into the amino acid thyroxine (the only molecule in the body which contains iodine). Therefore, iodine-131 is perfect for monitoring the thyroid gland.

Chromium, in the form of sodium chromate, attaches strongly to the hemoglobin of red blood cells. This makes radioactive chromium-151 an excellent isotope for determining the flow of blood through the heart. This isotope is also useful for determining the lifetime of red blood cells, which can be of great importance in the diagnosis of anemias.

Radioactive cobalt (cobalt-59 or cobalt-60) is used to study defects in vitamin B_{12} absorption. Cobalt is the metallic atom at the center of the B_{12} molecule. By injecting a patient with vitamin B_{12}, labeled with radioactive cobalt, the physician can study the path of the vitamin through the body and discover any irregularities.

We've already discussed how radiation therapy can be used to destroy cancer cells. Radiation can be delivered to a malignant area in three ways.

One method is *teletherapy*, in which a high-energy beam of radiation is aimed at the cancerous tissues (Fig. 8-16). A second method is *brachytherapy*, in which a radioactive isotope is placed into the area to be treated. This is usually done by means of a seed, which could be a glass bead containing the isotope. In this way the isotope delivers a constant beam of radiation to the affected area.

The third method is called *radiopharmaceutical therapy*. This method involves oral or intravenous administration of the isotope. The isotope then uses the normal body pathways to seek its target. This is the method that is used to get iodine-131 to the thyroid gland. A summary of various isotopes used in the diagnosis and treatment of various medical disorders can be found in Table 8-5.

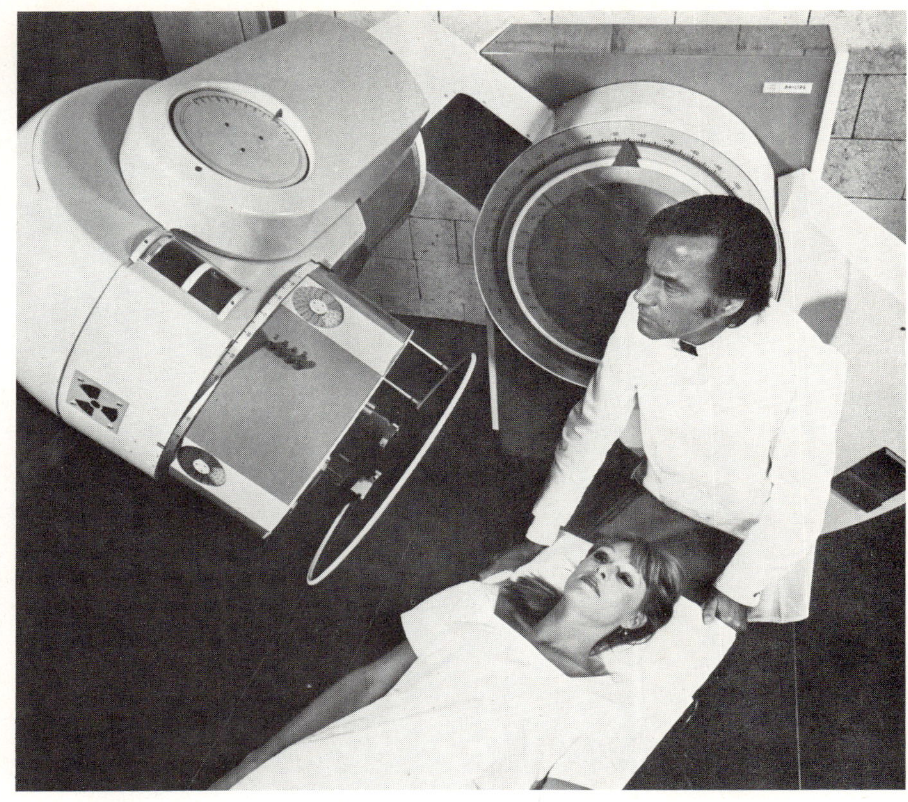

FIGURE 8-16
A rotational teletherapy unit for cobalt-60 therapy. (Philips Medical Systems)

TABLE 8-5   Radioactive isotopes used in medical therapy and diagnosis

NAME AND SYMBOL	PRINCIPAL NUCLEAR PROPERTIES[a]	FORM	USE	DOSAGE	MODE OF ADMINIS- TRATION
Americium ^{241}Am	Half-life 458Y $\alpha(5.49, 5.44)$ $\gamma(0.060)$; e^-	Encapsulated source	Diagnostic: In bone mineral analyzer		External
Cesium ^{137}Cs Daughter ^{137m}Ba	Half-life 30.0Y $\beta^-(1.176, 0.514)$ $\gamma(0.662)$; e^-	Cesium chloride or cesium sulfate (encased in needles and/or applicator cells	Therapeutic: As a tele- therapy source or intracavity or intersti- tial radiation source in treatment of malignancies	Depends on tumor sensitivity and type of applicator used	External, intracavity, or interstitial irradiation
Chromium ^{51}Cr	Half-life 27.8D K; $\gamma(0.32)$; e^-	Sodium chromate (labelled red blood cells)	Diagnostic: Study of blood volume and red cell survival; spleen imaging; placental localization	20 to 200 μCi; max in body at any time 390 μCi	i.v.

TABLE 8-5 Radioactive isotopes used in medical therapy and diagnosis (*cont.*)

NAME AND SYMBOL	PRINCIPAL NUCLEAR PROPERTIES[a]	FORM	USE	DOSAGE	MODE OF ADMINISTRATION
		Labeled human serum albumin	Diagnostic: Placental localization; loss of gastrointestinal proteins		
Cobalt ^{60}Co	Half-life 5.26Y β^-(0.314, 1.48) γ(1.173, 1.332)	Metallic cobalt	Therapeutic: As a source of teleo-roentgen therapy and as intracavitary or interstitial radiation source for treatment of malignancies known to be responsive to gamma radiation	Depends on tumor sensitivity and type of applicator used. One millicurie of ^{60}Co unfiltered gives 13.5 r/hr at 1 cm	External intracavitary or interstitial irradiation
		Radioactive vitamin B$_{12}$	Diagnostic: For absence of intrinsic factor (P.A.) or defect in absorption (sprue)	0.33–0.60 μCi	Orally
			Metabolic studies	Up to 3 μCi	i.v.
^{57}Co	Half-life 270D K; γ(0.122)		Same as ^{60}CoB$_{12}$	0.5 μCi	Orally or i.v.
Copper ^{64}Cu	Half-life 12.8H β^-(0.573) β^+(0.656)	Copper versenate Copper acetate	Brain scans for tumors Study Wilson's disease	1–2 mCi 1 mCi	i.v. i.v. or orally
Fluorine ^{16}F	Half-life 109.7M β^+(0.635)	Sodium fluoride (reactor produced)	Diagnostic: Bone scan	1–2 mCi	i.v. or orally
Gallium ^{67}Ga	Half-life 77.9H K; γ(0.93, 0.184, 0.296, 0.388)	Gallium citrate	Diagnostic: Tumor-seeking agent	1.5–3.0 mCi	i.v.
Gold ^{198}Au	Half-life 2.697D β^-(0.962) γ(0.412)	Colloidal gold or seeds	Therapeutic: Treatment of widespread abdominal carcinomatosis with ascites; carcinomatosis of pleura with effusion; lymphomas; interstitially in metastatic tumors	100–500 mCi when given intraperitoneally or intrapleurally. May be repeated if necessary 6–8 weeks later. No definite dosage established as yet for i.v. or interstitial use in treatment of lymphomas or localized metastatic tumors. 76 REP are obtained for complete decay of 1 microcurie ^{198}Au/g of tissue	Intraperitoneal, intrapleural, interstitial, or i.v.
		Colloidal gold	Diagnostic: Liver imaging	0.1–0.15 mCi (2 μCi/kg)	i.v.

NAME AND SYMBOL	PRINCIPAL NUCLEAR PROPERTIES[a]	FORM	USE	DOSAGE	MODE OF ADMINISTRATION
Indium ^{113m}In	Half-life 99.8M γ(0.393)	Indium-DTPA	Diagnostic: Brain imaging	5–10 mCi	i.v.
		Indium-transferrin	Diagnostic: Static cardiovascular blood pool imaging	1–3 mCi	i.v.
			Hepatic blood pool imaging; placental localization	1 mCi	
		Indium-Fe(OH)$_3$	Diagnostic: Perfusion lung scan	1–1.5 mCi (15–20 μCi/kg)	i.v.
		Indium-colloid	Diagnostic: Static liver imaging; spleen imaging	1–3 mCi	i.v.
^{111}In	Half-life 2.81D K; γ(0.173, 0.247)	Indium-DTPA	Diagnostic: Cisternography	0.5–2 mCi	Intrathecal instillation
		Indium chloride	Diagnostic: Hematopoietic bone marrow imaging; tumor-seeking agent	1–2 mCi	i.v.
Iodine ^{131}I	Half-life 8.05D β^-(0.606, 0.81, 0.335) γ(0.080, 0.284, 0.364, 0.637, 0.723)	Sodium iodide	Therapeutic: Hyperthyroidism	Dosage depends on size of gland and on percent of radioactive iodine retained by the thyroid; usually 3–10 mCi. May be necessary to repeat dosage	i.v. or orally
			Therapeutic: Cancer of thyroid	100–150 mCi. Repeat treatment at 8–12 week intervals until cancerous tissue no longer retains any ^{131}I	i.v. or orally
			Therapeutic: Severe heart disease	Dose depends on degree of thyroid suppression desired. 30–50 mCi usually necessary to produce myxedema	i.v. or orally
			Diagnostic: Thyroid scan	50 μCi	Orally
			Diagnostic: Study action of thyroid and antithyroid drugs; study of chloride space; aid in determining thyroid activity	5 μCi	i.v. or orally

NAME AND SYMBOL	PRINCIPAL NUCLEAR PROPERTIES[a]	FORM	USE	DOSAGE	MODE OF ADMINIS-TRATION
		Diiodofluorescein	Diagnostic: Diagnosis and localization of brain tumors	Recommended tolerance dose is 4.2 μCi/kg. May use up to 0.8 mCi if necessary	i.v.
		Iodinated serum albumin	Diagnostic: Determination of plasma volume, peripheral vascular flow, cardiac output, circulation time, and cerebral vascular flow	Approx. 5 μCi. Saturate thyroid gland with stable iodine prior to studies	i.v.
			Diagnostic: Diagnosis and localization of brain tumors	5 μCi/kg. Saturate thyroid gland with stable iodine prior to studies	i.v.
			Diagnostic: Placental localization	10 μCi. Protect thyroid from accumulation of radioactivity before using	i.v.
			Diagnostic: Cisternography	100 μCi	Intrathecal instillation
		Macroaggregated iodinated serum albumin	Diagnostic: Perfusion lung scan	200–350 μCi (4–5 μCi/kg). Protect thyroid from accumulation of radioactivity before using	i.v.
		Colloidal micro-aggregated iodinated serum albumin	Diagnostic: Hepatic blood pool imaging	3 μCi/kg	i.v.
		Iodinated fibrinogen	Diagnostic: Determination of fibrinolytic enzymes *in vitro*	*In vitro* diagnostic test	
		Iodinated rose bengal	Diagnostic: Liver function *in vivo*—hepatic excretion studies	3 μCi/kg (150–300 μCi)	i.v.
		Iodopyracet, Na iodohippurate, Na diatrizoate, diatrizoate	Diagnostic: Bilateral renal function test *in vivo*	30 μCi	i.v.
		methylglucamine, Na diprotrizoate, Na acetrizoate or Na iothalamate	Diagnostic: Kidney imaging	200–300 μCi. Protect thyroid from accumulation of radioactivity before using	i.v.

NAME AND SYMBOL	PRINCIPAL NUCLEAR PROPERTIES[a]	FORM	USE	DOSAGE	MODE OF ADMINISTRATION
		Iodinated fats or fatty acids (e.g., ^{131}I triolein	Diagnostic: Pancreatic function, intestinal fat absorption	25 μCi	Orally
		Copolymer of *p*-toluidine vinyl-pyrrolidone ^{131}I (Tolpovidone Abbott)	Diagnosis of exudative enteropathy	10–25 μCi	i.v.
^{125}I	Half-life 60D K; γ(0.035)	Sodium iodide	Diagnostic: Thyroid imaging	50–100 μCi	i.v.
		Iodinated serum albumin	Diagnostic: Determination of plasma volume	5 μCi	i.v.
		Iodinated rose bengal	Diagnostic: Liver function—hepatic excretion studies		i.v.
		Iodopyracet etc. See list for ^{131}I	Same as ^{131}I	50 μCi	i.v.
		Iodinated fats or fatty acids	Diagnostic: Intestinal fat absorption		i.v.
		Sealed source	For use in bone mineral analyzer		External
Iridium ^{192}Ir	Half-life 74.2D β^-(0.67) γ(0.296, 0.308, 0.317, 0.468, 0.589, 0.604, 0.612)	Seed encased in nylon ribbon	Therapeutic: Interstitial treatment of tumors		Interstitial irradiation
Iron ^{59}Fe	Half-life 45.6D β^-(1.573, 0.475, 0.273) γ(0.143, 0.192, 1.095, 1.292)	Ferrous citrate $FeSO_4$	Diagnostic: Determination of blood volume with red blood cells labeled *in vivo*; study of iron metabolism; blood transfusion studies	5–10 μCi	i.v. or orally
^{55}Fe also present	Half-life 2.60Y K				
Krypton ^{85}Kr	Half-life 10.76Y β^-(0.67) γ(0.514)	Gas	Diagnostic: Cardiac abnormalities; skeletal muscle, coronary or cerebral blood flow		Intramuscular or intraarterial injection
Lead RaD (^{210}Pb)	Half-life 20.4Y β^-(0.061, 0.015) γ(0.047)	Beta ray applicator	Same as strontium (^{90}Sr)	Same as strontium (^{90}Sr)	External irradiation

NAME AND SYMBOL	PRINCIPAL NUCLEAR PROPERTIES[a]	FORM	USE	DOSAGE	MODE OF ADMINISTRATION
Daughter RaE (^{210}Bi)	β^- (1.160)				
Mercury ^{197}Hg	Half-life 65H K; γ(0.077, 0.191, 0.268)	Chlormerodrin ^{197}Hg (Neohydrin)	Brain scans for tumors Renal studies for defects, clearance, etc.	700–10000 μCi 100–150 μCi	i.v.
^{203}Hg	Half-life 46.9D β^- (0.214) γ(0.279)	Chlormerodrin ^{203}Hg	Brain scans for tumors Renal scanning	700–900 μCi 100 μCi	i.v.
Phosphorus ^{32}P	Half-life 14-28D β^- (1.71)	Disodium hydrogen phosphate	Therapeutic: Chronic myeloid leukemia	1–2 mCi every 5–7 days until white blood cell count drops to 15,000 cells/mm^3. If given orally, dosage can be increased 50%	i.v. or orally
			Therapeutic: Polycythemia vera	100–175 μCi/kg given over a 10-day period. Maximum effect delayed up to 5 months	i.v. or orally
			Therapeutic: Lymphomas and widespread carcinomatosis	2 mCi every 5–7 days to a total of 14–18 mCi. May have to stop treatment early if marked depression of hematopoietic system occurs	i.v. or orally
			Diagnostic: Determination of blood volume with red blood cells labeled *in vitro*; study of peripheral vascular disease; localization of brain tumors; study of carcinomas of breast	Tolerance dose is 1.2 μCi/kg. In suspected malignancies, up to 500 μCi can be used	i.v. or orally
		Chromium phosphate	Pleural or peritoneal effusions from metastasis	5–7 mCi	Intrapleural or intraperitoneal
Potassium ^{42}K	Half-life 12.36H β^- (3.52) γ(1.524)	Potassium carbonate	Diagnostic: Localization of brain tumors; determination of intracellular fluid space	50–100 μCi	i.v. or orally
Radium ^{226}Ra	Half-life 1602Y α(4.78, 4.60)	Radium bromide; alpha and beta	Same as radon	5000 and 10,000 r of gamma radiation; de-	Interstitial irradiation

NAME AND SYMBOL	PRINCIPAL NUCLEAR PROPERTIES[a]	FORM	USE	DOSAGE	MODE OF ADMINISTRATION
	$\gamma(0.186)$	particles filtered by platinum		pends on uniform distribution of radium and tumor sensitivity. Point source of 1 mg radium filtered by 0.5 mm of platinum gives 8.4 r/hr of gamma radiation of 1 cm	
Radon (Radium Emanation) ^{222}Rn (Daughter of ^{226}Ra)	Half-life 3.8229D $\alpha(5.49)$ $\gamma(0.510)$	Gaseous radon; alpha and beta particles filtered by 0.3 mm of gold	Therapeutic: Treatment of malignancies by interstitial radiation. Used most often in cancer of uterine cervix and fundus, oral pharynx, urinary bladder, skin, and in metastatic cancer of lymp nodes	Usually 1–2 mCi, gold seeds with uniform distribution. Complete decay of 1 mCi radon is equivalent to 133 mg hr of radium. 0.3 mm of gold is required to remove the beta particles completely	Interstital irradiation
Ruthenium ^{106}Ru Daughter ^{106}Rh	Half-life 368D β-(0.039) β^-(3.53, 3.1, 2.4, 2.0) $\gamma(0.512, 0.622, 1.05)$	Beta ray applicator	Same as strontium (^{90}Sr)	Same as strontium (^{90}Sr)	External irradiation
Selenium ^{75}Se	Half-life 120D K; $\gamma(0.265, 0.136,$ also 0.066, 0.097, 0.121, 0.280, 0.401)	Seleno-methionine	Diagnostic: Pancreas imaging	250 μCi (3–4 μCi/kg)	i.v.
Sodium ^{24}Na	Half-life 14.96H β^-(1.389, 4.17) $\gamma(1.369, 2.754)$	Sodium chloride	Diagnostic: Study of peripheral vascular disease, extracellular space, circulation time, formation of cerebrospinal fluid, sodium metabolism	Tolerance dose is 13.5 μCi/kg; may be repeated 2–3 times at weekly intervals	i.v.
Strontium ^{85}Sr ^{87m}Sr	Half-life 64.0D K; $\gamma(0.514)$ Half-life 2.83H $\gamma(0.388)$	Strontium nitrate or chloride	Diagnostic: Bone imaging in patients with known or suspected malignancies	Maximum allowable dose (^{85}Sr): 100 μCi Dosage (^{87m}Sr): 1–3 mCi	i.v.
Strontium ^{90}Sr Daughter ^{90}Y	Half-life 27.7Y β^-(0.546) β^-(2.27)	Beta ray applicator	Therapeutic: Treatment of benign conditions of eye such as pterygia, traumatic	1500–5000 r given in divided doses	External irradiation

TABLE 8-5 Radioactive isotopes used in medical therapy and diagnosis (*cont.*)

NAME AND SYMBOL	PRINCIPAL NUCLEAR PROPERTIES[a]	FORM	USE	DOSAGE	MODE OF ADMINIS-TRATION
			corneal ulceration, corneal scars, vernal conjunctivitis, heman-gioma of eyelid, vascu-larization of cornea, and in preparation for a corneal transplant		
Technetium ^{99m}Tc	Half-life 6.049H γ(0.140)	Pertechnetate $NaTcO_4$	Diagnostic: Brain scan Blood pool Placental localization Thyroid scan	5–15 mCi 3–5 mCi 0.3–1 mCi 1–3 mCi	i.v. i.v. or orally
		Colloidal sulfate	Diagnostic: Liver, spleen, and bone marrow scans	1–3 mCi	i.v.
		Tc albuminate	Diagnostic: Heart scan Placental localization	3–5 mCi 1 mCi	i.v.
		Tc albuminate macro- or micro-aggregates	Diagnostic: Perfusion lung scan	1–1.5 mCi (15–20 μCi/kg)	i.v.
		Tc DTPA (iron ascorbate)	Diagnostic: Kidney scan	10 mCi	i.v.
		Tc DTPA (tin)	Diagnostic: Kidney and brains scans	10 mCi	i.v.
		Tc stannous polyphosphate Tc stannous etidronate	Diagnostic: Bone scan	10 mCi	i.v.
Xenon ^{133}Xe	Half-life 5.3D β^-(0.346) γ(0.081)	Gas or gas in saline solution	Diagnostic: Pulmonary function—ventilation studies	5–10 mCi 10–30 mCi	Inhalation i.v.
			Diagnostic: Cerebral blood flow; coronary abnormalities; skeletal muscle blood flow	0.5–1 mCi	Intraarterial or intramus-cular in-jection
Ytterbium ^{169}Yb	Half-life 31.8D K; γ(0.063, 0.110, 0.131, 0.177, 0.198, 0.308)	Yb-DTPA	Diagnostic: Cisternog-raphy Brain scan	1–2 mCi	Intrathecal instillation i.v.

[a] H = hours, D = days, M = months; Y = years.

References

1. Charles Behrns, E. Richard King, and James W. J. Carpenter, *Atomic Medicine* (The Williams & Wilkins Co., Baltimore, 5th ed., 1969).
2. *CRC Handbook of Radioactive Nuclides,* Yen Wang, Editor (The Chemical Rubber Co., Cleveland, Ohio, 1969).
3. *Physicians' Desk Reference for Radiology and Nuclear Medicine,* Leonard M. Freeman and M. Donald Blaufox, Editorial Consultants (Medical Economics Company, Oradell, 4th ed., 1974/75, 5th ed., 1975/76).
4. Sheldon Baum and Roland Bramlet, *Basic Nuclear Medicine* (Appleton-Century-Crofts, New York, 1975).

Source: The Merck Index, Merck & Co., Inc., Rahway, N.J., 9th. ed., 1976. By permission of the publisher.

SUMMARY

In this chapter we surveyed the topic of radiation. We began with the basics and read about the discovery of radiation. We discovered that there are three major types of ionizing radiation: alpha rays, beta rays, and gamma rays. We learned a little about nuclear transformation, and how one radioactive isotope can change into another isotope by emitting nuclear particles. We also read about something called half-life, which is the amount of time it takes for one-half of the radioactive material to decay. We also learned about the processes of nuclear fission and fusion and how these processes can be used for humankind's benefit or destruction.

In the second half of the chapter we looked at radiation and its medical uses. We discovered how radiation damages cells, and what effects this can have on the body. We examined the various methods used to detect radiation as well as the units used to express the dosages of radiation. Finally, we looked at various radioisotopes and how they are used in diagnosis and therapy of various disorders.

EXERCISES

1. Name the three types of nuclear radiation and give the Greek symbols for each of them.

2. Show the charge, mass, and symbol of
    (a) an alpha particle
    (b) a beta particle

3. When an isotope emits an alpha particle, what happens to its
    (a) mass number?
    (b) atomic number?

4. When an isotope emits a beta particle, what happens to its
    (a) mass number?
    (b) atomic number?

5. Complete the following nuclear equations.
    (a) $^{14}_{6}C \longrightarrow {}^{0}_{-1}e + ?$
    (b) $? \longrightarrow {}^{0}_{-1}e + {}^{24}_{12}Mg$
    (c) $? \longrightarrow {}^{4}_{2}He + {}^{234}_{90}Th$
    (d) $^{87}_{36}Kr \longrightarrow {}^{1}_{0}n + ?$
    (e) $^{212}_{84}Po \longrightarrow {}^{4}_{2}He + ?$

6. The isotope iodine-131 has a half-life of 8 days. If an individual is injected with $1\overline{00}$ mg of iodine-131 today, how much will remain in his body after 24 days (assuming that none is lost through normal body channels)?

7. The isotope strontium-90 has a half-life of 28 years. Suppose you have a sample of strontium-90 that originally weighed 64 g. How old is this sample if today it only weighs 8 g?

8. Explain the following terms.
   (a) chain reaction   (b) nuclear fission   (c) nuclear fusion

9. Explain the two ways that radiation can damage or destroy cells.

10. Write the Lewis dot structure for
    (a) a hydroxide free radical
    (b) a hydrogen free radical
    (c) hydrogen peroxide

11. Determine whether each of the following statements refers to a stage 1 or stage 2 effect.
    (a) Yttrium-90 encased in glass beads is placed into the pituitary gland, where it destroys some of its cells.
    (b) Iodine-131 therapy decreases the amount of thyroxine produced by the thyroid gland.
    (c) A patient with inoperable brain tumors receives cobalt-60 therapy, which shrinks the size of the tumors.

12. Name two types of cells that are very sensitive to radiation, and explain what makes them this way.

13. Name four types of radiation detectors, and explain what each one is used for.

14. Define the following terms.
    (a) curie   (b) roentgen   (c) rad   (d) rem   (e) LD_{50}^{30}

15. A $\overline{3000}$ curie cobalt-60 source is sometimes used to shrink inoperable tumors. What is the activity of this cobalt-60 source in disintegrations per second?

16. The LD_{50}^{30} for humans is about 400 roentgens; what would this be in rads?

17. The U.S. Environmental Protection Agency estimates that each of us receives about 4 millirems (0.004 rem) of radiation from fallout each year. If all of this were in the form of alpha rays, what would the dose be in rads?

18. Name the three ways that radiation can be delivered to malignant areas. Give an example showing how each method is used.

19. When iodine is administered to an individual, it goes straight to the thyroid. Why?

20. What type of shielding is needed to stop
    (a) alpha rays?   (b) beta rays?   (c) gamma rays?

CHAPTER 9

Introduction to Organic Chemistry or the Prolific Carbon Atom

You should be able to:

1. State whether a compound is organic or inorganic when given its chemical formula.
2. Write the electron dot diagram for carbon.
3. Write the electron dot diagram for a methane molecule (CH_4).
4. Define and give an example of a tetrahedron.
5. Show how two carbon atoms can bond to form a carbon–carbon single bond, double bond, or triple bond.
6. Define and give an example of a molecular formula and structural formula for an organic compound.
7. Define what is meant by an ''R group.''
8. State the properties of alkanes.
9. Name the first 10 straight-chain members of the alkane series and give the molecular formula for each.
10. State the properties of alkenes, and give some examples.
11. State the properties of alkynes, and give some examples.
12. Define and give an example of a saturated hydrocarbon, an unsaturated hydrocarbon, a cyclic hydrocarbon, and an aromatic hydrocarbon.
13. State what is meant by a homologous series, and give an example of such a series.
14. Define and illustrate the term isomer.
15. Name alkanes, alkenes, and alkynes using the IUPAC system.
16. Give an example of hydrocarbon combustion.
17. Write the products of alkene and alkyne hydrogenation given the initial reactants.

Introduction

During the 1700's and early 1800's, most chemists believed that there were two distinct classes of chemical compounds: organic and inorganic. They said that organic compounds were derived from living or once-living organisms and that inorganic compounds were part of the nonliving world. Organic compounds were substances such as sugar, fats, oils, and wood. Inorganic compounds were substances such as salt and iron. And although many chemists knew that organic substances could be converted into inorganic substances by heating, the reverse was not true—inorganic substances could not be converted into organic ones. It was these observations that gave rise to the theory of *vitalism.* This theory stated that life and substances associated with life were not bound by the laws of science and that these life-connected substances had a *vital force* that human beings could not control. This, of course, meant that chemists would never be able to synthesize organic compounds from inorganic ones.

But in 1828, the German chemist Friedrich Wöhler astounded the chemical world by having synthesized the organic compound urea from what was then considered the inorganic compound ammonium cyanate. Urea is an organic compound found in urine, and until Wöhler synthesized it in his laboratory, urea was thought by most scientists to be associated with living things. The belief in the theory of vitalism was on its way out.

Today, a little more than 150 years after Wöhler's discovery, the synthesis of organic compounds is commonplace. In fact, many of the complex molecules associated with life itself, for example insulin, have been artificially produced in the laboratory.

Organic Compounds Defined

Organic compounds are primarily those which contain carbon atoms. The old definition that organic substances are those from living or once-living sources does not apply any more. For example, plastics and pharmaceuticals are organic compounds, because their molecules contain numerous carbon atoms. However, not all carbon-containing compounds are organic. For example, carbonate compounds such as $CaCO_3$ and Na_2CO_3, as well as the oxides of carbon (CO and CO_2) are considered to be inorganic compounds. Although a precise definition of organic compounds is difficult to give, we can say that for the most part, *organic compounds are those which contain carbon along with elements such as hydrogen, oxygen, nitrogen, sulfur, and the group VIIA elements* (Fig. 9-1). After we've looked at some organic compounds, you'll see that we really don't need a precise definition anymore.

The Amazing Carbon Atom

Perhaps the most unique thing about the carbon atom is its ability to combine with itself. The other elements in the periodic table don't have this ability. But

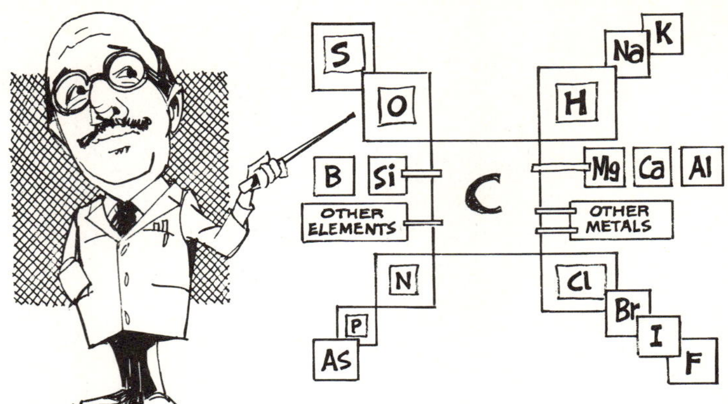

FIGURE 9-1
For the most part, organic compounds are those that contain carbon along with elements such as hydrogen, oxygen, nitrogen, sulfur, and the group VIIA elements.

it is this unusual ability that allows carbon atoms to form all kinds of chainlike and ring-shaped molecules, making the number of potential organic compounds almost infinite (Fig. 9-2). Just think, there are over 3 million organic compounds known today, and the possibility exists for countless more to be discovered or synthesized.

Carbon is a group IVA element, which means that a carbon atom has four electrons in its outermost energy level. The carbon atom can use its four electrons to form covalent bonds with other carbon atoms or atoms of other elements. This is exactly what happens in organic compounds.

FIGURE 9-2
Carbon atoms can form chainlike and ring-shaped molecules.

Pentane

Nonane

Cyclohexane

Cyclooctane

When a carbon atom shares each of its outer electrons with four other atoms, the result is a compound such as CH_4, the compound *methane*, which we know as natural gas (cooking gas). In this compound, the central carbon atom is bonded to each of the four hydrogen atoms. We can write the electron dot structure in the usual way.

$$\begin{array}{ccc} & H & \\ H\!:\!\overset{..}{\underset{..}{C}}\!:\!H & \quad or \quad & H{-}\!\!\overset{\textstyle H}{\underset{\textstyle H}{\,C\,}}\!\!{-}H \end{array}$$

Methane

As you can see, this diagram gives the impression that the carbon-hydrogen bonds are at 90° angles to each other.

However, this is not the way the molecule actually exists. Molecules are *three*-dimensional, and the four outer electrons in carbon move as far apart from each other as possible, because electrons repel each other. (Remember that particles with similar charges always repel one another.) The electrons can put the greatest distance between themselves if the carbon-hydrogen bonds are in the shape of a *tetrahedron*. The tetrahedral shape allows the four electrons to get as far apart as possible (Fig. 9-3). A tetrahedron ("tetra" is from the Greek

FIGURE 9-3
The tetrahedral structure of a CH_4 molecule.

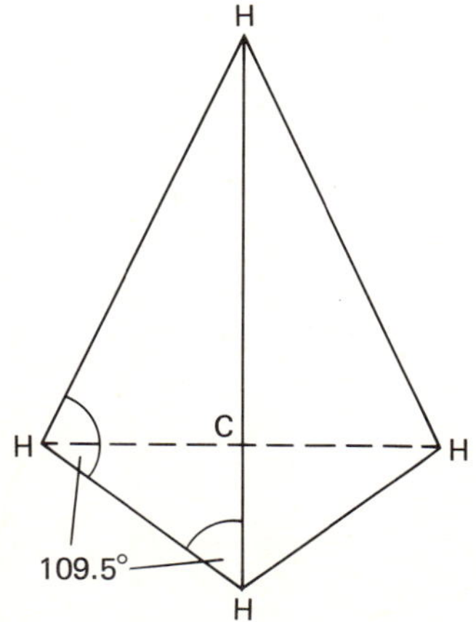

word "four") is a pyramid-shaped figure with three upper sides plus a bottom side. You can make a tetrahedron this way: Take six pencils of equal height. Place three of them on a table in the form of a triangle. Hold the other three pencils upright, placing one at each corner of the triangle. Now tilt the three upright pencils inward until they touch. You have formed a tetrahedron (Fig. 9-4). The tetrahedral shape means that the carbon-hydrogen bonds in methane are 109.5 degrees apart. Methane is only one example; this tetrahedral structure is common to many organic compounds.

The Bonding Between Carbon Atoms

When two carbon atoms bond with each other, they can do it in three ways:

1. They can form a single bond: ·C—C·

2. They can form a double bond: ·C═C·

3. They can form a triple bond: ·C≡C·

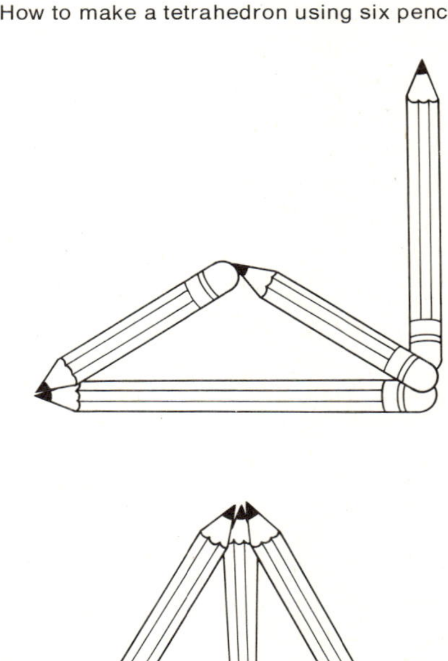

FIGURE 9-4
How to make a tetrahedron using six pencils.

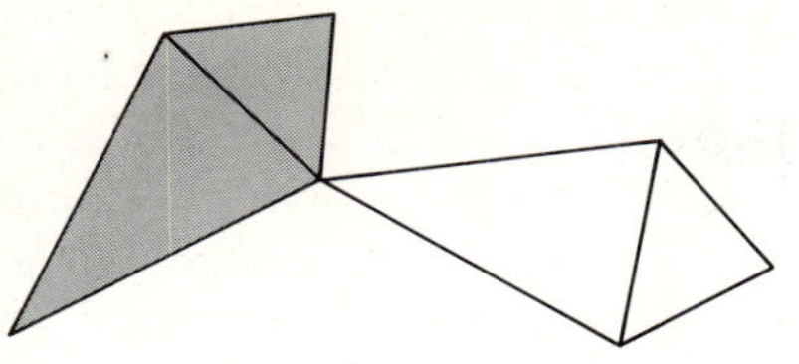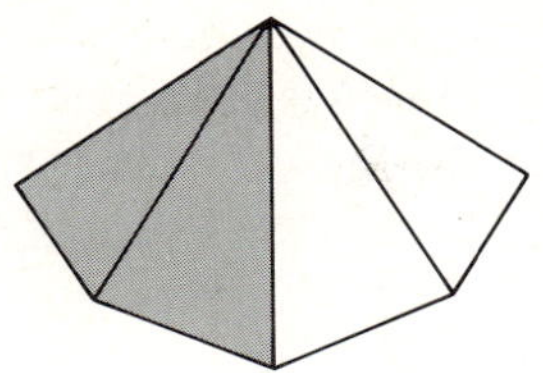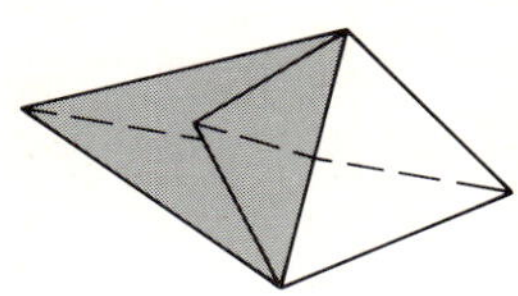

FIGURE 9-5

Two carbon atoms joined together: left, by a single bond; center, by a double bond; and right, by a triple bond.

In the case of a single bond, we can visualize the two carbon atoms joining in such a way that the compound looks like two tetrahedrons with their corners touching. The lines in the figure represent the bonds, and the corners represent the atoms. For the double bond, we can visualize two tetrahedrons with a common *edge*. For a triple bond, we can visualize two tetrahedrons with a common *face* (Fig. 9-5).

Structural Formulas of Organic Compounds

In the earlier chapters of this text we wrote the *molecular formulas* for various compounds. For example, barium chloride is $BaCl_2$. The molecular formula tells us the actual number of atoms of each element in a molecule of the compound.

For the organic chemist, however, the molecular formula is not enough. As we have seen, carbon atoms can bond together in different ways. So there can be many different compounds with the same molecular formula. For example, the molecular formula C_4H_{10} represents two distinct compounds; and the formula C_8H_{18} represents 18 distinct compounds! This is why, when organic chemists are writing the formula of an organic compound, they use a *structural formula*. The structural formula indicates *in what way* the carbon atoms are bonded to each other. We can write the two compounds represented by the formula C_4H_{10} in this way:

$$
\begin{array}{ccccc}
 & H & H & H & H \\
 & | & | & | & | \\
H- & C- & C- & C- & C-H \\
 & | & | & | & | \\
 & H & H & H & H
\end{array}
\quad \text{and} \quad
\begin{array}{ccc}
H & H & H \\
\backslash & | & / \\
H \;\; C & H \\
| & | & | \\
H-C- & C- & C-H \\
| & | & | \\
H & H & H
\end{array}
$$

Structural formulas are sometimes written in a more convenient form. For example, the two formulas we just wrote can also be written like this:

$$CH_3-CH_2-CH_2-CH_3 \quad \text{and} \quad CH_3-\underset{\underset{CH_3}{|}}{CH}-CH_3$$

Take a minute to see what we've done. Compare the two ways of writing the formulas and you'll soon get the idea. This second way—the more convenient one—still shows how each carbon atom is bonded to its neighbors.

EXAMPLE 9-1   Write the following structural formulas in the more convenient form, sometimes called the condensed form.

(a)

$$H\!-\!\overset{\displaystyle H}{\underset{\displaystyle H}{C}}\!-\!\overset{\displaystyle H}{\underset{\displaystyle H}{C}}\!-\!\overset{\displaystyle H}{\underset{\displaystyle H}{C}}\!-\!\overset{\displaystyle H}{\underset{\displaystyle H}{C}}\!-\!\overset{\displaystyle H}{\underset{\displaystyle H}{C}}\!-\!\overset{\displaystyle H}{\underset{\displaystyle H}{C}}\!-\!\overset{\displaystyle H}{\underset{\displaystyle H}{C}}\!-\!\overset{\displaystyle H}{\underset{\displaystyle H}{C}}\!-\!H$$

(b)

$$H\!-\!\overset{\displaystyle \overset{\textstyle H\;H\;H}{\diagdown|\diagup}}{\underset{\displaystyle \underset{\textstyle H\;H\;H}{\diagup|\diagdown}}{\overset{\displaystyle H\!-\!\overset{H}{\underset{H}{C}}\!-\!H}{\underset{\displaystyle H\!-\!\overset{H}{\underset{H}{C}}\!-\!H}{C}}}\!-\!H$$

(c)

SOLUTION   Check each structure carefully. Be sure that you have the same number of carbon atoms and hydrogen atoms as in the original structure.

(a) $CH_3\!-\!CH_2\!-\!CH_2\!-\!CH_2\!-\!CH_2\!-\!CH_2\!-\!CH_2\!-\!CH_3$

(b)
$$\begin{array}{c}CH_3\\ |\\ CH_3\!-\!C\!-\!CH_3\\ |\\ CH_3\end{array}$$

(c)
$$\begin{array}{c}H\\ |\\ CH_3\!-\!CH_2\!-\!C\!-\!CH_3\\ |\\ CH_3\end{array}$$

The Classification of Organic Compounds

We have said that there are more than 3 million organic compounds known to modern chemists. Is each of these compounds completely unique—or do they have similarities? In other words, can we classify this great number of compounds into groups?

Fortunately, yes, we can. We can classify most of them into a few simple groups. The categories are set up so that compounds that react alike are placed together. Let's take a general look at organic compounds, and see what we can learn about these different groups.

The R group

Suppose that an organic chemist begins by studying a compound with the following formula.

$$CH_3—C\equiv C—CH_3$$

The chemist then extends his study to cover other compounds with the same type of structure:

$$CH_3—CH_2—C\equiv C—CH_2—CH_3$$

$$CH_3—CH_2—CH_2—C\equiv C—CH_2—CH_2—CH_3$$

$$CH_3—CH_2—CH_2—CH_2—CH_2—C\equiv C—CH_2—CH_2—CH_2—CH_3$$

(By the way, this technique is used by many pharmaceutical companies to find new drugs that have certain similarities to existing ones.) The main point, however, is that the chemist's study shows that, in all these compounds, the *reactive* part of the molecule is the triple bond. Also, all these compounds behave alike. In other words, the chemist finds that all these compounds *belong in the same group.* We can write the general formula for this group as

$$R—C\equiv C—R'$$

The R and R' (pronounced "R prime") in this general formula are called *alkyl groups*, and they represent one, two, three, or any number of carbon atoms attached to the reactive part of the molecule. In Chapter 10 we'll use R-group notation for classes of organic compounds.

The Alkanes

Let's begin our study of particular organic compounds with the *alkanes*. This group of compounds is one of the simplest in composition and structure, for the following reasons.

1. All alkanes are *hydrocarbons*, which means that they contain only carbon and hydrogen.

2. The bonds between carbon atoms in alkanes are all single bonds. When all the carbon-carbon bonds in an organic compound are single bonds, we say that the compound is *saturated*.

The simplest member of this group of hydrocarbons is methane, CH_4, which we're already familiar with:

$$
\begin{array}{c}
H \\
| \\
H-C-H \\
| \\
H
\end{array}
$$

Methane

The next-to-the-simplest alkane has two carbon atoms and is called *ethane*, CH_3—CH_3, or

$$
\begin{array}{cc}
H & H \\
| & | \\
H-C-C-H \\
| & | \\
H & H
\end{array}
$$

Ethane

Then we come to the member of this group that has three carbon atoms—and many of us have had some experience with this one—*propane* (also known as bottled gas), CH_3—CH_2—CH_3, or

$$
\begin{array}{ccc}
H & H & H \\
| & | & | \\
H-C-C-C-H \\
| & | & | \\
H & H & H
\end{array}
$$

Propane

Note that each member of this group differs from the member before it by having one more CH_2 group. We can see this more clearly if we write the molecular formulas of these compounds.

$$CH_4 \qquad C_2H_6 \qquad C_3H_8$$

Methane   Ethane   Propane

So we say that these compounds are members of a *homologous series*. This means that all the compounds in this series are of the same chemical type, and differ only by fixed increments (an increment is an increase in quantity). In the alkane

TABLE 9-1   The first ten members of the alkane series

NUMBER OF CARBON ATOMS	MOLECULAR FORMULA	STRUCTURAL FORMULA	NAME	COMMENTS
1	CH_4	CH_4	Methane	Main ingredient of natural gas
2	C_2H_6	$CH_3—CH_3$	Ethane	Minor ingredient of natural gas
3	C_3H_8	$CH_3—CH_2—CH_3$	Propane	Minor ingredient of natural gas; usually separated from other ingredients and sold as bottled gas
4	C_4H_{10}	$CH_3—(CH_2)_2—CH_3$	Butane	Also found in natural gas; usually separated from other ingredients and used as a fuel—for example, in butane lighters
5	C_5H_{12}	$CH_3—(CH_2)_3—CH_3$	Pentane	
6	C_6H_{14}	$CH_3—(CH_2)_4—CH_3$	Hexane	
7	C_7H_{16}	$CH_3—(CH_2)_5—CH_3$	Heptane	Ingredients of gasoline
8	C_8H_{18}	$CH_3—(CH_2)_6—CH_3$	Octane	
9	C_9H_{20}	$CH_3—(CH_2)_7—CH_3$	Nonane	
10	$C_{10}H_{22}$	$CH_3—(CH_2)_8—CH_3$	Decane	Ingredient of kerosene

series, the fixed increment is CH_2. The general formula for this series of compounds can be written C_nH_{2n+2}, where n represents the number of carbon atoms (Table 9-1). Think about this formula for a minute and look at the formulas in Table 9-1, and you'll understand.

It's important that you learn the names of the alkanes we've just mentioned, because these terms are used in naming many other organic compounds.

Isomers

A little while ago we said that the molecular formula for an organic compound could actually represent many different compounds. As an example we showed that the formula C_4H_{10} represented two compounds. This phenomenon is known as *structural isomerism*. *Isomers* are compounds with the same *molecular* formula, but different *structural* formulas.

The title of Table 9-1 states that it shows the first 10 members of the alkane series. However, it really shows the first 10 *straight-chain* members of the alkane series. This is because, starting with C_4H_{10}, there are isomers for the rest of the alkane series (Table 9-2). Let's write the structures for the isomers of C_4H_{10} once again.

$$CH_3-CH_2-CH_2-CH_3 \qquad CH_3-\overset{\displaystyle CH_3}{\underset{\displaystyle H}{\overset{|}{\underset{|}{C}}}}-CH_3$$

Normal butane   Isobutane

The first structure is the *straight-chain* isomer, which we call normal butane (or *n*-butane). (The word "normal" or the letter *n* before the word "butane" signifies the straight-chain isomer.) The second structure is the *branched-chain* isomer, which is commonly called isobutane. Look at these two structures carefully and be sure that you understand the differences between them.

The three isomers of pentane (C_5H_{12}) would look this way.

$$CH_3-CH_2-CH_2-CH_2-CH_3 \qquad CH_3-CH_2-\overset{\displaystyle CH_3}{\underset{\displaystyle H}{\overset{|}{\underset{|}{C}}}}-CH_3$$

n-Pentane   Isopentane

$$CH_3-\overset{\displaystyle CH_3}{\underset{\displaystyle CH_3}{\overset{|}{\underset{|}{C}}}}-CH_3$$

Neopentane

Again, be sure that you see the differences among the three structures.

TABLE 9-2   Number of isomers of some compounds in the alkane series

MOLECULAR FORMULA	NUMBER OF ISOMERS
C_4H_{10}	2
C_5H_{12}	3
C_6H_{14}	5
C_7H_{16}	9
C_8H_{18}	18
C_9H_{20}	35
$C_{10}H_{22}$	75
$C_{15}H_{32}$	5000 (approx.)
$C_{40}H_{82}$	60 trillion (approx.)

At this point, you might be asking whether there are still more structures that can be written for the formula C_5H_{12}. You might try writing the following structure:

$$
\begin{array}{c}
CH_3 \\
| \\
H-C-CH_2-CH_3 \\
| \\
CH_3
\end{array}
$$

Is this structure different from that of isopentane? The answer is no! It is simply the molecule of isopentane flipped around. *The bonding sequence is still the same.*

EXAMPLE 9-2   Write the structural formulas for the five isomers of C_6H_{14}.

SOLUTION   Try to obtain five different bonding sequences. Remember that *each carbon atom can be bonded to only four other atoms.*

(a) $CH_3-CH_2-CH_2-CH_2-CH_2-CH_3$

(b) $CH_3-CH_2-CH_2-\overset{\displaystyle CH_3}{\underset{\displaystyle H}{\overset{\displaystyle |}{\underset{\displaystyle |}{C}}}}-CH_3$

(c) $CH_3-CH_2-\overset{\displaystyle CH_3}{\underset{\displaystyle H}{\overset{\displaystyle |}{\underset{\displaystyle |}{C}}}}-CH_2-CH_3$

(d) $CH_3-CH_2-\overset{\displaystyle CH_3}{\underset{\displaystyle CH_3}{\overset{\displaystyle |}{\underset{\displaystyle |}{C}}}}-CH_3$

(e) $CH_3-\overset{\displaystyle CH_3}{\underset{\displaystyle H}{\overset{\displaystyle |}{\underset{\displaystyle |}{C}}}}-\overset{\displaystyle CH_3}{\underset{\displaystyle H}{\overset{\displaystyle |}{\underset{\displaystyle |}{C}}}}-CH_3$

Before we can name compounds like these, we have to take up the topic of *organic nomenclature,* or how to name organic compounds.

FIGURE 9-6
Because of the huge number of organic compounds, scientists need a systematic way of naming them.

How to Name Organic Compounds

By the beginning of the twentieth century, chemists realized that there were a great number of organic compounds and that some type of systematic way of naming them was going to be required (Fig. 9-6). A system was developed in 1892 at an international meeting of chemists in Geneva. This system was revised at another international meeting of chemists—The International Union of Pure and Applied Chemistry, IUPAC, pronounced "I-you-pack." The IUPAC took on the job of standardizing chemistry so that chemists all over the world would use the same symbols for elements, the same units for measurements, and the same names for chemical formulas. The IUPAC continues to do this job today.

The chief feature of the IUPAC system is that it assigns numbers to carbon atoms to show where various R groups are located along the main carbon chain. You'll be able to see how this works as we introduce you to some simple rules for naming alkanes.

RULE 1.   Look for the longest hydrocarbon chain and use this as the base name of the compound.

EXAMPLE 9-3   For each of the compounds that follow, find the longest hydrocarbon chain and name it. (To make things easier to see, we'll show just the carbon atoms, keeping in mind that there are hydrogens attached to these carbon atoms.)

(a) C—C—C—C—C—C
            |
            C
            |
            C

(b) C—C—C—C—C—C
              |
              C
              |
              C
              |
              C

(c)   C
              |
              C
              |
    C—C—C—C
    |
    C
    |
    C

SOLUTION We'll use boldface to show the longest chain. Don't be misled by the way the structure is written; just look for the longest connecting chain of carbon atoms.

(a) **C—C—C—C—C—C**
              |
              C
              |
              C

The longest chain has six carbon atoms, so the base name of this compound is hexane.

(b) **C—C—C—C—C**—C
                |
                C
                |
                C
                |
                C

The longest chain has eight carbon atoms. This compound would be better written as

                      C
                      |
    C—C—C—C—C—C—C—C

So the base name of this compound is octane.

ALKANE	ALKYL GROUP
CH_4 Methane	CH_3— Methyl
CH_3—CH_3 Ethane	CH_3—CH_2— Ethyl
CH_3—CH_2—CH_3 Propane	CH_3—CH_2—CH_2— n-propyl
	CH_3—CH—CH_3 Isopropyl

(c)

$$
\begin{array}{c}
\quad\quad C \\
\quad\quad | \\
\quad\quad C \\
\quad\quad | \\
C-C-C-C \\
| \\
C \\
| \\
C
\end{array}
$$

The longest chain has seven carbon atoms. This compound would be better written as

$$C-C-C-C-C-C-C$$
$$|$$
$$C$$

So the base name of this compound is heptane.

RULE 2.   Next add the names of the side chains. Change the ending of the alkane side-chain group from *-ane* to *-yl* (Table 9-3).

EXAMPLE 9-4   Using the compounds in Example 9-3, add the names of the alkane side chains to the names of the compounds.

SOLUTION   This time we'll use boldface to denote the side chains.

(a) C—C—C—**C**—C—C
$$|$$
$$\mathbf{C}$$
$$|$$
$$\mathbf{C}$$

Ethylhexane

(b)

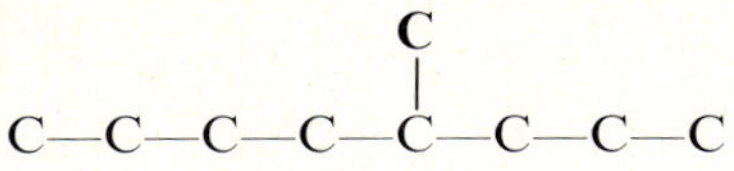

$$C-C-C-C-\underset{\underset{C}{|}}{C}-C-C-C$$

Methyloctane

(c) $C-C-C-C-\underset{\underset{C}{|}}{C}-C-C$

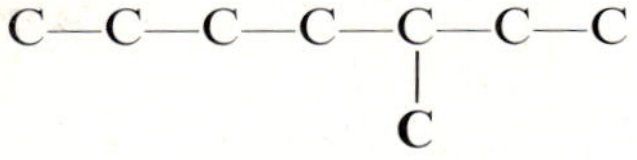

Methylheptane

RULE 3. Specify the location of the side chain by numbering the carbon atoms along the *longest* chain. And always number the longest chain so that the side-chain group will have the lowest possible number. For example, in this compound $C-C-C-C$, do we number from the

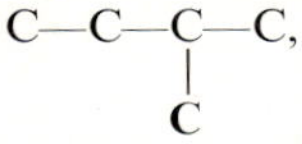

$$C-C-\underset{\underset{C}{|}}{C}-C$$

right or left? The answer is that we number from the side that will give the methyl side-chain group the lowest number possible. So in this compound we number from the right.

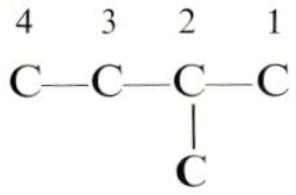

$$\overset{4\ \ \ 3\ \ \ 2\ \ \ 1}{C-C-\underset{\underset{C}{|}}{C}-C}$$

The name of this compound is 2-methylbutane.

EXAMPLE 9-5 Using the compounds in Example 9-3, give the entire name of the compound.

SOLUTION We already have the name of the side-chain group; therefore, all we have to do is number the main chain and add the position of the side-chain group to the name.

$$\overset{6\ \ \ \ 5\ \ \ \ 4\ \ \ \ 3\ \ \ \ 2\ \ \ \ 1}{\text{(a)}\ C-C-C-\underset{\underset{\underset{\underset{C}{|}}{C}}{|}}{C}-C-C}$$

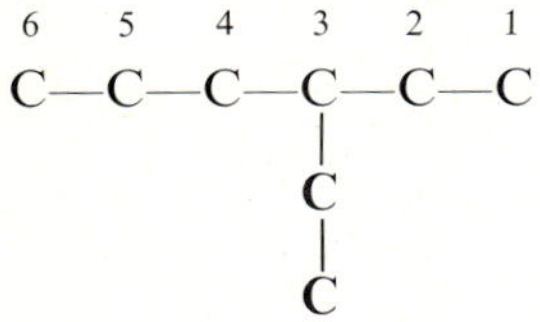

The name of this compound is 3-ethylhexane. In this compound we must number the carbon atoms from the right, so that the ethyl side-chain group has the lowest possible number, *three*.

$$\text{(b)}\ C-C-C-C-\underset{\underset{C}{|}}{\overset{\overset{C}{|}}{C}}-C-C-C$$
$$\ \ \ \ 8\ \ \ 7\ \ \ 6\ \ \ 5\ \ \ 4\ \ \ 3\ \ \ 2\ \ \ 1$$

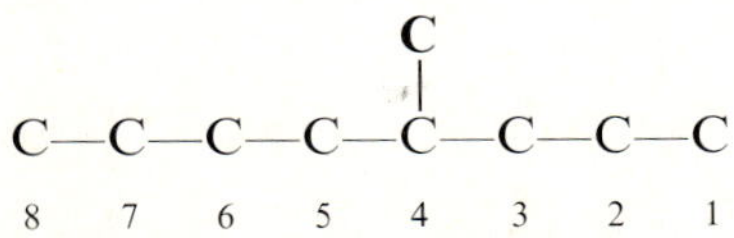

The name of this compound is 4-methyloctane. In this compound we also number from the right so as to give the methyl side chain the lowest possible number, *four*.

$$\overset{7}{C}-\overset{6}{C}-\overset{5}{C}-\overset{4}{C}-\overset{3}{C}-\overset{2}{C}-\overset{1}{C}$$

(c) with a C branching below the 4 position.

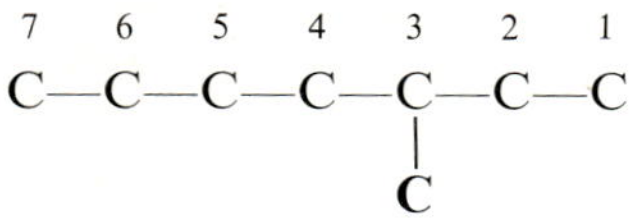

The name of this compound is 3-methylheptane. The chain was again numbered from the right, so as to give the methyl side-chain group the lowest possible number.

RULE 4.   If more than one side chain is present, use the following procedure to name the compound.
(a) Find the *longest chain*, as usual, to get the base name of the compound.
(b) Locate the *side chains*.
(c) Number the *longest chain* so that the *side chains* will have the lowest possible numbers.
(d) Add the name and position of the *longest side chain* TO THE BASE NAME.
(e) Now add the name and position of the *next-longest side chain*.

The following examples will show you how rule 4 is used.

EXAMPLE 9-6   Name the following compound.

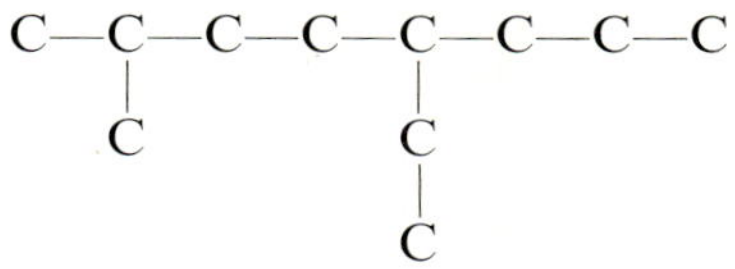

$$C-C-C-C-C-C-C-C$$
with a C branch below the second carbon, and a C—C branch below a later carbon.

SOLUTION   We'll name this compound in steps and use boldface to highlight the major points.

(a) The longest chain is octane.

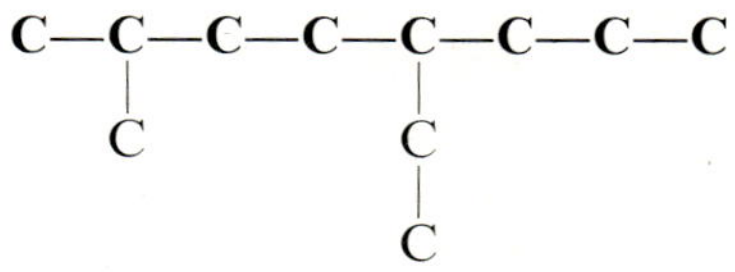

$$\mathbf{C-C-C-C-C-C-C-C}$$
with a C branch below the second carbon, and a C—C branch below a later carbon.

(b) The side chains are a methyl group and an ethyl group.

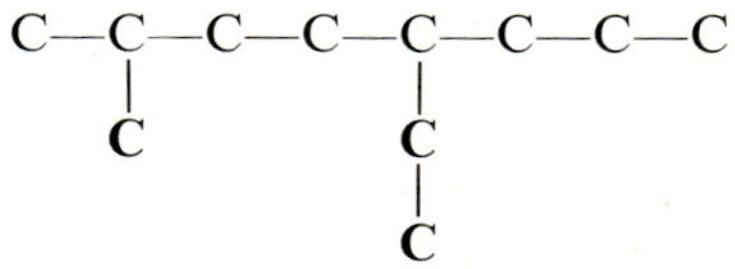

$$C-C-C-C-C-C-C-C$$
with a **C** branch below the second carbon, and a **C—C** branch below a later carbon.

(c) We number the longest chain from the left, so that the side chains have the lowest possible numbers: 2 and 5.

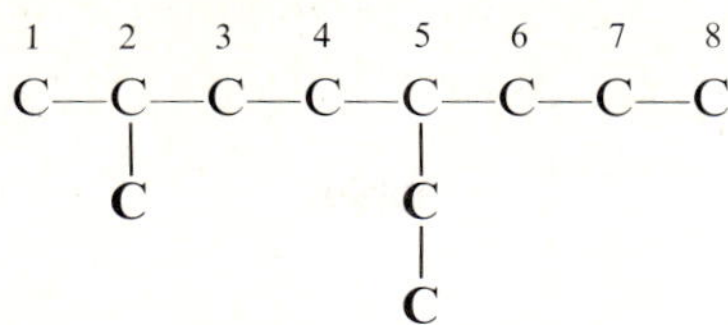

(If we had numbered from the right, the side-chain numbers would have been 4 and 7.)

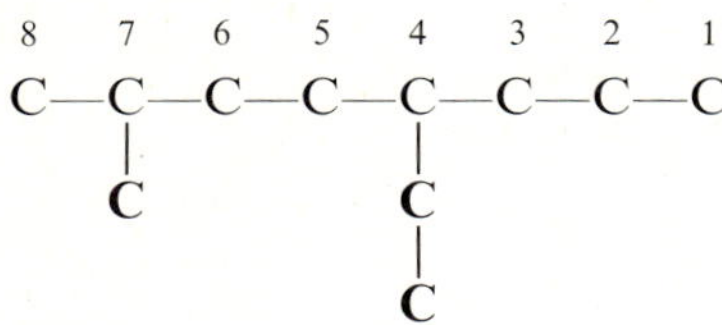

The numbers 2 and 5 are a lower combination than the numbers 4 and 7.

(d) We can now write the full name of the compound.

Longest chain	octane
Longest side chain and its position	5-ethyl
Next-longest side chain and its position	2-methyl
Full name of compound	2-methyl-5-ethyloctane

EXAMPLE 9-7   Name the following compound.

$$C-C-C-C-C$$

SOLUTION   We'll use the same procedure we used before.

(a) The longest chain is pentane.

$$C-C-C-C-C$$

(b) The side chains are two methyl groups.

$$C-C-C-C-C$$

(c) We number the longest chain from the right so that the side chains have the lowest possible numbers: 2 and 3.

$$\overset{5}{C}-\overset{4}{C}-\overset{3}{C}-\overset{2}{C}-\overset{1}{C}$$
$$\qquad\quad |\qquad\,\,\, |$$
$$\qquad\quad C\quad\,\, C$$

(d) We can now write the full name of the compound, but instead of saying 2-methyl-3-methyl, we say 2,3-dimethyl.

Longest chain	pentane
Longest side chain and its position	2,3-dimethyl
Full name of compound	2,3-dimethylpentane

The Alkenes and Alkynes

The alk*enes* and alk*ynes*, like the alkanes, are hydrocarbons. But the alk*enes* have a carbon–carbon double bond and the alk*ynes* have a carbon–carbon triple bond (Fig. 9-7).

The simplest *alkene* has the formula C_2H_4 and the structure

$$\begin{array}{c} H \\ \\ H \end{array} \,\,C{=}C\,\, \begin{array}{c} H \\ \\ H \end{array}$$

Ethylene

Notice that double bond linking the carbon atoms. The common name for this compound is ethylene. It is the substance from which we make polyethylene, a plastic used for containers, electrical insulation, and packaging. It is also a substance that has been used in the health area as an anesthetic.

The simplest *alkyne* has the formula C_2H_2 and the structure

$$H-C{\equiv}C-H$$

Acetylene

Notice that triple bond linking the carbon atoms. The common name for this compound is acetylene. Today it is used as a fuel, for example in the acetylene torch. But in the past it had been used as a surgical anesthetic.

Each of these compounds is the first compound in its homologous series (Table 9-4). Both of these series of hydrocarbons are said to be *unsaturated*. This means that there can be double or triple bonds in the compounds.

We can find the IUPAC name for the alkene or alkyne by changing the alkane ending -*ane* to -*ene* for alkenes or -*yne* for alkynes. This means that the

FIGURE 9-7
Alkanes have carbon-carbon single bonds; alkenes have a carbon-carbon double bond; and alkynes have a carbon-carbon triple bond.

$$-C-C-\qquad \text{Alkane}$$

$$C{=}C\qquad \text{Alkene}$$

$$-C{\equiv}C-\qquad \text{Alkyne}$$

	ALKENES	ALKYNES
Structural formula of the first member of the series	$\begin{array}{c} H \quad\quad H \\ \diagdown \quad\quad \diagup \\ C{=}C \\ \diagup \quad\quad \diagdown \\ H \quad\quad H \end{array}$	$H{-}C{\equiv}C{-}H$
Molecular formula	C_2H_4	C_2H_2
Common name	Ethylene	Acetylene
IUPAC name	Eth*ene*	Eth*yne*
General formula for the series	C_nH_{2n}	C_nH_{2n-2}

IUPAC name for C_2H_4 is eth*ene*, and the IUPAC name for C_2H_2 is eth*yne*. Both names are derived from the name of the two-carbon alkane compound, ethane; only the endings have changed.

EXAMPLE 9-8   Name the following compounds. (Again, in order to simplify writing the structures, we'll show only the carbon skeletons; keep in mind that there are hydrogens attached to these carbon atoms.)

(a) $C{-}C{\equiv}C$

(b) $C{-}C{=}C$

(c) $C{-}C{-}C{-}C{=}C$

(d) $C{-}C{-}C{=}C{-}C$

(e) $\begin{array}{c} C{-}C{-}C{-}C{\equiv}C{-}C \\ \quad | \quad\, | \\ \quad C \quad C \\ \quad\quad\quad | \\ \quad\quad\quad C \end{array}$

SOLUTION   (a) $C{-}C{\equiv}C$ is called prop*yne*. It is a three-carbon compound that is the second member of the alkyne series. It takes its name from the three-carbon alkane compound, except that it has a *-yne* ending.

(b) $C{-}C{=}C$ is called prop*ene*. It is a three-carbon compound that is the second member of the alkene series. It takes its name from the three-carbon alkane compound, except that it has an *-ene* ending.

(c) $\overset{5}{C}{-}\overset{4}{C}{-}\overset{3}{C}{-}\overset{2}{C}{=}\overset{1}{C}$ is called 1-pent*ene*. This compound takes its name from the five-carbon alkane compound, except that it has an

-ene ending. But there is one additional point: The position of the double bond must be specified. This is because there are other possible pentene structures with the double bond between other carbon atoms, as we'll see in part (d) of this example.

RULE 5.   Specify the position of the double bond by numbering the carbon atoms, as we did before, keeping the number of the double bond as low as possible. Then, in the name, we give the number of the carbon atom from which the double bond starts.

(d) $\overset{5}{C}-\overset{4}{C}-\overset{3}{C}=\overset{2}{C}-\overset{1}{C}$ is called 2-pent*ene*. This compound is similar to the compound in part (c), except the position of the double bond is different. Also, notice that we numbered the carbon chain from the right, to keep the number of the double bond as low as possible.

(e)
$$C-C-C-C\equiv C-C$$

We'll name this compound in steps. Note that when we number the carbon chain, we give preference to the double bond, and not to the side chains. In other words, we want to keep the number of the double bond as low as possible.

Longest chain and position of triple bond	2-hexyne
Longest side chain and its position	4-ethyl
Next-longest side chain and its position	5-methyl-
Full name of compound	5-methyl-4-ethyl-2-hexyne

The Cyclic Hydrocarbons

There is a whole series of hydrocarbons that form *cyclic structures*—that is, structures that join to form a closed chain, like a snake biting its tail. An example of this is:

$$\text{or} \quad C_3H_6$$

Cyclopropane

The name of this compound—cyclopropane—is derived from the name of the three-carbon alkane (propane) and the prefix "cyclo." Many cyclic hydrocarbons have useful properties. For example, cyclopropane is an anesthetic that is sometimes used in surgery. However, cyclopropane is very flammable, and great precautions must be taken in the operating room when it is used.

Another example of a cyclic hydrocarbon is

$$\text{or} \quad C_5H_{10}$$

Cyclopentane

The name of this compound—cyclopentane—is derived from the name of the five-carbon alkane (pentane) and the prefix "cyclo." In order to make the writing of cyclic hydrocarbons simpler, we usually do this:

we write as

The shape of the figure tells you the name of the compound. In this example the figure is a pentagon (five sides). Each corner of the pentagon represents a CH_2 group.*

EXAMPLE 9-9   Name the following compounds.

(a)   (b)

(c) $CH_2—CH_3$ (d) $CH_2—CH_2—CH_3$

CH_3

---

* In general, when writing cyclic hydrocarbons, each corner of the figure represents a C atom with the appropriate number of hydrogen atoms. Remember, each carbon atom can only have four bonds.

SOLUTION   (a) 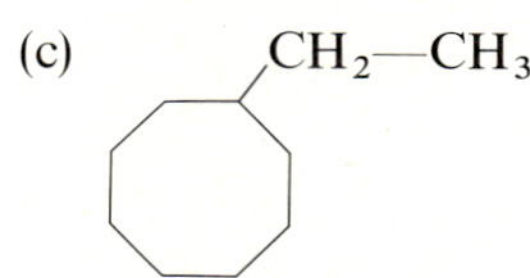 This is a cyclic hydrocarbon with four carbon atoms, and it is called cyclobutane.

(b) This is a cyclic hydrocarbon with six carbon atoms and it is called cyclohexane. It is used in some fungicidal preparations.

(c) CH_2—CH_3

This is a substituted cyclic hydrocarbon. The —CH_2—CH_3 side-chain group has replaced a hydrogen atom that was bonded to one of the carbon atoms. The compound is called ethylcyclooctane.

(d) CH_2—CH_2—CH_3

CH_3

This is a disubstituted cyclic hydrocarbon, which means that *two* hydrogens on the ring have been replaced by side-chain groups.

RULE 6.   When more than one side chain is attached to a cyclic hydrocarbon, you have to number the C atoms in the ring. Remember to number them so as to keep the numbers of the side chains as low as possible.

Name of cyclic structure	cyclopentane
Name of longest side chain and its position	1-propyl
Name of next-longest side chain and its position	3-methyl
Full name of compound	3-methyl-1-propylcyclopentane

Some cyclic hydrocarbons have double and triple bonds. These hydrocarbons take their names from the corresponding alkene or alkyne, with the prefix "cyclo" added.

EXAMPLE 9-10   Name the following compounds.

(a) 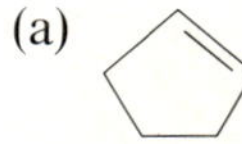 (b) 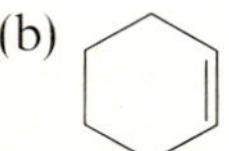 (c) 

SOLUTION

(a) 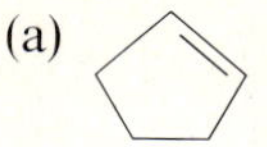 This is a five-carbon cyclic compound with a double bond. It takes its name from the corresponding *alkene*. The name of this compound is cyclopent*ene*.

(b) 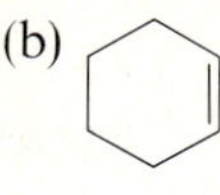 This is a six-carbon cyclic compound with a double bond. It takes its name from the corresponding alkene. The name of this compound is cyclohex*ene*. It is used as a stabilizer for high-octane gasolines.

(c) 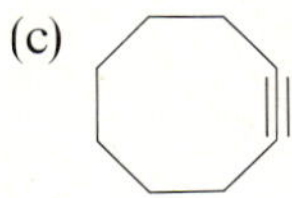 This is an eight-carbon cyclic compound with a triple bond. It takes its name from the corresponding alkyne. The name of this compound is cyclooct*yne*.

Aromatic Hydrocarbons: They Don't All Smell Sweet

One series of cyclic hydrocarbons has alternating single and double bonds. The members of this series are called *aromatic hydrocarbons*, because early organic chemists noticed that these compounds had pleasant smells. (However, later organic chemists found that some of these compounds stink!) The parent compound of the aromatic hydrocarbons is benzene, C_6H_6, whose structure is

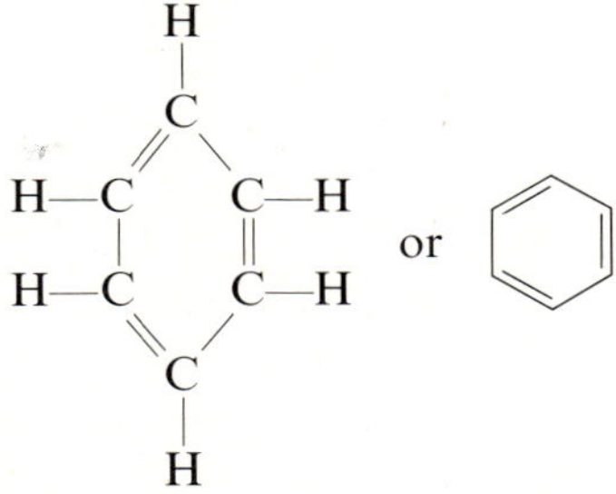

Benzene

Although only three of the carbon–carbon bonds in benzene are written as double bonds, in reality all the bonds seem to be equivalent. At one time chemists thought that the double bonds switched back and forth between the carbon atoms. If that were true, either of these two diagrams would give a valid picture of the benzene molecule:

Today, chemists still don't agree on how to represent the structure of benzene. Any one of the following representations can be used.

 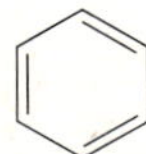 

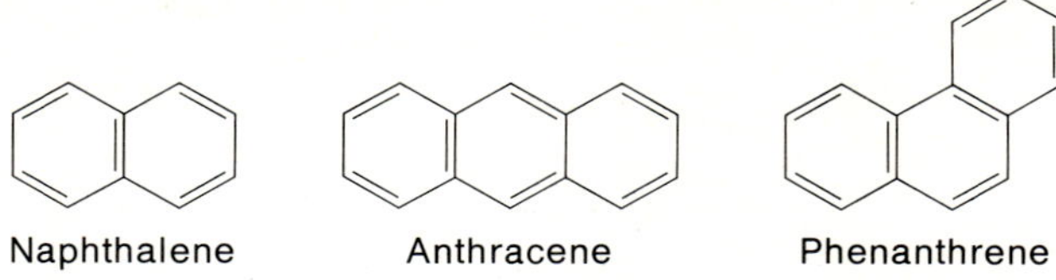

FIGURE 9-8
Some aromatic hydrocarbons.

The third diagram is meant to show that all the carbon–carbon bonds in benzene are equivalent.

Benzene, which can be obtained from coal tar, is one of the most important organic compounds. It is the starting material for countless products that make life more comfortable and pleasant, including many pesticides, pharmaceuticals, and synthetic fibers such as nylon, to say nothing of perfumes and synthetic dyes.

Other aromatic hydrocarbons seem to be made up of benzene rings joined together (Fig. 9-8). As you might have guessed, there are many compounds that consist of these aromatic compounds with various side chains. An example is

$$CH_3$$

Toluene

One name for this compound is methylbenzene, but organic chemists usually called it *toluene*. It is used in making explosives and dyes.

You'll be seeing a lot more of these aromatic compounds in subsequent chapters, as we look at organic compounds other than hydrocarbons.

Reactions of Hydrocarbons

Hydrocarbon compounds undergo many different kinds of chemical reactions to produce some very interesting and useful products. Many of these products are, in fact, useful to the health practitioner. Let's briefly examine some of the reactions of hydrocarbons.

Alkanes

Alkanes are a fairly inert group of organic compounds. However, they do react with oxygen in a process known as *combustion*. The resulting products of hydrocarbon combustion are carbon dioxide and water. For example, in the burning of natural gas, methane, the reaction looks something like this:

$$CH_4 + 2O_2 \xrightarrow{\text{spark}} CO_2 + 2H_2O$$

Now you can see why people who spend all day cooking with their gas oven turned on should also keep a window open. If they don't, the oxygen content in the kitchen may drop and the carbon dioxide level increase, causing hypoventilation.

Alkenes

Alkenes are more reactive than alkanes because these hydrocarbons have double bonds (they're unsaturated). The carbon atoms joined by double bonds can react with other substances in what is called an *addition reaction*. For example, an alkene can react with hydrogen gas to form an alkane. This process is known as *hydrogenation*.

$$H_2C{=}CH_2 + H_2 \longrightarrow H_3C{-}CH_3 \quad \text{(hydrogenation)}$$

Ethene   Ethane

We'll have more to say about hydrogenation when we study fats in Chapter 12.

Another type of addition reaction involves the reaction of an alkene with water to produce another type of organic compound known as an *alcohol*. This type of addition reaction is known as hydration.

$$H_2C{=}CH_2 + H{-}OH \longrightarrow CH_3CH_2{-}OH \quad \text{(hydration)}$$

Ethene   Ethanol

We'll learn more about alcohols in Chapter 10.

Alkynes

Alkynes are very reactive hydrocarbons. Like alkenes, they undergo addition reactions. In the case of hydrogenation, four atoms of hydrogen are added to the molecule.

$$HC{\equiv}CH + 2H_2 \longrightarrow CH_3{-}CH_3 \quad \text{(hydrogenation)}$$

Ethyne   Ethane

Cycloalkanes

Cyclic hydrocarbons are also reactive. This is especially true in the case of the smaller molecules, where bond angles are strained. For example, the hydrogenation of cyclopropane yields *n*-propane.

$$\underset{\underset{\text{CH}_2-\text{CH}_2}{}}{\overset{\text{CH}_2}{\diagup \diagdown}} + \text{H}_2 \longrightarrow \text{CH}_3\text{CH}_2\text{CH}_3$$

Cyclopropane also reacts very readily with oxygen in an explosive reaction. All that's necessary is a spark.

$$2\ \underset{\underset{\text{CH}_2-\text{CH}_2}{}}{\overset{\text{CH}_2}{\diagup \diagdown}} + 9\text{O}_2 \ \xrightarrow{\text{spark}}\ 6\text{CO}_2 + 6\text{H}_2\text{O}$$

We mention this because cyclopropane, a sweet-smelling gas, is used in hospitals as an anesthetic. Therefore, it's very important that precautions be taken to avoid sparks or static electricity in an operating room where this material is being used.

The list of organic reactions goes on and on, but we will not. Our purpose in the preceding pages was to give you a brief introduction into the reactions of hydrocarbons so that you could learn about some of the interesting reactions that they undergo.

SUMMARY

In this chapter we discussed the differences between organic and inorganic compounds. We looked at the properties of carbon and discovered why there are so many organic compounds. We learned how to write the structural formulas for organic compounds and how to name alkane, alkene, and alkyne hydrocarbons. We also learned about isomers and practiced writing the structures of isomers given the molecular formula of an organic compound. We learned the meaning of the words saturated, unsaturated, and homologous series. We also looked at some cyclic hydrocarbons. Finally, we looked at some of the reactions that hydrocarbons undergo.

In the following chapter we'll use much of the information that we learned here to study other classes of organic compounds.

EXERCISES

1. In your own words, define an organic compound.

2. State the three-dimensional shape of a methane molecule, and then write the Lewis dot structure for methane.

3. Describe in terms of three-dimensional geometry what happens when
   (a) two carbon atoms form a single bond
   (b) two carbon atoms form a double bond
   (c) two carbon atoms form a triple bond

4. Give an example of
   (a) a saturated hydrocarbon
   (b) an unsaturated hydrocarbon

5. Write the names and formulas for the first 10 straight-chain members of the alkane series. What is such a series called?

6. Write the following structural formulas in the more convenient form, sometimes called the condensed form.

(a)

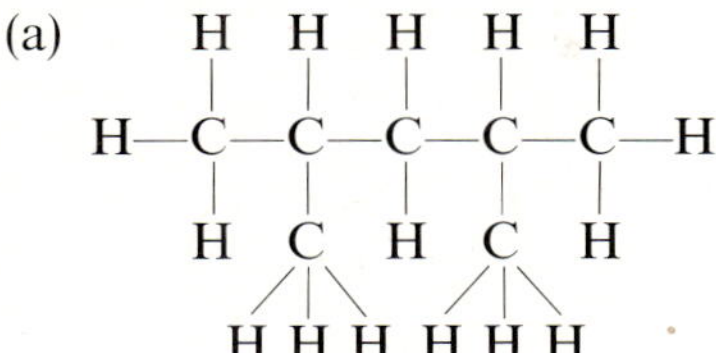

(b)

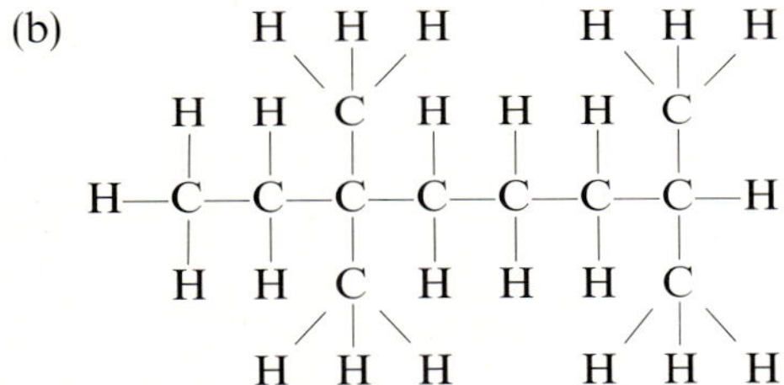

7. Write the structures for the nine isomers of heptane, C_7H_{16}.

8. Name the nine isomers of heptane from Exercise 7.

9. Rewrite each of the structures that follow by adding hydrogen atoms to the carbon skeleton.

(a) C—C—C—C—C—C
       | |
       C   C

(b) C—C—C=C—C—C
           |
           C

(c) C—C=C—C≡C
        |
        C

(d)

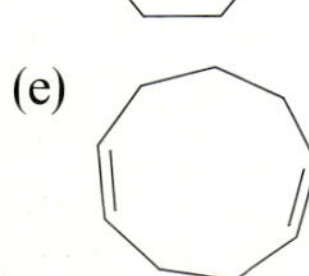

(e)

10. Name the following alkanes by the IUPAC system.

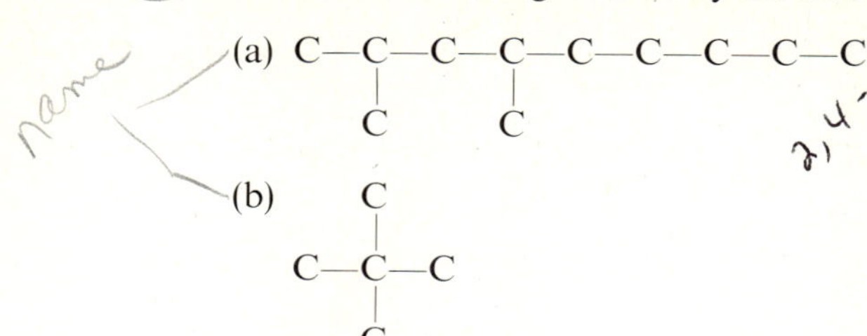

(a) C—C—C—C—C—C—C—C—C
          | |
          C   C

(b)   C
         |
     C—C—C
         |
         C

(c)   C
                 |
     C—C—C—C—C—C
             | |
             C   C
             |
             C
             |
             C

11. Name the following alkenes by the IUPAC system.

name

(a) C—C—C=C—C

(b) C—C=C—C—C

(c) C=C—C—C—C

(d) C—C—C—C=C—C—C
             | |
             C   C

12. Name the following alkynes by the IUPAC system

(a) C—C—C≡C—C—C
                   |
                   C
                   |
                   C

(b) C—C—C—C≡C—C
             |
             C
             |
             C

(c)   C
                     |
     C—C=C—C—C—C—C—C—C
                 | | |
                 C   C   C
                 |
                 C

13. Name the following cyclic hydrocarbons by the IUPAC system.

(a) 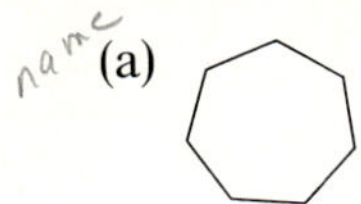

(b) 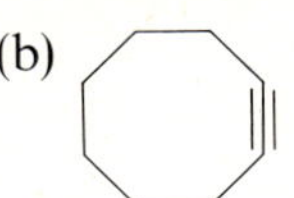

(c) 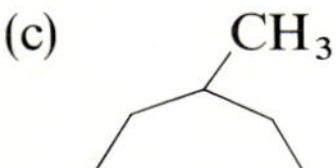

(d) 

(e) 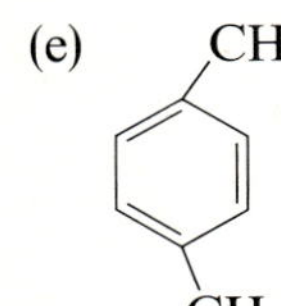

(*Hint:* This is an aromatic hydrocarbon.)

14. Write structures for the following hydrocarbons.
    (a) 2-methylbutane
    (b) 2,3,4-trimethylheptane
    (c) 2,2-dimethylpropane
    (d) 1-hexene
    (e) 2-hexene
    (f) 2-methyl-3-ethyl-1-pentene
    (g) 3-methyl-1-butyne

15. Write structures for the following cyclic hydrocarbons.
    (a) cyclopropane
    (b) cycloheptane
    (c) 1-methyl-3-ethylcyclohexane
    (d) cyclohexyne
    (e) cyclohexene
    (f) 1,2-dimethylbenzene
    (g) 1,3-dimethylbenzene

16. Write the products of the following hydrocarbon reactions.
    (a) $C_8H_{18} + O_2 \longrightarrow$
    (b) $CH_3CH_2-CH=CHCH_3 + H_2 \longrightarrow$
    (c) $CH=C-CH_2CH_2CH_3 + H_2 \longrightarrow$

CHAPTER 10

The Other Kinds of Organic Compounds

They're All Very Functional

Some Things You Should Know After Reading This Chapter

You should be able to:

1. Explain the meaning of *functional group*.
2. Write the general formula for an alcohol.
3. Name alcohols by their common and IUPAC names.
4. State whether an alcohol is primary, secondary, or tertiary when given its structural formula.
5. Give an example of at least one important alcohol and state some of its medical uses.
6. Write the general formula for an ether.
7. Name ethers by their common and IUPAC names.
8. Give an example of at least one important ether and state some of its medical uses.
9. Write the general formula for an acetal and hemiacetal.
10. Write the general formula for an aldehyde.
11. Name aldehydes by their common and IUPAC names.
12. Write an equation for the reduction of an aldehyde.
13. Write an equation for the oxidation of an aldehyde.
14. Give an example of at least one important aldehyde and state some of its medical uses.
15. Write the general formula for a ketone.
16. Name ketones by their common and IUPAC names.
17. Write an equation for the reduction of a ketone.
18. Give an example of at least one important ketone and state some of its medical uses.
19. Write the general formula for a carboxylic acid.
20. Name carboxylic acids by their common and IUPAC names.
21. Give an example of at least one important carboxylic acid and state some of its medical uses.
22. Write the general formula for an ester.
23. Write an equation for the formation of an ester.
24. Name esters by their common and IUPAC names.
25. Give an example of at least one important ester and state some of its medical uses.
26. Write the three general formulas for amines.
27. State whether an amine is primary, secondary, or tertiary when given its structural formula.
28. Name amines by their common and IUPAC names.
29. Give an example of at least one important amine and state some of its medical uses.
30. Write the three general formulas for amides.
31. Name amides by their common and IUPAC names.
32. Give an example of at least one important amide and state some of its medical uses.

Introduction

In the previous chapter we studied only hydrocarbon compounds. However, there are many classes of organic compounds. We can write the general formula of these compounds by using the R-group notation that we discussed in Chapter 9. Each class of organic compounds is denoted by a *functional group*. The *functional group* is what the R group is attached to; it is the reactive part of the molecule (Fig. 10-1). Table 10-1 lists some of the common classes of organic compounds, along with their general formulas.

Each class of compounds has very different properties and uses. In this chapter we will survey the various classes of organic compounds and learn about some of their properties and uses.

Alcohols: The —OH Functional Group

The general formula for an alcohol is

$$R\text{—}OH$$

The —OH is called the functional group. The R is usually a chain of carbon atoms. If the R group consists of only a few carbon atoms, the alcohol tends to be polar, because of the polarity of the —OH group (Fig. 10-2). This allows the alcohol to dissolve in polar solvents such as water. For example, the alcohol CH_3CH_2OH, ethanol, dissolves in water. However, if the R group consists

FIGURE 10-1
The functional group is what the R group is attached to; it is the reactive part of the molecule.

FIGURE 10-2
If the R group of an alcohol consists of only a few carbon atoms, the alcohol tends to be polar, as a result of the polarity of the —OH group.

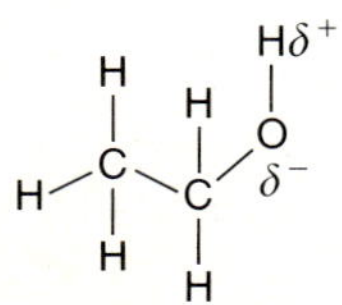

TABLE 10-1   Some common classes of organic compounds

CLASS	GENERAL FORMULA[a]	EXAMPLE	COMMON NAME	IUPAC NAME
Alkane	C_nH_{2n+2}	CH_4	Methane (swamp gas)	Meth*ane*
Alkene	C_nH_{2n}	$CH_2{=}CH_2$	Ethylene	Eth*ene*
	(n = 2 or more)	$CH_3{-}CH{=}CH_2$	Propylene	Prop*ene*
Alkyne	C_nH_{2n-2}	$H{-}C{\equiv}C{-}H$	Acetylene	Eth*yne*
	(n = 2 or more)	$CH_3{-}C{\equiv}C{-}H$	Methylacetylene	Prop*yne*
Alcohol	$R{-}OH$	$CH_3{-}OH$	Methyl alcohol (wood alcohol)	Methan*ol*
		$CH_3{-}CH_2{-}OH$	Ethyl alcohol (grain alcohol)	Ethan*ol*
Ether	$R{-}O{-}R$	$CH_3{-}O{-}CH_3$	Dimethyl ether	Methoxymethane
		$CH_3{-}CH_2{-}O{-}CH_2{-}CH_3$	Diethyl ether	Ethoxyethane
Aldehyde	$R{-}\overset{\displaystyle C=O}{\underset{\displaystyle H}{\vert}}$	$H{-}\overset{\displaystyle C=O}{\underset{\displaystyle H}{\vert}}$	Formaldehyde	Methan*al*
		$CH_3{-}CH_2{-}\overset{\displaystyle C=O}{\underset{\displaystyle H}{\vert}}$	Propionaldehyde	Propan*al*
Ketone	$R{-}\underset{\displaystyle O}{\overset{\displaystyle \vert}{C}}{-}R$	$CH_3{-}\underset{\displaystyle O}{\overset{\displaystyle \vert}{C}}{-}CH_3$	Dimethyl ketone (acetone)	Propan*one*
		$CH_3{-}CH_2{-}\underset{\displaystyle O}{\overset{\displaystyle \vert}{C}}{-}CH_2{-}CH_3$	Diethyl ketone	3-pentan*one*
Carboxylic acid	$R{-}\underset{\displaystyle O}{\overset{\displaystyle \vert}{C}}{-}OH$	$H{-}\underset{\displaystyle O}{\overset{\displaystyle \vert}{C}}{-}OH$	Formic acid	Methan*oic* acid
		$CH_3{-}\underset{\displaystyle O}{\overset{\displaystyle \vert}{C}}{-}OH$	Acetic acid	Ethan*oic* acid
Ester	$R{-}\underset{\displaystyle O}{\overset{\displaystyle \vert}{C}}{-}O{-}R$	$H{-}\underset{\displaystyle O}{\overset{\displaystyle \vert}{C}}{-}O{-}CH_3$	Methyl formate	Methylmethanoate
		$CH_3{-}\underset{\displaystyle O}{\overset{\displaystyle \vert}{C}}{-}O{-}CH_2{-}CH_3$	Ethyl acetate	Ethylethanoate
Amine	$R{-}NH_2$	$CH_3{-}NH_2$	Methylamine	Aminomethane
		$CH_3{-}CH_2{-}NH_2$	Ethylamine	Aminoethane
Amides	$R{-}\underset{\displaystyle O}{\overset{\displaystyle \vert}{C}}{-}NH_2$	$CH_3{-}\underset{\displaystyle O}{\overset{\displaystyle \vert}{C}}{-}NH_2$	Acetamide	Ethanamide
		$CH_3CH_2{-}\underset{\displaystyle O}{\overset{\displaystyle \vert}{C}}{-}NH_2$	Propionamide	Propanamide

[a] Remember that n = number of carbon atoms.

Source: Alan Sherman, Sharon Sherman, and Leonard Russikoff, *Basic Concepts of Chemistry*, Houghton Mifflin Company, Boston, 1976. By permission of the publisher.

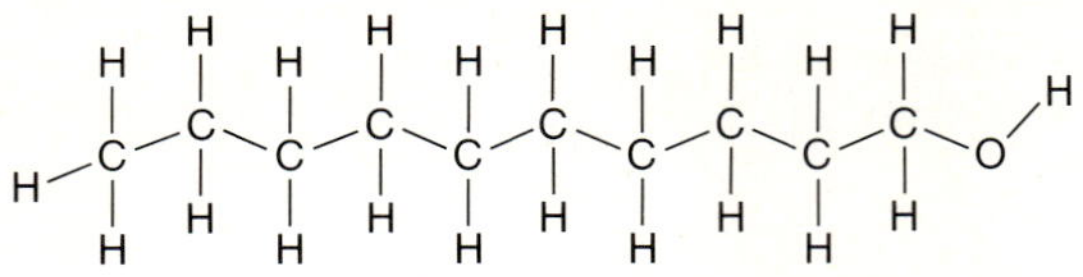

FIGURE 10-3
If the R group of an alcohol consists of a long chain of carbon atoms, the alcohol tends to be
nonpolar, because the compound is mostly an alkane chain and behaves like a nonpolar
alkane.

of a long chain of carbon atoms, the polarity of the —OH group is less pronounced and the alcohol appears to be nonpolar (Fig. 10-3). The alcohol will
dissolve in nonpolar solvents such as benzene and carbon tetrachloride. For
example, the alcohol $CH_3CH_2CH_2CH_2CH_2CH_2OH$, hexanol, dissolves well
in benzene.

Alcohols can be named in two ways:

1. By their common names, which usually involves adding the word
   alcohol to the alkyl group. For example, CH_3CH_2OH is called ethyl
   alcohol.

2. By their IUPAC names, which involves changing the alkane ending,
   -ane to the alcohol ending, *-ol*. For example, CH_3CH_2OH is also called
   etha*nol*.

In fact, you can think of simple alcohols as being derived from alkanes in which
one of the alkane hydrogens has been replaced by an —OH group. (The —OH
group is sometimes called the *hydroxyl* group.) For example,

$$CH_4 \qquad\qquad CH_3\text{—}OH$$

Methane   Methanol
(or methyl alcohol)

EXAMPLE 10-1   Name the following alcohols by their IUPAC names.

(a) $CH_3CH_2CH_2OH$

(b) $CH_3\underset{\displaystyle OH}{CH}CH_3$

(c) $CH_3CH_2\underset{\displaystyle CH_3}{CH}CH_2CH_2\underset{\displaystyle OH}{CH}CH_3$

(d) 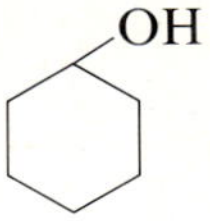   OH

SOLUTION

(a) $CH_3CH_2CH_2OH$ The IUPAC name is derived from the name of the three-carbon alkane, only with an *-ol* ending. Therefore, the name of this compound is 1-propan*ol*. (By the way, the common name of this compound is propyl alcohol, or more correctly, *n*-propyl alcohol. The *n* stands for normal, which means the straight-chain structure; in other words, the OH group comes off the end carbon atom. Does this mean there could be another propyl alcohol where the OH group does not come off the end carbon atom? The answer is in part (b).)

(b) $CH_3\overset{|}{C}HCH_3$ This is also a three-carbon alcohol, a propanol,
    OH

except that the OH group is not on the end carbon atom. Therefore, we have to designate its position, so that this compound can be differentiated from the one in part (a). The name of this compound is 2-propanol. (The common name of this compound is isopropyl alcohol. This is because the CH_3—CH—CH_3 alkyl group is known as the isopropyl group. Other alkyl groups are listed in Table 10-2.)

(c) $\overset{7}{C}H_3\overset{6}{C}H_2\overset{5}{C}H\overset{4}{C}H_2\overset{3}{C}H_2\overset{2}{C}H\overset{1}{C}H_3$ To name this compound we must
       CH_3 OH
number the main carbon chain so that the -OH group has the lowest possible number. We must also designate the position of the methyl side chain. The name of this compound is 5-methyl-2-heptanol.

(d) 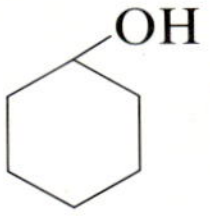   OH   The name of this compound comes from the name of the cycloalkane, but with an *-ol* ending. The name of this compound is cyclohexanol.

Alcohol Categories

Alcohols can be classified into three categories:

1.   PRIMARY ALCOHOLS in which the —OH group is attached to a carbon atom that is attached to only one other carbon atom. For example, CH_3CH_2OH, ethanol is a primary alcohol.

TABLE 10-2   Alkyl groups

NAME	FORMULA
Methyl	CH_3—
Ethyl	CH_3CH_2—
Propyl	$CH_3CH_2CH_2$—
Isopropyl	CH_3—$\overset{\displaystyle |}{CH}$—$CH_3$
Butyl	$CH_3CH_2CH_2CH_2$—
Isobutyl	CH_3—$\underset{\underset{\textstyle CH_3}{|}}{CH}$—$CH_2$—
sec-Butyl	CH_3CH_2—$\overset{\displaystyle |}{CH}$—$CH_3$
tert-Butyl	CH_3—$\overset{\overset{\textstyle CH_3}{|}}{\underset{\underset{\textstyle CH_3}{|}}{C}}$—
Phenyl	(benzene ring)

2. SECONDARY ALCOHOLS in which the —OH group is attached to a carbon atom that is attached to two other carbon atoms. For example, CH_3CHCH_3, 2-propanol (isopropyl alcohol), is a secondary alcohol.

$$\underset{\underset{\textstyle OH}{|}}{CH_3CHCH_3}$$

3. TERTIARY ALCOHOLS —in which the —OH group is attached to a carbon atom that is attached to three other carbon atoms. For example,

$$CH_3-\overset{\overset{\textstyle CH_3}{|}}{\underset{\underset{\textstyle OH}{|}}{C}}-CH_3$$

whose IUPAC name is 2-methyl-2-propanol, is a tertiary alcohol. (By the way, the common name for this alcohol is tert-butyl alcohol:

the word "tert" because it is a tertiary alcohol, and "butyl" because it involves four carbon atoms.)

EXAMPLE 10-2 State whether the alcohol is primary, secondary, or tertiary. Also, state its IUPAC or common name.

(a) CH_3CHCH_2OH
$|$
CH_3

(b) $CH_3CH_2CHCH_3$
$|$
OH

(c) $CH_3CH_2CH_2CH_2OH$

(d) 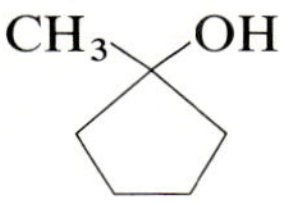 CH_3 OH

SOLUTION (a) CH_3CHCH_2OH This compound is a primary alcohol be-
$|$ cause the OH is attached to a carbon atom
CH_3 that is attached to only one other carbon
atom. The IUPAC name of this compound is 2-methylpropanol. (By the way, the common name of this alcohol is isobutyl alcohol. This is because the CH_3CHCH_2—group is called the isobutyl group.)
$|$
CH_3

(b) $CH_3CH_2CHCH_3$ This compound is a secondary alcohol be-
$|$ cause the OH is attached to a carbon atom
OH that is attached to two carbon atoms. The

FIGURE 10-4
The four butyl alcohols.

$CH_3—CH_2—CH_2—CH_2—OH$

n-Butyl alcohol or butanol

$CH_3—CH—CH_2—OH$
$|$
CH_3

Isobutyl alcohol or 2-methylpropanol

$CH_3—CH_2—CH—CH_3$
$|$
OH

sec-Butyl alcohol or 2-butanol

CH_3
$|$
$CH_3—C—CH_3$
$|$
OH

tert-Butyl alcohol or 2-methyl-2-propanol

IUPAC name of this compound is 2-butanol. (The common name of this compound is *sec*-butyl alcohol. The word "sec" indicates that it is a secondary alcohol, and the word "butyl" indicates that it contains four carbon atoms.)

(c) $CH_3CH_2CH_2CH_2OH$ This compound is a primary alcohol. The IUPAC name of this compound is butanol. The common name of this compound is just butyl alcohol (or, more correctly, *n*-butyl alcohol). The *n* stands for normal, which indicates the straight-chain alcohol (Fig. 10-4).

(d) This compound is a tertiary alcohol, because the OH is attached to a carbon atom that is attached to three other carbon atoms. Let's write out the structure so you can see what we mean.

The IUPAC name of this compound is 1-methylcyclopentanol.

Methanol, CH_3OH, whose common name is *methyl alcohol*, is the simplest alcohol. This very poisonous compound is also known as wood alcohol because it can be obtained from the burning of wood in the absence of air (a process called destructive distillation). Death can result from drinking just 30 ml of it: smaller amounts can produce nausea, convulsions, blindness, and respiratory failure. During the time of prohibition in this country, many people drank beverages containing methanol, not realizing the dangers involved.

Ethanol, CH_3CH_2OH, whose common name is *ethyl alcohol*, is the alcohol we all consume in our favorite wine, liqueur, or beer. The ethanol found in these beverages is produced through the process of fermentation, which involves the breakdown of the sugar glucose.

$$C_6H_{12}O_6 \xrightarrow{\text{yeast}} 2CH_3CH_2OH + 2CO_2 + \text{energy}$$

Glucose

A solution that is 70% ethanol by volume acts as an antiseptic by coagulating the proteins of bacteria. Ethanol is also used to make *tinctures*, which are medicinal substances dissolved in this alcohol. For example, tincture of iodine consists of 2% I_2, 2.4% NaI in a 50% weight/volume alcohol water solution.

2-Propanol, CH_3CHCH_3, whose common name is *isopropyl alcohol*,

is used to give sponge baths to reduce high fevers. It is for this reason that 2-propanol is also called rubbing alcohol.

1,2,3-Propantriol, $CH_2CH_2CH_2$, which is commonly known as *glycerol*

or *glycerine*, is a trihydroxy alcohol (in other words, it has three OH groups). Glycerol is a very thick, sweet-tasting compound that is found as part of natural fats and oils. It is nontoxic, which makes it an excellent carrier for pharmaceuticals. The cosmetics industry uses glycerol in the formulation of hand and skin creams because it acts as a good lubricant.

5-Methyl-2-isopropyl-1-cyclohexanol,

commonly called *menthol*, is a substituted cyclohexanol. This substance has many commercial uses, for example, in cough drops, nasal inhalers, shaving creams, liqueurs, and cigarettes. Menthol is found naturally in peppermint oil. It has a cooling sensation when rubbed on the skin. When placed in cough drops, menthol causes the mucous membranes of the throat to increase their secretions. This has a soothing effect on inflamed tissues.

Hydroxybenzene,

which is better known as *phenol*, is the simplest aromatic alcohol. Because of the nature of the benzene ring, phenol can act as an acid and ionize in the following manner:

It is for this reason that phenol is also known as carbolic acid. Although phenol can be quite corrosive in its pure form, many hospitals use a solution of 1 to 2% phenol as a germicide for linens and medical instruments.

5-Methyl-2-isopropyl-1-phenol,

commonly known as *thymol*, also has antiseptic properties. It is used in mouth-washes and by dentists for sterilizing a tooth before filling it.

2,6-di-tert-Butyl-4-methyl-1-phenol,

$$
\begin{array}{c}
\text{OH} \\
(CH_3)C\text{—}\!\!\bigcirc\!\!\text{—}C(CH_3)_3 \\
\text{CH}_3
\end{array}
$$

commonly known as *butylated hydroxytoluene*, or BHT for short, is used in many food substances as an antioxidant. This means that it doesn't allow the food to spoil. BHT is also put into certain types of rubbers and plastics to prevent them from undergoing oxidation.

As you can see by these brief examples, alcohols are an important class of compounds to the health practitioner. Some additional alcohols and their uses can be found in Table 10-3.

Ethers: The R—O—R Compounds

The general formula for an ether is

$$R\text{—}O\text{—}R'$$

If the R and R' groups are the same, we say that the ether is symmetrical. If the R and R' groups are different, we say that the ether is unsymmetrical.

Ethers can be named in either of two ways:

1. By their common names, which involves naming the two alkyl groups in alphabetical order and adding the name ether. For example, $CH_3CH_2OCH_3$ is called ethyl methyl ether.
2. By their IUPAC names, which involves naming one of the alkyl groups as an alkoxy compound. For example, $CH_3CH_2OCH_3$ is called methoxyethane. Notice that the shorter of the two R groups is named as the alkoxy compound.

EXAMPLE 10-3   Name the following ethers by their IUPAC names and their common names.

(a) $CH_3CH_2OCH_2CH_3$
(b) $CH_3CH_2CH_2CH_2OCH_2CH_3$
(c) $CH_3CH_2O\text{—}\bigcirc$

TABLE 10-3   Some alcohols and their uses

NAME	STRUCTURE	USE
ortho-Cresol	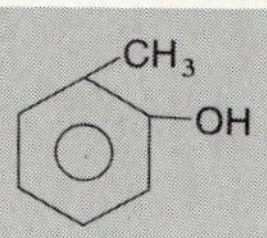	Antiseptic and disinfectant (found in products such as Lysol)
meta-Cresol	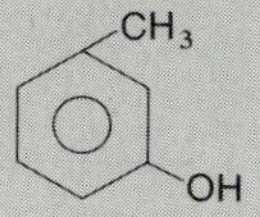	Antiseptic and disinfectant
para-Cresol	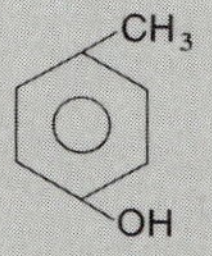	Antiseptic and disinfectant
Diethylstilbestrol	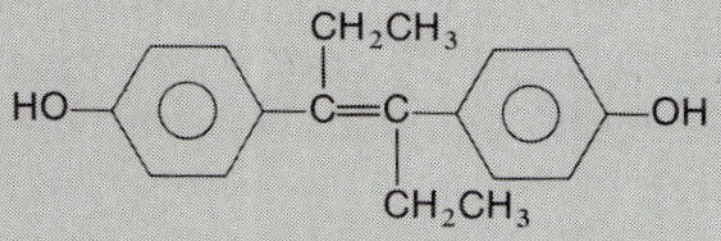	Estrogenic hormone therapy (this compound has been linked to uterine cancer in females whose mother's received this compound during pregnancy)
Hexylresorcinal	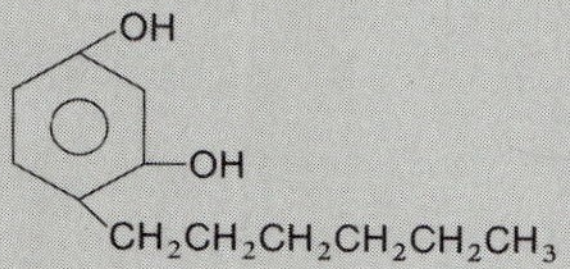	Antiseptic and germicide
Terpin hydrate	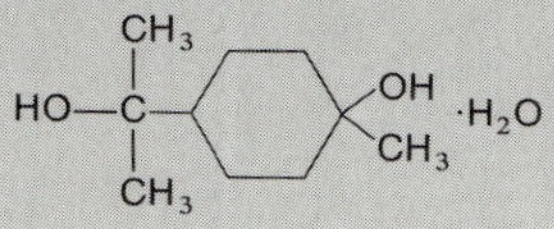	An expectorant

SOLUTION

(a) $CH_3CH_2OCH_2CH_3$ The IUPAC name of this compound is ethoxyethane. The common name of this compound is diethyl ether, or simply ethyl ether (it's not imperative to use the prefix *di* on symmetrical ethers).

(b) $CH_3CH_2CH_2CH_2OCH_2CH_3$ The IUPAC name of this compound is ethoxybutane. The common name of this compound is *n*-butyl-ethyl ether.

(c) CH_3CH_2O    The IUPAC name of this compound is ethoxybenzene. The common name of this compound is ethyl phenyl ether (Phenyl is the common name for the C_6H_5 aromatic group,    .)

Symmetrical ethers can be synthesized from alcohols, in the presence of sulfuric acid. For example, the simplest ether, methoxymethane (dimethyl ether) can be synthesized as follows:

$$CH_3OH + HOCH_3 \xrightarrow{H_2SO_4} CH_3OCH_3 + H_2O$$

Methyl   Dimethyl
alcohol   ether

The sulfuric acid acts as a dehydrating agent. Two molecules of methyl alcohol condense to form one molecule of dimethyl ether.

Some Properties of Ethers

Because ethers have two alkyl groups bonded covalently to an oxygen atom, ethers tend to have little polarity. This means that they don't dissolve well in water. Also, unlike their corresponding alcohols, ethers cannot hydrogen-bond. This means that they have much lower boiling points than the alcohols from which they are derived. Ethers tend to be very flammable, and many of them have some type of anesthetic properties.

Some Important Ethers and Their Uses

Ethoxyethane, $CH_3CH_2OCH_2CH_3$, also called diethyl ether, or sometimes just "ether," is probably the best known to medical personnel. It is used in hospitals as an anesthetic for long surgical procedures (Fig. 10-5). It is an extremely safe anesthetic because there is a large difference in dosages between that needed to anesthetize and that needed to kill. Also, it has little effect on blood pressure or respiration. Because of its high volatility, diethyl ether is easy to administer. But it does have some side effects, such as irritation to the respiratory membranes as well as nausea and vomiting upon awakening. Also, great care must be taken in the operating room, because of the high flammability of diethyl ether.

4-Allyl-2-methoxyphenol,

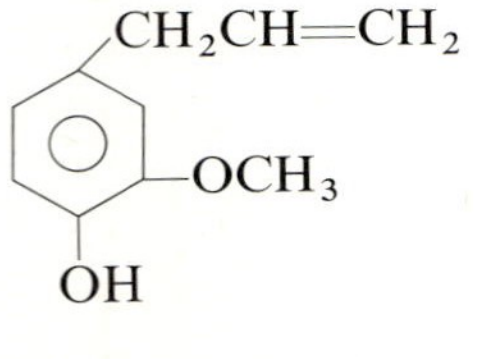

$$\begin{array}{c} CH_2CH{=}CH_2 \\ \text{(benzene ring)}\!\!-\!\!OCH_3 \\ OH \end{array}$$

whose common name is *eugenol,* is found in oil of cloves. It has a spicy, pungent taste and has many uses in dentistry; for example, in toothache preparations it acts as a pain reliever, and in temporary fillings it acts as an antiseptic.

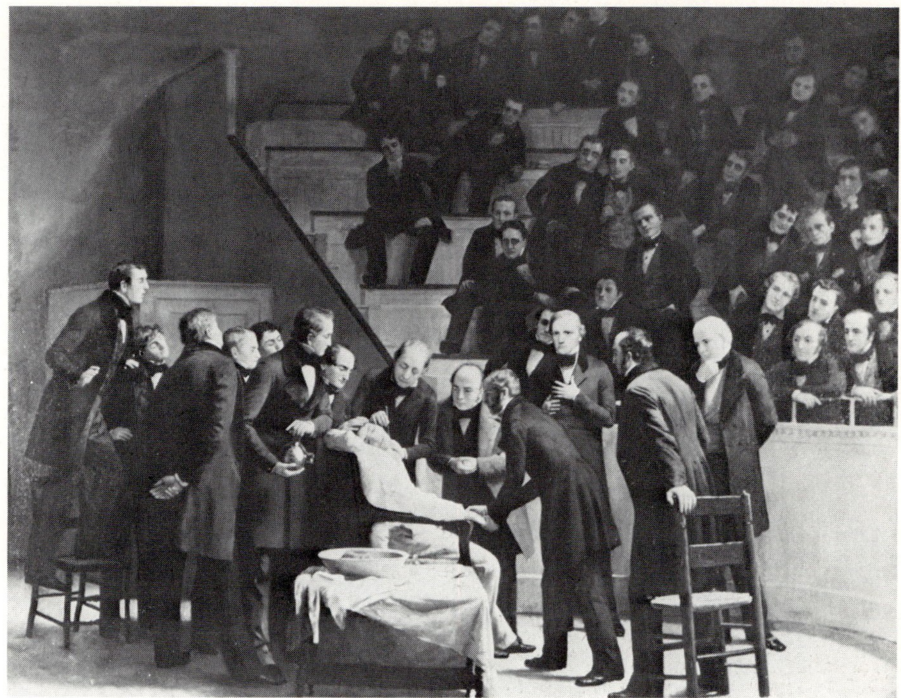

FIGURE 10-5
The Hinckley painting depicting the first public demonstration of the use of ether as a surgical anesthetic. (Massachusetts General Hospital)

Ethylene oxide,

$$\begin{array}{c} O \\ \diagup \diagdown \\ CH_2 \!\!-\!\!-\!\! CH_2 \end{array}$$

which is a gas at room temperature, is a cyclic ether, used as a fumigant (disinfectant) for foodstuffs and as a fungicide in agriculture. It is also used to sterilize surgical instruments.

Some additional ethers and their uses can be found in Table 10-4.

Acetals and Hemiacetals: They're Ether-Related

Consider a compound with the following structure:

$$\begin{array}{c} H \\ | \\ R\!\!-\!\!C\!\!-\!\!O\!\!-\!\!R' \\ | \\ O\!\!-\!\!R'' \end{array}$$

NAME	STRUCTURE	USE
Furazolidone	O_2N—(furan ring)—$CH{=}N$—N—(oxazolidinone ring with =O and O)	Antimicrobial
Divinyl ether	$CH_2{=}CH_2$—O—$CH{=}CH_2$	Quick-acting anesthetic
Mephenesin	(benzene ring)—OCH_2—CH—CH_2OH, with CH_3 OH	A skeletal muscle relaxant

This compound has two ether groups attached to the same carbon atom. Such a compound is known as an *acetal* (or if R′ and R″ are the same, a *ketal*).

If one of the ether groups is turned into an alcohol group (a hydroxyl group), we have the following structure:

$$
\begin{array}{c}
H \\
| \\
R{-}C{-}O{-}R' \\
| \\
O{-}H
\end{array}
$$

This compound is known as a *hemiacetal.* We bring this to your attention because in Chapter 11 you'll be learning about sugars and you will see molecules containing the hemiacetal structure. Now you'll know what they are!

Aldehydes: The $-\overset{\overset{\displaystyle O}{\|}}{C}-H$ *Functional Group*

The general formula for an aldehyde is

$$
R{-}\underset{\underset{\displaystyle O}{\|}}{C}{-}H
$$

The $-\underset{\underset{\displaystyle O}{\|}}{C}{-}H$ is called the *aldehyde group.* Many aldehydes have attractive odors and are used in perfumes. And some aldehydes have pleasant tastes and are used in flavoring agents.

Aldehydes may also be named in two ways:

1. By their common names, which in many cases is simply the name of the alkyl group plus the word "aldehyde." For example, CH_3CH_2CH
$\quad\quad\quad\quad\quad\quad\quad\quad\quad\quad\overset{\|}{O}$

is called propionaldehyde.

2. By their IUPAC names, which involves changing the alkane ending *-ane* to the aldehyde ending *-al*. For example, $CH_3CH_2\overset{\|}{\underset{O}{C}}H$ is called propan*al*.

EXAMPLE 10-4   Name the following aldehydes by their IUPAC names and their common names.

(a) $CH_3CH_2CH_2\overset{\|}{\underset{O}{C}}H$

(b) $CH_3CH_2CH_2CH_2CH_2\overset{\|}{\underset{O}{C}}H$

(c) $CH_3\overset{\|}{\underset{O}{C}}H$

SOLUTION   (a) $CH_3CH_2CH_2\overset{\|}{\underset{O}{C}}H$ The IUPAC name of this compound is butanal. The common name of this compound is butyraldehyde.

(b) $CH_3CH_2CH_2CH_2CH_2\overset{\|}{\underset{O}{C}}H$ The IUPAC name of this compound is hexanal. The common name of this compound is hexaldehyde. (This compound also has another common name, caproaldehyde, since it is the aldehyde of caproic acid, which is found in goat fat. *Caper* is Latin for goat.)

(c) $CH_3\overset{\|}{\underset{O}{C}}H$ The IUPAC name of this compound is ethanal. The common name of this compound is acetaldehyde.

The —C—H group makes aldehydes chemically reactive. For example,
     ‖
     O
many aldehydes can undergo the process of reduction and be converted to primary alcohols. For example,

$$CH_3CH_2\underset{\overset{\|}{O}}{C}H + H_2 \xrightarrow{Pt} CH_3CH_2CH_2OH$$

Propan*al*   Propan*ol*

Aldehydes can also undergo the process of oxidation and be converted to carboxylic acids. For example,

$$CH_3CH_2\underset{\overset{\|}{O}}{C}H \xrightarrow{KMnO_4} CH_3CH_2\underset{\overset{\|}{O}}{C}-OH$$

Propanal   Propanoic acid

Some Important Aldehydes and Their Uses

Methanal, H—C—H, commonly called *formaldehyde*, is the simplest and
           ‖
           O
best known aldehyde. Although it is a gas at room temperature, it is usually dissolved in water and commonly sold in a 40% weight/volume solution. In this form it is used as a germicide, disinfectant, and preservative, and is commonly called *formalin*. Biologists use formalin solution to preserve tissues of various organisms. Many of you are probably familiar with the pungent odor of this compound. Formaldehyde is also very irritating to your eyes, nose, and throat.

3-Methoxy-4-hydroxybenzaldehyde,

better known as *vanillin*, is used both commercially and medically as a flavoring agent. It has a very pleasant odor and taste that can cover up the harsh taste of many medicines. The vanilla taste in vanilla ice cream is due to the compound vanillin.

Some additional aldehydes and their uses can be found in Table 10-5.

TABLE 10-5   Some aldehydes and their uses

NAME	STRUCTURE	USE
Benzaldehyde	(benzaldehyde structure)	Almond flavoring
Cinnamalaldehyde	(cinnamaldehyde structure)	Cinnamon flavoring
Propionaldehyde	$CH_3CH_2{-}CH$ with $=O$	Disinfectant

Ketones: The R—C—R′ Compounds

The general formula for a ketone is

$$R{-}\underset{\underset{O}{\|}}{C}{-}R'$$

The —C— group (with $=O$) is called the *carbonyl* or *keto group*. Ketones are very similar in structure to aldehydes. Ketones, however, have two R groups attached to the carbonyl group, whereas aldehydes have only one.

$$R{-}\underset{\underset{O}{\|}}{C}{-}R' \qquad R{-}\underset{\underset{O}{\|}}{C}{-}H$$

Ketone   Aldehyde

Ketones may also be named in two ways:

1. By their common names, which involves simply naming each of the alkyl groups in alphabetical order plus the word "ketone." For example, $CH_3CH_2{-}C{-}CH_2CH_2CH_3$ (with $=O$) is called ethyl propyl ketone.

2. By their IUPAC names, which involves counting the number of carbon atoms in the compound, including the carbonyl group, only changing the alkane ending *-ane* to the ketone ending *-one*. For example,

$CH_3-\overset{\displaystyle O}{\underset{\displaystyle \|}{C}}-CH_3$ is called propan*one*. Sometimes it is necessary to state the carbon atom that has the oxygen. This is done when structural isomers exist for the compound. For example, $CH_3CH_2-\overset{\displaystyle O}{\underset{\displaystyle \|}{C}}-CH_2CH_3$ is 3-pentanone, whereas $CH_3-\overset{\displaystyle O}{\underset{\displaystyle \|}{C}}-CH_2CH_2CH_3$ is 2-pentanone.

EXAMPLE 10-5   Name the following ketones by their IUPAC names and their common names.

(a) $CH_3-\overset{\displaystyle O}{\underset{\displaystyle \|}{C}}-CH_3$

(b) $CH_3CH_2CH_2CH_2-\overset{\displaystyle O}{\underset{\displaystyle \|}{C}}-CH_2CH_3$

(c) cyclohexanone structure

SOLUTION

(a) $CH_3-\overset{\displaystyle O}{\underset{\displaystyle \|}{C}}-CH_3$ The IUPAC name of this compound is propanone. The common name of this compound is dimethyl ketone. However, this compound is probably best known as *acetone*.

(b) $CH_3CH_2CH_2CH_2-\overset{\displaystyle O}{\underset{\displaystyle \|}{C}}-CH_2CH_3$ The IUPAC name of this compound is 3-heptanone. The common name of this compound is butyl ethyl ketone.

(c) cyclohexanone structure   The IUPAC name of this compound is cyclohexanone. The common name for this compound does not follow the rules for straight-chain ketones, so we will omit it here.

Ketones are not as reactive as aldehydes. However, with proper catalysts, ketones can be reduced to secondary alcohols. For example,

$$CH_3-\overset{\displaystyle O}{\underset{\displaystyle \|}{C}}-CH_3 + H_2 \xrightarrow{\ Pt\ } CH_3-\underset{\displaystyle OH}{\underset{\displaystyle |}{CH}}-CH_3$$

Propanone   2-Propanol

Unlike aldehydes, ketones cannot be oxidized to carboxylic acids very readily. This, in fact, gives rise to a test to differentiate aldehydes from ketones. You will learn more about this test in the laboratory portion of your course.

Some Important Ketones and Their Uses

Propanone, $CH_3-\underset{\underset{O}{\|}}{C}-CH_3$, also called *dimethyl ketone* or *acetone*, is the simplest and best known ketone. Acetone is an excellent solvent for many organic substances. It is found in paint removers and in nail polish removers. However, small amounts of acetone are also found in the blood, where it is called a *ketone body*. Large amounts of acetone in the blood indicates a body malfunction in the breakdown of fats. This is usually a sign of diabetes mellitus. Acetone has a very unique odor and sweetish taste. An individual suffering from uncontrolled diabetes mellitus will give off an acetone smell when he or she exhales. This is called *acetone breath*, which a physician can easily detect in an examination of such an individual.

Methyl phenyl ketone,

also known as *acetophenone*, is used in perfumery to impart an orange-blossom-like odor. It has also been used in medicine as a hypnotic.

Progesterone,

also called the *pregnancy hormone*, is a diketone. This substance is one of the major female hormones. It is secreted during the second half of the menstrual cycle and prepares the uterine lining to receive the embryo. Progesterone will also prevent ovulation if it is given during the fifth to twenty-fifth days of the menstrual cycle. This, of course, is the basis of the oral contraceptives.

Some additional ketones and their uses can be found in Table 10-6.

TABLE 10-6   Some ketones and their uses

NAME	STRUCTURE	USE
Camphor		Anesthetic and mild antiseptic
Carvone		Oil of spearmint
Testosterone		Male sex hormone

Carboxylic Acids:

The Organic Acids with That $-\overset{\text{O}}{\overset{\|}{\text{C}}}-\text{OH}$ *Group*

The general formula for a carboxylic acid is

$$R-\overset{}{\underset{\overset{\|}{\text{O}}}{\text{C}}}-\text{OH}$$

The $-\overset{}{\underset{\overset{\|}{\text{O}}}{\text{C}}}-\text{OH}$ is called the *carboxyl group*. This group acts like an acid

because it releases hydrogen ions in aqueous solutions.

$$R-\overset{}{\underset{\overset{\|}{\text{O}}}{\text{C}}}-\text{OH} + H_2O \;\rightleftharpoons\; R-\overset{}{\underset{\overset{\|}{\text{O}}}{\text{C}}}-O^- + H_3O^+$$

This ionization occurs only to a slight degree, so carboxylic acids are weak acids. Some carboxylic acids are found in fats and are called *fatty acids*. We will discuss fatty acids in more detail when we study lipids in Chapter 12. Still other carboxylic acids contain an extra hydroxyl group and are called *hydroxy acids*. These hydroxy acids play an important part in certain metabolic reactions (see p. 446). Figure 10-6 shows an example of a fatty acid and a hydroxy acid.

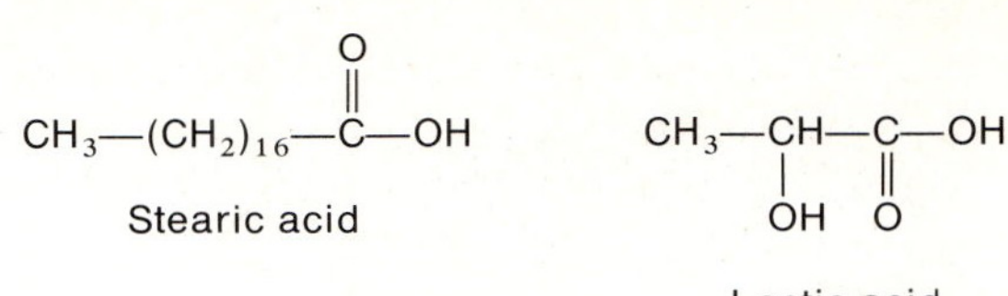

FIGURE 10-6
Stearic acid is a fatty acid used to make soaps, and lactic acid is a hydroxy acid found in sour milk.

Carboxylic acids can also be named in two ways:

1. By their common names, which involves (in most cases) the common alkyl root plus the ending -*ic*, followed by the word "acid." For example, CH_3CH_2—C—OH is called propion*ic* acid.

2. By their IUPAC names, which involves using the alkane name plus the ending -*oic*, followed by the word "acid." For example, CH_3CH_2—C—OH is called propan*oic* acid.

EXAMPLE 10-6   Name the following carboxylic acids by their IUPAC names and their common names.

(a) CH_3—C—OH

(b) $CH_3CH_2CH_2$—C—OH

(c) [benzoic acid structure] C—OH

SOLUTION   (a) CH_3—C—OH   The IUPAC name of this compound is ethanoic acid. The common name of this compound is acetic acid.

(b) $CH_3CH_2CH_2$—C—OH   The IUPAC name of this compound is butanoic acid. The common name of this compound is butyric acid.

(c) 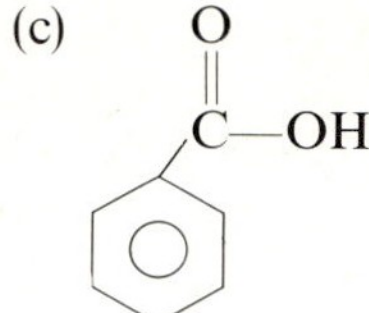

The IUPAC name of this compound is benzoic acid. There's no good name for this compound under the common naming system, except, perhaps, for the name benzenecarboxylic acid. But just about everybody calls this acid benzoic acid.

Some Important Carboxylic Acids and Their Uses

Methanoic acid, $H-\underset{\underset{O}{\|}}{C}-OH$, whose common name is *formic acid*, is the simplest of the carboxylic acids. It is the acid that is produced by ants and bees and is injected into their victims when they bite or sting. If you were ever stung by a bee, you may be interested in knowing that it was the formic acid that caused your skin to become inflamed.

By the way, it is formic acid that is formed by the liver when an individual consumes methanol.

$$CH_3-OH \xrightarrow[\text{by liver}]{\text{oxidation}} H-\underset{\underset{O}{\|}}{C}-OH$$

The formic acid, once formed, causes severe acidosis in the blood, which can lead to death.

Ethanoic acid, $CH_3-\underset{\underset{O}{\|}}{C}-OH$, also known as *acetic acid*, is a substance found in vinegar. We've already discussed some of the properties of this acid in Chapter 7. Acetic acid plays an important role in living cells, where it helps to produce long-chain fatty acids (see Chapter 12).

Butanoic acid, $CH_3CH_2CH_2-\underset{\underset{O}{\|}}{C}-OH$, also known as *butyric acid*, is produced when butter turns rancid. In its pure form it has a very pungent odor. It is also one of the substances that causes body odor.

2-Hydroxypropanoic acid, $CH_3\underset{\underset{OH}{\|}}{C}H-\underset{\underset{O}{\|}}{C}-OH$, better known as *lactic acid*, is a hydroxy acid. It is produced in your muscles when you exercise. Lactic acid is formed in milk when it turns sour, as a result of the action of lactic acid bacteria.

Benzoic acid,

NAME	STRUCTURE	USE
Citric acid	$\begin{array}{c} O{=}C{-}OH \\ \mid \\ HO{-}C{-}CH_2{-}C{-}CH_2{-}C{-}OH \\ \parallel \quad\quad \mid \quad\quad \parallel \\ O \quad\quad OH \quad\quad O \end{array}$	The taste of citrus fruits, also used for rickets and as a mild astringent
Lactic acid	$\begin{array}{c} CH_3{-}CH{-}C{-}OH \\ \mid \quad \parallel \\ OH \quad O \end{array}$	Found in sour milk
Nicotinic acid	(pyridine ring with $-C{-}OH$, $\parallel$ O)	Part of the essential B vitamins; also used as a vasodilator
Salicylic acid	(benzene ring with $-C{-}OH$ / $\parallel$ O and $-OH$)	Used extensively as an antimicrobial agent; also used to make aspirin
Stearic acid	$CH_3{-}(CH_2)_{16}{-}C{-}OH$, $\parallel$ O	Used for suppositories and for coating various types of pills, especially those which have a bitter taste; also used to make soap

which can be thought of as benzenecarboxylic acid, is the simplest aromatic acid. It is used as an antiseptic and antifungal agent, as well as a preservative. Other acids related to benzoic acid have numerous medical properties (Table 10-7).

Some additional carboxylic acids and their uses can be found in Table 10-7.

Esters: The R—C—OR' Compounds

$$R{-}\overset{\displaystyle O}{\underset{\displaystyle O}{\|}}{C}{-}OR'$$

The general formula for an ester is

$$R{-}C{-}OR' \\ \parallel \\ O$$

Esters can be formed by the reaction of an alcohol with a carboxylic acid.

For example:

$$CH_3CH_2-C-\boxed{OH + H}O-CH_3 \longrightarrow CH_3CH_2-C-OCH_3 + H_2O$$
$$\qquad\qquad\; \| \qquad\qquad\qquad\qquad\qquad\qquad\qquad \|$$
$$\qquad\qquad\; O \qquad\qquad\qquad\qquad\qquad\qquad\qquad O$$

    Propanoic acid   Methanol   An ester

Notice how the acid and alcohol react to produce the ester.

$$\boxed{R-C \qquad}\ \boxed{OR'}$$
$$\quad \|$$
$$\quad O$$

    Acid   Alcohol
    part   part

Water is also produced in this reaction, which is called *esterification*.

EXAMPLE 10-7   Write the products of the following esterification reaction.

$$CH_3CH_2CH_2CH_2-C-OH + HO-CH_2CH_3 \longrightarrow$$
$$\qquad\qquad\qquad\qquad\quad \|$$
$$\qquad\qquad\qquad\qquad\quad O$$

SOLUTION   We'll follow the example that we've just examined.

$$CH_3CH_2CH_2CH_2-C-OH + HO-CH_2CH_3 \longrightarrow CH_3CH_2CH_2CH_2-C-OCH_2CH_3 + H_2O$$
$$\qquad\qquad\qquad\quad \| \qquad\qquad\qquad\qquad\qquad\qquad\qquad\qquad\qquad \|$$
$$\qquad\qquad\qquad\quad O \qquad\qquad\qquad\qquad\qquad\qquad\qquad\qquad\qquad O$$

Some of the most common flavors and fragrances are due to esters (Table 10-8). Esters can also be named in two ways:

1. By their common names, which involves taking the alkane part of the alcohol group and adding it to the common name of the acid—but changing the *-ic* ending to *-ate*. For example, $CH_3CH_2-C-OCH_3$ is
$$\qquad\qquad\qquad\qquad\qquad\qquad\qquad \|$$
$$\qquad\qquad\qquad\qquad\qquad\qquad\qquad O$$

    called methyl propionate.

2. By their IUPAC names, which involves taking the alkane part of the alcohol group and adding it to the IUPAC name of the acid—but changing the *-ic* ending to *-ate*. For example, $CH_3CH_2-C-OCH_3$ is called
$$\qquad\qquad\qquad\qquad\qquad\qquad\qquad\qquad \|$$
$$\qquad\qquad\qquad\qquad\qquad\qquad\qquad\qquad O$$

    methylpropanoate.

NAME	FORMULA	FLAVOR OR FRAGRANCE
Amyl acetate	$CH_3\text{—}(CH_2)_4\text{—}O\text{—}\underset{\underset{O}{\|\|}}{C}\text{—}CH_3$	Banana
Amyl butyrate	$CH_3\text{—}(CH_2)_4\text{—}O\text{—}\underset{\underset{O}{\|\|}}{C}\text{—}CH_2CH_2CH_3$	Apricot
Ethyl butyrate	$CH_3CH_2\text{—}O\text{—}\overset{\overset{O}{\|\|}}{C}\text{—}CH_2CH_2CH_3$	Pineapple
Ethyl formate	$CH_3CH_2\text{—}O\text{—}\underset{\underset{O}{\|\|}}{C}\text{—}H$	Rum
Isoamyl acetate	$CH_3\text{—}\underset{\underset{CH_3}{\|}}{CH}\text{—}CH_2CH_2\text{—}O\text{—}\underset{\underset{O}{\|\|}}{C}\text{—}CH_3$	Pear
Isobutyl formate	$CH_3\text{—}\underset{\underset{CH_3}{\|}}{CH}\text{—}CH_2\text{—}O\text{—}\underset{\underset{O}{\|\|}}{C}\text{—}H$	Raspberry
Octyl acetate	$CH_3\text{—}(CH_2)_7\text{—}O\text{—}\underset{\underset{O}{\|\|}}{C}\text{—}CH_3$	Orange

EXAMPLE 10-8   Name the following esters by their IUPAC names and their common names.

(a) $CH_3\text{—}\underset{\underset{O}{\|\|}}{C}\text{—}OCH_2CH_3$

(b) $CH_3CH_2CH_2\text{—}\underset{\underset{O}{\|\|}}{C}\text{—}OCH_2CH_3$

(c) $\overset{\overset{O}{\|\|}}{C}\text{—}OCH_3$ attached to a benzene ring

SOLUTION   (a) $CH_3\text{—}\underset{\underset{O}{\|\|}}{C}\text{—}OCH_2CH_3$ The IUPAC name of this compound is ethylethanoate. The common name of this compound is ethyl acetate.

(b) 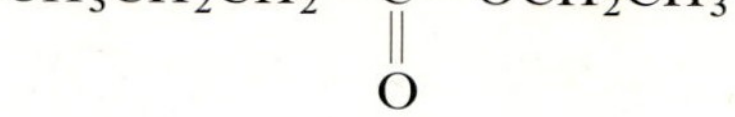 $CH_3CH_2CH_2\!-\!\overset{\displaystyle O}{\underset{\displaystyle \parallel}{C}}\!-\!OCH_2CH_3$ The IUPAC name of this compound is ethylbutanoate. The common name of this compound is ethyl butyrate. It has the smell of pineapple.

(c) 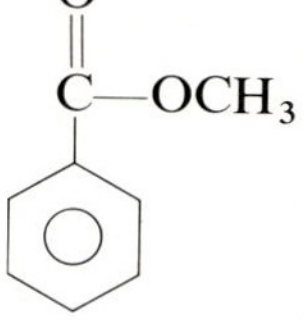 The IUPAC name of this compound is methylbenzoate; this is also its common name.

Some Important Esters and Their Uses

Methyl Salicylate,

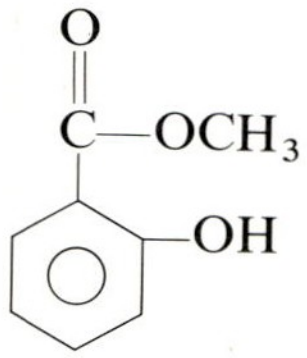

common known as *oil of wintergreen*, is the methyl ester of salicyclic acid,

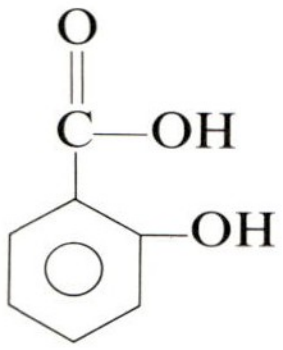

It has a pleasant spearmintlike odor, and for this reason it is used (in very low concentrations) in candies and perfumes. Because it also creates a mild burning sensation on your skin, which makes sore muscles feel good, methyl salicylate is found in many liniments.

Acetylsalicylic acid,

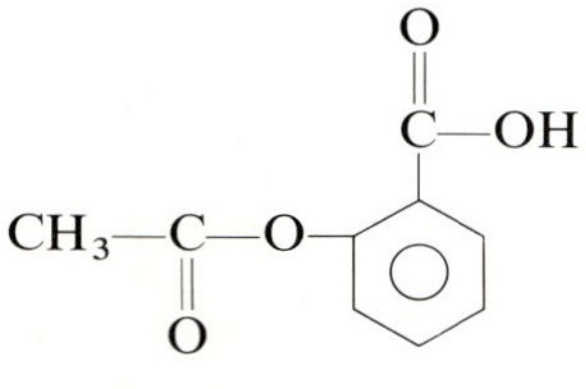

more commonly known as *aspirin*, is probably the best known and safest pain reliever (*analgesic*) and fever reducer (*antipyretic*). If you look closely at the structure of this compound, you'll see that it is the ester formed from acetic acid and salicylic acid.

NAME	STRUCTURE	USE
m-Cresyl acetate		Topical antiseptic and fungicide
Methylparaben		Used as a preservative for certain drugs
Nitroglycerin	H_2C-ONO_2 $HC-ONO_2$ H_2C-ONO_2	Nitro ester, used as a vasodilator for certain heart conditions

Some additional esters and their uses can be found in Table 10-9.

$$Amines: The \quad —N— \quad Compounds$$

Amines: The —N— Compounds

There are three general formulas for amines:

1. $R—NH_2$ (primary amine).

2. $R—NH$ (secondary amine).
      |
      R'

3. $R—N—R''$ (tertiary amine).
      |
      R'

Amines are related to the compound ammonia, NH_3, except that the hydrogen atoms in the ammonia molecule have been replaced by R groups. If one R group has replaced a hydrogen atom, the compound is a primary amine. If two R groups have replaced two hydrogen atoms, the compound is a secondary amine. And if three R groups have replaced three hydrogen atoms, the compound is a tertiary amine.

EXAMPLE 10-9 Determine whether each of the following compounds is a primary, secondary, or tertiary amine.

(a) $CH_3CH_2-N-CH_3$
$\quad\quad\quad\quad\quad\;|$
$\quad\quad\quad\quad\;\;CH_3$

(b) $CH_3CH_2CH_2-NH$
$\quad\quad\quad\quad\quad\quad\;\;|$
$\quad\quad\quad\quad\quad\;\;CH_2CH_3$

(c) $CH_3CH_2NH_2$

(d) $\quad\;\;NH_2$

SOLUTION

(a) $CH_3CH_2-N-CH_3$ This compound has three R groups attached
$\quad\quad\quad\quad\;\;|$ to the nitrogen atom, so it is a tertiary amine.
$\quad\quad\quad\;CH_3$

(b) $CH_3CH_2CH_2-NH$ This compound has two R groups,
$\quad\quad\quad\quad\quad\;|$ attached to the nitrogen atom, so it is
$\quad\quad\quad\quad CH_2CH_3$ a secondary amine.

(c) $CH_3CH_2NH_2$ This compound has one R group attached to the nitrogen atom, so it is a primary amine.

(d) $\quad NH_2$ This compound has one R group attached to the nitrogen atom, so it is a primary amine.

Amines may also be named in two ways:

1. By their common names, which involves using the alkane name plus the ending amine. For example, CH_3NH_2 is called methylamine. If two or three R groups are attached to the nitrogen, then each group is named in alphabetical order. For example, CH_3-NH is called di-

$\quad\quad\quad\quad\quad\quad\quad\quad\quad\quad\quad\quad\;\;CH_3$
methylamine, and CH_3CH_2-NH is called ethylmethylamine.
$\quad\quad\quad\quad\quad\quad\quad\quad\quad\quad\quad\quad\;\;|$
$\quad\quad\quad\quad\quad\quad\quad\quad\quad\quad\quad\quad CH_3$

2. By their IUPAC names, which involves, for primary amines, naming the —NH_2 group as an amino group along the alkane chain, and numbering the carbon atoms to locate its position. For example, CH_3—CH—CH_3 is called 2-aminopropane. For secondary and tertiary

$\qquad$ NH_2

amines the IUPAC rules become more complex and we will not deal with them here. Besides, most people use the common naming system for amines.

·EXAMPLE 10-10   Name the following amines by their common names, or, if possible, by their IUPAC names.

(a) CH_3CH_2—NH_2

(b) $CH_3CH_2CH_2$—N—CH_3
$\qquad$ CH_3

(c) $CH_3CH_2CH_2CH_2CH_2$—CH—CH_2CH_3
$\qquad$ NH_2

SOLUTION

(a) CH_3CH_2—NH_2 The IUPAC name of this compound is aminoethane. The common name of this compound is ethylamine.

(b) $CH_3CH_2CH_2$—N—CH_3 The common name of this compound
$\qquad$ CH_3 is dimethylpropylamine.

(c) $\overset{8}{C}H_3\overset{7}{C}H_2\overset{6}{C}H_2\overset{5}{C}H_2\overset{4}{C}H_2$—$\overset{3}{C}H$—$\overset{2}{C}H_2\overset{1}{C}H_3$ The IUPAC name of
$\qquad$ NH_2 this compound is 3-aminooctane.

Amines in general have very strong ammonialike odors. Many of them act as organic bases.

$$RNH_2 + H_2O \;\rightleftharpoons\; RNH_3^+ + OH^-$$

Amines are produced when living organisms die and decay. And amines are found in our bodies as amino acids. (Amino acids are compounds that have both an amine group and a carboxylic acid group. For example, H_2N—CH—C—OH is the amino acid alanine. We'll have more to say about

$\qquad$ CH_3 $\;O$

amino acids when we study proteins in Chapter 13.

Many important pharmaceutical compounds contain the amino group. One very important amine, acetylcholine, is involved in transmitting nerve impulses in the body. Let's take a look at some important amines.

Some Important Amines and Their Uses

1,5-Diaminopentane, $H_2N—CH_2CH_2CH_2CH_2CH_2—NH_2$, also called 1,5-pentanediamine, but better known as *cadaverine*, is a compound produced by decaying organisms. Such amines are called *ptomaines* and, in fact, can also be formed by the action of bacteria on meat and fish (thus the name ptomaine poisoning). Cadaverine has a very pungent odor; it also acts as a strong base.

4-Aminobenzoic acids,

also known as *para-aminobenzoic acid*, or *PABA* for short, is used today in suntan lotions as a sunscreen (in other words, it prevents ultraviolet light from reaching the skin). PABA has also been used as an antirickettsial agent (rickettsia are bacterialike organisms that cause diseases such as Rocky Mountain spotted fever). PABA is also found in foods as part of the B complex vitamins. For example, baker's yeast contains 5 to 6 parts per million PABA, and brewer's yeast contains from 10 to 100 parts per million.

1-Phenyl-2-aminopropane,

also called (phenylisopropyl)amine, but better known as *amphetamine* (or *benzedrine*), acts as a central nervous system stimulant. This compound and others related to it are called "uppers," because of their ability to keep you active and awake. Unfortunately, they can cause drug dependence, which means that after you've been taking them for a while, you need to continue to take them to stay alert. Amphetamine has also been used in nasal inhalers to relieve congestion in the nose.

Some additional amines and their uses can be found in Table 10-10.

TABLE 10-10   Some amines and their uses

NAME	STRUCTURE	USE
Ephedrine	(phenyl)—CH—CH—NH—CH$_3$ with OH and CH$_3$	Used as a vasoconstrictor, it also opens stuffed nasal passages
Methadone	CH$_3$CH$_2$—C(=O)—C—CH$_2$—CH—N—CH$_3$ with CH$_3$ CH$_3$ and two phenyl groups	Substitute for heroin in the treatment of heroin addiction
Piperazine	piperazine ring (N—H)	Kills pinworms and roundworms in many types of domestic animals

Amides: The R—C(=O)—N— *Compounds*

There are three general formulas for amides:

1. R—C(=O)—NH$_2$

2. R—C(=O)—NH—R′

3. R—C(=O)—N(R′)—R″

The first example represents the formula for simple amides. The only R group is off the carbon atom. This type of amide can be thought of as being analogous to a carboxylic acid with the OH group replaced by an NH$_2$ group.

R—C(=O)—OH   R—C(=O)—NH$_2$

Acid   Amide

The second and third types of amides are more complex, in that they have one or two R groups off the nitrogen atom.

Amides can be formed by the reaction of a carboxylic acid with an amine (or ammonia). The general formula for such a reaction is

$$R-\underset{\underset{O}{\|}}{C}-OH + H-\underset{\underset{R'}{|}}{N}-R'' \longrightarrow R-\underset{\underset{O}{\|}}{C}-\underset{\underset{R'}{|}}{N}-R'' + H_2O$$

An understanding of the amide group is very important in the study of proteins. Proteins are composed of amino acids, connected by an amide linkage (called a *peptide bond*). For example,

$$H_2N-CH_2-\underset{\underset{O}{\|}}{C}-OH + H_2N-\underset{\underset{CH_3}{|}}{C}-\underset{\underset{O}{\|}}{C}-OH \longrightarrow H_2N-CH_2-\underset{\underset{O}{\|}}{C}-\underset{\underset{H}{|}}{N}-\underset{\underset{CH_3}{|}}{C}-\underset{\underset{O}{\|}}{C}-OH + H_2O$$

Glycine   Alanine   Glycylalanine

Peptide bonds hold the amino acids together to form the long protein chains.

Simple amides can be named in two ways:

1. By their common names, which involves replacing the *-ic* ending of the corresponding acid with the ending *-amide*. For example, $CH_3-\underset{\underset{O}{\|}}{C}-NH_2$ is called acetamide.

2. By their IUPAC names, which involves replacing the *-oic* ending of the corresponding acid with the ending *-amide*. For example, $CH_3-\underset{\underset{O}{}}{C}-NH_2$ is called ethanamide.

The more complex amides may also be named by their common or IUPAC names. However, the groups substituted off the nitrogen atom must be designated by the capital letter N. For example, $CH_3CH_2-\underset{\underset{O}{\|}}{C}-\underset{\underset{CH_3}{|}}{N}-CH_3$ is called N,N-dimethylpropanamide (IUPAC name), or N,N-dimethylpropionamide (common name).

EXAMPLE 10-11 Name the following amides by their IUPAC names and common names.

(a) $CH_3CH_2CH_2-\underset{\underset{O}{\|}}{C}-NH_2$

(b) CH_3-C-NH
$\quad\quad\overset{\|}{O}\ \ \overset{|}{CH_3}$

(c) $CH_3-C-N-CH_3$
$\quad\quad\overset{\|}{O}\ \ \overset{|}{CH_3}$

SOLUTION   (a) $CH_3CH_2CH_2-C-NH_2$ The IUPAC name of this compound is butanamide. The common name of this compound is butyramide.
$\quad\quad\quad\quad\quad\quad\quad\quad\quad\overset{\|}{O}$

(b) CH_3-C-NH The IUPAC name of this compound is *N*-methylethanamide. The common name of this compound is *N*-methylacetamide.
$\quad\quad\overset{\|}{O}\ \ \overset{|}{CH_3}$

(c) $CH_3-C-N-CH_3$ The IUPAC name of this compound is *N,N*-dimethylethanamide. The common name of this compound is *N,N*-dimethylacetamide.
$\quad\quad\overset{\|}{O}\ \ \overset{|}{CH_3}$

Some Important Amides and Their Uses

Carbamide, $H_2N-\overset{\overset{\textstyle O}{\|}}{C}-NH_2$, better known as *urea*, is not a simple amide, but a diamide. Urea is the major endproduct of nitrogen metabolism in the body.

N-Phenylethanamide,

$$CH_3-C-NH$$

also called *N*-phenylacetamide, but better known as *acetanilide*, is a pain reliever and fever reducer. However, it has some serious side effects.

Para-ethoxy-N-phenylacetamide,

$$CH_3-C-NH$$
$$OCH_2CH_3$$

also called *para*-ethoxyacetanilide, but better known as *phenacetin*, is also a pain reliever and fever reducer. However, this compound seems to have fewer side effects than acetanilide, and is used as an aspirin substitute.

N,N-Diethyllysergamide,

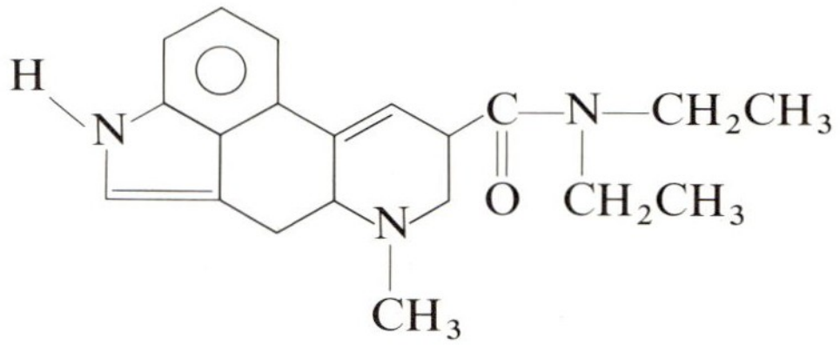

better known as *lysergic acid diethylamide* (LSD, for short), has been used in the study and treatment of mental disorders. LSD gained notoriety in the 1960's and 1970's as a powerful hallucinogen. It appears that this drug can block the action of a chemical in the brain called serotonin, which controls nerve impulses in the brain. When the function of the serotonin is blocked, nerve impulses in the brain become uncontrollable. This results in hallucinations and certain types of abnormal behavior.

Some additional amides and their uses can be found in Table 10-11.

TABLE 10-11   Some amides and their uses

NAME	STRUCTURE	USE
Acetominophen	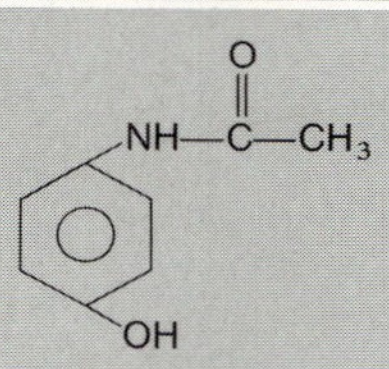	Aspirin substitute
Lidocaine	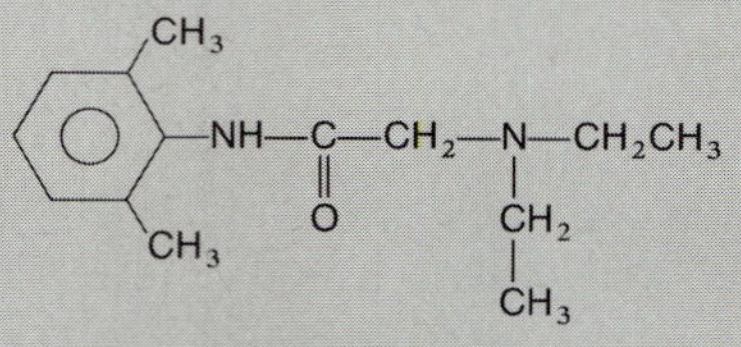	Local anesthetic
Oxanamide	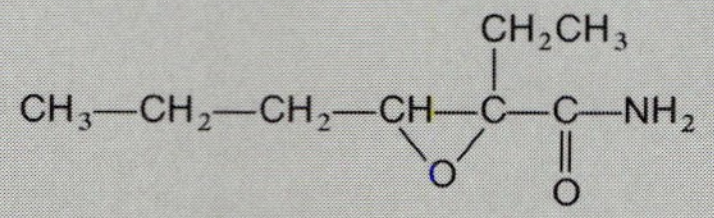	Tranquilizer
Xylocaine	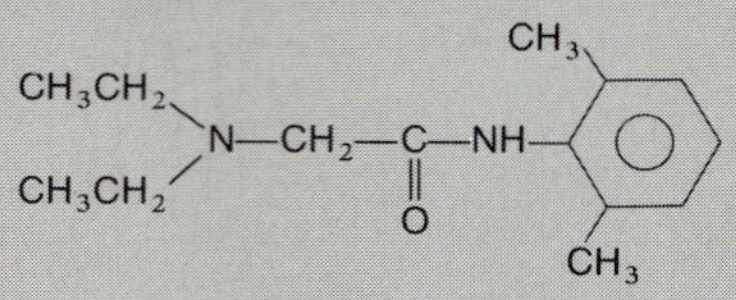	Local anesthetic

SUMMARY

In this chapter we looked at eight major functional groups of organic compounds. We studied a little about the chemistry of each group of compounds. We also learned how to name each group of compounds by the IUPAC system and common naming system. Finally, we surveyed some important members of each class of compounds and learned about their medical uses.

EXERCISES

1. Write the general formulas for
   (a) an alcohol
   (b) an ether
   (c) an aldehyde
   (d) a ketone
   (e) a carboxylic acid
   (f) an ester
   (g) an amine
   (h) an amide

2. Name the following alcohols; use either the IUPAC or common name.
   (a) $CH_3CH_2CH_2CH_2OH$
   (b) $CH_3CH_2CHCH_2CH_2CH_3$
                   |
                  OH

   (c) (cycloheptane with OH)

   (d) CH_3CHCH_2OH
            |
           CH_3

   (e) CH_3CH_2 (cyclohexane with OH)

3. State whether the alcohols in Exercise 2 are primary, secondary, or tertiary.

4. Name the following ethers; use either the IUPAC or common name.
   (a) CH_3OCH_3
   (b) $CH_3CH_2CH_2CH_2CH_2OCH_2CH_3$

   (c) CH_3O (benzene ring)

   (d) 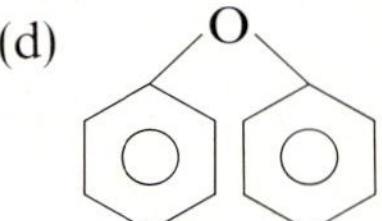 (diphenyl ether)

5. Write an equation for the synthesis of the following symmetrical ethers.
   (a) $CH_3CH_2OCH_2CH_3$
   (b) $CH_3CH_2CH_2CH_2OCH_2CH_2CH_2CH_3$

6. Write the general formula for an acetal and hemiacetal.

7. Name the following aldehydes; use either the IUPAC or common name.

(a) $CH_3CH_2CH_2CH_2CH_2CH$
       $\overset{\|}{O}$

(b)   benzaldehyde structure: phenyl ring with $\overset{\|}{O}$ — CH

(c) $CH_3CH_2CHCH_2CH$
       $\underset{CH_3}{|}$ $\overset{\|}{O}$

(d) $CH_3CH_2CH_2CH_2CH_2CHCH_2CH$
                  $\underset{CH_3}{|}$ O

8. Write an equation for the
(a) reduction of CH_3CH
                    $\overset{\|}{O}$

(b) oxidation of CH_3CH
                    $\overset{\|}{O}$

9. Name the following ketones; use either the IUPAC or common name.

(a) $CH_3CH_2—C—CH_2CH_3$
            $\overset{\|}{O}$

(b)   cyclopentanone structure (ring with O)

(c) $CH_3CH_2—C—CH_2CH_2CH_2CH_2CH_3$
            $\overset{\|}{O}$

(d)   benzophenone structure: phenyl—C(=O)—phenyl
            $\overset{\|}{O}$

10. Write an equation for the reduction of

(a) $CH_3CH_2—C—CH_3$
            $\overset{\|}{O}$

(b) $CH_3CH_2—C—CH_2CH_3$
            $\overset{\|}{O}$

11. Name the following carboxylic acids; use either the IUPAC or common name.

(a) $CH_3CH_2CH_2CH_2CH_2—C—OH$
                  $\overset{\|}{O}$

(b)   benzoic acid derivative structure: phenyl ring with $\overset{\|}{O}$ — C—OH and CH_3 substituent

(c) $CH_3CH_2CHCH_2$—$\overset{\displaystyle O}{\underset{\displaystyle \|}{C}}$—OH   (d) H—$\overset{\displaystyle O}{\underset{\displaystyle \|}{C}}$—OH
    |
    CH_3

12. Write an equation for the synthesis of the following esters.
    (a) CH_3CH_2—C—OCH_2CH_3 (b) 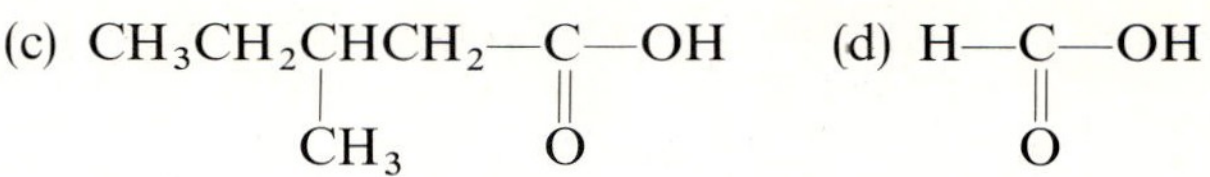

13. Name the following esters; use either the IUPAC or common name.
    (a) $CH_3CH_2CH_2CH_2$—C—OCH_3 (b)

    (c) $CH_3CH_2CH_2$—C—O—

14. Name the following amines; use either the IUPAC or common name.
    (a) $CH_3CH_2CH_2CHCH_2CH_3$
                  |
                 NH_2

    (b) CH_3CH_2—N—CH_2CH_3
              |
             CH_2CH_3

    (c) NH_2

    (d) $CH_3CH_2CH_2CH_2$—NH
                      |
                     $CH_2CH_2CH_3$

15. Write an equation for the synthesis of the following amides.
    (a) CH_3—C—NH_2 (b) CH_3CH_2—C—NH
              || || |
              O   O CH_3

16. Name the following amides; use either the IUPAC or common name.
    (a) $CH_3CH_2CH_2CH_2$—C—N—CH_2CH_3
                        || |
                        O CH_3

    (b) CH_3CH_2—C—NH
              || |
              O CH_2CH_3

(c) 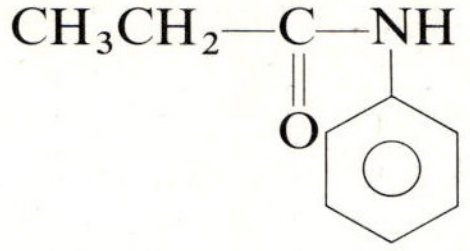 $CH_3CH_2-\overset{\underset{\|}{O}}{C}-NH$

17. State whether the amines in Exercise 14 are primary, secondary, or tertiary.

18. Write structural formulas for the following compounds.
    (a) 3-pentanol
    (b) 1-methylcyclohexanol
    (c) ethoxypentane
    (d) 3-heptanone
    (e) 2-methylpentanal
    (f) 4-ethylbenzoic acid
    (g) *N,N*-diethylpropanamide
    (h) 2-aminopentane
    (i) *n*-propylethanoate
    (j) trimethylamine

CHAPTER 11

The Carbohydrate Story

How Sweet It Is!

Some Things You Should Know
After Reading This Chapter

You should be able to:

1. Define and give some examples of carbohydrates.
2. Define monosaccharides, disaccharides, and polysaccharides and give an example of each.
3. Define aldose and ketose and give an example of each.
4. Write the products of disaccharide and polysaccharide hydrolysis.
5. Define and give an example of dehydration synthesis.
6. Define and give an example of a triose, tetrose, pentose, and hexose.
7. Define and give an example of a reducing sugar.
8. Define metabolism and give an example of this process.
9. Write the products of glucose metabolism.
10. Write the structural formulas of glucose, galactose, and fructose.
11. Explain the difference between the alpha and beta forms of a monosaccharide molecule.
12. Describe the Molisch and Benedict's tests for carbohydrates.
13. Explain and give examples of carbohydrate oxidation, fermentation, and hydrolysis.

Introduction

The foods we eat are composed of proteins, carbohydrates, lipids, minerals, vitamins, and water in varying amounts. Carbohydrates, fats, and proteins are the foods that supply us with energy to allow our bodies to function. For every gram of carbohydrate that we take in, 4 kcal of energy are produced for use by our bodies (Fig. 11-1). Sugars, starches, cellulose, glycogen, and heparin are all examples of carbohydrates (Table 11-1).

In plants, carbohydrates compose about three-fourths of the solid plant material. They actually make up the structures that support the plant, as well as the structures that store the plant's energy supply. The wood that supports

FIGURE 11-1
Every gram of carbohydrate that we take in gives us 4 kcal of energy for our bodies.

TABLE 11-1   Some foods that are high in carbohydrates

Chocolate cake	Oatmeal
Cookies	Raisins
Dates and figs	Rice
Rye flour	Sugar
White flour	Shredded wheat
Honey	Lima beans
Jellies	Bran cereal

a tree is made up of the carbohydrate molecule cellulose. A plant's energy is stored in the form of starch, which is found in the seeds, roots, and leaves of the plant.

The sugar found in our blood and tissues, the sugar found in plant sap, as well as the sugar found in fruit juices and honey are all examples of carbohydrate molecules.

As you can see, carbohydrates are pretty important compounds. So let's take a closer look at carbohydrates in this chapter, which we call The Carbohydrate Story.

What Are Carbohydrates?

Carbohydrate molecules are organic compounds which are generally composed of the elements carbon, hydrogen, and oxygen in the ratio $1:2:1$, or CH_2O. (There are exceptions to this definition, which will be discussed shortly.) The name *carbohydrate* means hydrated carbon or $C(H_2O)$. Carbohydrates are produced by plants and by certain microorganisms, both of which contain the compound *chlorophyll*. This compound is essential for the process of *photosynthesis* to take place. Photosynthesis involves the chemical reaction of carbon dioxide (from air) and water (from soil) in the presence of sunlight, to produce a simple sugar (a carbohydrate) and oxygen. The plant stores the energy taken from the sun in the form of an energy-rich carbohydrate molecule. The balanced chemical equation for the process of photosynthesis is

$$6CO_2 + 6H_2O \xrightarrow[\text{chlorophyll}]{\text{sunlight}} C_6H_{12}O_6 + 6O_2 \qquad \text{photosynthesis}$$

It is interesting to see how plants and animals rely on each other to provide an oxygen–carbon dioxide balance in nature. Plants can use the carbon dioxide from the air and water from the soil to synthesize carbohydrates. Animals can't form carbohydrates in this way, so they must rely on plants for their carbohydrate supply. But animals have the ability to utilize carbohydrates in their bodies in a process called metabolism. *Metabolism* is the process by which the food we eat is transformed in the body by a series of complex chemical reactions, catalyzed by enzymes, to provide material for cellular function and energy to power the body. Many of the chemical processes that take place in the body fall under the name "metabolism." In this overall reaction, a carbohydrate molecule reacts with oxygen to produce carbon dioxide, water, and energy. The carbon dioxide and water are returned to the environment and the energy is used to power the body. The balanced equation for this reaction is

$$C_6H_{12}O_6 + 6O_2 \longrightarrow 6CO_2 + 6H_2O + \text{energy} \qquad \text{animal metabolism}$$

You will notice that this is the reverse of the equation for photosynthesis, where carbohydrates are produced. So you see that a carbon dioxide–oxygen cycle exists in nature. Plants take in carbon dioxide from the air during photosynthesis and give off oxygen. Animals take in oxygen from the air and give off carbon dioxide during metabolism. This is a mutually beneficial cycle.

How Are Carbohydrates Classified?

Chemically speaking, carbohydrates are defined as *polyhydroxyaldehydes* or *polyhydroxyketones*. You remember from our discussion of organic chemistry that polyhydroxy means that several alcohol (—OH) groups are present. Aldehydes and ketones both contain the carbonyl group ($-\overset{\text{O}}{\underset{\|}{C}}-$). So carbohydrates are alcohols and are also aldehydes or ketones. Substances that form polyhydroxyaldehydes or polyhydroxyketones upon reaction with water (hydrolysis) are also classified as carbohydrates.

Generally speaking, we say that carbohydrates are composed of the elements carbon, hydrogen, and oxygen in the ratio 1:2:1. But using this generalization we find that there are exceptions to the rule. For example, rhamnose has the formula $C_6H_{12}O_5$, but is a carbohydrate. Other compounds have formulas that should classify them as carbohydrates, yet they are not. For example, acetic acid has the empirical formula CH_2O but is not a carbohydrate. So the chemical classification of polyhydroxyaldehydes and polyhydroxyketones or compounds that form these groups upon hydrolysis is more correct and is the definition that we will refer to in our discussions. The ending *-ose* signifies that a compound is a carbohydrate.

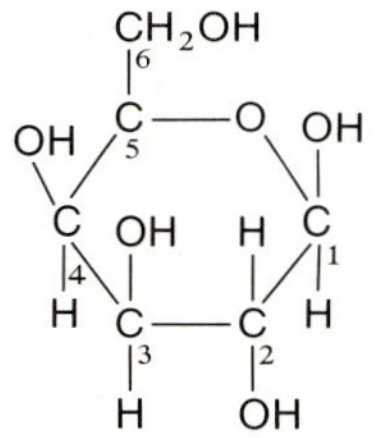

Carbohydrates are classified according to the size of the molecule. There are three main categories based on the number of sugar (saccharide) groups in the molecule. These are monosaccharides, disaccharides, and polysaccharides. *Monosaccharides*, which are also called *simple sugars*, contain only one saccharide group in the molecule (Fig. 11-2). *Disaccharides* contain two saccharide groups in the molecule (Fig. 11-3), and *polysaccharides* contain many saccharide groups in the molecule (Fig. 11-4).

FIGURE 11-2
Galactose: a monosaccharide molecule.

FIGURE 11-3
Lactose: a disaccharide molecule.

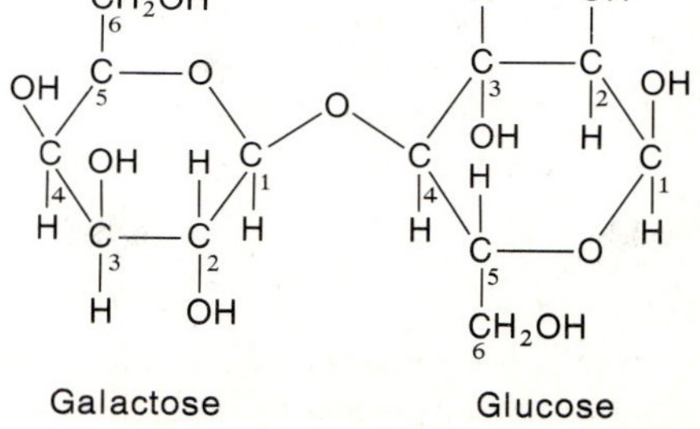

Galactose   Glucose

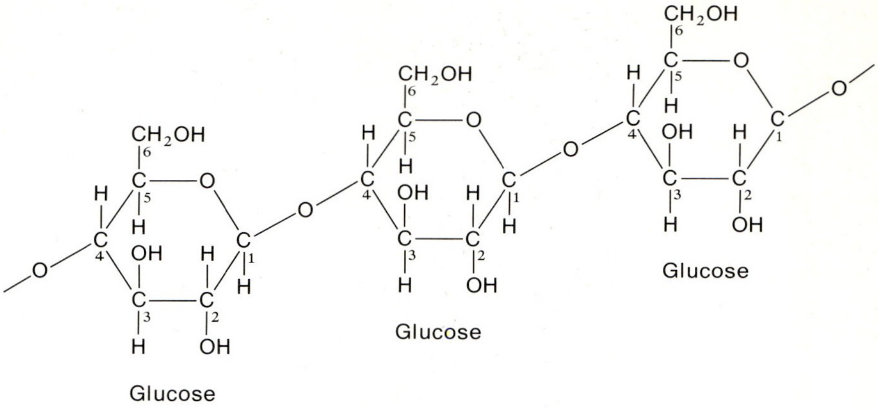

CH$_2$OH
Glucose
CH$_2$OH
Glucose
CH$_2$OH
Glucose

FIGURE 11-4
Cellulose: a polysaccharide molecule.

TABLE 11-2   Hydrolysis products

Disaccharides	$\xrightarrow{\text{hydrolysis (H}_2\text{O) (acid)}}$	2 monosaccharides
Polysaccharides	$\xrightarrow{\text{hydrolysis (H}_2\text{O) (acid)}}$	many monosaccharides

Monosaccharides are the simplest sugars that exist and can't be changed into simpler sugars upon reaction with water. However, disaccharides and polysaccharides can be changed into simpler sugars upon reaction with water. On hydrolysis (reaction with water), disaccharides, which are double sugars, form two simple sugars. On hydrolysis, polysaccharides, which are complex sugars, form many simple sugars. Table 11-2 shows the hydrolysis products of disaccharides and polysaccharides.

A Closer Look at Monosaccharides

The monosaccharide molecule has the general formula $(CH_2O)_x$, where x is 3, 4, 5, or 6. A monosaccharide molecule with three carbon atoms is called a *triose* molecule; four carbon atoms, *tetrose*; five carbon atoms, *pentose*; and six carbon atoms, *hexose* (Table 11-3).

A further, more specific classification system also exists. You remember that carbohydrates are polyhydroxyaldehydes or polyhydroxyketones. There is an —OH group bonded to each carbon atom—with the exception of one oxygen atom that is double-bonded to a carbon atom that forms the carbonyl group (—C—). If the oxygen happens to be bonded to the first carbon atom,

$$\underset{O}{\overset{\|}{C}}$$

TABLE 11-3   Molecular formulas of monosaccharides

NUMBER OF CARBON ATOMS	NAME	MOLECULAR FORMULA
3	Triose	$C_3H_6O_3$
4	Tetrose	$C_4H_8O_4$
5	Pentose	$C_5H_{10}O_5$
6	Hexose	$C_6H_{12}O_6$

then the monosaccharide would have an aldehyde group ($-\overset{\|}{\underset{O}{C}}-H$) and is called *aldose*. If the oxygen is double-bonded to the second carbon atom, then the monosaccharide would have a ketone group ($-\overset{\|}{\underset{O}{C}}-$) and is called *ketose*.

Aldoses and ketoses are further classified according to the number of carbon atoms in the molecule. For example, three-carbon-atom molecules are *trioses*, four are *tetroses*, and five are *pentoses*. To understand how these compounds are named in total, let's look at a monosaccharide molecule that has six carbon

TABLE 11-4   An aldohexose and a ketohexose

NUMBER OF CARBON ATOMS	MOLECULAR FORMULA	STRUCTURAL FORMULA (Fischer formula)	NAME
6	$C_6H_{12}O_6$	H—C=O (1); HO—C—H (2); H—C—OH (3); HO—C—H (4); HO—C—H (5); HO—C—H (6); H	Aldohexose
6	$C_6H_{12}O_6$	H; HO—C—H (1); C=O (2); H—C—OH (3); HO—C—OH (4); HO—C—H (5); HO—C—H (6); H	Ketohexose

atoms and an aldehyde group. This compound is an *aldohexose*. If we had a molecule composed of six carbon atoms and a ketone group, the compound would be a ketohexose (Table 11-4).

EXAMPLE 11-1 Determine whether each compound is an aldose or ketose. Then classify each compound according to the number of carbon atoms in the molecule.

(a)

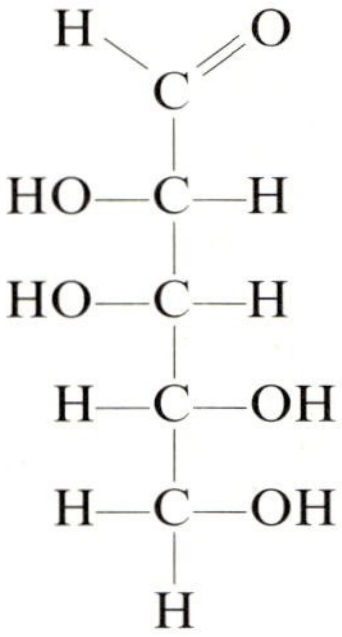

$$
\begin{array}{c}
\text{H}\diagdown \!\!\!\!\! _{\displaystyle C}\!\!\!\!\!\diagup\!\!^{\text{O}} \\
\text{HO—C—H} \\
\text{HO—C—H} \\
\text{H—C—OH} \\
\text{H—C—OH} \\
\text{H}
\end{array}
$$

Lyxose

(b)

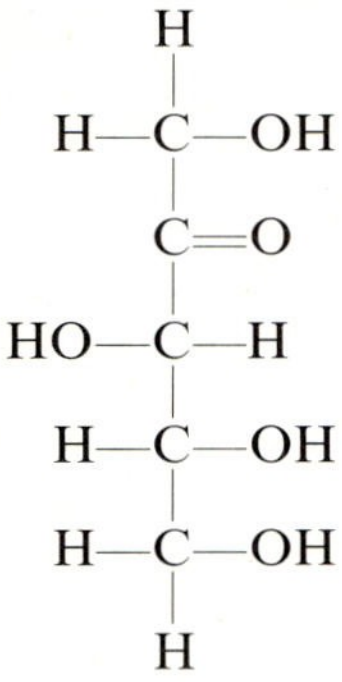

$$
\begin{array}{c}
\text{H} \\
\text{H—C—OH} \\
\text{C}=\text{O} \\
\text{HO—C—H} \\
\text{H—C—OH} \\
\text{H—C—OH} \\
\text{H}
\end{array}
$$

Xylulose

(c)

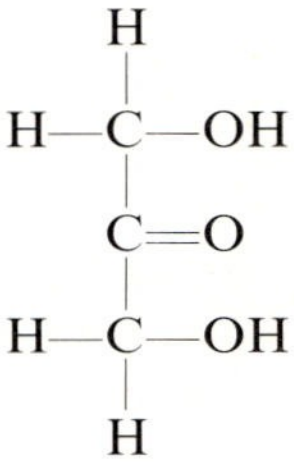

$$
\begin{array}{c}
\text{H} \\
\text{H—C—OH} \\
\text{C}=\text{O} \\
\text{H—C—OH} \\
\text{H}
\end{array}
$$

Dihydroxyacetone

(d)

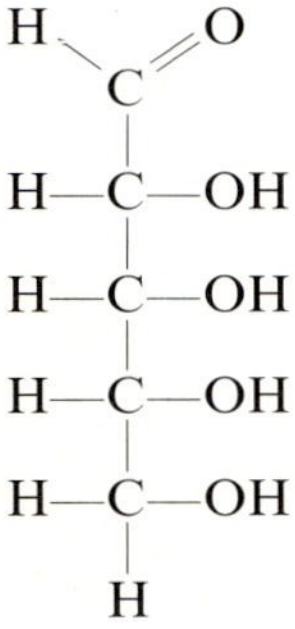

$$
\begin{array}{c}
\text{H}\diagdown \!\!\!\!\! _{\displaystyle C}\!\!\!\!\!\diagup\!\!^{\text{O}} \\
\text{H—C—OH} \\
\text{H—C—OH} \\
\text{H—C—OH} \\
\text{H—C—OH} \\
\text{H}
\end{array}
$$

Ribose

SOLUTION

(a)

$$
\begin{array}{c}
\text{H}\diagdown \!\!\!\!\! _{\displaystyle C}\!\!\!\!\!\diagup\!\!^{\text{O}} \\
\text{HO—C—H} \\
\text{HO—C—H} \\
\text{H—C—OH} \\
\text{H—C—OH} \\
\text{H}
\end{array}
$$

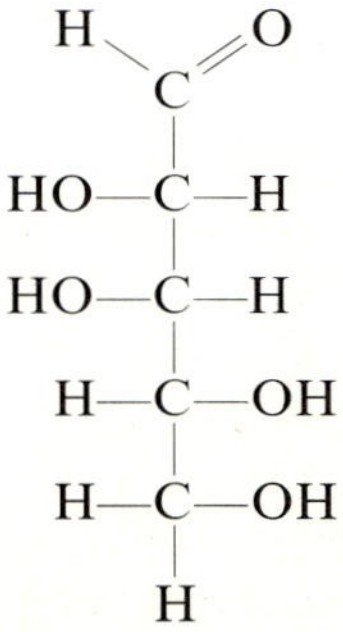

Lyxose

The carbonyl group is attached to the first carbon atom. The compound is therefore an aldose. The compound also has five carbon atoms; therefore, it is an aldopentose.

(b)

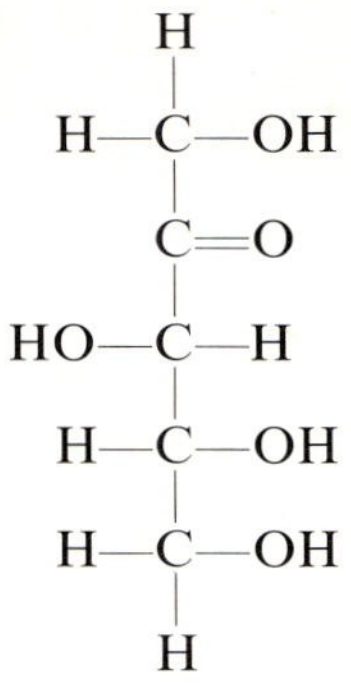

$$H-C-OH$$
$$C=O$$
$$HO-C-H$$
$$H-C-OH$$
$$H-C-OH$$
$$H$$

Xylulose

The carbonyl group is attached to the second carbon atom. The compound is therefore a ketose. The compound also has five carbon atoms; therefore, it is a ketopentose.

(c)

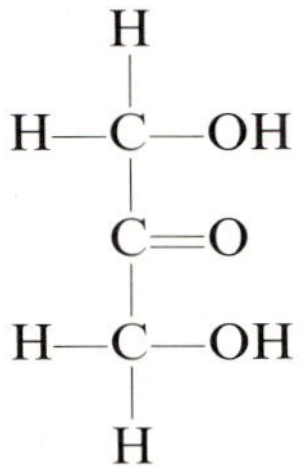

$$H-C-OH$$
$$C=O$$
$$H-C-OH$$
$$H$$

Dihydroxyacetone

The carbonyl group is attached to the second carbon atom. The compound is therefore a ketose. The compound also has three carbon atoms; therefore, it is a ketotriose.

(d)

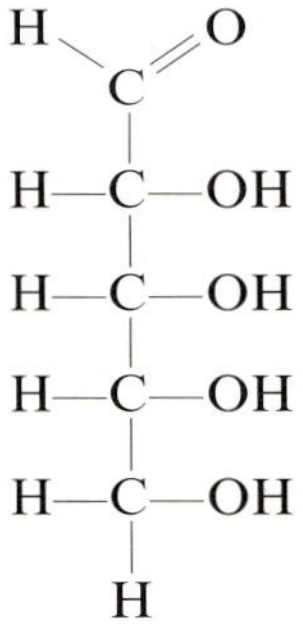

$$H\!-\!C\!=\!O$$
$$H-C-OH$$
$$H-C-OH$$
$$H-C-OH$$
$$H-C-OH$$
$$H$$

Ribose

The carbonyl group is attached to the first carbon atom. The compound is therefore an aldose. The compound also has five carbon atoms; therefore, it is an aldopentose.

FIGURE 11-5
Glyceraldehyde: an aldotriose.

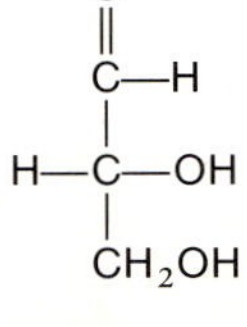

$$O$$
$$C-H$$
$$H-C-OH$$
$$CH_2OH$$

Let's look at some of the more important monosaccharides. *Glyceraldehyde*, a triose, is formed from the breakdown of hexose molecules in muscle tissue. Glyceraldehyde is an aldotriose (Fig. 11-5). The phosphate esters of the trioses glyceraldehyde and dihydroxyacetone are found in the body during glucose metabolism (Fig. 11-6).

Ribose and *deoxyribose* are important pentoses. These five carbon monosaccharides compose part of the genetic molecules ribonucleic acid and deoxyribonucleic acid (Fig. 11-7). Some other pentose molecules are *xylulose*, which is found in various fruits, and *lyxose*, which is found in the heart muscle.

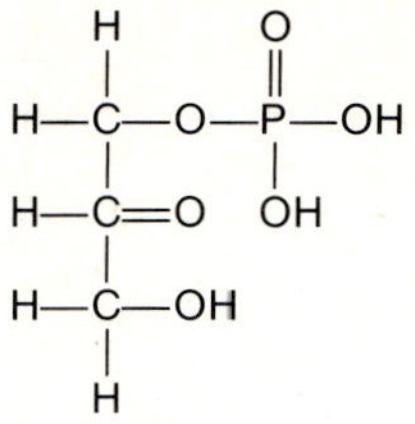

Dihydroxyacetone phosphate

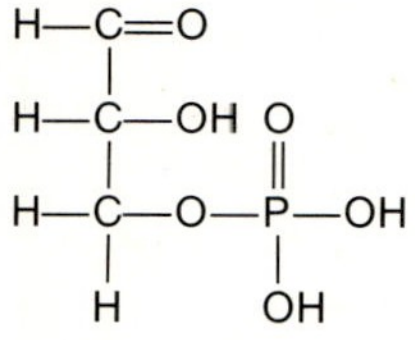

Glyceraldehyde 3-phosphate

FIGURE 11-6
The phosphate esters of glyceraldehyde and dihydroxyacetone are found in the body during glucose metabolism.

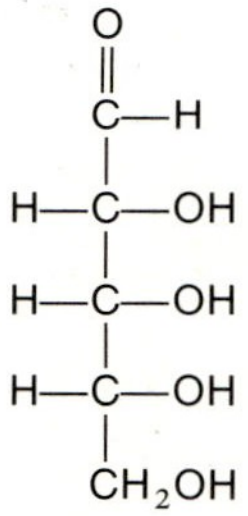

Ribose 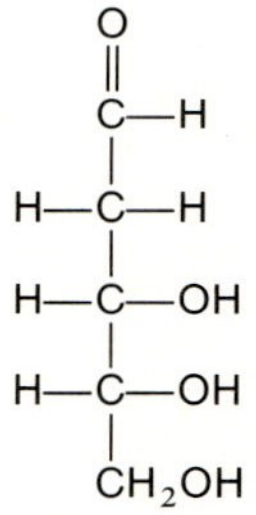   Deoxyribose (contains one
                less oxygen than ribose)

FIGURE 11-7
Ribose and deoxyribose are five carbon monosaccharides that compose part of the genetic molecules ribonucleic acid and deoxyribonucleic acid.

The most important monosaccharides are the hexoses, or six-carbon sugars. *Glucose, fructose,* and *galactose* are the three most important hexoses found in the body. They all have the molecular formula, $C_6H_{12}O_6$, but have different structural formulas and are therefore isomers (Fig. 11-8).

Glucose

In water solution, the *glucose* molecule is found in three forms. Only a small number of glucose molecules is found in the straight-chain form shown

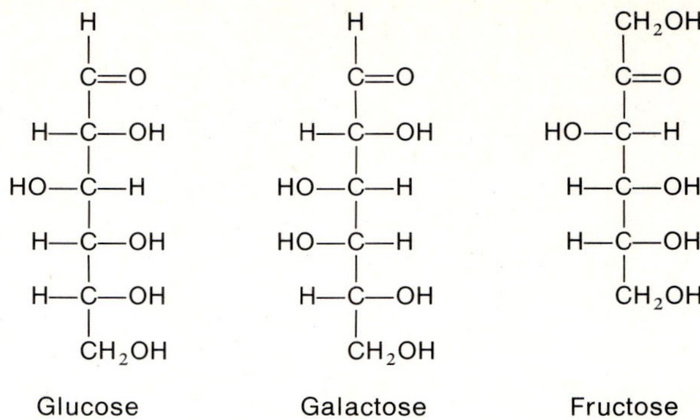

Glucose   Galactose   Fructose

FIGURE 11-8
Glucose, fructose, and galactose are the three most important hexoses found in the body. They all have the molecular formula $C_6H_{12}O_6$.

in Figure 11-8. In water solution, the three forms of glucose are easily converted from one form to another and are said to be in dynamic equilibrium. The three forms include the straight-chain compound and two ring compounds. These ring compounds are called *cyclic hemiacetals* and are formed from the addition of an alcohol group to an aldehyde group (Fig. 11-9). In glucose, the hemiacetal forms between the aldehyde group on carbon 1 and the alcohol group on carbon 5 (Fig. 11-10).

When a six-membered ring is formed by a sugar, it is called *pyranose*. When a five-membered ring is formed by a sugar, it is called *furanose*. The compounds under discussion are called *heterocyclic* compounds, meaning that they are cyclic structures composed of carbon plus an additional element in the ring. In this case one oxygen atom becomes a part of the ring. The pyran and furan rings shown in Figure 11-11 are examples of heterocyclic rings.

The ring can form in two ways. One is called alpha (α) and the other, beta (β). The difference between the alpha and beta forms of the hemiacetal is the orientation of the hydroxyl group on carbon 1 (for a *ketohexose* you look at carbon 2). This is most easily understood when drawing the structural formulas of the ring compounds. Two methods of drawing these heterocyclic

FIGURE 11-9
The addition of an alcohol group to an aldehyde group forms a hemiacetal. If R and R′ are connected, a cyclic hemiacetal is formed.

Aldehyde + Alcohol ⟶ Hemiacetal

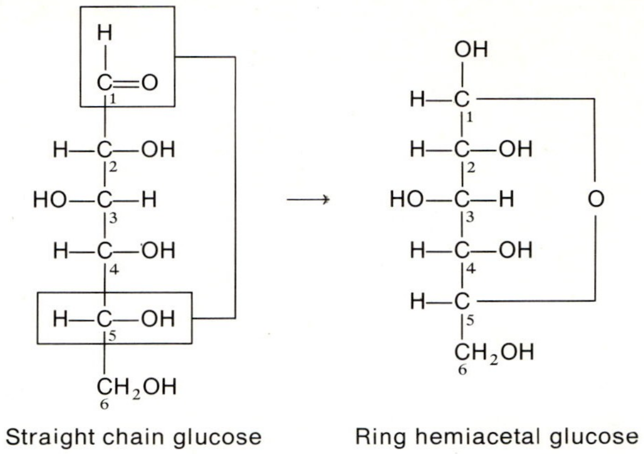

Straight chain glucose $\longrightarrow$ Ring hemiacetal glucose

FIGURE 11-10
In glucose the hemiacetal forms between the aldehyde group on carbon 1 and the alcohol group on carbon 5.

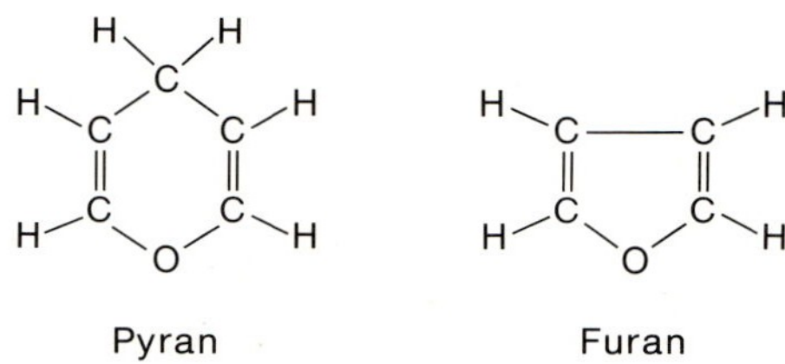

FIGURE 11-11
Pyran and furan are examples of heterocyclic rings.

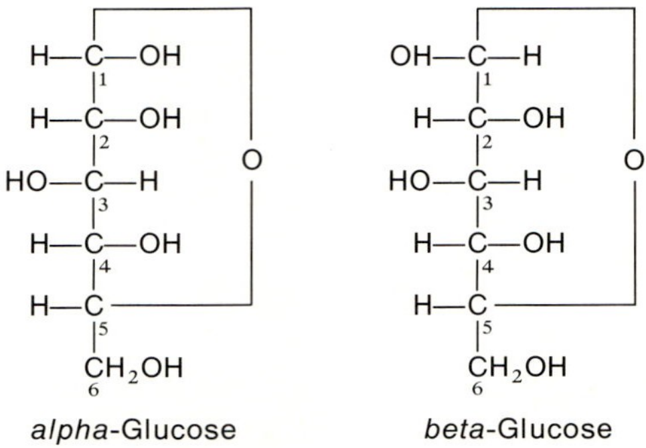

FIGURE 11-12
Fischer formulas for glucose. In the alpha form, the OH group on carbon 1 is written on the right side; in the beta form, the OH on carbon 1 is written on the left side.

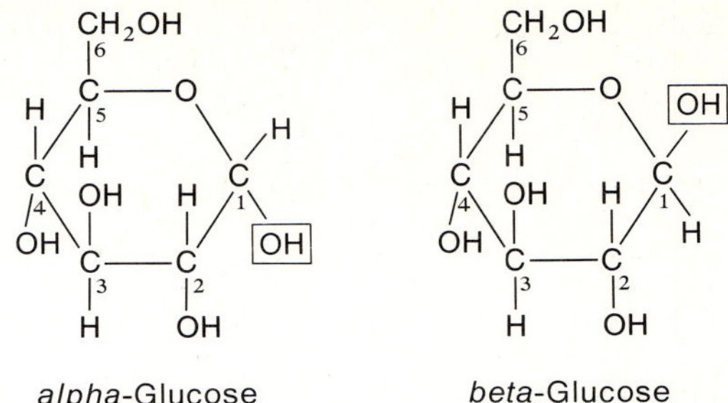

alpha-Glucose   *beta*-Glucose

FIGURE 11-13
Haworth formulas for glucose. In the alpha form, the OH group on carbon 1 is written below the plane of the ring; in the beta form, the OH group on carbon 1 is written above the plane of the ring.

compounds are used. These are the Fischer formula and the Haworth formula. In the Fischer formula the carbon atoms are written vertically and the hydroxyl group is written on the right side of carbon 1 to represent the alpha form. The beta form is written with the hydroxyl group on the left side of carbon 1 (Fig. 11-12). In the Haworth formula, the carbon atoms are drawn in a ring. In the alpha form the hydroxyl group of carbon 1 is written below the plane of the ring and in the beta form it is written above the plane of the ring (Fig. 11-13). About two-thirds of the glucose molecules found in water solution exist in the beta form, about one-third exist in the alpha form and less than 1% exist in the straight-chain form.

EXAMPLE 11-2 Decide whether the following structures are alpha or beta forms.

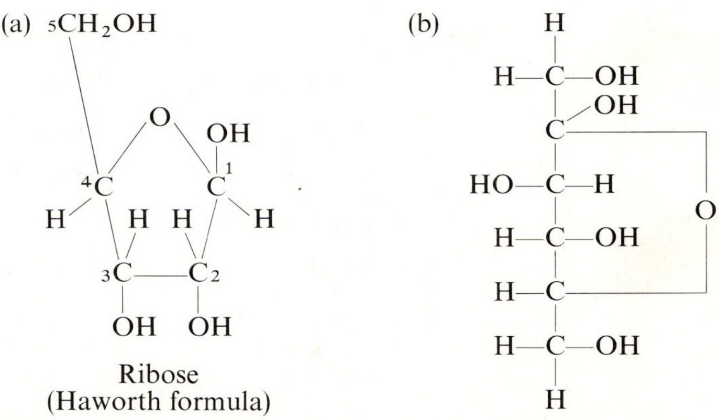

(a) Ribose
(Haworth formula)

(b) Fructose (a ketohexose)
(Fischer formula)

SOLUTION

(a) The hydroxyl group of carbon 1 is *above* the plane of the ring; therefore, it is the *beta* form of ribose.

(b) This is a ketohexose, so we look at carbon-2. The hydroxyl group of carbon 2 is written on the *right side*; therefore, it is the *alpha* form of fructose.

Glucose is found in the bloodstream and is also known as *dextrose* or *blood sugar*. It doesn't have to be digested to be used by the body, so it is given intravenously to people who can't take food orally. Blood glucose level can vary. Eight to 12 hours after a meal there is from 70 to 100 mg of glucose present for every 100 ml of blood. This is the *normal fasting level* of blood sugar. After a meal the level increases and can range from 100 to 160 mg per 100 ml of blood. The body normally deals with this increased blood sugar level by producing a hormone called insulin. This hormone facilitates the passage of simple sugars through cell membranes. Most cells, with the exception of nerve cells, kidney cells, and red blood cells, require insulin to bring sugar across the cell membrane and into the cell, where it serves as a source of energy. Insulin helps keep the blood glucose level normal by moving glucose out of the bloodstream and into the cells. In addition, insulin helps to speed up enzyme-catalyzed reactions in the liver, where glucose molecules are converted into larger glycogen molecules. Glycogen, then, is stored sugar found in the liver. A lack of insulin means that the blood sugar level can continue to increase above the level of 160 mg per 100 ml of blood. At this point, which is called the *renal threshold*, the kidneys cannot reabsorb the glucose into the bloodstream and excess glucose is excreted in the urine. The condition caused by a lack of the hormone insulin in the body is called *diabetes mellitus*. A diabetic must watch his or her sugar intake, take oral drugs to stimulate insulin production, or take daily insulin injections, depending upon the case and the treatment required. If untreated, the blood glucose level could increase until it becomes toxic and the person could fall into a coma or die.

When the blood sugar level is higher than normal, *hyperglycemia* exists. Diabetes mellitus is a hyperglycemic condition. There are other conditions characterized by severe hyperglycemia, where *glycosuria* (sugar in the urine) results. Kidney failure or administration of certain drugs could lead to the presence of sugar in the urine.

Hypoglycemia, or low blood sugar, is another condition involving blood glucose level. Overproduction or overinjection of insulin could result in *insulin shock* when the blood sugar level falls dangerously low. This is characterized by convulsions or a coma. Such a condition would be indicative of *severe hypoglycemia*. Eating a diet too rich in carbohydrates can cause the pancreas to produce too much insulin, in which case too much sugar would be removed from the blood. A person who suffers from this condition, called *mild hypoglycemia*, would be subject to dizziness, fainting, excessive tiredness, and irritability. This condition can usually be alleviated by eating a proper diet. Tumors in the cells that produce insulin can also result in overproduction of insulin and hypoglycemia.

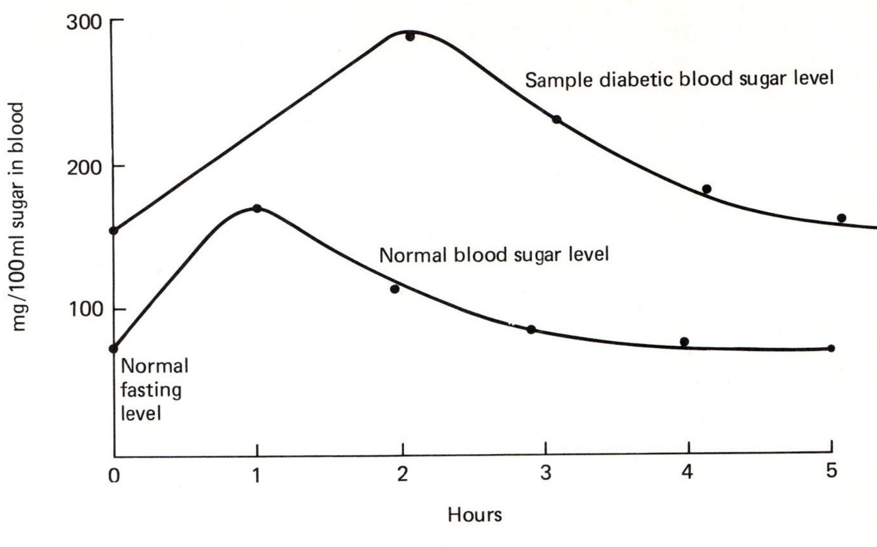

FIGURE 11-14
Glucose-tolerance test curves.

The *glucose tolerance test* is helpful in diagnosing abnormalities in blood sugar level. A person fasts for at least 8 hours but not more than 16 hours, and should have a minimum carbohydrate intake of 150 g per day for 3 days preceding the test. Caffeine and nicotine should be avoided before and during the test, as well as other drugs which modify glucose tolerance. The patient has his or her blood sugar level checked at the start of the test. Fifty to 100 g of glucose is then taken orally, and the blood sugar level is tested regularly for the next few hours to determine the amount of glucose present in the blood. Normal readings should show the blood sugar level at a maximum during the first hour and then it should begin to fall. A diabetic will show a higher-than-normal level at the start and throughout the test (Fig. 11-14 and Table 11-5).

TABLE 11-5   Oral glucose tolerance tests

Diagnostic criteria for diabetes mellitus
Wilkerson point load system: glucose load, 100 g

TIME	Mg/100 ml BLOOD	POINTS
Fasting	≥ 110	1
1 hour	≥ 170	$\frac{1}{2}$
2 hours	≥ 120	$\frac{1}{2}$
3 hours	≥ 110	1
Diabetes $\geq$ 2 points		

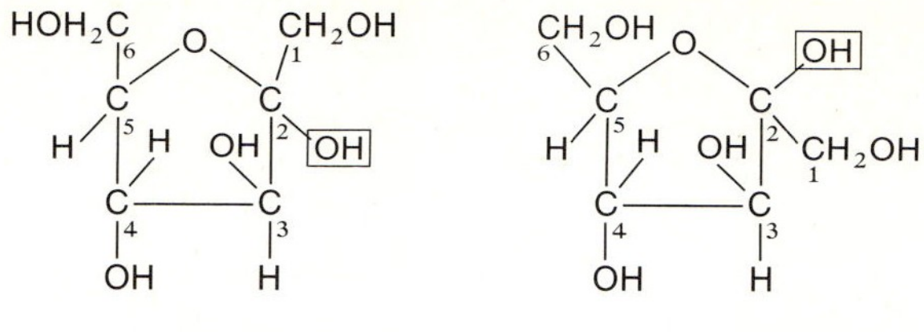

FIGURE 11-15
The cyclic structures of alpha and beta fructose.

Fructose

The sweetest sugar known is found in honey and in many fruits. It is called *frutose* and combines with a molecule of glucose to produce the disaccharide sucrose, or table sugar. Therefore, the hydrolysis of sucrose produces fructose. Fructose is also called *levulose* or *fruit sugar*. It is a ketohexose and has the molecular formula $C_6H_{12}O_6$. It exists as the straight-chain form seen in Figure 11-8 and in two ring forms, alpha and beta (Fig. 11-15).

An inability to metabolize fructose in the body due to the lack of the enzyme 1-phosphofructoaldolase exists in some people. This hereditary fructose intolerance is transmitted genetically. Persons who lack this enzyme may exhibit nausea, vomiting, excessive sweating, malaise, tremors, confusion, convulsions, or coma when fructose is ingested. Sometimes mental retardation or cirrhosis of the liver can develop with continued fructose ingestion. A series of blood tests following injection of a fructose solution is used to diagnose this condition. Elimination of fructose and sorbitol from the diet treats this abnormality.

Galactose

Galactose is another important monosaccharide. It isn't found free in nature but is formed on hydrolysis of certain disaccharides and polysaccharides. The galactose molecule is a component of such compounds as lactose (milk sugar), agar, and pectine. *Agar* is a polysaccharide prepared from seaweed. It is used to treat constipation since it absorbs water and therefore becomes bulky. It can't be digested, so it adds bulk to the feces. Agar is also used in growing bacteria. *Pectin* is used in the preparation of jellies. When added to fruit juices and sugar, pectin causes the thickening or setting of the jelly.

The galactose molecule has the formula $C_6H_{12}O_6$ and is an aldohexose. It exists in the straight-chain form and in two ring forms, alpha and beta galactose (Fig. 11-16).

Galactosemia is a disease that appears in some infants. It is the inability to metabolize galactose to form glucose in the body, caused by a lack of the

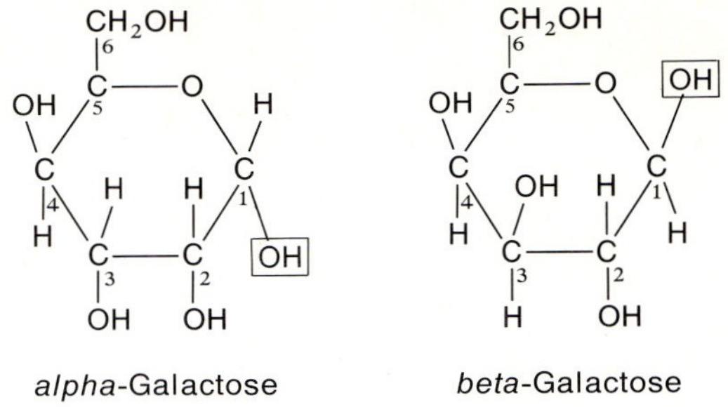

FIGURE 11-16

The two ring forms of alpha and beta galactose.

enzyme 1-P uridyl transferase. Although the infant seems to be normal at birth, there is some difficulty with feeding, followed by vomiting. It can be diagnosed by the presence of galactose in the urine and the absence of the enzyme in the red blood cells. Galactosemia is treated by elimination of milk and milk products from the diet. The infant can eat certain proteins, fruits, vegetables, sugar, vitamins, and minerals, but must eliminate all lactose- and galactose-containing foods.

Tests for Carbohydrates

The *Molisch test* is used to determine if a carbohydrate is present. The unknown solution is combined with an alcoholic solution of alpha naphthol, and cold concentrated sulfuric acid is added carefully down the side of the test tube. The formation of a red-violet ring at the point where both solutions meet indicates the presence of a carbohydrate.

The presence of glucose in urine can be detected by testing for a free aldehyde group. This is because less than 1% of glucose molecules in water exist in the straight-chain form and not as the hemiacetal ring. Small carbohydrate molecules such as glucose, which have a free aldehyde or ketone group in solution, can act as reducing agents. This is why these substances are called *reducing sugars.* Most monosaccharides are reducing sugars. This is true because most of them have an open chain with a free aldehyde or ketone group that can provide electrons during a reaction. Such carbohydrates will reduce alkaline solutions containing mild oxidizing agents.

The *Benedict's test* detects reducing sugars. An alkaline solution of copper(II) sulfate is mixed with the unknown solution. The copper(II) sulfate solution contains the Cu^{+2} ion and is blue. The mixture is heated. If a reducing sugar is present, the Cu^{+2} ions are reduced to Cu^{+1} ions. The aldehyde group is oxidized to a carboxylic acid group. Brick-red copper(I) oxide solid, Cu_2O, forms. The quantity of copper(I) oxide solid formed depends upon the quantity of reducing sugar present (Fig. 11-17).

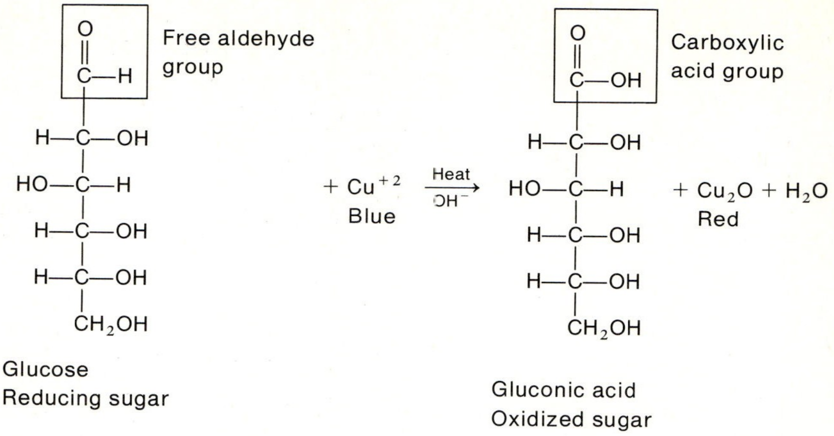

FIGURE 11-17

Benedict's test detects reducing sugars by reacting an alkaline solution of copper(II) sulfate with the suspected reducing sugar. If a reducing sugar is present, the blue copper(II) ions are reduced to copper(I), forming a brick-red copper(I) oxide.

Clinitest Reagent Strips are often used as a test for glucose in the urine since they are quite specific for glucose. A freshly voided urine sample is dipped into the test area of the strip. After 10 seconds the color of the test area is compared with a color chart and the test is read as positive or negative. No attempt is made to determine the amount of glucose present. Hydrogen peroxide and hypochlorite will give false positive readings and ascorbic acid will give false negative readings.

Clinitest Reagent Tablets are also available to test for reducing substances. Like the Benedict's test it is a nonspecific test for reducing substances. Five drops of urine and 10 drops of water are added to a test tube along with one Clinitest tablet. After the reaction stops the solution color can be compared with a color chart which approximates the amount of reducing agent present.

Disaccharides

Two monosaccharide molecules bonded together by an oxygen atom make up a disaccharide. The bond joining the two monosaccharides is called a *glycosidic linkage*. The reaction by which a disaccharide is formed from two monosaccharide molecules by means of the loss of a water molecule is called a *dehydration synthesis*. The three most important disaccharides have the molecular formula $C_{12}H_{22}O_{11}$. They are maltose, lactose, and sucrose. Maltose and lactose have a free aldehyde group and are therefore reducing sugars. Sucrose does not have a free aldehyde or ketone group and is not a reducing sugar. Figure 11-18 shows the Haworth forms of maltose, lactose, and sucrose.

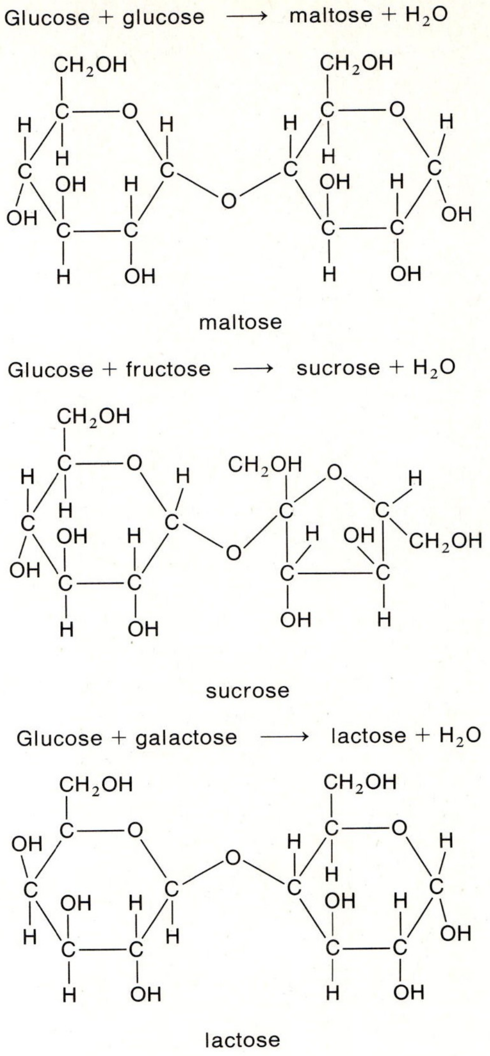

Glucose + glucose $\longrightarrow$ maltose + H_2O

maltose

Glucose + fructose $\longrightarrow$ sucrose + H_2O

sucrose

Glucose + galactose $\longrightarrow$ lactose + H_2O

lactose

FIGURE 11-18
The Haworth formulas of maltose, lactose, and sucrose.

Sucrose

Sucrose is found in sugar cane and sugar beets. It is also called *table sugar* and is used for sweetening foods at home. Sucrose is composed of a molecule of fructose and a molecule of alpha glucose which undergo loss of a water molecule (*dehydration synthesis*). A solution of sucrose treated properly with the enzyme invertase (from yeast) can be broken down to glucose and fructose and is called *invert sugar*. Honey is mostly invert sugar, where the invertase is supplied by the bees. Invert sugar is used by bakers and candy makers.

Lactose

Lactose is known as *milk sugar* and is produced by the mammary glands of nursing mothers. It is a component of baby formula, which aims to imitate mother's milk. Mammals produce lactose from the sugar which is found in the blood. Cow's milk contains 4 to 5% lactose. Human milk also contains about 5% lactose. Lactose is obtained commercially as a by-product of cheese manufacture. It is found in whey, which is the water solution that remains after the coagulation of the milk proteins in cheese manufacture. There are bacteria which can act upon lactose, causing it to become lactic acid. This reaction causes milk to become sour. Lactose is not very sweet. It is composed of a molecule of alpha glucose combined with a molecule of galactose.

Maltose

Maltose is known as *malt sugar*. It does not occur free in nature and is produced by hydrolysis of the polysaccharides starch, dextrin and glycogen. It can be produced from grain germinated under special conditions and this product, called malt, is used in the production of beer. Maltose is composed of two glucose molecules bonded together.

Polysaccharides

Polysaccharides are molecules composed of three or more monosaccharide groups bonded together by dehydration synthesis. They are actually polymers of monosaccharides. Upon complete hydrolysis of polysaccharides, several monosaccharide molecules are formed.

Polysaccharides are high-molecular-weight units. They are insoluble in water, and since the molecules are so large, they can't pass through a membrane. Two types of polysaccharides are possible; those formed from pentoses (called *pentosans*) and those formed from hexoses (called *hexosans* or *glucosans*).

Starch

Carbohydrates are important in our diets and starch is the most important carbohydrate that we eat. Plants store their energy in the form of starch. Such foods as vegetables, potatoes, wheat, corn, and rye contain large amounts of starch.

Two polymers, amylose and amylopectin, compose starch. Amylose is a straight-chain alpha glucose polymer and amylopectin is a branched polymer (Fig. 11-19).

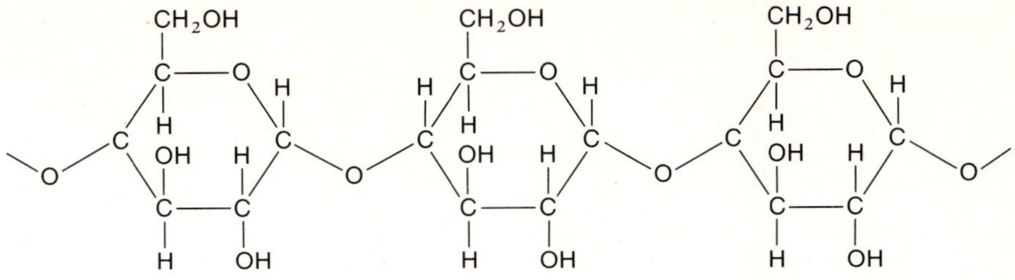

Amylose: a straight chain glucose polymer consisting of from
250 to 500 alpha-glucose molecules linked together

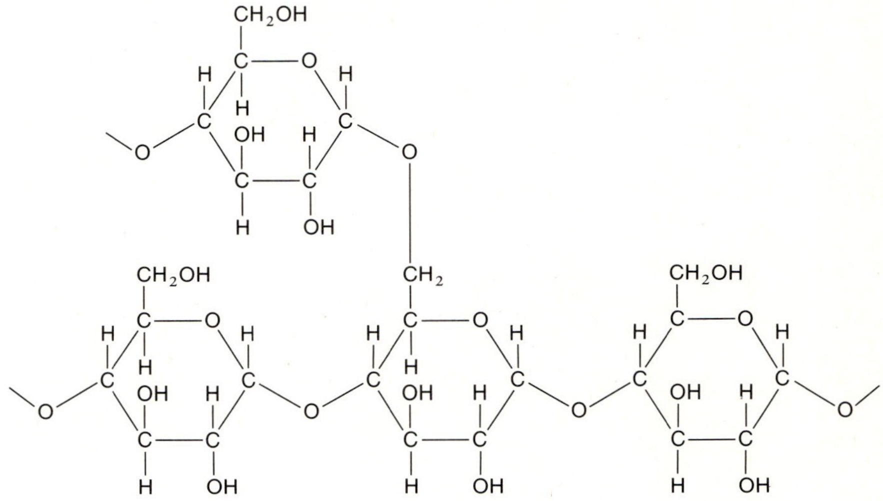

Amylopectin: a highly branched glucose polymer consisting of
about 1000 alpha-glucose molecules

FIGURE 11-19
Starch is a mixture of amylose and amylopectin.

On hydrolysis, starch first forms dextrins (which will soon be discussed) and is further broken down into maltose and then glucose. Dextrins can be used on postage stamps since they are sticky substances.

Cellulose

Cellulose is important in the diet since it can't be digested and therefore serves as bulk. It aids in the excretion of solid wastes, as it helps move foods through the intestinal tract.

Cellulose molecules are very large and consist of beta glucose units (Fig. 11-20). It is because cellulose consists of beta glucose, which can't be digested by the body, that cellulose is undigestible.

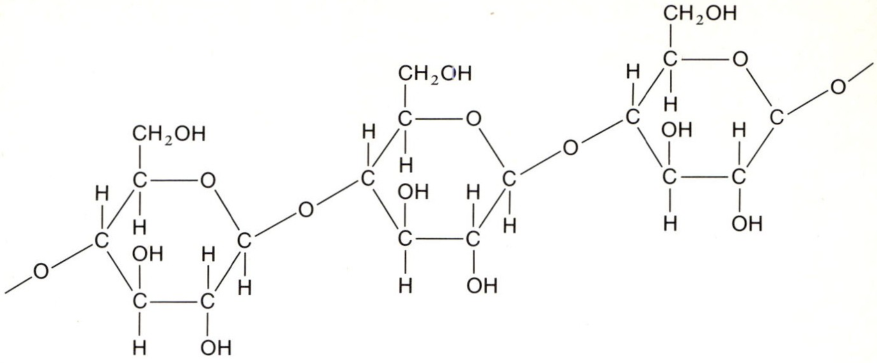

Cellulose consists of from 900 to 6000 glucose units linked together in a straight chain. This polymer consists of beta-glucose units.

FIGURE 11-20
Cellulose molecules are very large and consist of beta glucose units.

Cellulose is important in making fabrics. Cotton is composed almost totally of cellulose. Mercerized cotton is made by treating cotton with a quite concentrated sodium hydroxide solution and then expanding and drying the fibers. Rayon is made of cellulose treated with carbon disulfide and sodium hydroxide. Certain explosives, as well as movie film, are made of cellulose treated with nitric acid under the proper conditions. Carboxymethylcellulose is a thickener and is used in ice cream, salad dressing, and cosmetics.

Glycogen

Glycogen is stored sugar and is found in the liver and muscles. It is formed when glucose molecules bond to form glycogen in a process called *glycogenesis*. When the glucose molecules are needed again by the body, glycogen breaks down to form glucose in a process called *glycogenolysis* (Fig. 11-21).

Dextrin

The glue found on postage stamps is made of the polysaccharide called *dextrin*. Dextrin is formed during starch hydrolysis, before it breaks down to form maltose. When added to water, dextrins become sticky. During baking

FIGURE 11-21
The processes of glycogenesis and glucogenolysis.

$$\text{glucose} \xrightarrow{\text{glycogenesis}} \text{glycogen}$$

$$\text{glycogen} \xrightarrow{\text{glucogenolysis}} \text{glucose}$$

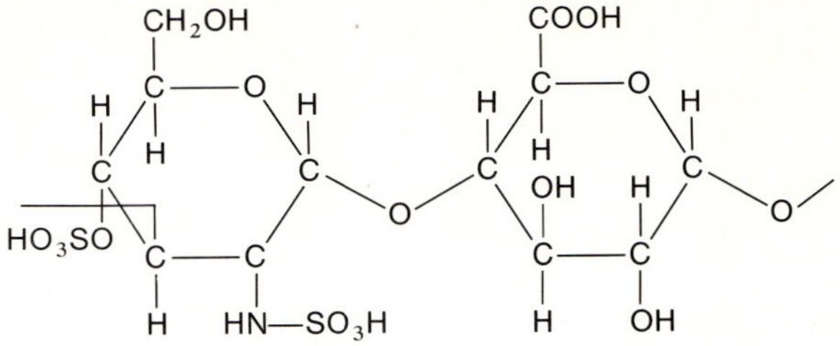

Heparin

FIGURE 11-22
Heparin is a polysaccharide that inhibits the formation of blood clots.

FIGURE 11-23
An onic acid, formed by the oxidation of the aldehyde group of an aldose.

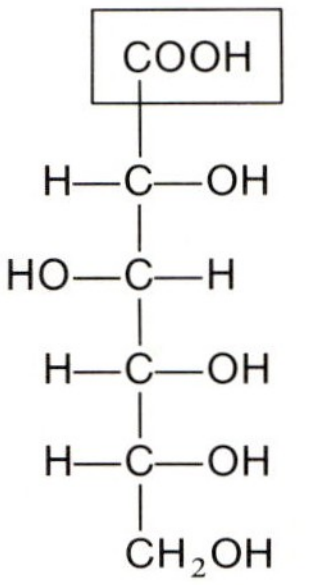

COOH

H—C—OH

HO—C—H

H—C—OH

H—C—OH

CH$_2$OH

Gluconic acid

FIGURE 11-24
A uronic acid, formed by the oxidation of the —OH group of an aldose.

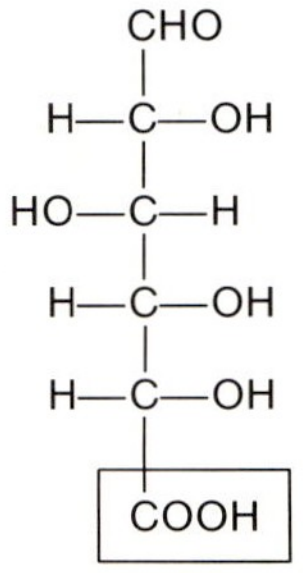

CHO

H—C—OH

HO—C—H

H—C—OH

H—C—OH

COOH

Glucuronic acid

of bread dextrins form and give bread its golden brown color. When bread is toasted, some of the starch turns to dextrins, changing the texture and taste of the bread.

Heparin

Heparin is an important polysaccharide since it serves as a blood anti-coagulant. It inhibits the formation of blood clots, since it interferes with the biochemistry of clot formation. This is important in patients who have undergone surgery and are in danger of forming clots in blood vessels that have been cut. Patients who have had heart attacks and strokes are often given this drug to prevent clotting. If a blood clot forms in a blood vessel, it may break loose and clog up a blood vessel leading to the heart or brain. This causes brain damage or death (Fig. 11-22).

Reactions of Carbohydrates

There are many types of reactions that carbohydrate molecules can take part in. The type of reaction depends upon which functional group is present.

Oxidation

Upon complete oxidation, carbohydrates form carbon dioxide, water, and energy. The equation for the oxidation of glucose is as follows:

$$C_6H_{12}O_6 + 6O_2 \longrightarrow 6CO_2 + 6H_2O + energy$$

Oxidation of Aldoses

Aldoses contain an aldehyde group and many hydroxyl groups. Oxidation of the aldehyde group forms an *onic acid* (Fig. 11-23). Oxidation of the alcohol group at the other end of the molecule forms a *uronic acid* (Fig. 11-24).

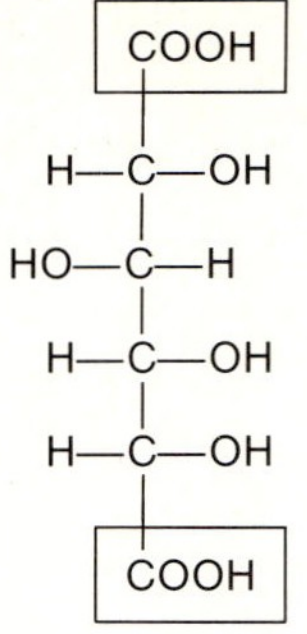

Saccharic acid

FIGURE 11-25
Saccharic acid, formed by the oxidation of both the aldehyde group and the —OH group of glucose.

FIGURE 11-26
Reduction of an aldohexose to an alcohol.

Oxidation of both the aldehyde and hydroxyl groups at the same time forms a *saccharic acid* (Fig. 11-25).

Reduction of Aldohexoses

Aldohexoses can be reduced to alcohols. Sorbitol is formed from the reduction of glucose. Sorbitol and mannitol are formed from the reduction of fructose (Fig. 11-26).

Reducing Agents: Reducing Sugars

Those carbohydrate molecules with free aldehyde or ketone groups are able to reduce alkaline metallic hydroxides. This is the basis of Benedict's test for sugar. The reaction is as follows:

$$2Cu(OH)_2 + \text{a reducing sugar} \longrightarrow Cu_2O + 2H_2O + \text{an oxidized sugar}$$

Figure 11-17 shows the oxidation of glucose to gluconic acid.

Fermentation

Certain microorganisms secrete enzymes. When the proper enzymes are present, they can act upon carbohydrates to decompose them. When simple sugars are treated with yeast, alcohol and carbon dioxide are produced. This is called *alcoholic fermentation.* The reaction is as follows:

$$C_6H_{12}O_6 \xrightarrow[\text{enzymes}]{\text{yeast}} 2C_2H_5OH + 2CO_2$$

Glucose   Ethyl alcohol

Alcoholic beverages such as beer and wine are made by the fermentation process. Crushed grapes contain sugars and they are acted upon by certain enzymes to produce wine. In the presence of certain enzymes, glucose is fermented to produce lactic acid, which makes cheese.

Bacteria that live in the alimentary canal secrete certain enzymes. When a diet rich in carbohydrates is eaten, these enzymes can act on the carbohydrates to produce gas. This often takes place in the intestines and causes pain and discomfort.

Hydrolysis

Polysaccharide and disaccharide molecules can break down to produce monosaccharides in the presence of certain enzymes, or when boiled with dilute acids. The reactions are as follows:

$$\left. \begin{array}{l} \text{Polysaccharides} \\ \text{Disaccharides} \end{array} \right\} + H_2O \xrightarrow[\text{dilute acids}]{\text{enzymes}} \text{monosaccharides}$$

SUMMARY

Carbohydrates are foods that provide our bodies with energy. Foods containing sugars, starches, and cellulose are carbohydrates. Breads, cereals, vegetables, spaghetti, cakes, and candies are largely carbohydrate in composition.

Plants, too, contain a great percentage of carbohydrate molecules. The plant support structure is composed of cellulose and the plant's food is stored as starch in the leaves, seeds, and root of the plant.

Carbohydrates are produced by plants during the process of photosynthesis. Carbon dioxide from the air and water from the soil react to form a simple carbohydrate molecule, and water, in the presence of sunlight and the compound chlorophyll. Human beings can metabolize carbohydrates during reaction with oxygen, producing carbon dioxide and water. There is an oxygen-carbon dioxide balance which exists between plants and animals in nature.

Carbohydrates are classified as polyhydroxyaldehydes and polyhydroxyketones or substances that form these products on hydrolysis. Carbohydrates are classified according to size. Monosaccharides are the simplest sugars, disaccharides contain two saccharide (sugar) units, and polysaccharides contain three or more saccharide units.

Monosaccharides have the general formula $(CH_2O)_x$, where x is 3, 4, 5, or 6. Three carbon monosaccharides are called triose, followed by tetrose, pentose, and hexose for 4, 5, and 6 carbon monosaccharides, respectively. The *-ose* ending indicates that the molecule is a carbohydrate. Monosaccharide molecules are also called aldose or ketose, depending upon whether they contain aldehyde or ketone groups. The hexoses are the most important monosaccharides. Glucose, galactose, and fructose are the most important hexoses. These molecules have the same molecular formula but different structural formulas. They exist in three forms; a straight-chain and two ring forms. The ring forms are heterocyclic compounds formed from the addition of an aldehyde to an alcohol group and are called hemiacetal rings. These hexoses form an alpha and a beta hemiacetal, depending upon the orientation of the hydroxyl group on carbon 1 in the ring.

Glucose, or blood sugar, is an important compound found in the body. It is carried in the bloodstream to the cells, where it is burned up to supply energy for the body. Insulin, a hormone, helps carry glucose into the cells. If this hormone isn't produced, high levels of sugar collect in the blood and hyperglycemia exists. Diabetes mellitus is characterized by high blood sugar levels. Certain drugs, as well as kidney failure, also result in glycosuria, as does diabetes mellitus. Hypoglycemia is a condition characterized by low blood sugar levels. This, too, varies in severity and can be caused by over-production or overinjection of insulin, a diet too rich in carbohydrates, or a tumor in the insulin-producing cells. A glucose tolerance test is useful in monitoring blood sugar levels.

Carbohydrates can be tested for with the Molisch test. Benedict's test is used to test for reducing sugars, which are those sugars containing a free aldehyde or ketone group that can cause reduction in alkaline solutions containing mild oxidizing agents.

Maltose, lactose, and sucrose are important disaccharide molecules. When specific monosaccharides undergo *dehydration synthesis*, disaccharides are formed. Disaccharides can be hydrolized to form monosaccharides.

Polysaccharides are polymers of monosaccharides. They, too, can undergo dehydration synthesis and hydrolysis reactions. Glycogen, cellulose, dextrin, heparin, and starch are important polysaccharides.

Carbohydrates can undergo such reactions as oxidation, reduction, fermentation, and hydrolysis.

EXERCISES

1. Name six essential dietary components.

2. What is the caloric value per gram of carbohydrate?

3. List five examples of foods that are highly carbohydrate in composition.

4. What plant structures store carbohydrates?

5. Briefly explain how carbohydrates are produced by photosynthesis.

6. Explain the mechanism of the oxygen–carbon dioxide balance in nature.

7. How are carbohydrates classified with regard to the saccharide group content?

8. Write the hydrolysis products of
   (a) a disaccharide   (b) a polysaccharide

9. Write the dehydration synthesis products of monosaccharides and disaccharides.

10. What elements compose
(a) the hydroxyl group?   (b) the carbonyl group?
(c) an aldehyde group?   (d) a ketone group?

11. What suffix is used to signify a carbohydrate?

12. Give an example of a molecular formula for
(a) a triose   (b) a tetrose   (c) a pentose   (d) a hexose

13. Define the following terms.
(a) aldose   (b) ketose   (c) aldohexose   (d) ketohexose

14. Name the three most important hexoses and draw their structural formulas in the open-chain form.

15. How is a hemiacetal formed?

16. Draw the pyran and furan rings.

17. Explain the difference between alpha and beta glucose
(a) using the Fischer formulas   (b) using the Haworth formulas

18. What is the importance of insulin in glucose transport?

19. What is the normal fasting level of blood sugar?

20. Explain hyperglycemic and hypoglycemic levels of blood sugar.

21. Explain the importance of the Benedict's test.

22. Explain what is meant by a reducing sugar.

23. Which monosaccharides form
(a) lactose?   (b) maltose?   (c) sucrose?

24. What is the importance of glycogen in the body?

25. Explain what is meant by the fermentation process?

26. Determine whether each compound is an aldose or ketose. Then classify each compound according to the number of carbon atoms in the molecule.

(a)
$$\begin{array}{c} H \\ | \\ C=O \\ | \\ H-C-OH \\ | \\ HO-C-H \\ | \\ HO-C-H \\ | \\ H-C-OH \\ | \\ H \end{array}$$

(b)
$$\begin{array}{c} H \\ | \\ H-C-OH \\ | \\ C=O \\ | \\ H-C-OH \\ | \\ H \end{array}$$

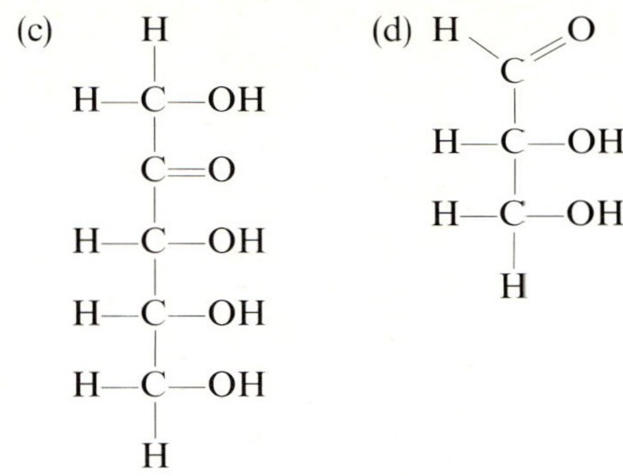

27. Decide whether these structures are alpha or beta forms.

(a)

```
            H
            |
        H—C—OH
            |
       HO—C————————┐
            |       |
       HO—C—H       |
            |        O
        H—C—OH      |
            |       |
        H—C—————————┘
            |
        H—C—OH
            |
            H
```

(b)

```
        H     OH
          \  /
           C——————————┐
           |          |
        H—C—OH        |
           |          |
       HO—C—H          O
           |          |
        H—C—OH        |
           |          |
        H—C———————————┘
           |
        H—C—OH
           |
           H
```

28. Decide whether the following structures are alpha or beta forms.

(a)

```
                    O
   HOH₂C₆ ╱     ╲  ¹CH₂OH
        C₅        C₂
     H ╱ \ H  OH╱ \ OH
         C₄————C₃
         OH     H
```

(b)

```
            ₆CH₂OH
             |
            C₅———O
       H     |        OH
       |     H         |
       C₄             C₁
       |    OH    H    |
       OH    |    |    H
            C₃————C₂
             |     |
             H     OH
```

CHAPTER 12

A Look at Lipids

A Heavy Experience

Some Things You Should Know After Reading This Chapter

You should be able to:

1. State at least two general characteristics of *all* lipids.
2. Identify each of the following as major lipid components: glycerol, fatty acids, nitrogen-containing compounds, and phosphoric acid.
3. Identify each of the following as a major class of lipids: fats, waxes, phospholipids, glycolipids, steroids, and terpenes.
4. Identify a fatty acid as saturated or unsaturated.
5. Identify an unsaturated fatty acid as a member of the oleic, linoleic, linolenic, or arachidonic acid series.
6. Identify the prostaglandins as cyclic fatty acids.
7. Write the structure of glycerol.
8. Identify a compound as a mono-, di-, or triglyceride.
9. Write the products of fat hydrolysis, complete hydrogenation, and saponification with NaOH.
10. Identify waxes as esters of fatty acids and high-molecular weight-alcohols other than glycerol.
11. Identify the three most important types of phospholipids as lecithins, cephalins, and sphingomyelins.
12. Identify the three major components of a glycolipid as a sugar, which is usually galactose; a fatty acid; and an alcohol, which is usually glycerol, sphingosine, or dihydrosphingosine.
13. Identify steroid structures.
14. Define sterols as steroid alcohols and name a specific example.
15. Identify the structure of terpenes.

Lipids are an important group of biochemical substances, found in both plant and animal tissues (Fig. 12-1). Lipids include fats and other compounds that resemble fats in physical properties (Table 12-1). Fats have been recognized since ancient times as an important category of foods. Using fats as foods probably seemed logical to early man. And applying fats and waxes as lubricants as well as using them in medicines and cosmetics must have also seemed like a good thing to do.

Lipids are found in almost all parts of a plant. The seeds and the fruit contain most of the fat found in plants. *Vegetable fats*, called *oils*, are found mainly in the seeds. Some seeds are very high in oil content. For example, the coconut seed is about 65% oil. *Waxes*, which are also lipids, serve as protective agents on plant leaves. Waxes are there to prevent or reduce the lost of moisture from the plant. Plants found in dry desert regions develop a very thick wax coat on the under side of the leave to reduce water loss. Carnauba wax, which is found on the leaves·of a special type of Brazilian palm tree, is widely used in floor polishes, shoe creams, car wax, and in the manufacture of carbon paper.

Let's look at the properties of lipids and see how this important group of organic molecules is classified.

TABLE 12-1   Some examples of lipids

Fats	Cholesterol	Oils
Waxes	Vitamin A	Cortisone

FIGURE 12-1
Lipids are an important group of biochemical substances that are found in both plant and animal tissues.

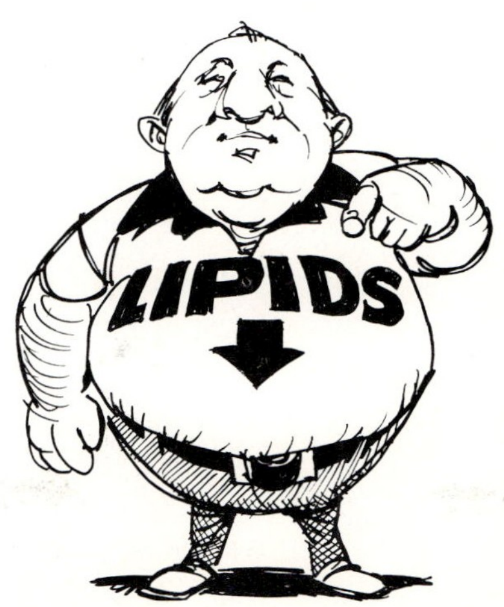

Lipids are organic compounds that have the following characteristics:

1. They are insoluble in water and soluble in one or more organic solvents, such as ethanol, acetone, benzene, chloroform, and carbon tetrachloride. These organic solvents are called "fat solvents," and they are nonpolar organic molecules.

2. They contain carbon, hydrogen, and oxygen atoms, and sometimes phosphorus or nitrogen atoms.

3. They are almost all esters that form *fatty acids* on hydrolysis. (Fatty acids are usually straight-chain carboxylic acids that generally contain an even number of carbon atoms. We'll have more to say about fatty acids shortly.)

4. They are important components of all living matter.

Classification of Lipids

Lipids vary considerably in their composition and structure. Because of this, most texts don't agree on how to classify lipids. Some texts divide lipids into two major groups, other texts use three major groups, and still other texts use four. For our purposes it will be easier to use the first classification scheme. This scheme divides lipids into two major groups, those that have fatty acids in their molecules and those that don't. Each of these two major groups can then be subdivided further to yield the major classes of lipids. Table 12-2 summarizes the classification of lipids and Table 12-3 lists the major lipid components. In the next few pages we will take a closer look at each of these lipid components, starting with fatty acids and glycerol.

TABLE 12-2   A scheme for classifying lipids

1. Lipids that contain fatty acids
   (a) Fats
   (b) Waxes
   (c) Mono and diglycerides
   (d) Phospholipids
       (1) Lecithins
       (2) Cephalins
       (3) Sphingomyelins
   (e) Glycolipids

2. Lipids that don't contain fatty acids
   (a) Steroids
           Sterols
   (b) Terpenes

TABLE 12-3   Some major lipid components

FATS: fatty acids plus glycerol

WAXES: fatty acids plus high-molecular-weight alcohols other than glycerol

PHOSPHOLIPIDS: fatty acids, a nitrogen-containing compound, phosphoric acid, and a polyhydroxyalcohol

Lecithins: fatty acids, choline, phosphoric acid, and glycerol

Cephalins: fatty acids, ethanolamine (or serine or inositol), phosphoric acid, and glycerol

Sphingomyelins: fatty acids, a nitrogen-containing compound, phosphoric acid, and sphingosine

GLYCOLIPIDS: a fatty acid, a sugar (usually galactose), and an alcohol (usually glycerol, sphingosine, or dihydrosphingosine)

STEROIDS: multicyclic ring structures

Sterols: steroid alcohols

TERPENES: polymers of isoprene

EXAMPLE 12-1   Identify the major class of each lipid from its components.

(a) fatty acids + sphingosine + a carbohydrate
(b) fatty acids + glycerol
(c) a multicyclic ring structure

SOLUTION   We use Table 12-3.

(a) *Glycolipids* contain fatty acids + sphingosine + carbohydrate.
(b) *Fats* contain fatty acids + glycerol.
(c) *Steroids* contain multicyclic ring structures.

The Fatty Acids: A Lipid Component

Most *fatty acids* are found as essential components of other lipid molecules. They can be obtained by the hydrolysis of lipid molecules such as fats. *Fats* are all *esters* of fatty acids and an alcohol, glycerol (Fig. 12-2). The general formula for a fat includes three fatty acid chains, R_1, R_2, and R_3, which may or may not be the same. These are bonded to a molecule of glycerol, in which all three of the glycerol —OH groups have undergone an esterification reaction (Fig. 12-3). This particular type of fat is called a *triglyceride* because *all* three of the glycerol —OH groups have undergone the esterification reaction. Fats can be broken

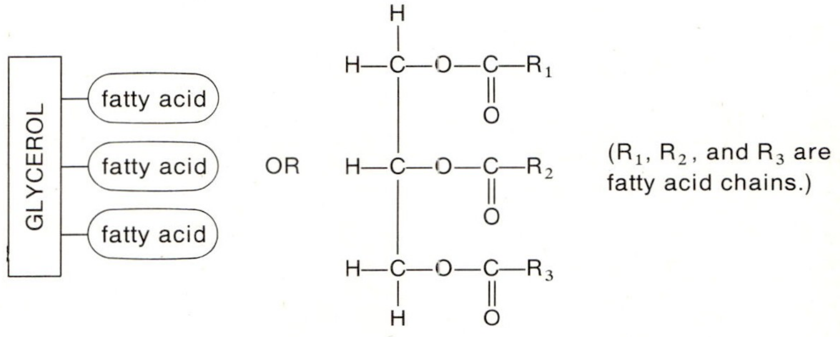

FIGURE 12-2
Fats are esters of fatty acids and glycerol.

FIGURE 12-3
The general formula for a fat (a triglyceride).

TABLE 12-4   The fatty acid series

NAME	GENERAL FORMULA	NUMBER OF DOUBLE BONDS
Saturated fatty acids	$C_nH_{2n+1}COOH$	0
Oleic acid series	$C_nH_{2n-1}COOH$	1
Linoleic acid series	$C_nH_{2n-3}COOH$	2
Linolenic acid series	$C_nH_{2n-5}COOH$	3
Arachidonic acid series	$C_nH_{2n-7}COOH$	4

down to fatty acids and glycerol by hydrolysis. We will discuss the hydrolysis of fats, in more detail, later in this chapter.

Most fatty acids are composed of a straight chain of carbon atoms connected to a carboxyl group, —C—OH, which is found at one end of the mole-

$$\overset{\|}{\underset{O}{}}$$

cule. Most of the natural fatty acids have an even number of carbon atoms in their molecules. The fatty acids that compose natural fats and oils can be classified into several series. The *saturated fatty acid series* contains fatty acids that have no double bonds. The *oleic acid series* contains fatty acids with one double bond. The *linoleic acid series* contains fatty acids with two double bonds. The *linolenic acid series* contains fatty acids with three double bonds. The *arachidonic acid series* contains fatty acids with four double bonds (Table 12-4). Other straight-chain fatty acids exist, but we will limit our discussion to those already mentioned (Table 12-5).

The most common straight-chain fatty acids are stearic, palmitic, and oleic acids (Fig. 12-4). Stearic and palmitic acids are *saturated* fatty acids.

TABLE 12-5   Some important fatty acids

ACID SERIES	NAME	TOTAL NUMBER OF CARBON ATOMS	FORMULA	OCCURRENCE
Saturated fatty acid series	Butyric	4	C_3H_7COOH	Milk fat
	Caproic	6	$C_5H_{11}COOH$	Butter, coconut oil, palm nut oil
	Caprylic	8	$C_7H_{15}COOH$	Butter, coconut oil, palm nut oil
	Capric	10	$C_9H_{19}COOH$	Butter, coconut oil, palm nut oil
	Lauric	12	$C_{11}H_{23}COOH$	Laurel oil, coconut oil
	Myristic	14	$C_{13}H_{27}COOH$	Nutmeg oil
	Palmitic	16	$C_{15}H_{31}COOH$	Animal and vegetable fats
	Stearic	18	$C_{17}H_{35}COOH$	Animal and vegetable fats
	Arachidic	20	$C_{19}H_{39}COOH$	Peanut oil
Oleic acid series	Crotonic	4	C_3H_5COOH	Croton oil
	Oleic	18	$C_{17}H_{33}COOH$	Animal and vegetable fats
Linoleic acid series	Linoleic	18	$C_{17}H_{31}COOH$	Vegetable oils such as linseed and Cottonseed oil
Linolenic acid series	Linolenic	18	$C_{17}H_{29}COOH$	Linseed oil
Arachidonic acid series	Arachidonic	20	$C_{19}H_{31}COOH$	Nervous tissue, lecithin, cephalin

$$CH_3-(CH_2)_{14}-\overset{\displaystyle O}{\underset{\displaystyle \|}{C}}-OH \qquad \text{Palmitic acid}$$

$$CH_3-(CH_2)_{16}-\overset{\displaystyle O}{\underset{\displaystyle \|}{C}}-OH \qquad \text{Stearic acid}$$

$$CH_3-(CH_2)_7-CH=CH-(CH_2)_7-\overset{\displaystyle O}{\underset{\displaystyle \|}{C}}-OH \qquad \text{Oleic acid}$$

FIGURE 12-4
The three most common straight-chain fatty acids.

They contain no double bonds and are unreactive. At room temperature they are waxy solids. Oleic acid is an *unsaturated* acid. The *unsaturated fatty acids* have one or more double bonds and are liquids at room temperature. In general, the presence of one double bond results in a considerable decrease in the melting point of a fatty acid. The presence of additional double bonds lowers the melting point even more (Table 12-6).

EXAMPLE 12-2 For each of the fatty acid structures given, state whether it is a member of the oleic, linoleic, linolenic, or arachidonic acid series. Also, state whether it is saturated or unsaturated and whether it will be a liquid or solid at room temperature.

$$\text{(a)} \quad CH_3-(CH_2)_{10}-\overset{\displaystyle \|}{\underset{\displaystyle O}{C}}-OH$$

$$\text{(b)} \quad CH_3-(CH_2)_4-CH=CHCH_2CH=CH-(CH_2)_7-\overset{\displaystyle \|}{\underset{\displaystyle O}{C}}-OH$$

TABLE 12-6   Melting points of several fatty acids

NAME	MELTING POINT (°C)	NUMBER OF DOUBLE BONDS
Palmitic	63.1	0
Stearic	69.6	0
Oleic	13.4	1
Linoleic	−5	2
Linolenic	−11	3
Arachidonic	−49.5	4

$$\text{(c)} \quad CH_3-(CH_2)_4-CH=CHCH_2CH=CHCH_2CH=CHCH_2CH$$
$$HO-\underset{\underset{O}{\|}}{C}-(CH_2)_3-CH$$

SOLUTION   (a) This acid, known as lauric acid, is a saturated acid, so it is not a member of the oleic, linoleic, linolenic, or arachidonic acid series. Because it is saturated, it should be a solid at room temperature.

(b) This compound is linoleic acid. It has two double bonds. This also means it should be a liquid at room temperature.

(c) This compound is arachidonic acid. It has four double bonds. It, also, is a liquid at room temperature.

Vegetable fats, called *oils*, generally contain more unsaturated fatty acids then saturated fatty acids, so they are liquids at room temperature. Animal fats are usually solids at room temperature because they contain an abundance of saturated fatty acids (Fig. 12-5).

Three of the fatty acids *can't* be synthesized by the body and are called the *essential fatty acids*. These fatty acids must be included in the diet. They are

FIGURE 12-5
A comparison of saturated and unsaturated fatty acids in some foods.

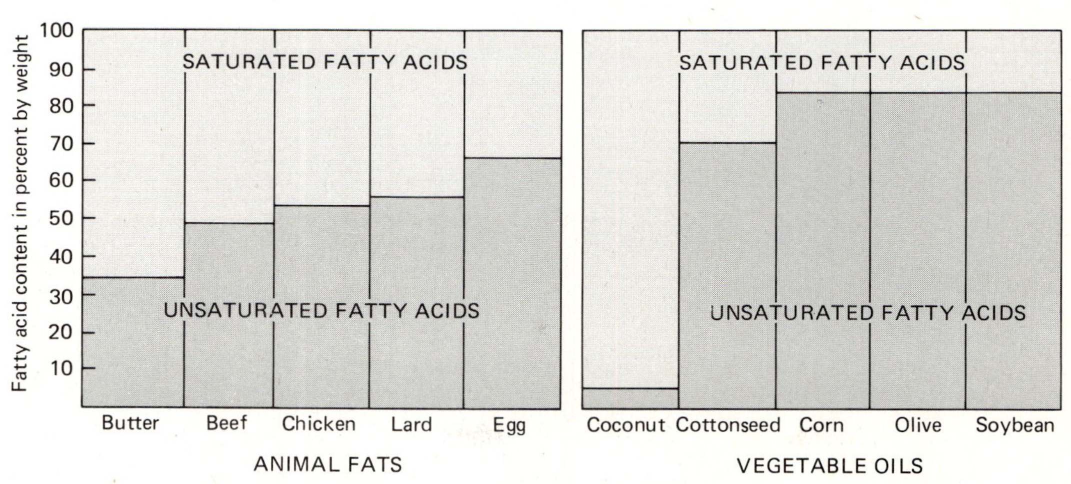

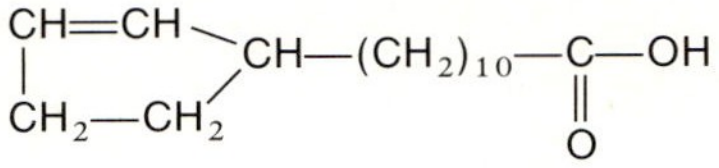

$CH_3-(CH_2)_4-CHOH-CH=CH-\overset{\displaystyle H}{\underset{\displaystyle CH}{\underset{\displaystyle \diagdown HO}{\overset{\diagup}{C}}}}...$

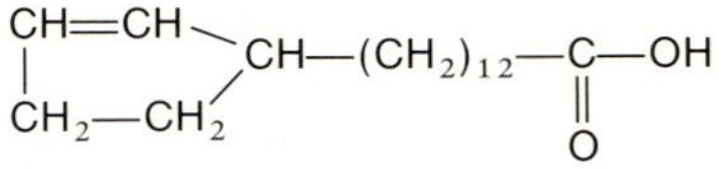

FIGURE 12-6
Some cyclic fatty acids.

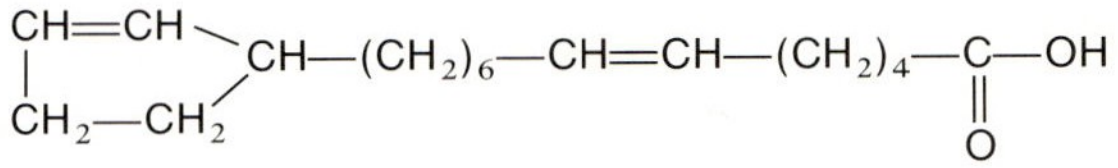

linoleic, *linolenic*, and *arachidonic* acids. Babies that don't consume these acids develop *eczema*, a skin condition that causes inflamed and itching skin. These babies also tend to lose weight. The three essential fatty acids are found in both animal and vegetable fats.

Besides the straight-chain fatty acids, there are also *cyclic fatty acids*. Three such acids are *hydnocarpic*, *chaulmoogric*, and *gorlic acids*, all of which contain a cyclopentene ring (Fig. 12-6). These cyclic fatty acids are specific for the treatment of leprosy, but have been replaced by more effective drugs in recent years.

Prostaglandins are cyclic fatty acids containing 20 carbon atoms that form a five-membered ring in the center of the molecule (Fig. 12-7). Prostaglandins are found in animal tissues and are known to affect endocrine, digestive, cardiovascular, reproductive, and respiratory functions. These molecules have

FIGURE 12-7
Prostaglandin E_2, a cyclic fatty acid containing 20 carbon atoms, with a five-membered ring in the center of the molecule.

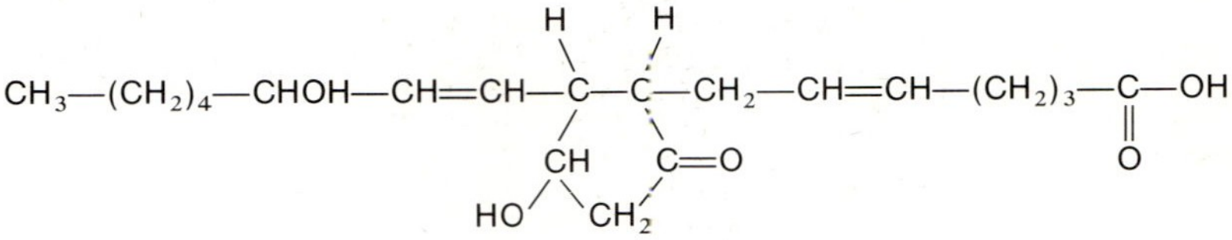

FIGURE 12-8
Arachidonic acid is used by the body to synthesize a prostaglandin molecule.

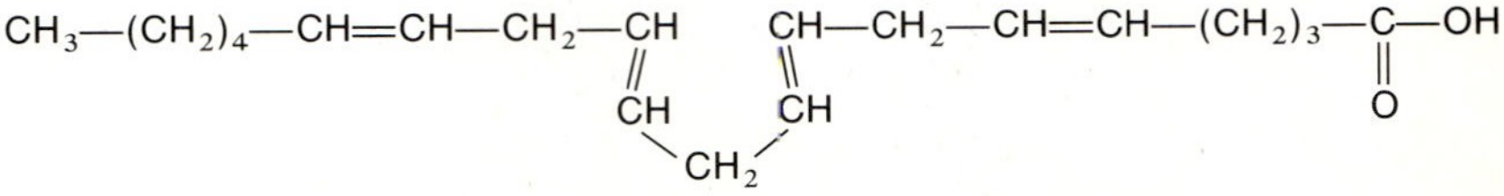

Arachidonic acid

$$
\begin{array}{ccc}
\text{CH}_2\text{OH} & & \text{H}-\text{C}-\text{OH} \\
| & & | \\
\text{CHOH} & \text{OR} & \text{H}-\text{C}-\text{OH} \\
| & & | \\
\text{CH}_2\text{OH} & & \text{H}-\text{C}-\text{OH}
\end{array}
$$

FIGURE 12-9
The structure of the polyhydroxy alcohol, glycerol.

important physiological properties, including the ability to raise and lower blood pressure and control gastric secretions. They are being studied intensively to understand their functions and possible uses. The essential fatty acids, linoleic, linolenic, and arachidonic acids, are used by the body to produce the prostaglandins (Fig. 12-8).

Glycerol: A Lipid Component

Glycerol is an alcohol containing three —OH groups (Fig. 12-9). It is commonly called *glycerin*. It has a sweet taste and is a colorless, oily liquid. It mixes readily with water and ethanol, but is practically insoluble in ether. When reacted with *dehydrating agents*, glycerol forms *acrolein*, a substance that is easily detected by its pungent odor (Fig. 12-10). (By the way, a dehydrating agent is a substance that removes water from a molecule.) When treated with nitric acid, glycerol forms nitroglycerin, which is sometimes called glyceryl trinitrate (Fig. 12-11). Nitroglycerin is an important pharmaceutical, because it is a vasodilator (it opens up blood vessels) and is useful in treating patients with certain circulatory disorders. Nitroglycerin is also an explosive and is used in making dynamite.

Mono-, Di-, and Triglycerides

As we already learned, fats are esters of fatty acids and glycerol. Glycerol is an alcohol containing three —OH groups. In our discussion so far, we have

FIGURE 12-10
The dehydration of glycerol produces acrolein.

$$
\begin{array}{ccc}
\text{CH}_2\text{OH} & & \text{CH}_2 \\
| & & \| \\
\text{CHOH} & \xrightarrow[\text{H}_2\text{SO}_4]{\text{heat}} & \text{CH} \qquad + 2\text{H}_2\text{O} \\
| & & | \\
\text{CH}_2\text{OH} & & \text{C}=\text{O} \\
& & | \\
\text{Glycerol} & & \text{H}
\end{array}
$$

Acrolein

$$
\begin{array}{ccccc}
CH_2OH & & & & CH_2(ONO_2) \\
| & & & & | \\
CHOH & + HNO_3 & \longrightarrow & & CH(ONO_2) \\
| & & & & | \\
CH_2OH & & & & CH_2(ONO_2) \\
\\
\text{Glycerol} & & & & \text{Nitroglycerin}
\end{array}
$$

FIGURE 12-11
Glycerol, treated with nitric acid, forms nitroglycerin.

only mentioned fats which are *triglycerides*, that is, fats in which all *three* of the glycerol —OH groups have undergone esterification and are bonded to the fatty acid chains. However, it is also possible for just one or two of the glycerol —OH groups to be combined with fatty acids in ester linkages. Monoglycerides are lipids in which *one* glycerol —OH group is combined with a fatty acid in an ester linkage. Diglycerides are lipids in which *two* glycerol —OH groups are combined with two fatty acids in ester linkages. The —OH groups that have not undergone esterification remain unchanged (Fig. 12-12).

EXAMPLE 12-3   State whether each of the following compounds are mono-, di-, or triglycerides.

(a)

$$
\begin{array}{c}
\quad\quad\quad\quad\quad\quad O \\
\quad\quad\quad\quad\quad\quad \| \\
CH_3-(CH_2)_{14}-C-O-CH_2 \\
\\
\quad\quad\quad\quad\quad\quad O \\
\quad\quad\quad\quad\quad\quad \| \\
CH_3-(CH_2)_{16}-C-O-CH \\
\\
\\
\quad\quad\quad\quad\quad\quad HO-CH_2
\end{array}
$$

FIGURE 12-12
The general formula for a monoglyceride and a diglyceride.

A monoglyceride

A diglyceride

(b)

$$CH_3-(CH_2)_{16}-\overset{\overset{\displaystyle O}{\|}}{C}-O-CH_2$$
$$HO-CH$$
$$HO-CH_2$$

SOLUTION

(a) This compound is a diglyceride because it has *two* fatty acid groups.

(b) This compound is a monoglyceride because it has *one* fatty acid group.

Fats: A Closer Look at a Major Class of Lipids

There are several reactions which fats undergo. Let's take a closer look at fats and see what reactions they take part in.

1. IODINE NUMBER Fats and oils are composed of saturated and unsaturated fatty acids. As we stated before, oils generally contain more unsaturated fatty acids than saturated fatty acids, whereas animal fats usually contain more saturated fatty acids than unsaturated fatty acids (Fig. 12-5). The unsaturated fatty acids absorb iodine and other halogens. Iodine is taken up at the double bonds and forms saturated compounds that are derivatives of iodine. The greater the number of double bonds in the molecule, the greater the quantity of iodine that is absorbed by the molecule. The *iodine* number is defined as the number of grams of iodine absorbed by 100 g of fat or oil. The iodine number is used to determine the purity of an oil. Values have been established for various oils, and if the oils are pure, their iodine numbers should match the standard values (Table 12-7). The higher the number of unsaturated fatty acids in the oil, the higher is its iodine number. Diets high in polyunsaturated fats, in other words, those with very high iodine numbers, appear to lower the risk of heart attacks in humans.

2. ACROLEIN TEST The presence of glycerol can be detected by the *acrolein test*. At high temperatures, fats undergo hydrolysis. Glycerol is

TABLE 12-7 Iodine numbers for some oils

NAME	IODINE NUMBERS
Cottonseed oil	103–111
Olive oil	79–88
Linseed oil	175–202

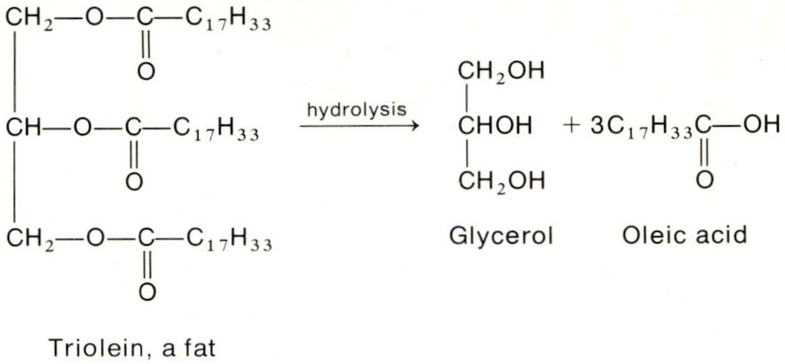

$$CH_2-O-\underset{\underset{O}{\parallel}}{C}-C_{17}H_{33}$$
$$CH-O-\underset{\underset{O}{\parallel}}{C}-C_{17}H_{33} \xrightarrow{\text{hydrolysis}} \quad \begin{matrix} CH_2OH \\ | \\ CHOH \\ | \\ CH_2OH \end{matrix} \quad + 3C_{17}H_{33}\underset{\underset{O}{\parallel}}{C}-OH$$
$$CH_2-O-\underset{\underset{O}{\parallel}}{C}-C_{17}H_{33}$$

Triolein, a fat

Glycerol   Oleic acid

FIGURE 12-13
Fat hydrolysis produces glycerol plus fatty acids.

produced and under these conditions it reacts to form acrolein. Acrolein has a pungent and irritating odor. If you've ever smelled burnt fat or oil, you've smelled the odor of acrolein.

3. HYDROGENATION   Margarine and vegetable shortenings are made by a process called *hydrogenation*. When hydrogen atoms are added to the double bonds of unsaturated fatty acids, oils can be converted to solid fats. Partial hydrogenation of corn oil or soybean oil is used to produce shortenings such as Crisco or Spry. Margarine is made by mixing partially hydrogenated oils with unsaturated oils, vitamins A and D, carotenoids (used for coloring), and artificial flavoring.

4. HYDROLYSIS   Fats can be broken down to form fatty acids and glycerol under the proper conditions. Superheated steam, alkalies, or fat-splitting enzymes, called lipases, can cause fat hydrolysis (Fig. 12-13).

5. SAPONIFICATION   Soaps can be formed by fat hydrolysis when alkalies are used in the reaction. This is called *saponification*. Glycerol and the salts of the fatty acids are produced (Fig. 12–14). In other words, soaps

FIGURE 12-14
Soaps can be formed by fat hydrolysis when alkalies are used in the reaction.

$$CH_2-O-\underset{\underset{O}{\parallel}}{C}-C_{17}H_{35}$$
$$CH-O-\underset{\underset{O}{\parallel}}{C}-C_{17}H_{35} + 3NaOH \longrightarrow \quad \begin{matrix} CH_2OH \\ | \\ CHOH \\ | \\ CH_2OH \end{matrix} \quad + 3C_{17}H_{35}-\underset{\underset{O}{\parallel}}{C}-Na$$
$$CH_2-O-\underset{\underset{O}{\parallel}}{C}-C_{17}H_{35}$$

Tristearin, a fat

Glycerol   Sodium stearate, a soap

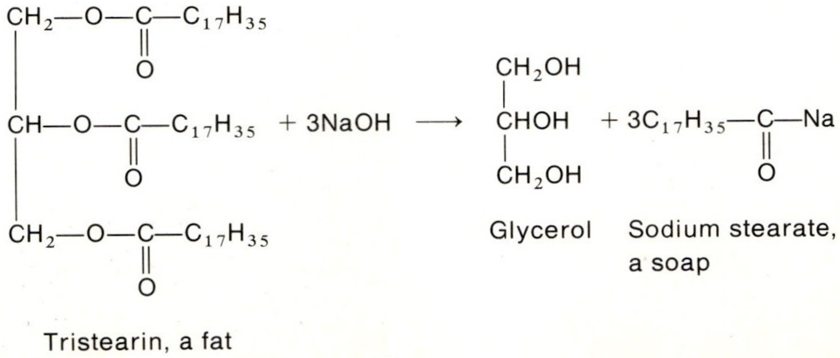

are salts of fatty acids. They are commonly salts of sodium or potassium ions. The sodium salts of fatty acids tend to produce solid soaps that are hard and can be shaped into bars. The potassium salts of fatty acids tend to produce liquid soaps or soaps that are not very hard. The hardness of a soap is also related to the degree of unsaturation of the fatty acids. Fats having a smaller degree of unsaturation tend to produce harder soaps.

Soap molecules have a dual polarity: the fatty acid portion, containing the long carbon chain, is nonpolar, whereas the other end of the molecule, containing the ionic salt, is polar (Fig. 12-15). This is what accounts for the cleaning action of soaps. The nonpolar end of the molecule dissolves nonpolar greases and dirt. The polar end of the molecule allows the soap to dissolve in polar water. So the nonpolar greases and dirt get attached to the nonpolar end of the soap molecule and get carried away with the water.

6. RANCIDITY   When fats are exposed to air, they undergo two reactions, hydrolysis and oxidation. This causes them to have poor taste and odors, or what is known as *rancidity*. *Slight* hydrolysis breaks down fats to form fatty acids that have unpleasant odors. If butter is left out of the refrigerator too long, the microorganisms in the air produce enzymes that break down the fat into butyric acid. This produces the odor of rancid butter. Unsaturated fats or oils can also be *oxidized* to produce aldehydes and short-chain fatty acids that have unpleasant odors. Chemicals which are called *antioxidants* can be added to help stop the oxidation of fats. Vitamin E is a natural antioxidant (we'll have more to say about vitamin E later). Certain fats keep better than others because they contain higher concentrations of natural antioxidants. It is important to slow down the production of rancidity in fats, otherwise, some of the important dietary substances in the fats, such as carotene, vitamin A, and linoleic acid, could also undergo oxidation. It is wise to keep foods such as peanut butter in the refrigerator to slow down rancidity.

This ends our discussion of the reaction of fats, but see what you've learned by trying the following example.

FIGURE 12-15
The fatty acid portion of the soap is nonpolar, whereas the other end of the molecule, containing the ionic salt, is polar.

$$CH_3 - - - - - (CH_2)_{18} - - - - - - \overset{\displaystyle O}{\overset{\displaystyle \|}{C}} - O^- \quad Na^+$$

Nonpolar portion of molecule

Polar portion
of molecule

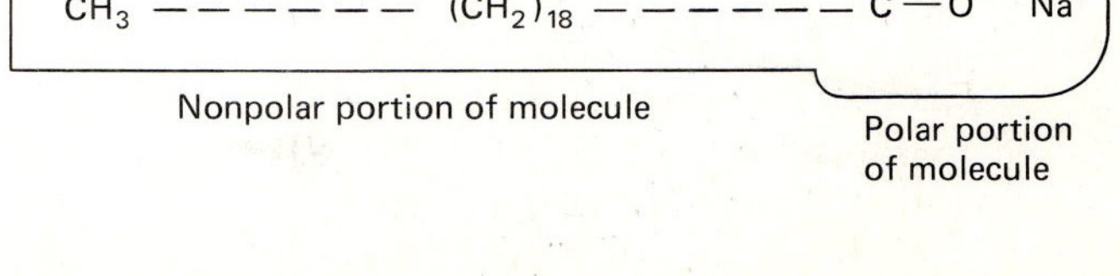

EXAMPLE 12-4 Show what happens to the following fat when it undergoes (a) hydrolysis, (b) complete hydrogenation, and (c) saponification with NaOH:

$$CH_3CH_2CH{=}CHCH_2CH{=}CHCH_2CH{=}CH-(CH_2)_7-\overset{\displaystyle O}{\overset{\|}{C}}-O-CH_2$$

$$CH_3-(CH_2)_7-CH{=}CH-(CH_2)_7-\overset{\displaystyle O}{\overset{\|}{C}}-O-CH$$

$$CH_3-(CH_2)_{16}-\underset{\displaystyle O}{\overset{\displaystyle}{\underset{\|}{C}}}-O-CH_2$$

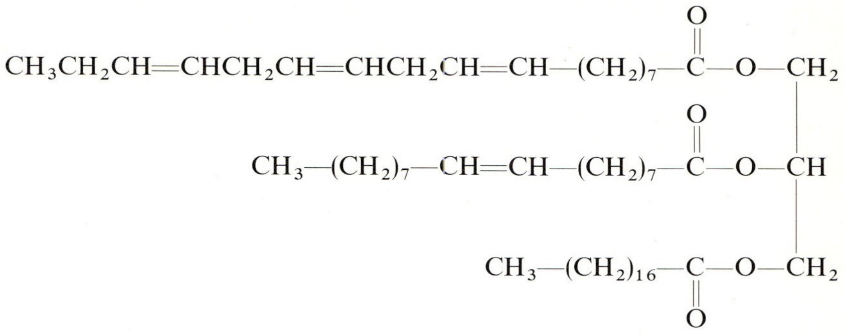

SOLUTION

(a) The hydrolysis of a fat produces the fatty acids plus glycerol. So for our example we get the following products:

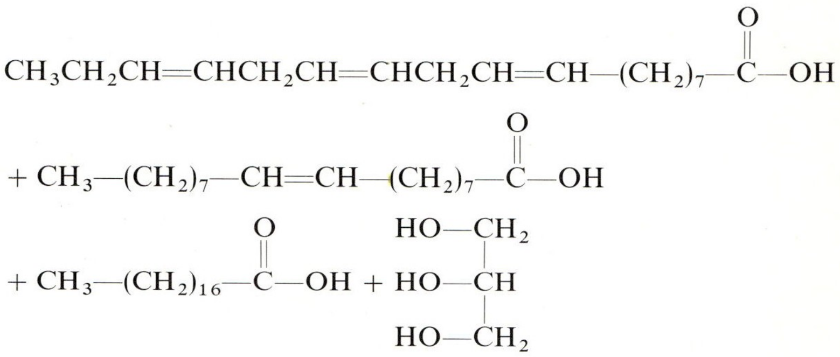

$$CH_3CH_2CH{=}CHCH_2CH{=}CHCH_2CH{=}CH-(CH_2)_7-\overset{\displaystyle O}{\overset{\|}{C}}-OH$$

$$+\ CH_3-(CH_2)_7-CH{=}CH-(CH_2)_7-\overset{\displaystyle O}{\overset{\|}{C}}-OH$$

$$+\ CH_3-(CH_2)_{16}-\overset{\displaystyle O}{\overset{\|}{C}}-OH\ +\ \begin{matrix} HO-CH_2 \\ HO-CH \\ HO-CH_2 \end{matrix}$$

(b) Complete hydrogenation of a fat means that hydrogen atoms are added to each side of the double bonds. So for our example we get the following saturated fat:

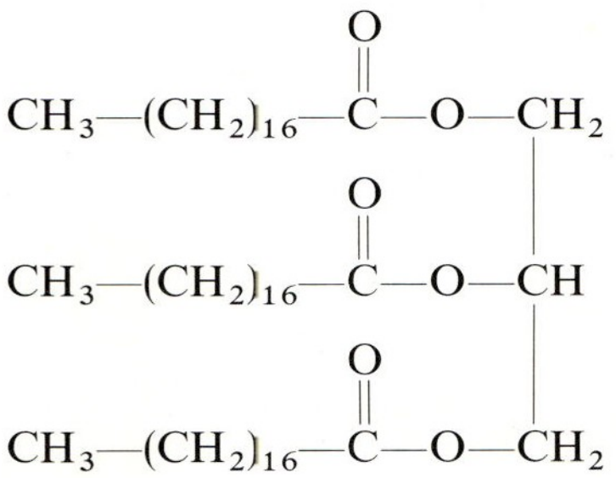

$$CH_3-(CH_2)_{16}-\overset{\displaystyle O}{\overset{\|}{C}}-O-CH_2$$

$$CH_3-(CH_2)_{16}-\overset{\displaystyle O}{\overset{\|}{C}}-O-CH$$

$$CH_3-(CH_2)_{16}-\overset{\displaystyle O}{\overset{\|}{C}}-O-CH_2$$

(c) Saponification with NaOH produces the sodium salt of each fat plus glycerol. So for our example we get the following products:

$$CH_3CH_2CH{=}CHCH_2CH{=}CHCH_2CH{=}CH{-}(CH_2)_7{-}\overset{\displaystyle O}{\overset{\|}{C}}{-}O^- \; Na^+$$

$$+ \; CH_3{-}(CH_2)_7{-}CH{=}CH{-}(CH_2)_7{-}\overset{\displaystyle O}{\overset{\|}{C}}{-}O^- \; Na^+$$

$$+ \; CH_3{-}(CH_2)_{16}{-}\overset{\displaystyle O}{\overset{\|}{C}}{-}O^- \; Na^+ \; + \;
\begin{array}{l} HO{-}CH_2 \\ | \\ HO{-}CH \\ | \\ HO{-}CH_2 \end{array}$$

Body Fat

One of the major problems of urbanized society today is conquering obesity. When we eat more food than we need, and consume too many calories, our bodies produce fat. The fat is stored in special cells in the body called *adipose cells*. Adipose cells compose *adipose tissue*, which is found under the skin, around all the organs of the body, around joints, and in the yellow marrow of certain bones. The adipose tissue serves as a storehouse of reserve food for the body. When the body needs energy, the fat stored in adipose tissue can be used to produce this energy. Fats are a very good source of energy since they produce 9 kcal of energy for every gram of fat. (If you remember from Chapter 11, we said that carbohydrates produce 4 kcal of energy per gram, so fats produce more than twice the energy of carbohydrates per gram.) In addition to being an energy reservoir, fats also insulate the body against excessive heat loss through the skin. The fat deposited around our organs helps to protect these organs from mechanical injury.

In diseases, such as certain gallbladder disorders, or conditions where fats can't be absorbed by the body, a low-fat diet must be eaten. Such a diet would eliminate egg yolks, fatty meats, cream, fried foods, fish canned in oil, gravy, cheeses (except cottage cheese), peanut butter, olives, potato chips, chocolate, and coconut. With all that good "stuff" out of the diet, what's left? Well, individuals on low-fat diets can eat fruits, vegetables, margarine, eggs (in limited amounts), bread, cottage cheese, lean meats, and skim milk.

Waxes: A Major Class of Lipids

Waxes are esters of fatty acids and high-molecular-weight alcohols other than glycerol. Waxes are not affected by the fat-splitting enzymes, the lipases, and are not as easily hydrolyzed as fats. They have *no* nutritional value. Waxes are soluble in the fat solvents but are *not* soluble in water. They are found in nature in the animal and vegetable worlds. Waxes serve as protective coverings for both animals and plants. Carnauba wax, as mentioned earlier, is a leaf secretion from the carnauba palm tree. Candelilla wax protects the

NAME	SOURCE	USES	MAJOR CONSTITUENTS
Beeswax	Honeycomb	Preparation of ointments and plasters	Esters of straight-chain monohydric alcohols with even-numbered carbon chains from C_{24} to C_{36}, esterified with straight-chain acids also having an even number of C atoms up to C_{36}
Carnauba wax	Carnauba palm tree	Floor polish, car wax, and the like	Esters of hydroxylated unsaturated fatty acids, having about 12 carbon atoms in the acid chain
Lanolin	Wool of sheep	Base for ointments	Complex mixture of esters and polyesters of 33 high-molecular-weight alcohols and 36 fatty acids. The alcohols are of thee types: aliphatic, steroid, and triterpenoid alcohols.
Spermaceti	Sperm whale	Making candles and cosmetics and as a base for ointments	Mostly cetyl palmitate; also esters of lauric, stearic, and myristic acids.

candelella plant found in Mexico. Lanolin provides a protective coating for the wool fibers and skin of sheep and other animals. Beeswax comes from the honeycomb and is used to make dental impressions. Spermaceti comes from the sperm whale and is used to make candles and cosmetics. In our own bodies we produce *cerumen*, ear wax, which protects our inner ear from infection. Table 12-8 lists major constituents of some of the waxes we've just discussed.

Phospholipids: A Major Class of Lipids

The *phospholipids* or *phosphatides* are important compounds which are a part of every plant and animal tissue. They are complex lipids which contain the element phosphorus. Most phospholipids are composed of fatty acids, a nitrogen-containing compound, phosphoric acid, and a polyhydroxy alcohol. *Phosphoglycerides* are phospholipids based on the alcohol, glycerol. Instead of being esterified with a fatty acid, one of the glycerol —OH groups is esterified with a phosphoric acid group, which in turn may be esterified with another group, such as a nitrogen-containing compound or a sugar (Fig. 12-16).

Lecithins: A Type of Phospholipid

Lecithins are phospholipids composed of glycerol, fatty acids, phosphoric acid, and choline (Fig. 12-17). *Choline* is a nitrogen-containing compound which is a base. (When choline reacts with acetic acid, acetylcholine is formed

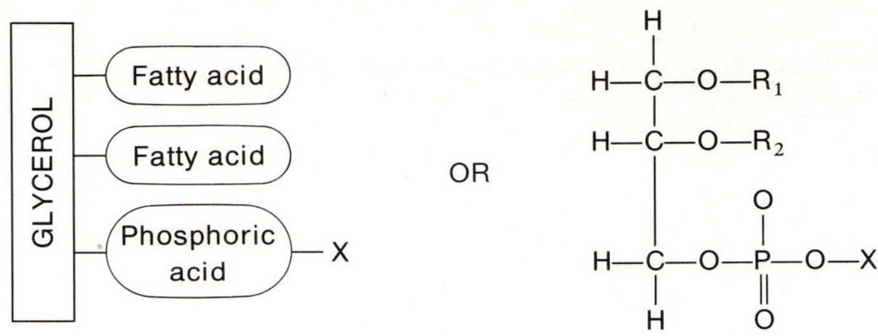

FIGURE 12-16
The general formula of a phosphoglyceride.

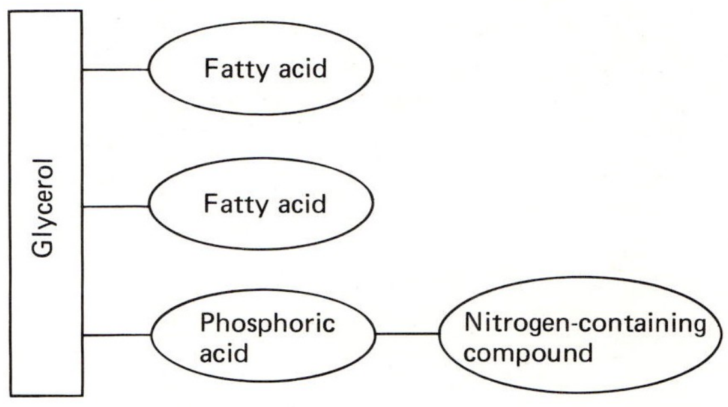

FIGURE 12-17
The general formula of a lecithin.

(Fig. 12-18). This compound is an important neurotransmitter which helps nerve impulses move quickly through the nervous system.) Lecithins are found in animal tissues and in soybeans. They are good emulsifying agents, which means that they can be used to hold fatty or oily substances in suspension. When fats are moved from one place to another in the body, they are partly converted into lecithins.

FIGURE 12-18
The structures of choline and acetylcholine.

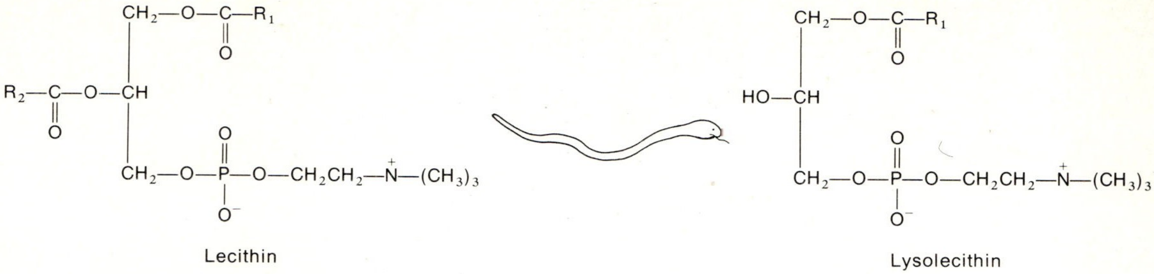

Lecithin

Lysolecithin

FIGURE 12-19
Cobra venom contains a lecithinase that is capable of removing an unsaturated fatty acid from a lecithin, producing lysolecithin.

Lecithins are soluble in hot alcohol, benzene, chloroform, and ether, but not in acetone. *Lecithinases* are enzymes that hydrolyze lecithins. Cobra venom contains a lecithinase which is capable of removing an unsaturated fatty acid from a lecithin, producing *lysolecithin*. This compound can destroy red blood cells. Certain stinging insects and poisonous spiders contain such deadly chemicals (Fig. 12-19).

Cephalins: A Type of Phospholipid

Cephalins are another type of phospholipid. They are involved in blood clotting. They are soluble in hot alcohol, benzene, and chloroform, but not in acetone, ethyl ether, or methyl ether. The cephalins are similar in structure to the lecithins, except that they contain *no* choline. The compounds ethanolamine, serine, and inositol replace the choline (Fig. 12-20).

Sphingomyelins: A Type of Phospholipid

Sphingomyelins belong to the class of lipids called *sphingolipids*. Instead of glycerol, these lipids are based on the polyhydroxyalcohol *sphingosine* (Fig. 12-21). Sphingomyelins are found in the brain and nervous tissue, where they form part of the protective coating of the nerve—the *myelin sheath* (Fig. 12-22). In certain inherited disorders, such as *Niemann-Pick disease*, sphingomyelins may build up in the brain, spleen, and liver, resulting in mental retardation and early death.

EXAMPLE 12-5   Determine whether each of the following phospholipids is a cephalin, sphingomyelin, or lecithin.

    (a) glycerol + fatty acids + phosphoric acid + ethanolamine
    (b) glycerol + fatty acids + phosphoric acid + choline
    (c) fatty acids + sphingosine + phosphoric acid + a nitrogen-containing compound

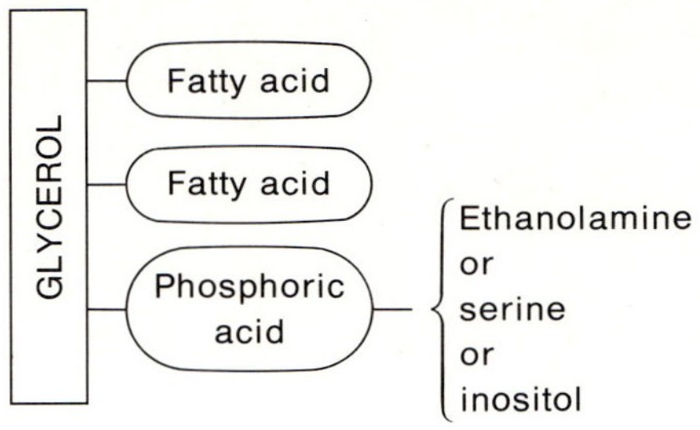

The general structure of a cephalin

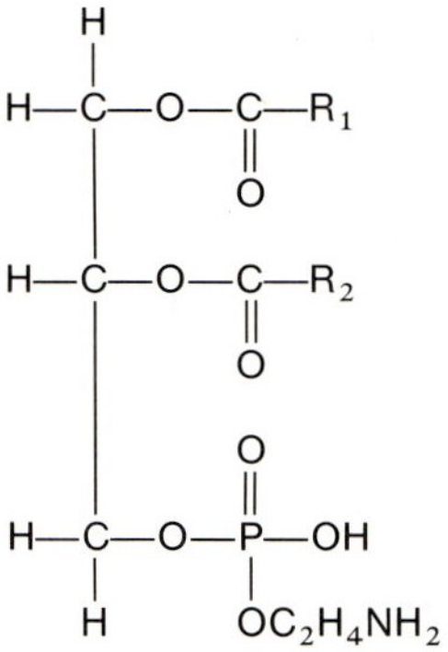

A cephalin containing
ethanolamine

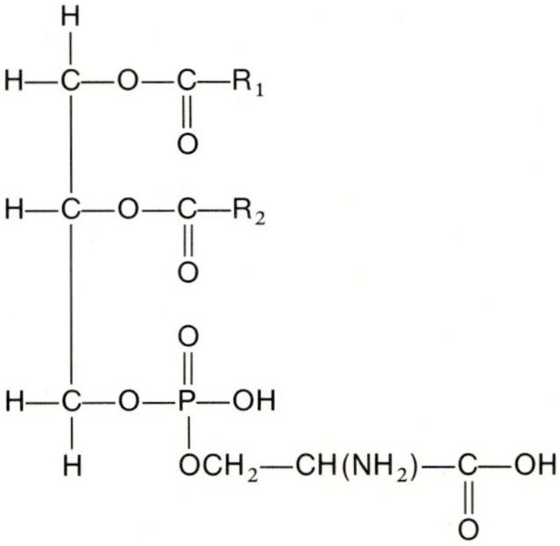

A cephalin containing serine

FIGURE 12-20
The cephalins.

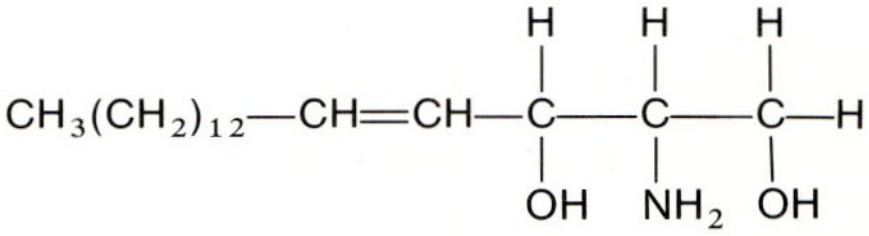

FIGURE 12-21
Sphingosine, a polyhydroxyalcohol.

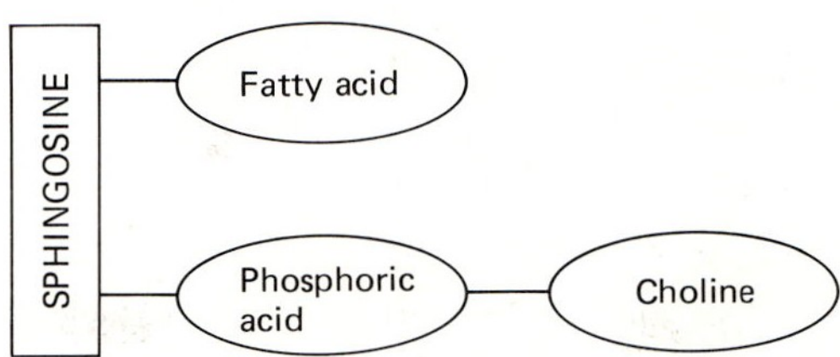

FIGURE 12-22
The general structure of a sphingomyelin.

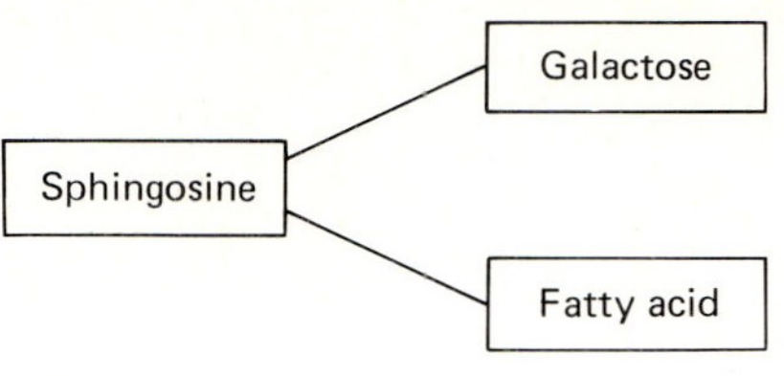

FIGURE 12-23
The general structure of a glycolipid.

SOLUTION

(a) The components in this example refer to a cephalin.

(b) The components in this example refer to a lecithin.

(c) The components in this example refer to a sphingomyelin.

Glycolipids: A Major Class of Lipids

Glycolipids are composed of:

1. A sugar, which is usually galactose.

2. A fatty acid.

3. An alcohol, which is usually glycerol, sphingosine, or dihydrosphingo-sine (Fig. 12-23). (By the way, the difference between sphingosine and dihydrosphingosine is that sphingosine is an unsaturated compound, whereas dihydrosphingosine is saturated.)

Glycolipids are also found in the brain and nerve tissue. They are called *cerebrosides* when they contain sphingosine. *Gaucher's disease* is a hereditary disorder in which glucose is substituted for galactose in the glycolipid molecule. This glucose-containing glycolipid accumulates in the liver, spleen, lymph nodes, and bone marrow. Infants affected with this disease usually die within a year. However, patients who survive to adolescence may live for many years. *Tay-Sach's disease* is a hereditary disorder in which infants can't break down glycolipids because of the lack of a specific enzyme. Glycolipids then build up in the tissues of the eyes and brain resulting in mental retardation, seizures, muscular weakness, blindness, and death by the third year of life.

Steroids: A Major Class of Lipids

Steroids are organic molecules based on an interlocking four-ring structure consisting of one cyclopentane ring and three cyclohexane rings (Fig. 12-24). Although this is the basic design of steroids, there are numerous variations that exist. For example, many steroids have various side chains attached to the

FIGURE 12-24
The general structure of a steroid.

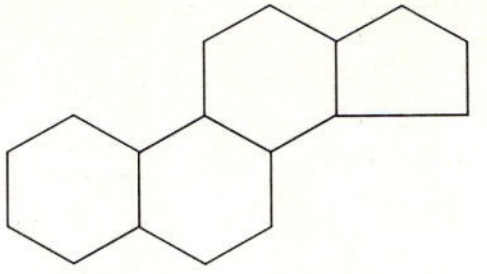

four rings. And many of the rings, themselves, may have one or more double bonds.

Steroids have very important functions in the body. *Testosterone* is a male hormone secreted from the testes. This hormone is responsible for the development of secondary sex characteristics such as male muscle structure and beard development (Fig. 12-25). *Estrogens* are female hormones secreted by the ovaries which cause the development of sex characteristics such as female structure formation (Fig. 12-26). *Progesterone* is a female hormone that prepares the lining of the uterus for pregnancy and is present throughout pregnancy (Fig. 12-27). *Digitoxigenin* is a steroid that comes from the digitalis plant. It is used in certain types of heart disease to help make the heart pump better (Fig. 12-28).

Another important steroid is vitamin D (Fig. 12-29). This vitamin is necessary for the proper absorption of calcium from the small intestine. It

FIGURE 12-25
The structure of testosterone.

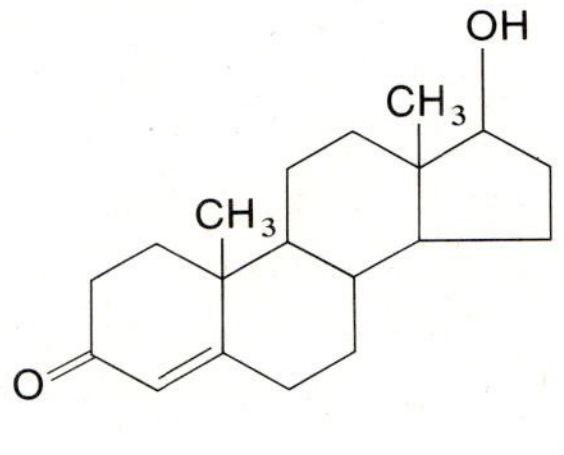

FIGURE 12-26
The structure of the estrogen, estradiol.

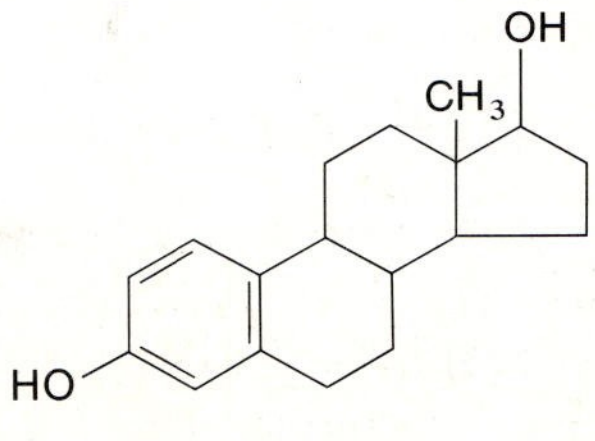

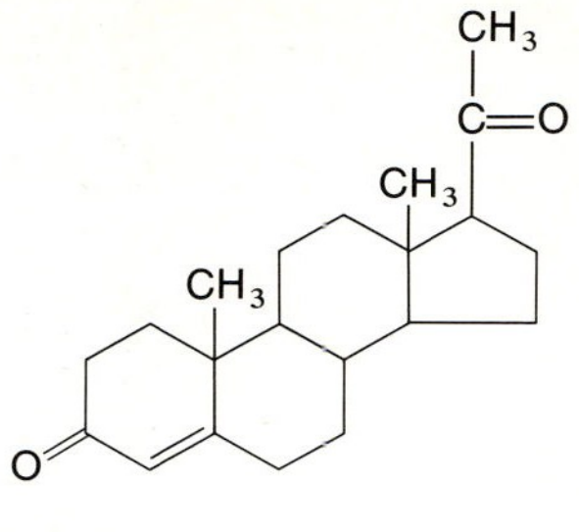

FIGURE 12-27
The structure of progesterone.

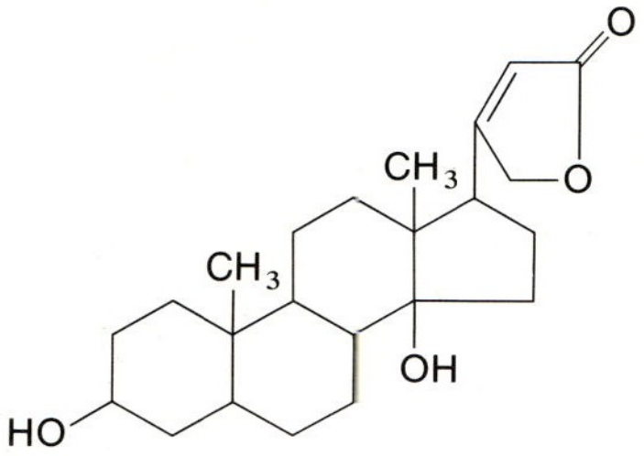

FIGURE 12-28
The structure of digitoxigenin.

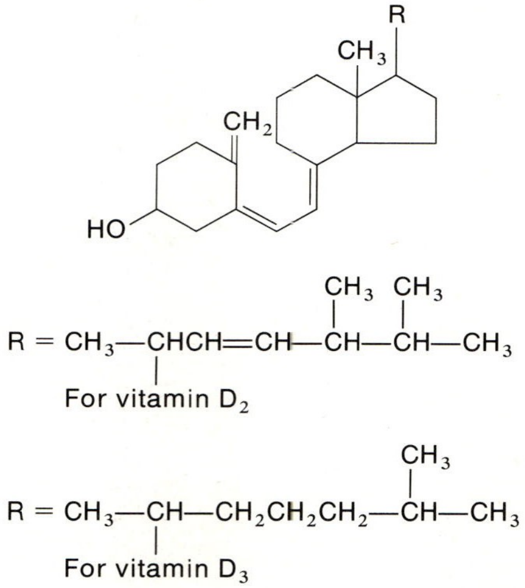

R = CH$_3$—CHCH=CH—CH—CH—CH$_3$
(with CH$_3$ and CH$_3$ branches)
For vitamin D$_2$

R = CH$_3$—CH—CH$_2$CH$_2$CH$_2$—CH—CH$_3$
(with CH$_3$ branch)
For vitamin D$_3$

FIGURE 12-29
The structure of vitamin D$_2$ and vitamin D$_3$

FIGURE 12-30
The structure of cholesterol.

also plays an important part in phosphate metabolism. A lack of vitamin D leads to the disease known as *rickets*, which causes the improper development of bones and teeth because of a lack of necessary building materials such as calcium. Vitamin D is found in cod liver oil and other fish liver oil. Milk that has been irradiated with ultraviolet radiation is also a good source of the vitamin.

Sterols: The Steroid Alcohols

Sterols are steroid alcohols. Cholesterol is the most important sterol (Fig. 12-30). The name *cholesterol* comes from the Greek words meaning "solid bile." Cholesterol plays a role in the development of *atherosclerosis*, which is the deposition of large amounts of cholesterol and other lipids on the walls of the arteries. These deposits form plaques, which reduce the flow of blood through the arteries. Blood must then be pumped much harder through the narrowed arteries, causing high blood pressure or *hypertension* (Fig. 12-31).

FIGURE 12-31
The progress of atherosclerosis in an artery. (a) A normal artery. (b) A partially blocked artery. (c) An almost completely blocked artery. (American Heart Association)

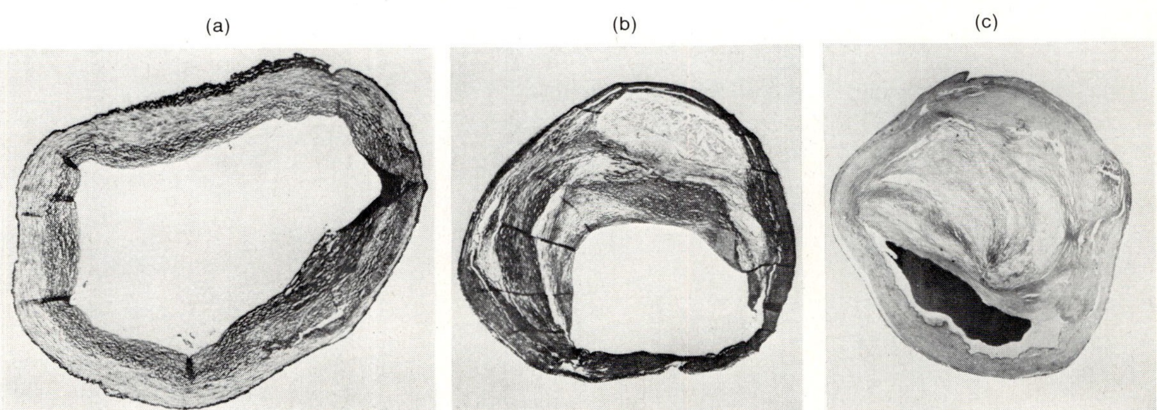

The passage of nutrients to the cells is also affected. The plaque deposits on the arterial walls increase the tendency for blood clots to form. If a clot forms, it has the potential to block a blood vessel and cause a heart attack or stroke. Foods high in cholesterol are animal fats and egg yolks, and should be eaten in moderation or avoided in those individuals who have to restrict their cholesterol intake. A diet low in cholesterol would include low-fat, high-quality protein foods such as fat-free milk, cottage cheese, egg whites (not yolks), liver, veal, fish or poultry.

Terpenes: A Major Class of Lipids

Terpenes are polymers of isoprene (Fig. 12-32). Polymers are large molecules composed of many smaller units linked together. The smaller units, called *monomers*, are repeated over and over again in the polymer. An example of a terpene is *carotene* (Fig. 12-33). It was isolated from carrots in 1826, and is the chemical that gives carrots their orange coloring. Sweet potatoes are also colored orange because of the presence of carotene, So too, is the yellow color of eggs yolks. Orientals have a pale, yellowish tone to their skins because of the presence of small quantities of carotene found beneath the skin. Carotene is also responsible for many of the beautiful colors of autumn leaves.

Vitamins A, E, and K are terpenes (Fig. 12-34). These three vitamins, along with vitamin D (which is a steroid, not a terpene), are known as the *fat-soluble vitamins*, because they are stored and carried by fats in the body. Because of their ability to be stored, an excess of these vitamins, known as *hypervitaminosis*, can occur. This, in turn, can lead to toxic concentrations of these vitamins in the body.

The major role of vitamin A in the body is the regeneration of the protein pigment, *rhodopsin*, found in the retina of the eye. Lack of vitamin A causes night blindness, which is the inability to see objects clearly in dim light. Vitamin A is also necessary for certain metabolic and reproductive processes. Foods containing vitamin A are cod liver oil, butter, eggs, cheese, and animal livers.

The major role of vitamin E in the body is to prevent the oxidation of unsaturated fatty acids and to add stability to biological membranes. A number of claims have also been made that this vitamin slows down the aging

FIGURE 12-32
A molecule of isoprene.

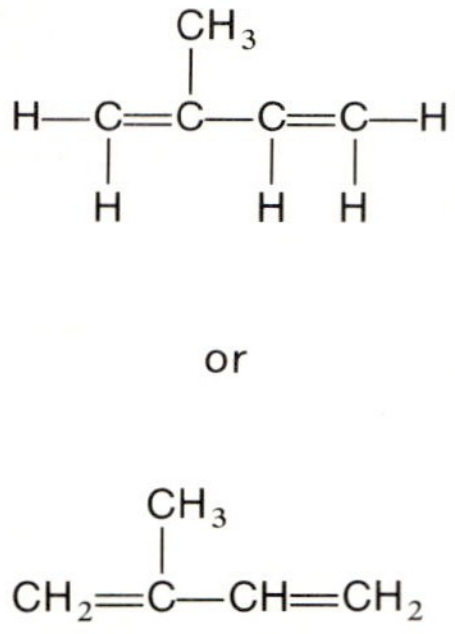

FIGURE 12-33
The structure of carotene.

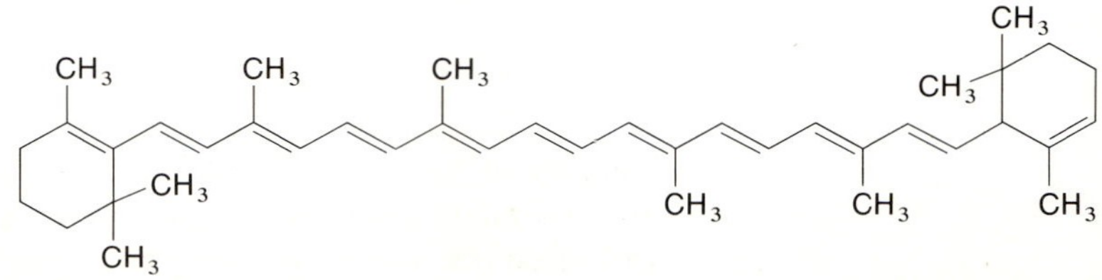

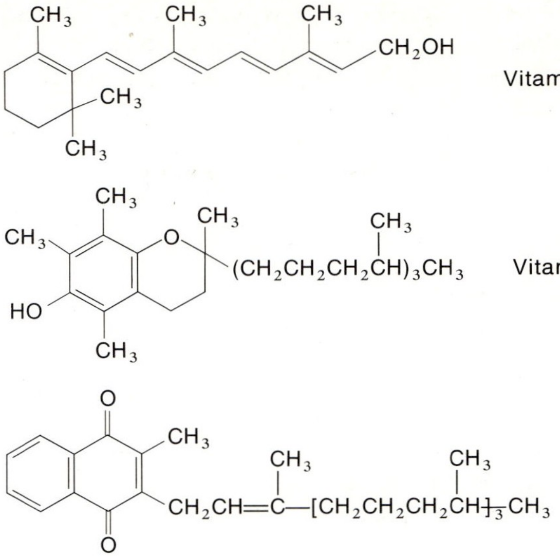

FIGURE 12-34
The structures of vitamins A, E, and K.

process. Vitamin E is found in milk, eggs, fish, meats, cereal, lettuce, parsley, spinach, cottonseed oil, corn oil, peanut oil, wheat germ oil, and at your natural food store.

The major role of vitamin K in the body is to aid in the coagulation of blood in the case of a body injury. A lack of vitamin K means that it will take longer for the blood to clot, which could be a serious matter if the injury is extensive. Vitamin K is found in cabbage, alfalfa, cauliflower, spinach, liver, eggs, and putrified fish meal.

EXAMPLE 12-6   State whether each of the following compounds is a steroid or terpene.

(a) testosterone
(b) vitamin A
(c) vitamin D

SOLUTION   (a) Testosterone is a steroid.
(b) Vitamin A is a terpene.
(c) Vitamin D is a steroid.

SUMMARY

In this chapter we learned about the various kinds of lipids. We learned that there are two major groups of lipids, those that have fatty acids in their molecules and those that don't. We saw that these two major groups could be further subdivided into six major classes of lipids. We then reviewed the components of these classes of lipids.

One of the major classes of lipids that we discussed was fats. We looked at the different types of fats as well as some of the important reactions that fats undergo. We then went on to look at some of the other major classes of lipids, including waxes, phospholipids, glycolipids, steroids, and terpenes. In our discussion of these major lipid classes we studied many of the important compounds which are representative of these lipids, and we frequently saw how they relate to life.

EXERCISES

1. State at least two general characteristics of all lipids.

2. Name the four major lipid components.

3. Identify the major class of each lipid from its components.
(a) fatty acids + glycerol
(b) fatty acids + sphingosine + a carbohydrate
(c) fatty acids + high-molecular-weight alcohols other than glycerol
(d) glycerol + fatty acids + phosphoric acid + choline
(e) fatty acids + sphingosine + phosphoric acid + a nitrogen compound

4. Write the general formula for members of the
(a) saturated fatty acids
(b) oleic acid series
(c) linoleic acid series
(d) linolenic acid series
(e) arachidonic acid series

5. Fill in the blank.
The _________ (saturated or unsaturated) fatty acids have one or more double bonds and are liquids at room temperature.

6. For each of the fatty acid structures given, state the following information:
(1) whether it is saturated or unsaturated
(2) whether it is a solid or liquid at room temperature
(3) whether it is a member of the oleic, linoleic, linolenic, or arachidonic acid series

$$\text{(a) } CH_3(CH_2)_6CH{=}CHCH_2CH{=}CHCH_2CH{=}CHCH_2CH{=}CH(CH_2)_3{-}\overset{\displaystyle O}{\overset{\|}{C}}{-}OH$$

$$\text{(b) } CH_3{-}(CH_2)_{18}{-}\underset{\displaystyle O}{\underset{\|}{C}}{-}OH$$

(c) $CH_3(CH_2)_7CH{=}CH(CH_2)_7{-}\overset{\displaystyle O}{\underset{\displaystyle \|}{C}}{-}OH$

7. Name the three essential fatty acids.

8. Write the structure of glycerol. What test can be used to detect glycerol in a fat?

9. Define each of the following terms.
   (a) prostaglandins
   (b) monoglyceride
   (c) diglyceride

10. State whether each of the following compounds are mono-, di-, or tri-glycerides:

(a)
$$CH_3{-}(CH_2)_{18}{-}\overset{O}{\overset{\|}{C}}{-}O{-}CH_2$$
$$CH_3{-}(CH_2)_{16}{-}\overset{O}{\overset{\|}{C}}{-}O{-}CH$$
$$HO{-}CH_2$$

(b)
$$CH_3{-}(CH_2)_{16}{-}\overset{O}{\overset{\|}{C}}{-}O{-}CH_2$$
$$CH_3{-}(CH_2)_{16}{-}\overset{O}{\overset{\|}{C}}{-}O{-}CH$$
$$CH_3{-}(CH_2)_{16}{-}\overset{O}{\overset{\|}{C}}{-}O{-}CH_2$$

(c)
$$CH_3{-}(CH_2)_{18}{-}\overset{O}{\overset{\|}{C}}{-}O{-}CH_2$$
$$HO{-}CH$$
$$HO{-}CH_2$$

11. Show what happens to the following fat when it undergoes
    (a) hydrolysis
    (b) complete hydrogenation

(c) saponification with NaOH

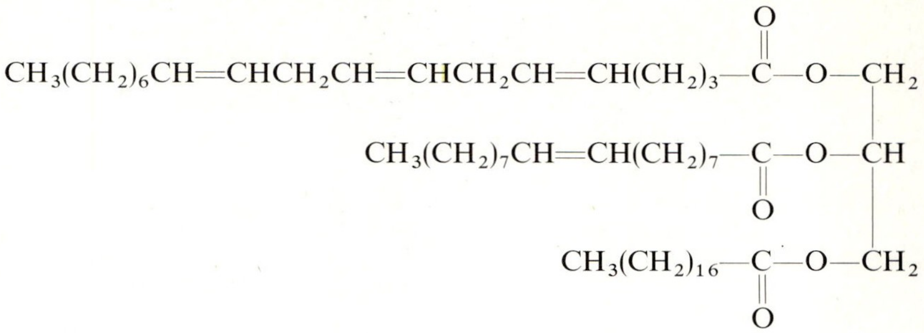

$$CH_3(CH_2)_6CH=CHCH_2CH=CHCH_2CH=CH(CH_2)_3-\overset{\overset{\textstyle O}{\|}}{C}-O-CH_2$$

$$CH_3(CH_2)_7CH=CH(CH_2)_7-\overset{}{C}-O-CH$$

$$CH_3(CH_2)_{16}-\overset{}{C}-O-CH_2$$

12. Match the lipid on the left with its description on the right.
    (1) lecithin
    (2) sphingomyelin
    (3) wax
    (4) cephalin
    (5) glycolipid

    (a) An ester of a fatty acid and a high-molecular-weight alcohol other than glycerol.
    (b) A lipid that contains sphingosine, fatty acids, phosphoric acid, and a nitrogen compound.
    (c) A lipid that contains fatty acids, a sugar, and glycerol.
    (d) A lipid that contains glycerol, fatty acids, phosphoric acid, and choline.
    (e) A lipid that contains glycerol, fatty acids, phosphoric acid, and ethanolamine.

13. State whether each of the following compounds is a steroid or terpene.

(a)

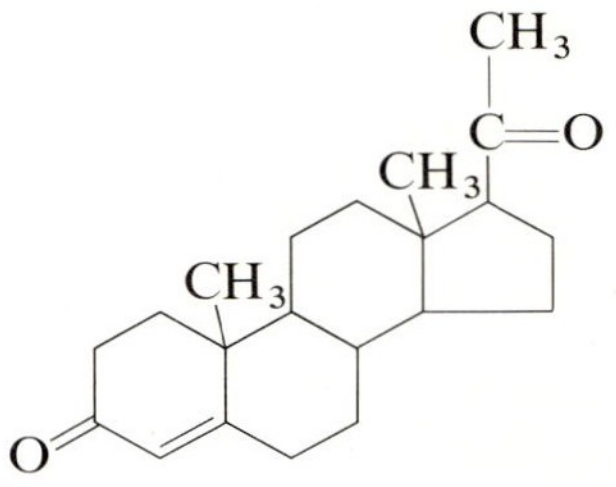

(b)

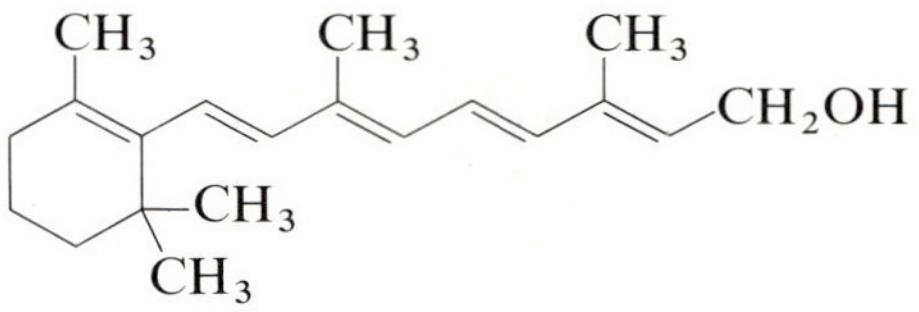

(c)

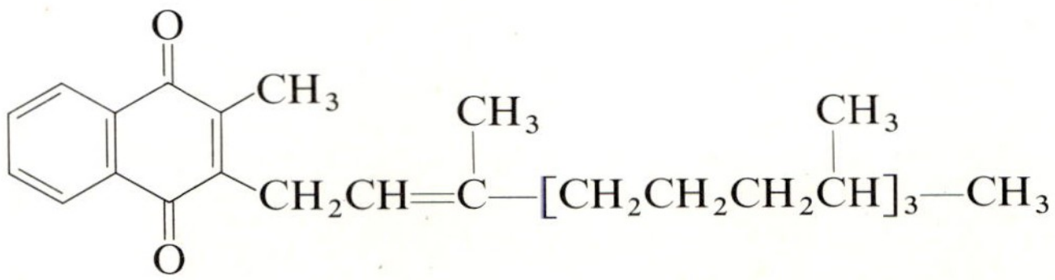

(d)

$$CH_3(CH_2)\text{—}CH_2\text{—}CH_2CH_2CH_2\text{—}CH\text{—}CH_3$$

(Steroid structure with CH$_3$ groups, HO— at ring)

14. Complete the following statement.
    The higher the iodine number, the more __________ (saturated or un-saturated) fatty acids there are in the fat.

15. Write the product of the triglyceride formed from stearic acid and glycerol.

$$3CH_3(CH_2)_{16}\text{—}\underset{O}{\overset{\|}{C}}\text{—}OH + \begin{array}{l} HO\text{—}CH_2 \\ HO\text{—}CH \\ HO\text{—}CH_2 \end{array} \longrightarrow$$

16. Match the vitamin with its role in the body.
    (1) vitamin A   (a) Aids in the coagulation of blood.
    (2) vitamin E   (b) Helps prevent night blindness.
    (3) vitamin D   (c) Prevents the oxidation of unsaturated fatty acids.
    (4) vitamin K   (d) Necessary for the proper absorption of calcium from the small intestine.

17. Define each of the following terms.
    (a) rancidity
    (b) iodine number
    (c) sterol
    (d) terpene
    (e) steroid

18. Nondairy creamers are often made from hydrogenated vegetable oils. This means that they are high in (saturated or unsaturated) fatty acids.

19. Explain the cleansing action of soaps from a molecular point of view.

20. Write the formula of the mixed triglyceride containing glycerol, butyric acid, oleic acid, and stearic acid.

CHAPTER 13

All About Proteins

It's Sometimes Denaturing

Some Things You Should Know After Reading This Chapter

You should be able to:

1. Write the general formula of an amino acid as $H_2N-\overset{\displaystyle R}{\underset{\displaystyle H}{C}}-\overset{\displaystyle }{\underset{\displaystyle O}{C}}-OH$.

2. Describe proteins as being composed of various combinations of amino acids.
3. Define what is meant by an essential amino acid.
4. Describe what is meant by complete and incomplete proteins.
5. List two sources of complete protein and incomplete protein.
6. Define and give an example of a polar, nonpolar, and ionic R group found in amino acids.
7. State how an amino acid can act as an acid or a base.
8. Define and write the structure of a zwitterion.
9. Show how two amino acids are joined together by a peptide bond.
10. Define the terms dipeptide, tripeptide, and polypeptide.
11. Name dipeptides, tripeptides, or polypeptides using their full or abbreviated names.
12. Describe and give an example of a fibrous protein and a globular protein.
13. Tell what is meant by the primary, secondary, tertiary, and quaternary structure of a protein.
14. Define what is meant by denaturation of proteins.
15. State at least three ways proteins can be denatured.
16. Name the three protein tests and describe each one.

Proteins are important molecules in the living cell. Without them the human being would not exist. Proteins are essential in sustaining normal life processes. Some of the functions of proteins include catalyzing reactions in the body, defending the body against disease, helping with the digestion of foods, transporting oxygen in the blood, aiding in the clotting of blood, regulating cellular activities, and participating in genetic functions. About one-third of all the proteins in the body are structural proteins which make up the membranes, musculature, and connective tissues of the body (Fig. 13-1).

Proteins are synthesized by living matter. Plants produce the greatest amount of proteins. They synthesize carbohydrates and then combine these carbohydrates with nitrogen-containing compounds from the soil to produce proteins. Animals have a very limited ability to produce proteins, so they must include them in the diet and transform them into other types of proteins in the body.

Proteins are composed of the elements carbon, hydrogen, oxygen, and nitrogen. Some proteins contain sulfur, iodine, phosphorus, or iron. Proteins are large molecules with weights ranging from 5000 to millions of grams per mole. They are such large molecules because they are polymers composed of simpler units called *amino acids*. In fact, it is the amino acids that are the key to understanding the functions and compositions of proteins.

FIGURE 13-1

The structure of collagen, the major protein of connective tissue and bone. Three polypeptide chains form a triple helix in the basic unit of this protein, called tropocollagen.

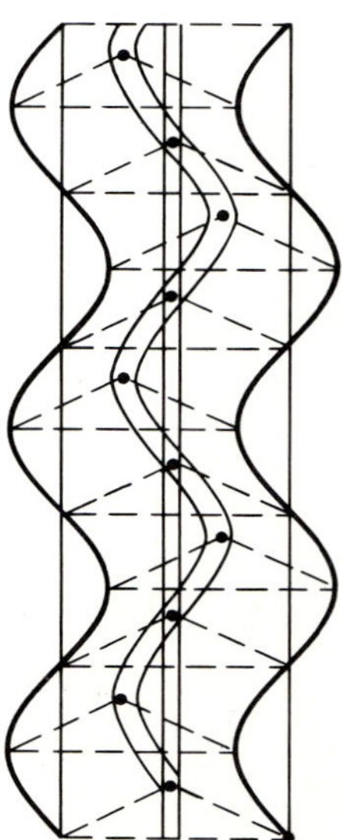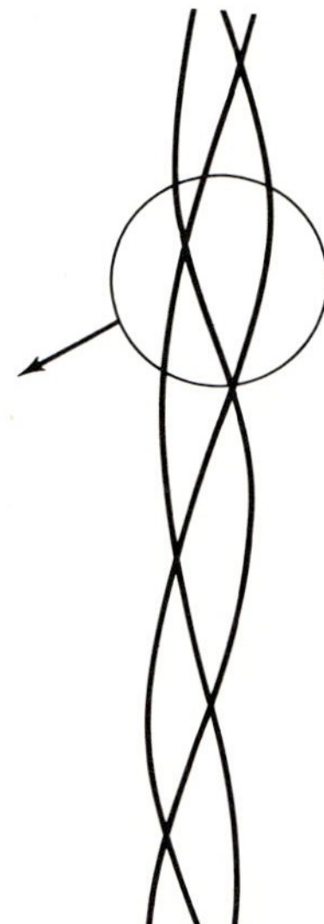

Amino Acids: The Building Blocks of Proteins

As we just stated, amino acids are the building blocks of proteins. It is the union of amino acids in the proper sequence that gives each protein its special structure and function. To understand proteins, we must first understand amino acids. So let's study the general properties of amino acids and then look into the *peptide linkage*, which is the bond that joins amino acids together to form proteins.

All About Amino Acids

Amino acids are the molecules that make up proteins. They are difunctional organic compounds that contain a carboxyl group ($-\overset{\text{O}}{\underset{\text{}}{\text{C}}}-$OH) on one end of the molecule and an amino group ($-$NH$_2$) on the other end of the molecule. The general structure of an amino acid can be seen in Figure 13-2.

Each amino acid has a different R group, which is what makes each amino acid unique. There are 20 major amino acids found in proteins. Table 13-1 lists these amino acids. Because amino acids are the building blocks of proteins, they are very important to us biologically. Some amino acids can by synthesized by the body and others can be interconverted from closely related amino acids. There are also eight amino acids that can't be synthesized by the body in any way and must be included in the diet. These are called the *essential amino acids* and they include lysine, tryptophan, phenylalanine, methionine, threonine, leucine, isoleucine, and valine. Another amino acid, histidine, is sometimes listed with the essential amino acids because it is required by infants and children for proper growth and development (along with the other eight essential amino acids, of course). Consumption of these essential amino acids is important for the maintenance of a healthy body (Table 13-2).

FIGURE 13-2
The general structure of an amino acid.

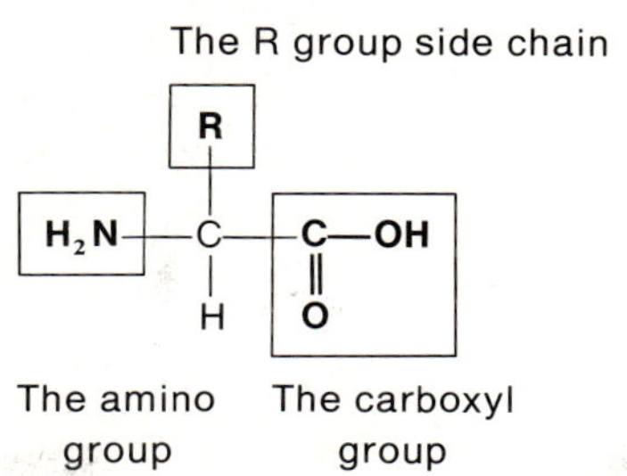

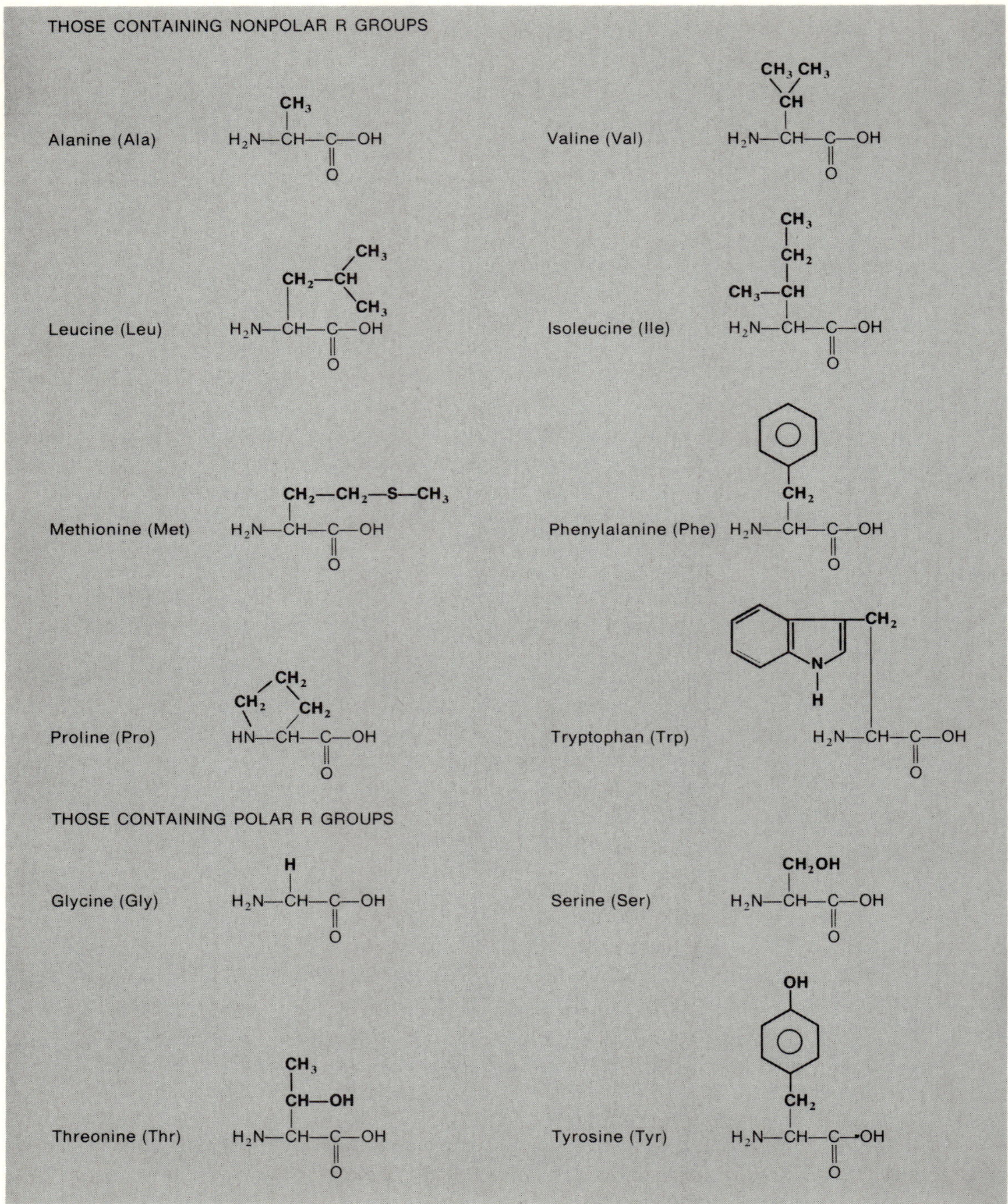

THOSE CONTAINING NONPOLAR R GROUPS
Alanine (Ala)
Valine (Val)
Leucine (Leu)
Isoleucine (Ile)
Methionine (Met)
Phenylalanine (Phe)
Proline (Pro)
Tryptophan (Trp)
THOSE CONTAINING POLAR R GROUPS
Glycine (Gly)
Serine (Ser)
Threonine (Thr)
Tyrosine (Tyr)

TABLE 13-1   The 20 amino acids commonly found in protein (*cont.*)

THOSE CONTAINING POLAR R GROUPS (*cont.*)

Cysteine (Cys)

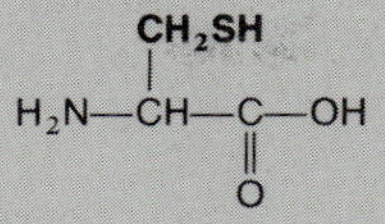

Asparagine (Asn)

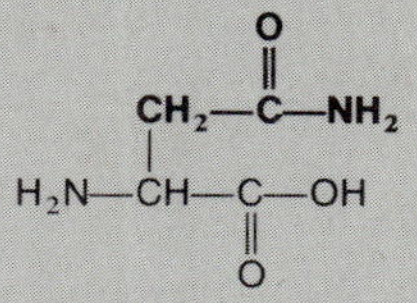

Glutamine (Gln)

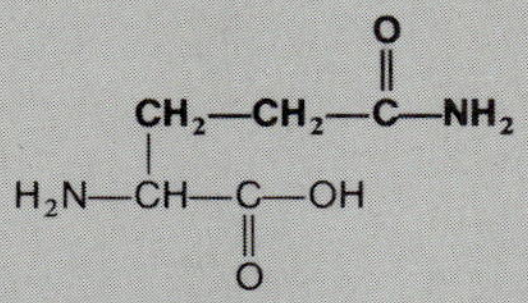

THOSE CONTAINING IONIC R GROUPS

Aspartic acid (Asp)

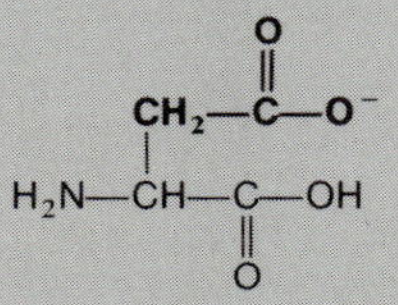

Glutamic acid (Glu)

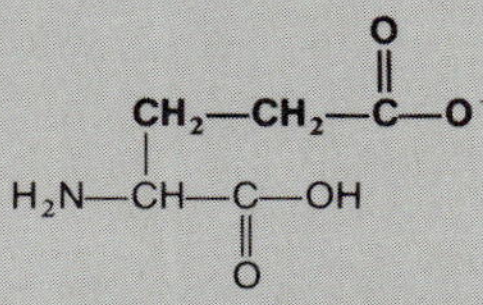

Lysine (Lys)

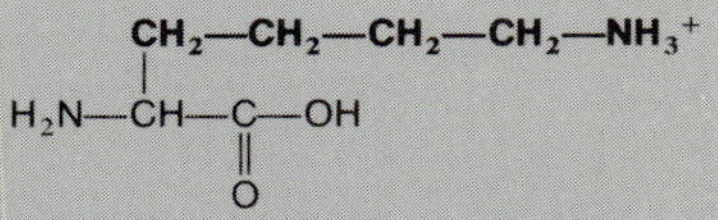

Arginine (Arg)

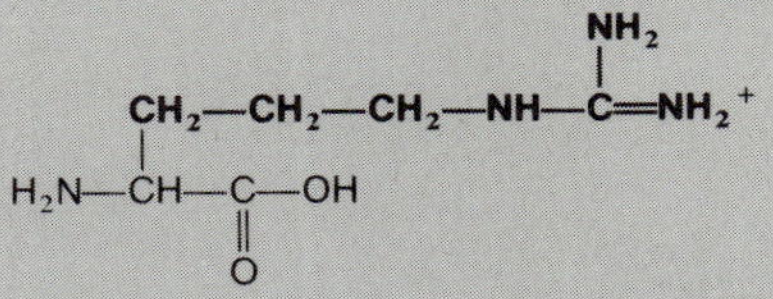

Histidine (His)

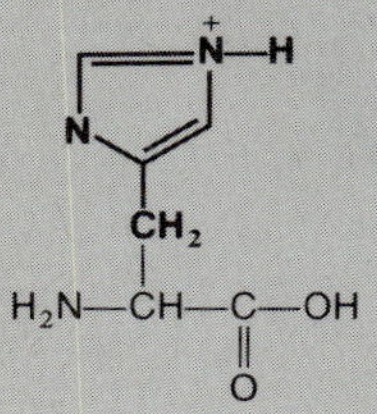

Complete proteins contain all the amino acids essential for maintenance of good health. Eggs, milk, meat, fish, poultry, and soybean are the best sources of complete protein in the diet. Cereal grain, nuts, and vegetables are not sources of complete protein, since they may be low in one or more of the amino acids. Proteins found in such foods are called *incomplete proteins.* However, mixtures of whole grain cereals and vegetables, or mixtures of whole grains and milk, meat, or eggs are goods sources of complete protein. Generally speaking, proteins that come from vegetable sources are incomplete and those that come from animal sources are complete proteins.

TABLE 13-2   Essential amino acid requirements

ESSENTIAL AMINO ACID	GRAMS/DAY (for a 150-pound adult)
Isoleucine	0.84
Leucine	1.12
Lysine	0.84
Methionine[a]	0.70
Phenylalanine[b]	1.12
Threonine	0.56
Tryptophan	0.21
Valine	0.98

[a] Includes cystine.
[b] Includes tyrosine.
Source: The National Research Council—National Academy of Sciences, 1974.

EXAMPLE 13-1   List two sources of complete protein and two sources of incomplete protein.

SOLUTION   Complete protein: milk, eggs, cheese, meat, fish, or soybean.
Incomplete protein: vegetables, cereal grains, or nuts.

Classification of Amino Acids

Amino acids are classified by looking at the polarity of their R groups. (Remember that the different R groups are what distinguish one amino acid from another.) There are three classes of R groups: *polar*, *nonpolar*, and *ionic*. *The polar R groups* contain a carbonyl group ($-\overset{\displaystyle \|}{\underset{\displaystyle O}{C}}-$), a hydroxyl group ($-OH$), or a mercaptan group ($-SH$). Because water is a polar molecule, and polar compounds dissolve in polar solvents, the amino acids with polar R groups are extremely water-soluble. The *nonpolar R groups* are generally hydrocarbons, for example, the isopropyl R group ($CH_3-\overset{\displaystyle |}{\underset{\displaystyle CH_3}{CH}}-$) found in valine. The amino acids that contain these nonpolar R groups are not as soluble in water. The *ionic R groups* contain carboxyl groups ($-\overset{\displaystyle \|}{\underset{\displaystyle O}{C}}-OH$) or amino groups ($-NH_2$).

They ionize in water and therefore are very soluble (Table 13-3).

NONPOLAR	POLAR	IONIC
Alanine	Glycine	Aspartic acid
Valine	Serine	Glutamic acid
Leucine	Threonine	Lysine
Isoleucine	Tyrosine	Arginine
Methionine	Cysteine	Histidine
Phenylalanine	Asparagine	
Proline	Glutamine	
Tryptophan		

Acid-Base Properties of Amino Acids

Amino acids have important acid-base properties because they contain both an amino group ($-NH_2$), which is basic, and a carboxyl group ($-\overset{\underset{\|}{O}}{C}-OH$), which is acidic. This allows amino acids to act as either acids or bases in water. Substances that can act as either acids or bases in water are called *amphoteric*.

EXAMPLE 13-2

(a) Show how an amino acid acts as a base when added to a solution of nitric acid.

(b) Show how an amino acid acts as an acid when added to a solution containing potassium hydroxide.

SOLUTION

(a) The amino part of the amino acid acts as a base because it reacts with the H^+ ions from the nitric acid.

$$H^+NO_3^- + H_2N-\underset{\underset{H}{|}}{\overset{\overset{R}{|}}{C}}-\underset{\underset{O}{\|}}{C}-OH \longrightarrow H_3\overset{+}{N}-\underset{\underset{H}{|}}{\overset{\overset{R}{|}}{C}}-\underset{\underset{O}{\|}}{C}-OH + NO_3^-$$

(b) The carboxyl part of the amino acid acts as an acid because it reacts with the OH^- ions from the potassium hydroxide.

$$K^+OH^- + H_2N-\underset{\underset{H}{|}}{\overset{\overset{R}{|}}{C}}-\underset{\underset{O}{\|}}{C}-OH \longrightarrow H_2N-\underset{\underset{H}{|}}{\overset{\overset{R}{|}}{C}}-\underset{\underset{O}{\|}}{C}-O^-K^+ + H-OH$$

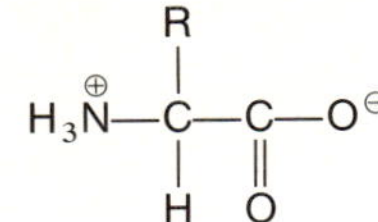

FIGURE 13-3
The general structure
of a zwitterion or
dipolar ion.

Amino acids can also act as *buffers*. Buffers can neutralize small quantities of acids and bases and maintain the pH of a solution. This is an important property of amino acids because maintenance of the proper chemical balance in the body is due in part to the buffering ability of proteins. For example, maintenance of blood pH is due partly to protein buffers (composed, of course, of amino acids) found in the blood.

The Zwitterion

Amino acids have high melting points and they are soluble in water. These observations have led biochemists to conclude that in the solid state, amino acids *don't* exist as neutral molecules, but instead as dipolar ions (Fig. 13-3). The dipolar ion is formed by the transfer of a hydrogen ion from the carboxyl group of the amino acid to the amine group. The dipolar form of an amino acid is called a *zwitterion*. The zwitterion form of an amino acid is also found when the amino acid is placed in a solution at a certain pH.

Peptide Bonds

Now that we have a basic knowledge of amino acids, let's look at the bonds that hold two or more amino acids together to form proteins.

Peptide bonds, or *peptide linkages* as they are often called, hold amino acids together in a protein molecule. The carboxyl group of one amino acid bonds to the amino group of another amino acid. When two amino acids combine to form a peptide bond, one molecule of water is eliminated in the reaction (Fig. 13-4).

EXAMPLE 13-3 Show how the amino acids glycine and alanine bond to form the dipeptide called glycylalanine. (*Hint*: The carboxyl group of glycine combines with the amino group of alanine in this compound.)

FIGURE 13-4
When two amino acids combine to form a peptide bond, one molecule of water is eliminated in the reaction.

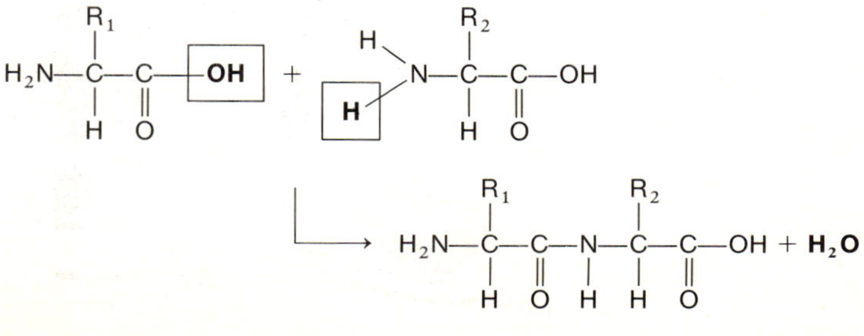

SOLUTION   We'll write out the reaction.

$$H_2N-CH_2-\underset{O}{C}-\boxed{OH} + \boxed{H}-N-\underset{H}{\underset{|}{C}}-\underset{O}{\overset{CH_3}{C}}-OH$$

Glycine   Alanine

$$\longrightarrow H_2N-CH_2-\underset{O}{C}-N-\underset{H}{C}-\underset{O}{\overset{CH_3}{C}}-OH + H_2O$$

Peptide linkage

Glycylalanine

Amino acids can form long protein chains because each amino acid has an amino end and a carboxyl end available for bonding. Let's look at the following situation to understand this better. Suppose that you were going to tie many pairs of sneakers together. Each pair of sneakers has a left lace and a right lace. You would tie the right lace of one pair of sneakers to the left lace of the next pair of sneakers. You would always end up with a free left lace at one end of the sneaker chain and a free right lace at the other end of the sneaker chain (Fig. 13-5). It's the same with protein chains. You bond the carboxyl group of one amino acid to the amino group of the next amino acid. You always end up with a free amino group at one end of the protein chain and a free carboxyl group at the other end of the protein chain.

FIGURE 13-5
The sneaker chain.

Compounds formed by the bonding of amino acids have special names. *Dipeptides* are formed when two amino acids form a peptide bond. *Tripeptides* are formed when three amino acids form peptide bonds. *Polypeptides* are formed when more than three amino acids form peptide bonds. Peptide bonds are covalent bonds. They can be broken by the action of certain enzymes, or by acid or base hydrolysis.

When dipeptides form between *two* amino acids, they can combine in two ways. For example, when glycine and alanine bond, they can do the following:

1. The carboxyl group of the glycine can combine with the amino group of the alanine, to form the dipeptide called glycylalanine.

$$H_2N-CH_2-\underset{\underset{O}{\|}}{C}-\underset{\underset{H}{|}}{N}-\underset{\overset{CH_3}{|}\ \underset{H}{|}}{C}-\underset{\underset{O}{\|}}{C}-OH$$

Glycylalanine

2. The carboxyl group of the alanine can combine with the amino group of the glycine, to form the dipeptide called alanylglycine.

$$H_2N-\underset{\overset{CH_3}{|}\ \underset{H}{|}}{C}-\underset{\underset{O}{\|}}{C}-\underset{\underset{H}{|}}{N}-CH_2-\underset{\underset{O}{\|}}{C}-OH$$

Alanylglycine

The compound formed depends upon which amino acid contains the free amino group and which contains the free carboxyl group.

Dipeptides are named using the following rules:

1. Write the name of the amino acid containing the free-amino end first; however, change the end of its name from *ine* to *yl*. For example, alan*ine* becomes alan*yl* and glyc*ine* becomes glyc*yl*.

2. Add the name of the amino acid containing the free carboxyl end.

Sometimes abbreviations are used to name the amino acids in a protein (Table 13-1). The same order of writing the amino acids is followed as when writing the full names, but each three-letter abbreviation is followed by a hyphen. In other words, alanylglycine becomes Ala-Gly.

EXAMPLE 13-4 Name the dipeptides that can be formed from lysine and serine. Write out their full names and abbreviated names.

SOLUTION   There are two ways that these amino acids can combine with each other:

> (a) lysylserine   (Lys-Ser)
> (b) seryllysine   (Ser-Lys)

Tripeptides are composed of three amino acids. Therefore, there are six ways in which these amino acids can arrange themselves. For example, if we have three amino acids, A, B, and C, they could combine in the following ways:

$$A\text{-}B\text{-}C \qquad B\text{-}A\text{-}C \qquad C\text{-}A\text{-}B$$
$$A\text{-}C\text{-}B \qquad B\text{-}C\text{-}A \qquad C\text{-}B\text{-}A$$

Just imagine the number of ways that you could arrange a polypeptide containing 20 amino acids!

The rules for naming tripeptides and polypeptides are the same as those for dipeptides. The amino acid with the free amino group is named first (along with its *-yl* ending). The other amino acids follow in order (with the endings of their names also changed to *yl*). The amino acid with the free carboxyl group is named last.

EXAMPLE 13-5   Name the six tripeptides that can be formed from the amino acids proline (Pro), valine (Val), and lysine (Lys). Write out their full names and abbreviated names.

SOLUTION

Pro-Val-Lys	prolylvalyllysine
Pro-Lys-Val	prolyllysylvaline
Val-Pro-Lys	valylprolyllysine
Val-Lys-Pro	valyllysylproline
Lys-Pro-Val	lysylprolylvaline
Lys-Val-Pro	lysylvalylproline

EXAMPLE 13-6   Show the structure of the tripeptide Ala-Ser-Val.

SOLUTION

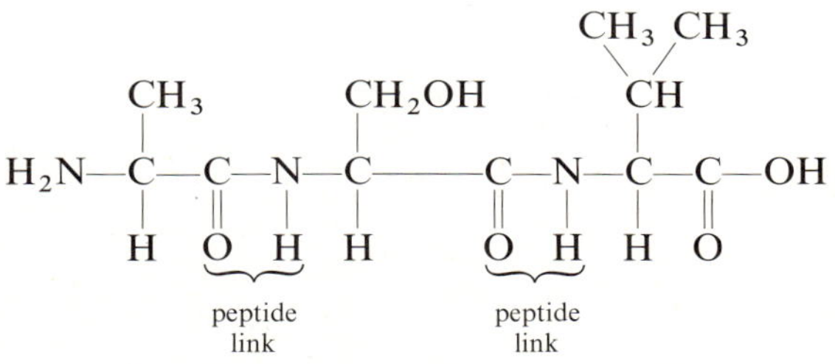

$$H_2N-\overset{\overset{\displaystyle CH_3}{|}}{\underset{\underset{\displaystyle H}{|}}{C}}-\overset{\overset{}{}}{\underset{\underset{\displaystyle O}{||}}{C}}-\overset{}{\underset{\underset{\displaystyle H}{|}}{N}}-\overset{\overset{\displaystyle CH_2OH}{|}}{\underset{\underset{\displaystyle H}{|}}{C}}-\overset{}{\underset{\underset{\displaystyle O}{||}}{C}}-\overset{}{\underset{\underset{\displaystyle H}{|}}{N}}-\overset{\overset{\displaystyle CH}{\overset{\diagup\diagdown}{CH_3\ CH_3}}}{\underset{\underset{\displaystyle H}{|}}{C}}-\overset{}{\underset{\underset{\displaystyle O}{||}}{C}}-OH$$

peptide link   peptide link

Ala   Ser   Val

Classification of Proteins

Now that we see how proteins are formed, let's see how they're classified. There are several ways in which proteins are classified. This is because they are so complex. The scheme we will use classifies proteins according to their solubilities in water. *Fibrous proteins* are those which are insoluble in water. These proteins have a protective or structural function in the body. Some examples of fibrous proteins are elastin, which is found in the large blood vessels; keratin, which is found in hair nails and wool; and collagen, which is found in tendons. *Globular proteins* are those which are soluble in water. These proteins generally carry out the chemical processes of the body. Some examples of globular proteins are the enzymes; hemoglobin, which transport oxygen in the blood; lipoprotein, which transports fatty acids in the blood; and prothrombin, which causes blood clotting.

EXAMPLE 13-7 Name the two classes of proteins, then define them and give an example of each.

SOLUTION

(a) Fibrous proteins are those which are insoluble in water and have a protective or structural function in the body. Some examples of these proteins are elastin, keratin, and collagen.

(b) Globular proteins are those which are water-soluble. These proteins carry out many of the chemical processes of the body. Some examples

TABLE 13-4 The classification of proteins by function

CLASS	EXAMPLE
Blood proteins	Fibrinogen, to help blood clot
Catalytic proteins	Enzymes, to speed up and regulate cellular reactions
Contractile proteins	Myosin, which is found in muscle tissues
Digestive proteins	Pepsin, an *enzyme* that splits other proteins into smaller molecules
Hormonal proteins	Insulin, which regulates activities in cells
Natural-defense proteins	Antibodies, which protect the body from disease
Repressor proteins	Proteins that regulate expression of genetic information
Respiratory proteins	The cytochromes, which help transport electrons to electron acceptors such as oxygen
Structural proteins	Collagen, which is found in connective tissue of vertebrates
Transport proteins	Hemoglobin, which moves oxygen through the bloodstream

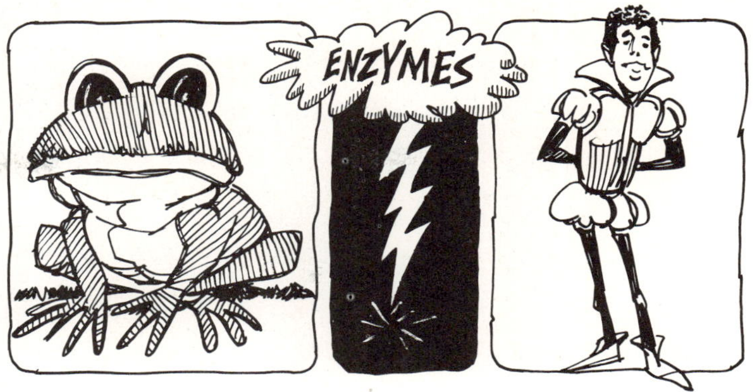

FIGURE 13-6
Enzymes speed up or catalyze chemical reactions in the body.

of these proteins are enzymes, hemoglobin, lipoprotein, and prothrombin.

The Functions of Proteins

Proteins have a wide range of functions, which makes them so important in
nature. This gives rise to another method of classification. There are 10 categories into which proteins are divided according to their functions. Table 13-4
summarizes the functional classification of proteins. Since all proteins are
important in sustaining life, no one functional classification ranks above another. Proteins that function as enzymes, however, are of special importance.
This is because enzymes catalyze all the chemical reactions that take place
in the cell. Enzymes speed up these reactions and determine how far each
reaction will proceed. This enables the cell to adapt to its changing needs in an
efficient and economical manner (Fig. 13-6).

Protein Structure: The Big Picture

Now that we have a basic understanding of proteins, let's take a look at a
protein's total structure. You probably think that we've already done this.
After all, we've learned that proteins are combinations of amino acids, arranged
in a specific order. But this is only part of a protein's structure, called its
primary structure. It is not unusual for a protein to have a secondary, tertiary,
and quaternary structure as well. Let's see what all this means.

Primary Structure of Proteins

The *primary structure* of a protein is the basic sequence or order of amino
acids which make up the protein (Fig. 13-7). We've already learned that there
are 20 different amino acids that make up most of the known proteins. A

Val—His—Leu—Thr—Pro—Glu—Glu—Lys—Ser—Ala—Val—Thr—Ala—Leu—Trp—Gly—Lys—Val—

 1   2   3   4   5   6   7   8   9   10   11   12   13   14   15   16   17   18

Asp—Val—Asp—Glu—Val—Gly—Gly—Glu—Ala—Leu—Gly—Arg—Leu—Leu—Val—Val—Tyr—Pro—

19   20   21   22   23   24   25   26   27   28   29   30   31   32   33   34   35   36

Trp—Thr—Glu—Arg—Phe—Phe—Glu—Ser—Phe—Gly—Asp—Leu—Ser—Thr—Pro—Asp—Ala—Val—

37   38   39   40   41   42   43   44   45   46   47   48   49   50   51   52   53   54

Met—Gly—Asp—Pro—Lys—Val—Lys—Ala—His—Gly—Lys—Lys—Val—Leu—Gly—Ala—Phe—Ser—

55   56   57   58   59   60   61   62   63   64   65   66   67   68   69   70   71   72

Asp—Gly—Leu—Ala—His—Leu—Asp—Asp—Leu—Lys—Gly—Thr—Phe—Ala—Thr—Leu—Ser—Glu—

73   74   75   76   77   78   79   80   81   82   83   84   85   86   87   88   89   90

Leu—His—Cys—Asp—Lys—Leu—His—Val—Asp—Pro—Glu—Asp—Phe—Arg—Leu—Leu—Gly—Asp—

91   92   93   94   95   96   97   98   99 100 101 102 103 104 105 106 107 108

Val—Leu—Val—Cys—Val—Leu—Ala—His—His—Phe—Gly—Lys—Glu—Phe—Thr—Pro—Pro—Val—

109 110 111 112 113 114 115 116 117 118 119 120 121 122 123 124 125 126

Glu—Ala—Ala—Tyr—Glu—Lys—Val—Val—Ala—Gly—Val—Ala—Asp—Ala—Leu—Ala—His—Lys—

127 128 129 130 131 132 133 134 135 136 137 138 139 140 141 142 143 144

Tyr—His

145 146

FIGURE 13-7

The primary structure of a beta chain of hemoglobin.

survey of these known proteins shows that the average *chain length* (or number of amino acids bonded together) is somewhere between 100 and 150 amino acids per protein molecule. Needless to say, proteins are very complex molecules, and there are numerous methods employed by chemists for determining the sequence of amino acids in proteins. In 1956, Sanger determined the exact amino acid sequence of insulin. Once the amino acid sequence is known, chemists attempt to synthesize these molecules in the laboratory. This is not an easy task, but some peptide chains and some proteins have been synthesized. Vasopressin and oxytocin, which are hormones produced by the posterior pituitary gland, have been synthesized. In 1965, a Chinese group of scientists combined the research of others and produced the first all-synthetic insulin.

Secondary Structure of Proteins

The *primary structure* tells us the sequence of amino acids in a protein. But a protein is more than just a straight chain of amino acids. It has a special geometrical arrangement. The *secondary structure* of a protein tells us how the polypeptide chain is arranged in space. It tells us, for example, if the chain is in an ordered helix, which is a coiled spiral-like arrangement, or if it is in an orderly pleated sheet arrangement, or if it is in a disordered random arrangement (Fig. 13-8).

FIGURE 13-8
Some secondary structures of proteins.

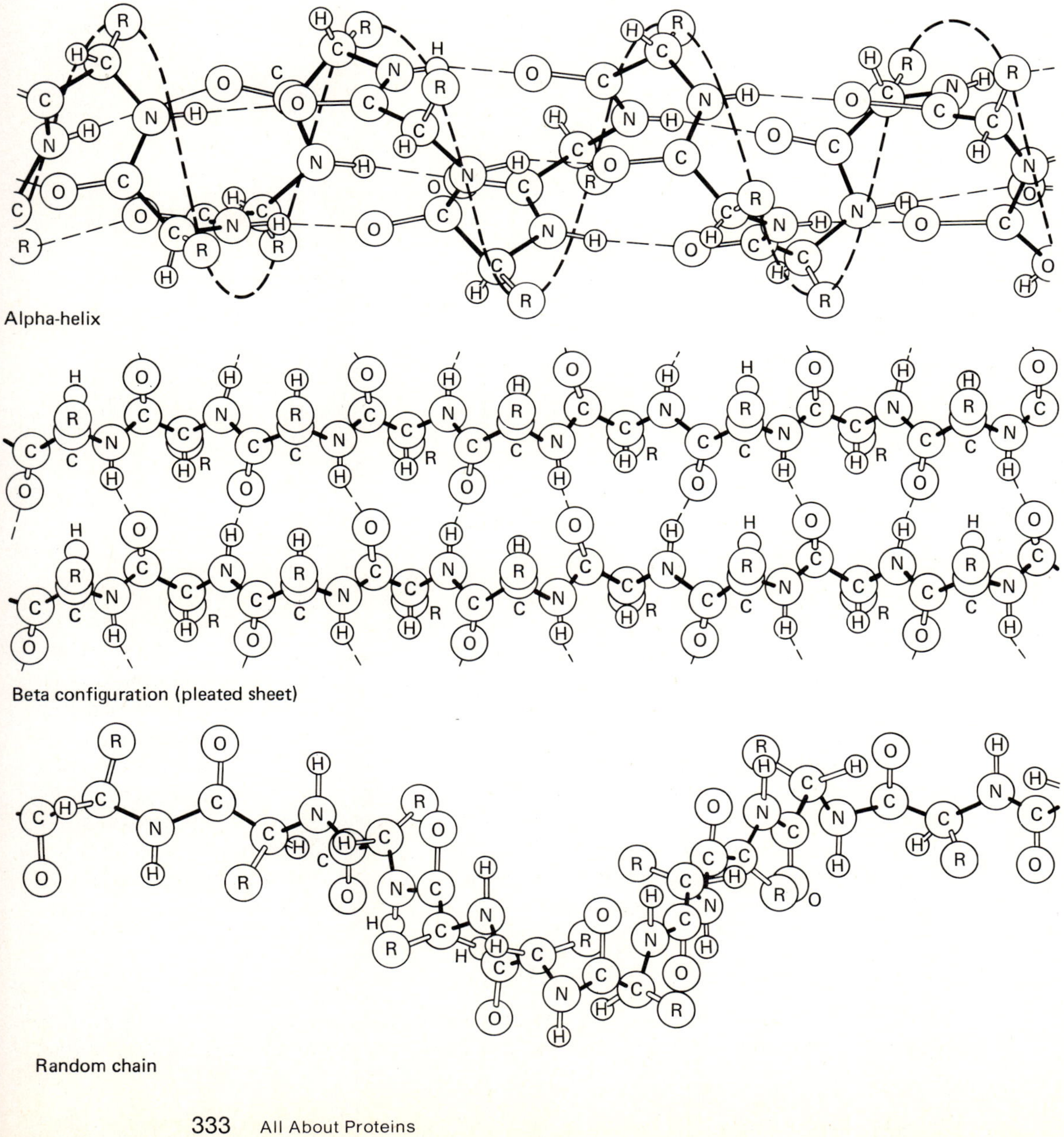

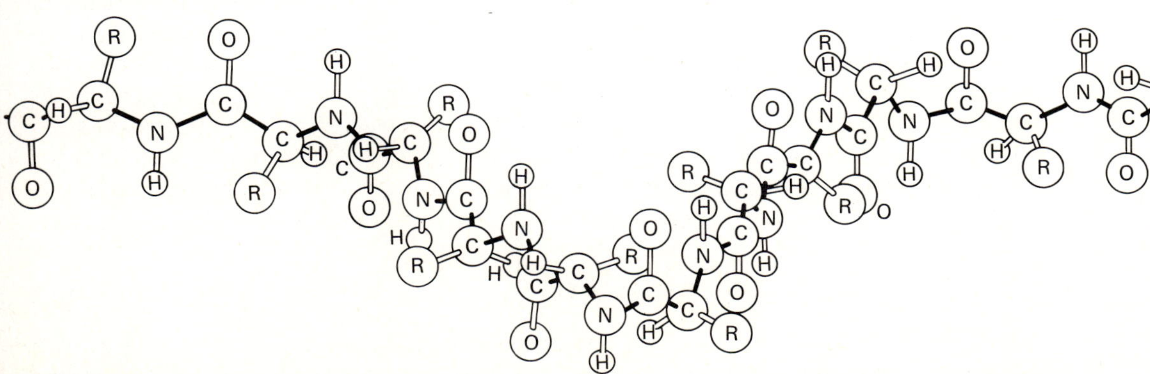

In the early 1950's, Linus Pauling and coworkers discovered that the polypeptide chains in the protein keratin are coiled in what is called an *alpha-helix*. In an alpha-helix there are 3.6 amino acids in every turn of the helix. The helix is held together by hydrogen bonds between a hydrogen atom (belonging to the amino part of a peptide bond) and an oxygen atom (belonging to the carboxyl part of a peptide bond). The hydrogen bonding takes place between atoms found along the turns of the helix, and since peptide bonds occur regularly along the helix, the entire helix is held together by these bonds (Fig. 13-9).

FIGURE 13-9
Hydrogen bonding in the alpha-helix. (From M. M. Bloomfield, *Chemistry and the Living Organism*, 1977. By permission of John Wiley & Sons.)

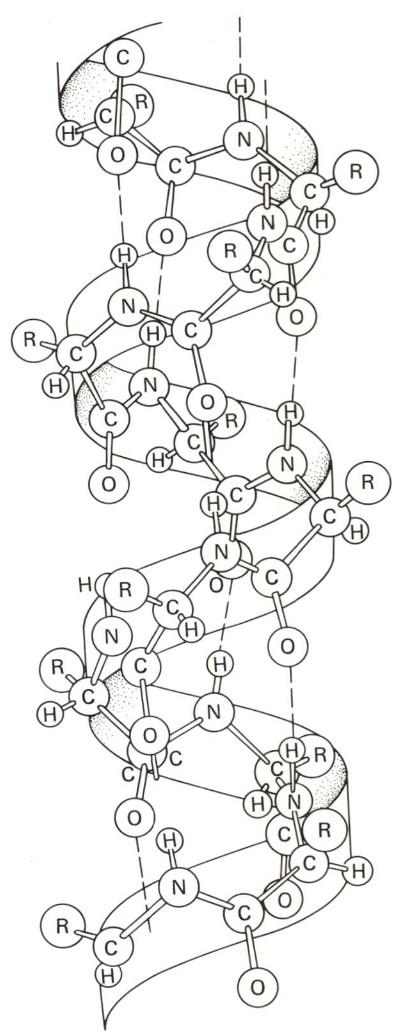

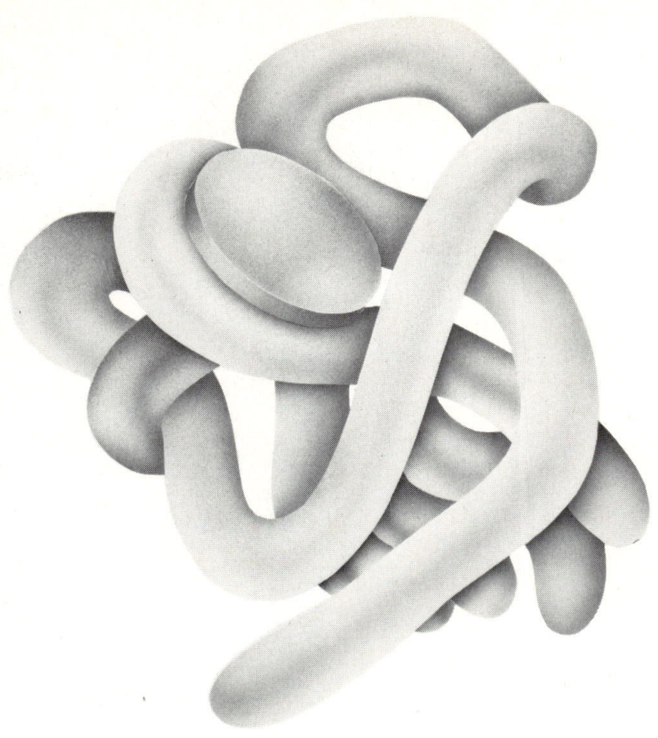

FIGURE 13-10
A model of the myoglobin molecule, showing its tertiary structure. The disk represents the heme group.

Tertiary Structure of Proteins

The *three-dimensional structure* of a protein is called its *tertiary structure*. It specifically pertains to the way the secondary structure is folded in space. Experiments with certain proteins, such as myoglobin, keratin, myosin, and albumin, have shown that the alpha-helix does not remain as a long extended coil. In such molecules the helix folds upon itself several times to produce a compact molecule. The protein myoglobin, which is an oxygen-carrying protein found in the muscle tissue of mammals, if extended would be an alpha-helix whose length is 20 times its width. But experiments show that it is a compact molecule which looks almost like a football because of the folding of the alpha-helix (Fig. 13-10).

Quaternary Structure of Proteins

Some proteins have a *quaternary structure*. This means that they exist as groups of two or more polypeptide chains. The polypeptide chains are tertiary structures, so a quaternary structure is really made up of groups of two

C
N
N
C

PROTEIN	NUMBER OF SUBUNITS	MOLECULAR WEIGHT
Insulin	2	11,466
Hemoglobin	4	64,500
Alcohol dehydrogenase	4	80,000
Apoferritin	20	480,000
Tobacco mosaic virus	2130	40,000,000

or more tertiary structures. Each tertiary structure is called a *subunit* (Table 13-5).

Hemoglobin is the classical example of a protein with a quaternary structure. Four peptide chains (which are tertiary structures, or subunits) make up the hemoglobin molecule. Each chain is able to combine with one oxygen molecule, so the hemoglobin molecule can combine with four oxygen molecules (Fig. 13-11). Hemoglobin is found in the blood and it is the molecule that carries oxygen to all the cells of the body.

Denaturation of Proteins: The Breakdown of Protein Structure

When the structure of a protein is altered in some way, and peptide bonds have *not* been broken, the protein is said to be *denatured*. This can occur if the tertiary structure is unfolded, or if disulfide bonds, which sometimes hold tertiary structures together, are broken, or if the primary structure becomes a randomly shaped coil (Fig. 13-12). Sometimes the protein structure unfolds and can be refolded. This is called *renaturation*, or reversal of the denaturation process. At other times an irreversible denaturation takes place, such as in the case of a hard-cooked egg. Once the egg is boiled in water, the protein coagulates and

FIGURE 13-11
The quaternary structure of hemoglobin. Hemoglobin molecule, as deduced from X-ray diffraction studies, is shown from above (top) and side (bottom). The drawings follow the representation scheme used in three-dimensional models built by M. F. Perutz and his co-workers. The irregular blocks represent electron-density patterns at various levels in the hemoglobin molecule. The molecule is built up from four subunits: two identical alpha chains (white blocks) and two identical beta chains (dark gray blocks). The letter "N" in the top view identifies the amino ends of the two alpha chains; the letter "C" identifies the carboxyl ends. Each chain enfolds a heme group (light gray disk), the iron-containing structure that binds oxygen to the molecule. (From M. F. Perutz, The hemoglobin molecule, *Scientific American*, Nov. 1964, *211*, (5), pp. 64–76. Copyright © 1964 by Scientific American, Inc.)

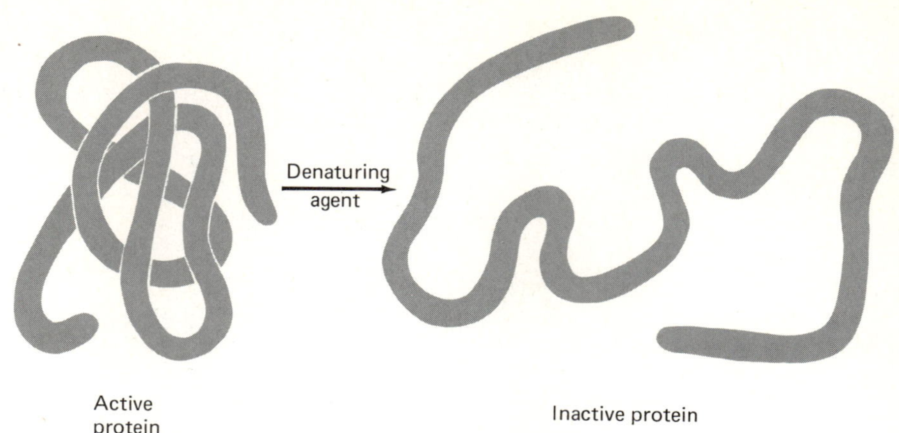

FIGURE 13-12
The denaturation of a protein.

the change cannot be reversed. This is because the unfolded chains group together, forming insoluble precipitates. At this point the protein has lost its biological activity.

Heat also causes proteins to denature. This is because as the temperature of the protein increases, the atoms making up the protein move faster and faster, and the bonds that hold atoms together in the tertiary structure break. The tertiary structure then unfolds and is lost. Heat is used to kill bacteria by destroying the bacterial protein during sterilization in boiling water or in an autoclave.

There are also many organic solvents that denature proteins by destroying their secondary and tertiary structures. Alcohol, for example, breaks the hydrogen bonds in proteins, and is used as an antiseptic to kill bacteria around wounds by denaturing the bacterial proteins.

Embalming is a method of disinfecting and preserving dead bodies until final disposition of the body occurs. Disease-producing microorganisms that survive in the dead tissues are killed by the organic solvents used in embalming. This is an important public health measure. In cases where a body must be preserved for a while, these chemicals serve to alter the structure of the tissue cell proteins by denaturation, preventing rapid decomposition of the body. The enzymes and bacteria which naturally exist in a living body begin to decompose the body after death. The chemicals used in embalming destroy these enzymes and bacteria and therefore slow down the decomposition process.

Proteins are also denatured by changing their pH. Most proteins are active at their optimum pH, which is in the 7.0 to 7.5 range. (Some exceptions to this are *pepsin*, which works in the stomach and has an optimum pH of 2; and *trypsin*, which works in the small intestine and has an optimum pH of 8.) When a protein is placed in a solution whose pH is *not* in the normal range, the protein may be denatured. This denaturation can result in coagulation when the protein is placed in a strong acid or base. This is how cheese is made. The proteins in the milk are acted upon by acid-producing bacteria. The milk protein, called *casein*, coagulates and curdles, and is made into cheese.

Some Tests for Proteins: A Brief Description

There are three tests which are used to determine the presence of amino acids, peptide bonds, and certain R groups. These tests produce different colors, which identify the substance that is present.

The *ninhydrin test* detects the presence of amino acids, peptides, or proteins. A blue color that forms when ninhydrin is heated with the test solution indicates a positive result.

The *biuret test* is used to detect the presence of two or more peptide bonds. The test solution is added to a small amount of copper sulfate and sodium hydroxide. The solution turns purple if a dipeptide or a larger peptide is present.

The *xanthoproteic test* detects the presence of the amino acids tryptophan, phenylalanine, and tyrosine. The test solution is combined with nitric acid. The solution turns yellow if these amino acids are present.

EXAMPLE 13-8   From each of the descriptions given, name the protein or amino acid test.

    (a) Detects peptides with two or more peptide bonds. The test requires copper(II) sulfate and sodium hydroxide.

    (b) This test is specific for the amino acids tryptophan, phenylalanine and tyrosine.

    (c) This test detects the presence of amino acids, peptides, or proteins. A blue color indicates a positive test.

SOLUTION   (a) The description given corresponds to the biuret test.

    (b) The description given corresponds to the xanthoproteic test.

    (c) The description given corresponds to the ninhydrin test.

SUMMARY

In this chapter we studied the major properties and characteristics of proteins. We began by learning about amino acids, the major building blocks of proteins. We discovered that most proteins are composed of various combinations of 20 amino acids. We next looked at the many properties of amino acids, including their acid–base properties, buffering ability, and their ability to exist in the form of zwitterions. We also learned how amino acids form peptide bonds, which was our key to understanding the basic structure of proteins.

In the second part of the chapter we learned that proteins can be divided into two major classes; fibrous and globular. We also learned that proteins may be classified according to their functions into 10 categories. Next we looked at the primary, secondary, tertiary, and quaternary structures of proteins. Finally, we concluded the chapter with a short discussion of some tests for proteins.

EXERCISES

1. Write the general formula of an amino acid.

2. List five functions of proteins.

3. What is the R group for each of the following amino acids?

   (a)
   $$H_2N-\underset{\underset{H}{|}}{C}-\underset{\underset{O}{\|}}{C}-OH \quad (\text{with } H \text{ on top})$$

   (b)
   $$H_2N-\underset{\underset{H}{|}}{C}-\underset{\underset{O}{\|}}{C}-OH \quad (\text{with } CH_3 \text{ on top})$$

   (c)
   $$O{=}C-OH$$
   $$CH_2$$
   $$CH_2$$
   $$H_2N-\underset{\underset{H}{|}}{C}-\underset{\underset{O}{\|}}{C}-OH$$

4. Define what is meant by an essential amino acid. Write the names and structures of each essential amino acid.

5. Define what is meant by complete protein and incomplete protein. Give an example of each type.

6. Write the structure of an amino acid containing a(n)
   (a) polar R group   (b) nonpolar R group   (c) ionic R group

7. Show how the amino acid phenylalanine reacts with
   (a) HCl   (b) NaOH

8. Show the zwitterion form of the amino acids
   (a) alanine   (b) valine   (c) leucine

9. Write the structure of the tripeptide alanylvalylproline.

10. Name the following polypeptide, using its abbreviated name.

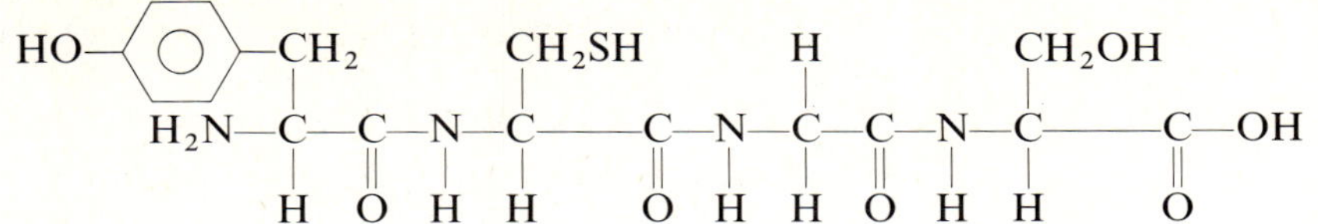

11. Name the six tripeptides that can be formed from the amino acids methionine (Met), threonine (Thr), and asparagine (Asn). Use their abbreviated names.

12. Name the two classes of proteins, define them, and give an example of each.

13. Tell whether each of the following statements describes a protein's primary, secondary, tertiary, or quaternary structure.
    (a) In sickle cell anemia, the protein hemoglobin is defective because out of about 300 amino acids, glutamic acid is replaced in one position by the amino acid valine.
    (b) In bovine ribonuclease the disulfide linkages between cysteine groups cause the protein to fold.
    (c) The protein collagen exists in the form of a triple helix.
    (d) The four subunits of hemoglobin must be present for this protein to carry out its function as an oxygen carrier.

14. Explain how each of the following processes denature proteins.
    (a) heating the protein, for example, in boiling water
    (b) placing the protein in ethyl alcohol
    (c) placing the protein in a strong acid or base

15. Describe each of the following protein tests and tell what they detect.
    (a) ninhydrin test   (b) biuret test   (c) xanthoproteic test

CHAPTER 14

Enzymes
A Very Special Group of Proteins

Some Things You Should Know After Reading This Chapter

You should be able to:

1. Define enzymes as biological catalysts, and define catalysts as substances that affect the rate at which chemical reactions occur.
2. Decide whether an enzyme is a simple enzyme or conjugated enzyme when given its components.
3. Define the terms cofactor and coenzyme.
4. Define the terms apoenzyme, activator, prosthetic group, holoenzyme, zymogen, substrate, and active site.
5. State that the name of an enzyme usually ends with the suffix *-ase*.
6. State the names of the six major classes of enzymes.
7. State that an enzyme provides an alternative reaction pathway for the formation of product from a substrate, and that this pathway requires less energy.
8. Discuss how the four factors, temperature, pH, concentration of substrate, and concentration of enzyme, affect enzyme activity.
9. Discuss why an enzyme has no effect on the amount of substrate remaining and product formed at equilibrium.
10. Define substrate specificity.
11. Discuss the lock-and-key theory of enzymes.
12. Discuss how the three types of enzyme inhibitors—competitive, noncompetitive, and irreversible—work.
13. Discuss how the isoenzymes of LDH serve as diagnostic tools.
14. State where the digestion of carbohydrates, fats, and proteins occurs.
15. Name the two types of vitamins as water-soluble vitamins and fat-soluble vitamins, and discuss the major difference between them.
16. Define the terms avitaminosis, hypovitaminosis, hypervitaminosis, deficiency diseases, and megavitamin therapy.
17. Associate the diseases night blindness, rickets, scurvy, pellagra, and beriberi with the lack of specific vitamins.

Introduction

Enzymes are a group of proteins that regulate biological reactions in our bodies. They are biological *catalysts* which drive the chemical reactions that take place inside our cells. *Catalysts* are substances that affect the rate at which chemical reactions occur. They take part in chemical reactions, but are not changed chemically in the overall reaction. *Biological catalysts* or enzymes, as they are called, are very specific in their actions. A particular enzyme will act only upon certain substances, and will only catalyze certain types of reactions.

There are over 1000 enzymes known to us. Enzymes work in our bodies to help us digest our food, conduct nerve impulses throughout our bodies, and break down carbohydrates, lipids, and proteins into smaller molecules. If even one enzyme is missing in the body, the results could be disastrous. Enzymes are very important, indeed, and in this chapter we're going to take a look at this very special group of proteins.

Enzyme Composition

There are two types of enzymes: simple enzymes and conjugated enzymes. *Simple enzymes* are proteins that are composed entirely of amino acids. Examples of simple enzymes are pepsin, trypsin, and chymotrypsin. They are entirely protein in composition. *Conjugated enzymes* are composed of a protein group and a nonprotein group. The nonprotein group is called a *cofactor*. Cofactors can be metal ions such as Fe^{+2}, Fe^{+3}, Mn^{+2}, Mg^{+2}, Zn^{+2}, or Cu^{+2}, or they can be complex organic molecules. When the cofactor of an enzyme is a complex organic molecule, it is called a *coenzyme*. The cofactor is essential for the biological activity of the conjugated enzyme. Some enzymes require both kinds of cofactors, a metal ion and a coenzyme, to be biologically active. Many of the vitamins are coenzymes, and we will look at them later in this chapter.

EXAMPLE 14-1   Decide whether the following enzymes are simple enzymes or conjugated enzymes.

(a) catalase, which contains the cofactor iron and the coenzyme porphyrin
(b) trypsin, which contains only protein
(c) ascorbic acid oxidase, which contains the cofactor copper

SOLUTION   (a) Catalase is a conjugated enzyme because it contains protein, a cofactor, and a coenzyme.
(b) Trypsin is a simple enzyme because it contains only protein.
(c) Ascorbic acid oxidase is a conjugated enzyme because it contains protein and a cofactor.

Some Enzyme Terminology

There are several terms that are used in the study of enzymes. We will review them here (Fig. 14-1).

1. APOENZYME The term *apoenzyme* refers to the protein part of the enzyme.

2. COFACTOR The term *cofactor* refers to the nonprotein part of a conjugated enzyme. A cofactor can be a metal ion or a complex organic molecule.

3. COENZYME The term *coenzyme* is used to describe a cofactor that is a complex organic molecule.

4. ACTIVATOR The term *activator* is used to describe a cofactor that is a metal ion.

5. PROSTHETIC GROUP The term *prosthetic group* refers to a cofactor that is tightly bound to the protein part of a conjugated enzyme.

6. HOLOENZYME The term *holoenzyme* refers to the apoenzyme and cofactor (the whole enzyme) that functions together as a biologically active molecule.

7. ZYMOGEN The term *zymogen* refers to the inactive form of an enzyme. Sometimes enzymes are produced in an inactive form and are transported in the body to the place where they will work. They are then changed to the biologically active form.

8. SUBSTRATE The term *substrate* refers to the substance that the enzyme will work on.

FIGURE 14-1
The parts of an enzyme-substrate reaction.

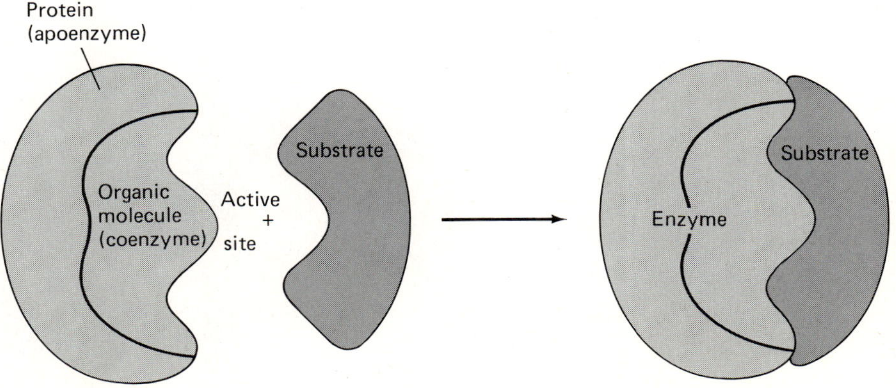

9. ACTIVE SITE   The term *active site* refers to the place where the substrate attaches to the enzyme during the chemical reaction.

Naming and Classifying Enzymes

Before we look at the six major classes into which enzymes are grouped, let's first see how enzymes are named. The name of the enzyme is obtained by adding the suffix *-ase* to the name of the substrate. (Remember that the substrate is the substance that the enzyme acts upon.) For example, a carbohyd*rase* is an enzyme that acts upon a carbohydrate molecule and breaks it down into smaller units. The specific enzyme that breaks down the carbohydrate maltose into two glucose molecules is called malt*ase*. As you read further in this chapter you will see which enzymes break down various substrates (Fig. 14-2).

[*Important note:* Another system of naming enzymes has been used in the past, and parts of this system are still in use today. In this system, the enzyme name ended with the suffix *-in*. Peps*in*, renn*in*, and tryps*in* are enzymes that were named in this way, and these names are still used today. The problem with this naming system is that there is no indication of what substrate is being acted upon, or what type of reaction is taking place. We will use the current naming system in most of our discussions.]

EXAMPLE 14-2   Decide which is the enzyme and which is the substrate.

    (a) lipid, lipase
    (b) protease, protein
    (c) esterase, ester

FIGURE 14-2
Specific enzymes break down specific substrates.

SOLUTION
    (a) Lipid is the substrate and lipase is the enzyme.
    (b) Protein is the substrate and protease is the enzyme.
    (c) Ester is the substrate and esterase is the enzyme.

Now let's look at the six major classes of enzymes.

1. OXIDOREDUCTASES They catalyze oxidation–reduction reactions. There are two types of oxidoreductases. *Oxidases* add oxygen to a substrate. *Dehydrogenases* remove hydrogen from a substrate.

2. TRANSFERASES They catalyze reactions in which a functional group is transferred. Transphosphatases transfer

$$
\begin{array}{c}
OH \\
| \\
-O-P-OH \\
\| \\
O
\end{array}
$$

groups. *Transaminases* transfer $-NH_2$ groups. *Transmethylases* transfer $-CH_3$ groups.

3. HYDROLASES They are enzymes that catalyze hydrolysis reactions to form smaller molecules by addition of H_2O. *Carbohydrases* break down carbohydrates into monosaccharides. *Proteases* break down proteins into amino acids and peptides. *Lipases* break down lipids into smaller molecules. *Amylases* break down starches into smaller molecules.

4. LYASES They are enzymes that catalyze the addition of a group to a double bond, or the removal of a group, leaving a double bond.

5. ISOMERASES They catalyze reactions involving the rearrangement of molecular structures to form isomers.

6. LIGASES They catalyze reactions involving the joining together of two molecules, using energy from the cleaving of phosphate bonds.

Enzyme Action

We know that enzymes are biological catalysts that affect the rates of chemical reactions. Let's see what happens when a chemical reaction takes place, so that we can understand how enzymes work. When a chemical reaction takes place, the molecules that compose the reactants collide with each other. The existing bonds between the atoms making up each molecule are broken and new bonds are formed. The reactants are now rearranged to form products. But it takes a certain amount of energy to break the existing bonds. This is called the *activation energy*. A catalyst, such as an enzyme, provides an alternative reaction pathway for the formation of the product that requires less

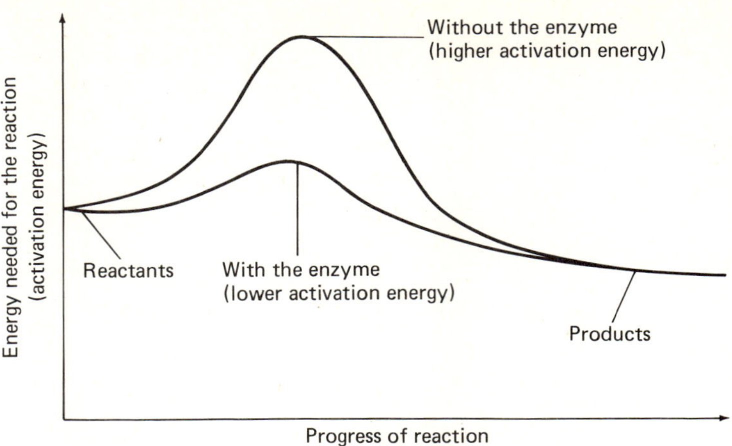

FIGURE 14-3

An enzyme lowers the activation energy of a chemical reaction and speeds up the reaction rate, forming more product in a shorter period of time.

energy. Since the activation energy is lowered when a catalyst such as an enzyme is used, there is less energy needed per molecule and more molecules can react. This increases the rate of the reaction because more product can be formed in a shorter amount of time (Fig. 14-3).

The Factors That Affect Enzyme Activity

There are various factors that affect enzyme activity, for example, temperature, pH, the concentration of the substrate, and the concentration of the enzyme. Let's see how each of these factors affect enzyme activity.

1. TEMPERATURE   Most enzymes that are found in our bodies work best at temperatures ranging from 35 to 45°C. At temperatures below 35°C, the reaction rate of an enzyme-catalyzed reaction slows down because enzyme activity decreases as temperature decreases. The enzymes do not become denatured at low temperatures; they just become inactive. As the temperature is raised, they resume their activity. This is important in the study of cryogenics, in preserving human organs for transplant, and in preserving bacteria and cell cultures for study. At high temperatures enzymes do become denatured. When you run a high temperature, your enzyme activity is inhibited by the high temperature.

2. pH   Enzymes are also affected by pH. They show the greatest activity at their optimum pH, which is usually in the range 6 to 8. Certain enzymes work best at specific pH's. Pepsin, which begins protein

digestion in the stomach, works best at a pH between 1 and 2, while trypsin, which works on protein digestion in the small intestine, works best at a pH between 8 and 9.

3. CONCENTRATION OF SUBSTRATE As the concentration of the substrate molecules increases, the reaction rate increases and more product is formed. There is a point at which the enzyme becomes saturated and the rate of reaction levels off.

4. CONCENTRATION OF ENZYME The reaction rate is also affected by the concentration of the enzyme. Only a small amount of enzyme is required for a reaction. But as the enzyme concentration is increased in the presence of excess substrate, the rate of reaction increases proportionately.

EXAMPLE 14-3 State what effect the following conditions will have on the rate of a reaction.

(a) increasing the temperature
(b) decreasing the temperature
(c) increasing the substrate concentration
(d) removing the enzyme
(e) changing the pH of the fluid surrounding the enzyme so that it is no longer optimum

SOLUTION (a) The reaction rate will increase as long as the temperature is not great enough to denature the enzyme.
(b) The reaction rate will decrease.
(c) The reaction rate will increase.
(d) The reaction rate will decrease.
(e) The reaction rate will decrease.

Enzymes and Reaction Equilibrium

An enzyme catalyzes the formation of products from the breakdown of the substrate molecules. But it may also help reform the substrate from the products. In other words, an enzyme catalyzes both the forward and reverse reactions, as long as the chemical reaction is reversible. When the *rates* at which the forward and reverse reactions occur is the same, then *equilibrium* exists. The enzyme has no effect on the amount of substrate remaining and product formed at equilibrium, but it does speed up the reaction rate so that equilibrium is reached faster. (Please remember that equilibrium means that the rates of the forward and reverse reactions are the same, not that there are equal amounts of substrate and products. The equilibrium may be in favor of the substrate or the products.)

FIGURE 14-4

The lock-and-key theory explains how enzymes and substrates react. The enzyme is like a lock, and the substrate is like a key that fits the lock.

A Theory of How Enzymes Work

Enzymes are biological catalysts that provide a reaction pathway that requires less energy than would otherwise be needed. But how do they do this?

An enzyme combines with a reactant called a *substrate*. Each enzyme strongly binds with a specific substrate. The attraction of an enzyme to a specific substrate is called *substrate specificity*. You remember that proteins have a tertiary structure which give them a three-dimensional shape. Since enzymes are proteins they have a unique tertiary structure. Only a certain substrate can react with a particular enzyme because of the unique structure.

The *lock-and-key theory* explains how enzymes and substrates react. The enzyme is like a lock that has a special structure. The substrate is like a key which also has a special structure that will fit the lock (Fig. 14-4). Some enzymes are very specific in their actions and others are more general. Some enzymes act on specific groups, for example, a *transaminase* transfers amino groups and a *methyltransferase* transfers methyl groups. A *decarboxylase* removes carboxyl groups. Some enzymes catalyze reactions that break specific types of bonds. An esterase cleaves ester linkages and a protease cleaves peptide bonds. Some enzymes catalyze very specific reactions. *Urease* has only one substrate: urea. *Maltase* works only on the maltose molecule. So we see that enzymes are very selective molecules.

Let's summarize what happens during an enzyme-catalyzed reaction using the letters E to represent the enzyme, S to represent the substrate, and P to represent the products (Fig. 14-5).

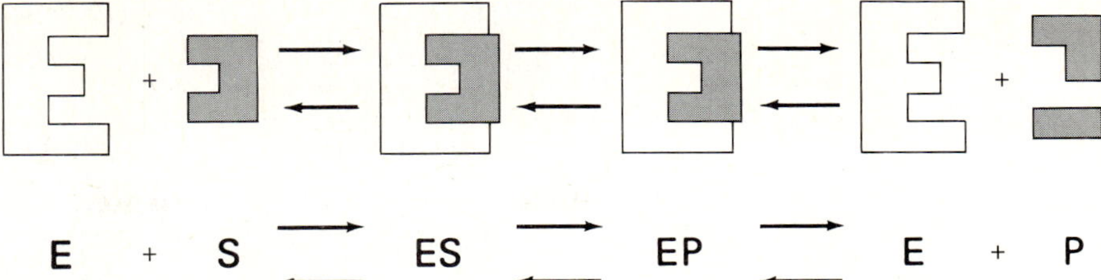

FIGURE 14-5
What happens during an enzyme-catalyzed reaction?

1. $E + S \rightleftharpoons ES$ The substrate combines with the enzyme to form an enzyme-substrate complex (ES).

2. $ES \rightleftharpoons EP$ The active site is the part of the enzyme where the reactants are converted to the products (EP).

3. $EP \rightleftharpoons E + P$ The products are released from the surface of the enzyme and the enzyme is now free to catalyze another reaction. The enzyme can be used over and over again.

Enzyme Inhibitors

Inhibitors are substances that slow down or stop enzymes from reacting with substrates. We will look at three types of enzyme inhibitors: competitive inhibitors, noncompetitive inhibitors, and irreversible inhibitors.

Competitive inhibitors are molecules that have structures that are similar to the substrate. They combine with the enzyme at the active site and stop the enzyme from catalyzing reactions involving the substrate. If the concentration of the substrate is increased, more substrate molecules than inhibitor molecules will react with the enzyme, and the inhibition reaction will be reversed. In a competitive inhibition reaction, the substrate and the inhibitor compete for the same active site on the enzyme.

Let's take a look at the sulfa drugs, which act through competitive inhibition. *Para*-aminobenzoic acid is a substrate needed by some bacteria to synthesize folic acid, which is needed for bacterial growth. The chemical structures of the sulfa drugs sulfanilamide and *para*-aminosalicylic acid are similar to the structure of *para*-aminobenzoic acid (Fig. 14-6). The sulfa drugs react with the enzyme at the active site and make it unavailable for reaction with the substrate, thereby stopping the growth of the bacteria.

Another group of enzyme inhibitors are called *noncompetitive inhibitors*. They bind to the enzyme somewhere on its surface other than the active site and change the structure of the enzyme. The substrate can no longer attach itself to the enzyme. Noncompetitive inhibitors do not have to be structurally similar to the substrate since they are not competing with the substrate for

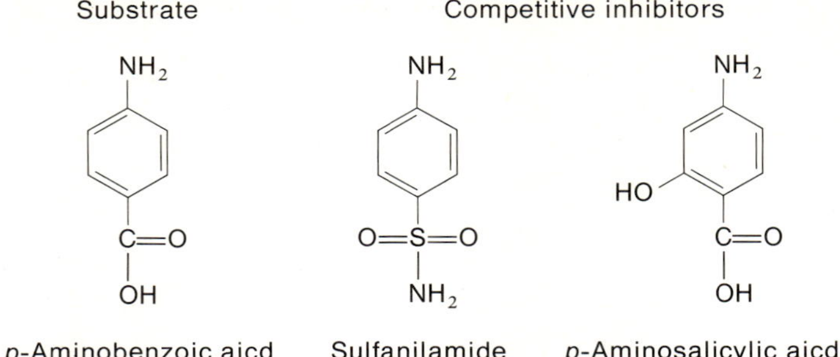

FIGURE 14-6

Competitive inhibitors have chemical structures that are similar to the substrate, such as the sulfa drugs, shown here, which are competitive inhibitors of para-aminobenzoic acid, a substance needed for bacterial growth.

the active site. Cyanide poisoning is caused by noncompetitive inhibition. The cyanide group (CN^-) binds to iron and inhibits the work of respiratory enzymes (Fig. 14-7).

Irreversible inhibition is another way of inhibiting enzymes. Certain molecules can make permanent changes in the structure of the enzyme causing the loss of biological activity. Poison nerve gases often act by destroying the enzyme acetylcholinesterase, which is vital to the nervous system (Fig. 14-8).

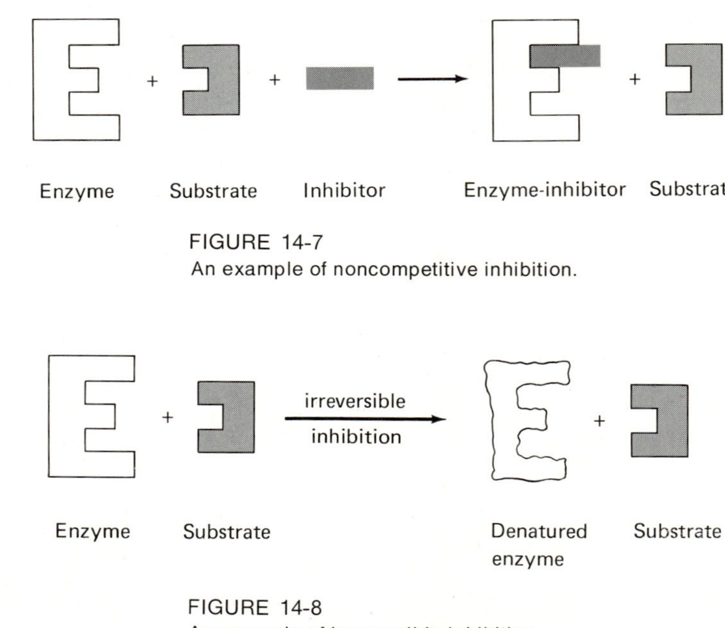

FIGURE 14-7

An example of noncompetitive inhibition.

FIGURE 14-8

An example of irreversible inhibition.

Medical Applications of Enzymes

Enzymes have wide use in medicine as diagnostic and therapeutic tools. When certain diseases are present, specific enzymes may appear in the blood or urine. This can be used to diagnose diseases such as cancer, heart or kidney disease, liver or bone disease, heart attack, or hepatitis. Let's see how such diseases can be diagnosed using enzymes.

Isoenzymes are isomeric forms of a particular enzyme, for example, lactate dehydrogenase (sometimes abbreviated LDH), comes in five isoenzyme forms. The five isoenzymes of LDH are very useful as diagnostic tools. Let's explain how they're used. First, you must know that the enzyme LDH has a quaternary structure. It is composed of four subunits. The subunits are either one of two different types of polypeptide chains, called H and M. The LDH found in heart tissue contains a greater amount of the H chain, and the LDH found in muscle tissue contains a greater amount of the M chain. The five isomeric forms of LDH are:

1. HHHH or H_4.

2. HHHM or H_3M.

3. HHMM or H_2M_2.

4. HMMM or HM_3.

5. MMMM or M_4.

The enzyme LDH is widely distributed in various tissues and body fluids. In normal, healthy tissues the LDH stays inside the cells and there are low levels of LDH in the blood. When cells aren't functioning properly, the LDH spills out into the blood, elevating the level of LDH, which can be detected. By determining which particular isoenzyme is present, the part of the body which has malfunctioned can be identified. For example, an increased level of LDH of the H_4 variety in the blood means that some cells in the heart have been damaged. This is an indication of a heart attack. Liver disease is accompanied by an increase in the level of the M_4 variety of LDH in the blood (Table 14-1).

Heart attack can also be detected by an increase in concentration of another enzyme which is not normally found in the blood. The enzyme glutamate transaminase is found in the blood in high levels about 6 to 12 hours after heart damage has occurred, and returns to the normal level in 36 to 48 hours. The blood level of the enzyme creatine phosphokinase also increases after a heart attack.

Liver damage caused by the disease called hepatitis is diagnosed by increased levels of the enzyme glutamic-oxalacetic transaminase in the blood. There is an enzyme called acid phosphatase which is normally found in the intestine and the prostate gland. When this enzyme is found in the blood it is an indication of the presence of cancer of these parts of the body.

TABLE 14-1 Concentrations of LDH isoenzymes before and after a heart attack (%)

ISOENZYME	BEFORE	AFTER
H_4	20.8	37.4
H_3M	36.4	41.2
H_2M_2	25.2	8.6
HM_3	12.0	12.8
M_4	5.6	0.0

TABLE 14-2 The enzymes of digestion

FOOD	WHERE DIGESTION OCCURS	ENZYME	PRODUCTS
Carbohydrates	Mouth	Amylase	Polysaccharides and disaccharides
Carbohydrates	Small intestine	Pancreatic amylase	Polysaccharides and disaccharides
Carbohydrates	Small intestine	Sucrase, lactase, and maltase	Monosaccharides glucose, fructose, and galactose
Fats	Stomach	Lipase	Fatty acids and glycerol
Fats	Small intestine	Bile salts and lipase	Fatty acids and glycerol and mono- and triglycerides
Proteins	Stomach	Pepsin and rennin	Polypeptides
Proteins	Small intestine	Trypsin and chymotrypsin	Dipeptides
Proteins	Small intestine	Carboxypeptidase, aminopeptidase, and dipeptidase	Amino acids

The Enzymes of Digestion

There are several enzymes that work in our bodies to digest our food. Table 14-2 summarizes the enzymes of digestion.

Vitamins: The Coenzymes

Earlier in this chapter we discussed two classes of enzymes. We said that simple enzymes are composed entirely of protein, and conjugated enzymes are composed of a protein and a nonprotein part. The nonprotein part is called a *cofactor.* A cofactor can be an activator, which is a metal ion, or a coenzyme, which is a complex organic molecule. Many coenzymes are vitamins or derivatives of vitamins. Most vitamins serve as catalysts that take part in chemical reactions.

There are two types of vitamins, *water-soluble vitamins* and *fat-soluble vitamins.* The water-soluble vitamins are polar molecules. They are not stored in the body and must be replenished daily. The fat-soluble vitamins (also called *lipid-soluble vitamins*) are absorbed from the intestine with the lipids from the foods we eat. The fat-soluble vitamins can be stored in the liver for a period of time.

It is important for us to have the proper amounts of vitamins in our diets each day. Lack of certain vitamins causes *deficiency diseases* such as *beriberi, pellagra,* and *scurvy* (Figs. 14-9 and 14-10). The term *avitaminosis* is used to describe disease that results from a missing vitamin in the diet. A condition known as *hypovitaminosis* occurs when an insufficient amount of a vitamin is present in the diet. Another condition, known as *hypervitaminosis,* can occur if too much of a vitamin is taken into the body and builds up to toxic levels. This can occur with the fat-soluble vitamins because they can be stored in the body. *Megavitamin therapy* is used by some medical practitioners to treat diseases. This could be dangerous if the concentrations of the fat-soluble vitamins are too high, causing them to reach toxic levels.

Let's see how vitamins work, to understand a little about their importance. We said that most vitamins are coenzymes. The coenzymes as a group have several chemical functions in the body. *Niacin,* which is known as part of the vitamin B_2 complex, is composed of nicotinic acid and nicotinamide. These

FIGURE 14-9
A child suffering from pellagra. This disease is caused by a niacin deficiency. (Center for Disease Control, Atlanta, Georgia)

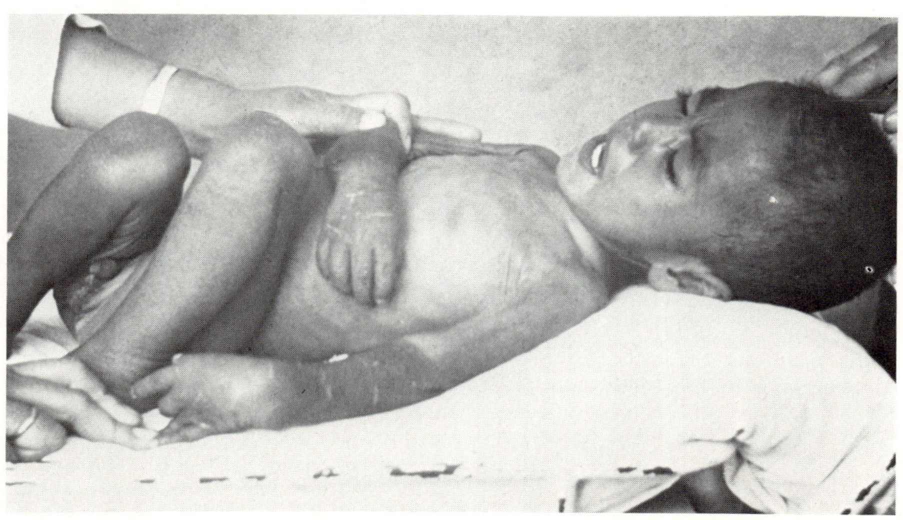

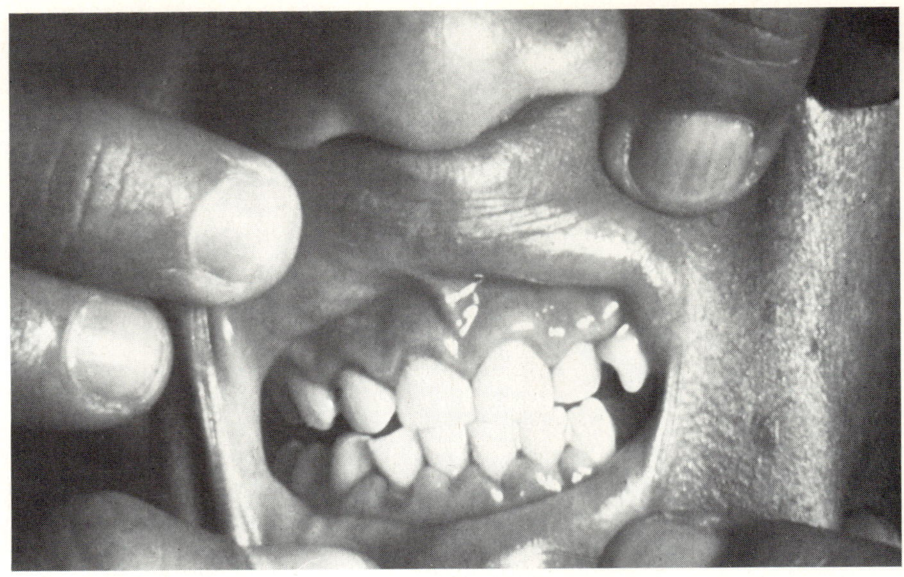

FIGURE 14-10
A victim of scurvy has sore gums and loose teeth. (Center for Disease Control, Atlanta, Georgia)

two compounds produce two important coenzymes. The coenzymes of niacin do such things as catalyze carbohydrate hydrolysis reactions, synthesize triglycerides in the body, and help supply oxygen to many tissues in the body. A person whose diet is short of niacin will suffer from a deficiency disease known as *pellagra*. A pellagra victim feels weak and suffers with inflammation of skin, inflammation of the mouth and tongue, inability to digest food, and disturbances of the central nervous system. A diet containing niacin cures pellagra in human beings. This is one example of how important vitamins are to our bodies. Table 14-3 summarizes the importance of vitamins in the diet.

SUMMARY

In this chapter we looked at the importance of enzymes in our bodies. We learned that enzymes are biological catalysts that drive the reactions that take place inside our cells. We discussed the two types of enzymes, simple enzymes and conjugated enzymes. We also learned something about enzyme terminology, and we discussed how enzymes are named and classified. We also talked about the factors that affect enzyme activity and we discussed the lock-and-key theory of how enzymes work. We also learned about the various types of enzyme inhibitors, including competitive inhibitors, noncompetitive inhibitors, and irreversible inhibitors.

TABLE 14-3   Some vitamins that function as coenzymes

NAME	MINIMUM DAILY ADULT REQUIREMENT	FUNCTION OF VITAMIN	DEFICIENCY DISEASE OR SYMPTONS	SOURCE OF VITAMIN IN THE DIET
FAT SOLUBLE				
Vitamin A (retinol and dehydro-retinol)	5000 I.U.[a] (1.5 mg)	Helps to form visual pigments and maintains the normal epithelial structure	Night blindness	Fish liver oils, liver, egg yolk, green leafy or yellow vegetables
Vitamin D (ergocalciforal and cholecalciforal)	400 I.U.	Needed for good bone and teeth formation	Rickets, which results in defective bone formation	Fish liver oil, butter, egg yolk, along with ultraviolet light (i.e., sunlight)
Vitamin E (tocopherol)	30 mg as synthetic dl-α-tocopherol acetate	Stabilizes biological membranes and acts as an antioxidant	Red blood cells will have a greater tendency for hemolysis	Vegetable oil, wheat germ, leafy vegetables
Vitamin K (K_2-phylloquinone)	About 0.03 mg/kg of body weight	Needed for prothrombin formation and normal blood coagulation	Hemorrhage	Leafy vegetables, vegetable oils; also is produced by intestinal flora after the fourth day of life.
WATER SOLUBLE				
Vitamin C (ascorbic acid)	60 mg	Needed for maintenance of connective tissues, vascular function, tissue respiration, and wound healing	Scurvy, which includes loose teeth and bleeding gums	Citrus fruits, tomatoes, and potatoes
Folic acid (folacin)	0.1–0.5 mg, depending on source	Needed for the synthesis of purines and pyrimidines and for the maturation of red blood cells	Anemia	Fresh green leafy vegetables and fruit, organ meats, liver, and dried yeast

NAME	MINIMUM DAILY ADULT REQUIREMENT	FUNCTION OF VITAMIN	DEFICIENCY DISEASE OR SYMPTONS	SOURCE OF VITAMIN IN THE DIET
WATER SOLUBLE (*cont.*)				
Niacin (nicotinic acid)	15–20 mg	Acts as a coenzyme in hydrogen transport; also used in carbohydrate and tryptophan metabolism	Pellagra (see text for discussion of this disease)	Dried yeast, liver, meat, fish, and whole grains
Vitamin B_1 (thiamine)	1–1.5 mg	Needed for carbohydrate metabolism and for nerve cell function	Beriberi (includes acute cardiac symptoms and heart failure, also weight loss and nerve inflammation)	Whole grains, pork, liver, nuts, and enriched cereal products
Vitamin B_2 (riboflavin)	1.0–1.7 mg	Coenzyme in energy transport; also involved in protein metabolism	Visual problems and skin fissures	Milk, cheese, liver, meat, and eggs
Vitamin B_6 group (pyridoxine, pyridoxal, pyridoxamine)	2 mg	Necessary for cellular function and for the metabolism of certain amino and fatty acids	Convulsions in infants and skin disorders in adults	Dried yeast, liver, organ meats, fish, and whole-grain cereals
Vitamin B_{12} (cyanocobalamin)	5 μg	Necessary for DNA synthesis related to folate coenzymes, also for maturation of red blood cells	Pernicious anemia and certain psychiatric syndromes	Liver, beef, pork, eggs, milk, and milk products
Vitamin H (biotin)	0.15–0.30 mg	Necessary for protein synthesis, transamination, and CO_2 fixation	Skin disorders	Liver, dried peas and lima beans, egg white, and by bacteria in the alimentary canal
Pantothenic acid	10 mg	Forms part of coenzyme A	Fatigue, malaise, neuromotor, and digestive disorders	Dried yeast, liver, and eggs

[a] I.U. stands for international unit. One international unit equals 0.3 μg of retinol.

In the final part of the chapter we discussed the medical applications of enzymes, as well as the enzymes of digestion, and vitamins—the coenzymes. In ending, we can truly say that enzymes are remarkable compounds.

EXERCISES

1. Define each of the following terms.
   (a) enzyme   (b) catalyst   (c) cofactor   (d) coenzyme

2. Identify each of the following enzymes as simple or conjugated.
   (a) the enzyme ascorbic acid oxidase needs Cu^{+2} for its activity
   (b) the enzyme glucose oxidase requires the coenzyme FAD for its activity
   (c) the enzyme maltase contains only protein

3. For each of the following substrates, write the name of the enzyme.
   (a) sucrose   (b) maltose   (c) lactose
   (d) lipids   (e) proteins

4. Match the word on the left with the definition on the right.
   (1) apoenzyme   (a) The apoenzyme and cofactor function together as a biologically active molecule.
   (2) coenzyme
   (3) activator   (b) The protein part of the enzyme.
   (4) holoenzyme   (c) A cofactor that is a metal ion.
   (5) zymogen   (d) The inactive form of the enzyme.
   (e) A cofactor that is a complex organic molecule.

5. Decide which is the enzyme and which is the substrate.
   (a) urea, urease
   (b) protease, protein
   (c) maltose, maltase
   (d) carbohydrate, carbohydrase

6. Match the word on the left with the definition on the right.
   (1) oxidases   (a) They catalyze reactions involving the joining together of two molecules.
   (2) dehydrogenases
   (3) transferases   (b) They catalyze hydrolysis reactions.
   (4) hydrolases   (c) They add oxygen to a substrate.
   (5) lyases   (d) They catalyze reactions in which a functional group is transferred.
   (6) isomerases
   (7) ligases   (e) They catalyze the addition of a group to a double bond.
   (f) They remove hydrogen from a substrate.
   (g) They catalyze reactions involving the rearrangement of molecular structure.

7. Explain how a catalyst works.

8. Tell what effect the following will have on the rate of a reaction.
    (a) Change the pH of the fluid surrounding the enzyme to 12.
    (b) Increase the enzyme concentration above normal.
    (c) Decrease the enzyme concentration below normal.
    (d) Decrease the temperature of the enzyme below 35°C.

9. Fill in the missing words.
    Enzymes catalyze both the _________ and _________ reactions.

10. Explain the lock-and-key theory of enzymes.

11. Complete the following equation for an enzyme reaction.

$$E + S \rightleftharpoons \; ? \; \rightleftharpoons \; ? \; \rightleftharpoons \; E + P$$

12. Match the word on the left with the definition on the right.
    (1) competitive inhibitors   (a) They bind to the enzyme,
    (2) noncompetitive inhibitors   somewhere on its surface, other
    (3) irreversible inhibitors   than the active site.
                                   (b) They make permanent changes
                                       in the structure of the enzyme.
                                   (c) They have structures that are
                                       similar to the substrate.

13. Explain how the isoenzymes of LDH serve as diagnostic tools.

14. Write the five isomeric forms of LDH.

15. Complete the following table on enzyme digestion.

FOOD	WHERE DIGESTION OCCURS	ENZYME	PRODUCTS
Carbohydrates	_________	Pancreatic amylase	_________
Fats	Stomach	_________	_________
Proteins	_________	_________	Polypeptides

16. Name the two types of vitamins. What is the major difference between them?

17. Complete the following statements using the words *avitaminosis, hypovitaminosis,* or *hypervitaminosis.*

    (a) A disease that results from an *insufficient* amount of a vitamin in the diet is due to _________.
    (b) A disease that results from the *lack* of a vitamin in the diet is due to _________.
    (c) A condition that occurs if *too much* of a vitamin is taken into the body is _________.

18. Vitamins that are parts of enzymes are called _________.

19. Match the disease with the vitamin deficiency.
    (1) night blindness   (a) Lack of vitamin C.
    (2) rickets   (b) Lack of vitamin A.
    (3) scurvy   (c) Lack of vitamin B_1 (thiamine).
    (4) pellagra   (d) Lack of vitamin D.
    (5) beriberi   (e) Lack of the vitamin niacin.

20. Using Table 14-3, tell what foods you think are the best sources of
    vitamins.

CHAPTER 15

Nucleic Acids
The Basis of Life

Some Things You Should Know After Reading This Chapter

You should be able to:

1. State that nucleic acids are the major components of chromosomes.
2. Name the two types of nucleic acids as DNA and RNA.
3. State that nucleic acids are composed of nucleotides.
4. Name the three subgroups that compose nucleotides.
5. Name the five-carbon sugar found in RNA and the five-carbon sugar found in DNA.
6. List the two types of nitrogen-containing bases and state whether they are single-ring or double-ring compounds.
7. Name the five different nitrogen-containing bases and state whether they are purines or pyrimidines.
8. State which nitrogen-containing bases are found in DNA and which nitrogen-containing bases are found in RNA.
9. Describe the Watson-Crick model of DNA.
10. State the complementary nitrogen base pairs for DNA and the complementary nitrogen base pairs for RNA.
11. Describe the replication of DNA.
12. Define the terms cytoplasm, ribosomes, *m*RNA, *t*RNA, transcription, codon, and anticodon.
13. State the complementary nitrogen base pairs between DNA and RNA.
14. Write the proper *m*RNA code for a nucleotide sequence on DNA, and vice versa.
15. Write the *m*RNA codons that correspond to *t*RNA anticodons, and vice versa.
16. Use Table 15-3 to write the amino acid sequence that corresponds to *m*RNA codons.
17. Explain how protein synthesis is regulated.
18. Define the terms repressor protein, regulatory gene, mutagen, and mutation.
19. State the major symptoms of sickle cell anemia and phenylketonuria. State the specific breakdown in the DNA code for each disease.

In the past four chapters we have learned about carbohydrates, lipids, proteins, and enzymes. We now know something about the foods we eat and how they are broken down by the body and used as the raw materials for the chemical reactions that take place in our cells. The foods we eat are metabolized, providing the energy and material for growth of new cells and replacement of old, worn-out, damaged cells. But what tells the body how to carry out these functions?

All human beings require the same basic nutrients which are used by the body to produce the correct set of proteins that make up each individual. The information needed for the synthesis of these proteins is found in the nucleus of each cell. The cell nucleus contains *nucleic acids,* which are responsible for hooking amino acids together in specific sequences to make the various proteins found in the body. Nucleic acids are the major components of *chromosomes,* Chromosomes hold all the genetic information that we inherit from our parents. The nucleic acids found in our cells hold the information for the maintenance, growth, and reproduction of our cells. There are two types of nucleic acids: *deoxyribonucleic acid,* or *DNA,* and *ribonucleic acid,* or *RNA.* DNA is the genetic material found in the nucleus of each cell, which contains all the information needed for the development of an individual. It tells how a person will look and how that person's body will function. It determines what chemical reactions will take place in the body, and which proteins and enzymes must be produced in order to carry out these chemical reactions. RNA brings the information contained in the cell nucleus to another part of the cell, called the *ribosomes,* where the actual synthesis of protein occurs. It is the DNA and RNA, combined with some protein molecules, which make up our chromosomes. Each human cell contains 46 chromosomes, 23 from the mother and 23 from the father. (The sex cells contain only 23 chromosomes, so that when fertilization occurs the embryo has 46 chromosomes.) *Genes,* which carry the information for each specific characteristic that the person will have, are found on the chromosomes.

Let's now take a closer look at the composition of nucleic acids.

Components of Nucleic Acids

Nucleic acids are composed of many small units called *nucleotides.* Just as amino acids are the basic building blocks of proteins, and monosaccharides combine to form polysaccharides, *nucleotides* join together to form *nucleic* acids (Fig. 15-1).

Nucleotides are composed of three subgroups:

1. A five-carbon sugar.

2. A phosphate group.

3. A nitrogen-containing base (Table 15-1).

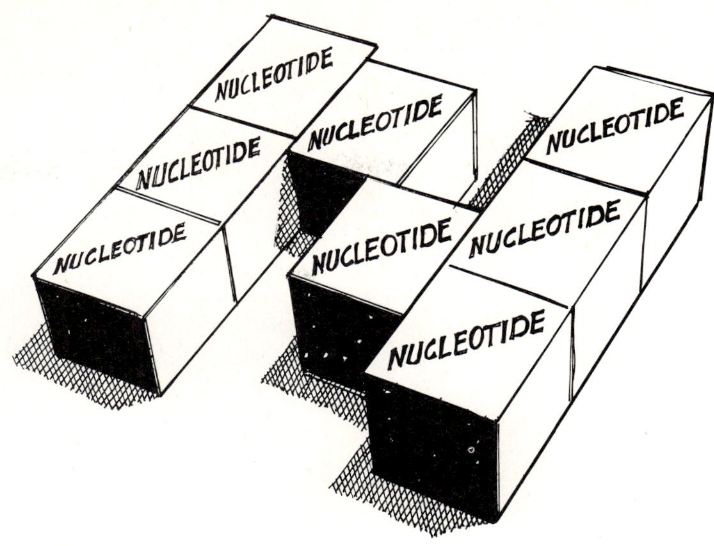

FIGURE 15-1
Nucleotides join together to form nucleic acids.

TABLE 15-1   The composition of nucleotides

SUBUNIT	SUBSTANCES FOUND IN RNA	SUBSTANCES FOUND IN DNA
Five-carbon sugar	Ribose	Deoxyribose
Phosphate group	$HO-\overset{\overset{\textstyle OH}{\textstyle \|}}{\underset{\underset{\textstyle O}{\textstyle \|}}{P}}-O-$	$HO-\overset{\overset{\textstyle OH}{\textstyle \|}}{\underset{\underset{\textstyle O}{\textstyle \|}}{P}}-O-$
Nitrogen-containing bases	Adenine, guanine, cytosine, and uracil	Adenine, guanine, cytosine, and thymine

EXAMPLE 15-1   List the components of a nucleotide.

SOLUTION   A nucleotide consists of a five-carbon sugar, a phosphate group, and a nitrogen-containing base.

The *five-carbon sugars* found in nucleotides are either *ribose* or *deoxyribose* (Fig. 15-2). Ribose is found in ribonucleic acid and deoxyribose is found in deoxyribonucleic acid.

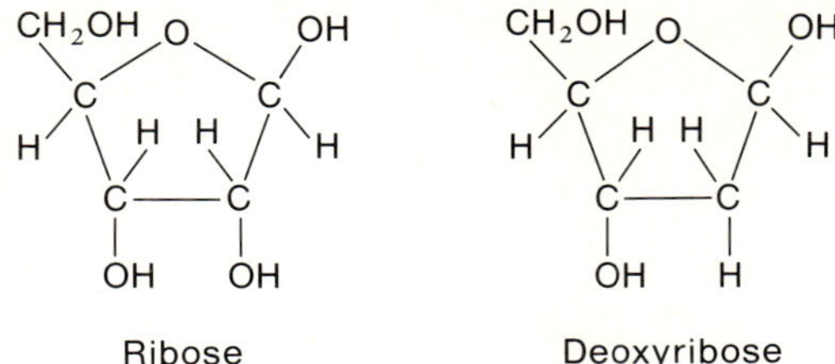

FIGURE 15-2
The five-carbon sugars found in nucleotides. Ribose is found in the nucleotides of ribonucleic acid, and deoxyribose is found in the nucleotides of deoxyribonucleic acid.

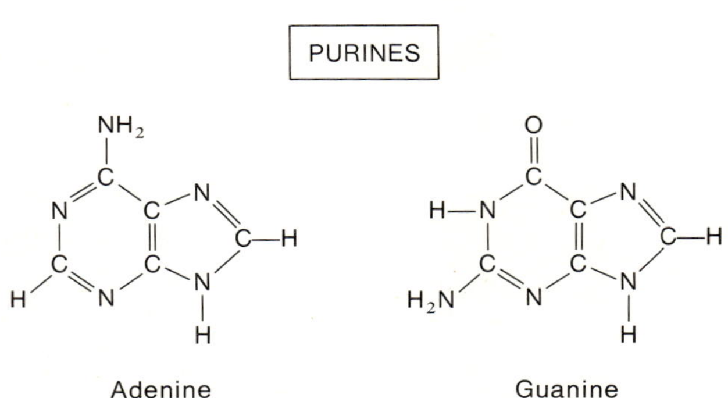

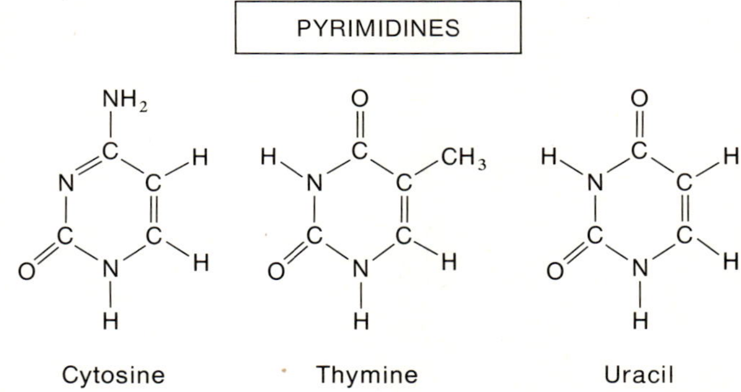

FIGURE 15-3
The nitrogen-containing bases.

EXAMPLE 15-2   Name the two five-carbon sugars and state the nucleic acids with which they are associated.

SOLUTION   The two five-carbon sugars are ribose and deoxyribose. Ribose is found in ribonucleic acid and deoxyribose is found in deoxyribonucleic acid.

The *phosphate group* bonds nucleotides together. The structure of the phosphate group is as follows:

$$\text{HO}-\overset{\displaystyle \text{OH}}{\underset{\displaystyle \text{O}}{\overset{|}{\underset{\|}{\text{P}}}}}-\text{O}-$$

A *nitrogen-containing base* is a cyclic compound containing both nitrogen and carbon in the ring. The ring can have either one or two cyclic parts. There are five different nitrogen-containing bases which can be found in DNA or RNA. They fall into two classes: the *purines* and *pyrimidines*. The purines are double-ring structures and the pyrimidines are single-ring structures. The purines are *adenine* and *guanine* and the pyrimidines are *cytosine, thymine,* and *uracil* (Fig. 15-3).

The *DNA nucleotides* are made of the nitrogen-containing bases adenine, guanine, cytosine, and thymine. The *RNA nucleotides* are made of the nitrogen-containing bases adenine, guanine, cytosine, and uracil. In other words, DNA can't contain uracil and RNA can't contain thymine (Table 15-2).

EXAMPLE 15-3   List the two types of nitrogen-containing bases, and state whether they are single-ring or double-ring compounds. Also, name the five nitrogen-containing bases.

SOLUTION   Purines are double-ring compounds. Adenine and guanine are the two purines. Pyrimidines are the single-ring compounds. Cytosine, thymine, and uracil are the three pyrimidines.

TABLE 15-2   The nitrogen-containing bases of DNA and RNA

NUCLEIC ACID	NITROGEN-CONTAINING BASES
DNA	Adenine, guanine, cytosine, and *thymine*[a]
RNA	Adenine, guanine, cytosine, and *uracil*[a]

[a] Both DNA and RNA contain the bases adenine, guanine, and cytosine. However, only DNA contains thymine and only RNA contains uracil.

EXAMPLE 15-4   State which nitrogen-containing bases are found in DNA and which are found in RNA.

SOLUTION   The nitrogen-containing bases in DNA are adenine, guanine, cytosine, and thymine. The nitrogen-containing bases in RNA are adenine, guanine, cytosine, and uracil.

The Structure of Nucleic Acids

DNA

In 1953, J. Watson and F. Crick proposed a structure for the DNA molecule, for which they received a Nobel prize. The Watson-Crick model of DNA shows the nucleotides arranged as a double-coiled chain, or what is better known as a *double helix* (Fig. 15-4). Let's look at the nucleotides of DNA.

We said earlier that a nucleotide is made up of a nitrogen-containing base, a five-carbon sugar, and a phosphate group. The bonding sequence in the nucleotide has the phosphate group bonded to the five-carbon sugar, which in turn is bonded to the nitrogen-containing base (Fig. 15-5). A nucleic acid is a long chain of nucleotides bonded together (Fig. 15-6). The phosphate group of one nucleotide bonds to the sugar group on the next nucleotide. This is how we explain the bonding in each of the nucleotide chains of DNA. But what holds the two chains together to make the double helix? The two

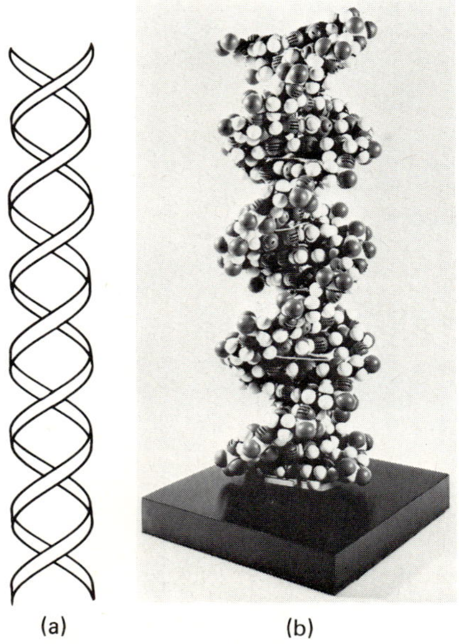

(a)   (b)

FIGURE 15-4
The double helix of DNA.   (a) A drawing of a double helix.   (b) A photograph of an actual model of DNA (The Ealing Corporation).

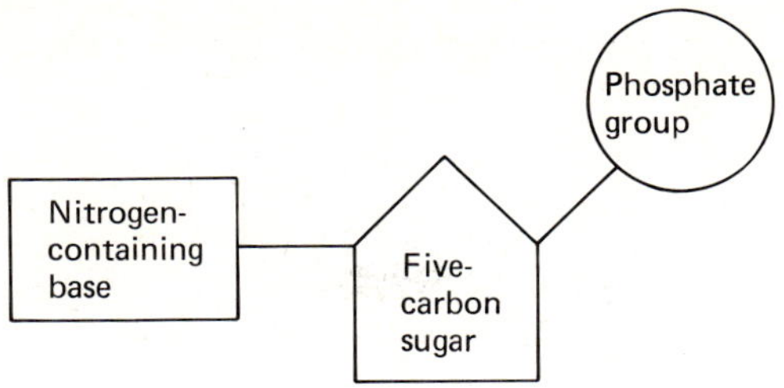

FIGURE 15-5
A nucleotide is composed of a phosphate group bonded to a five-carbon sugar, which in turn is bonded to a nitrogen-containing base.

chains are held together by hydrogen bonding between the nitrogen-containing bases. A nitrogen-containing base from one chain bonds with a nitrogen-containing base from the other chain. The bonding between the nitrogen-containing bases in DNA is very specific. Adenine will bond only with thymine, and cytosine will bond only with guanine. This is known as *complementary*

FIGURE 15-6
A nucleic acid is a long chain of nucleotides bonded together. The phosphate group of one nucleotide bonds to a sugar group on the next nucleotide.

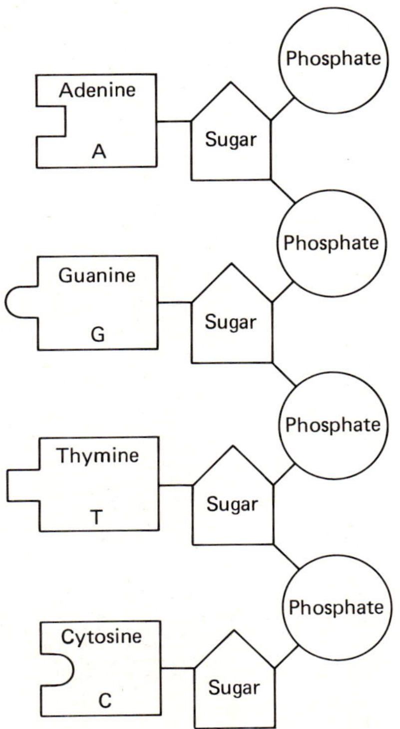

nitrogen base pairing. It means that the two strands of DNA are not identical, but are complementary. In other words, where adenine appears on one strand of DNA, thymine appears on the other, and where guanine appears on one strand of DNA, cytosine appears on the other. The adenine and thymine are held together by two hydrogen bonds and the cytosine and guanine are held together by three hydrogen bonds. Figure 15-7 shows what this hydrogen bonding would look like if we could uncoil the two nucleotide chains. But remember, the two chains actually exist in a helical arrangement (Fig. 15-8).

Another question we might ask is: What makes the DNA molecules that produce one type of cell in the body different from the DNA molecules that produce another type of cell? Each DNA molecule must contain phosphate groups, five-carbon sugars, and nitrogen-containing bases. But the number of base pairs and the sequence of base pairs can vary from molecule to molecule. Also, the number of nucleotides can vary from molecule to molecule. The range is from 5000 to 5 million nucleotides per DNA molecule. It is these variations in DNA molecules that allow one type of cell in the body to be

FIGURE 15-7

Hydrogen bonding between complementary base pairs holds together the two nucleotide chains that compose DNA.

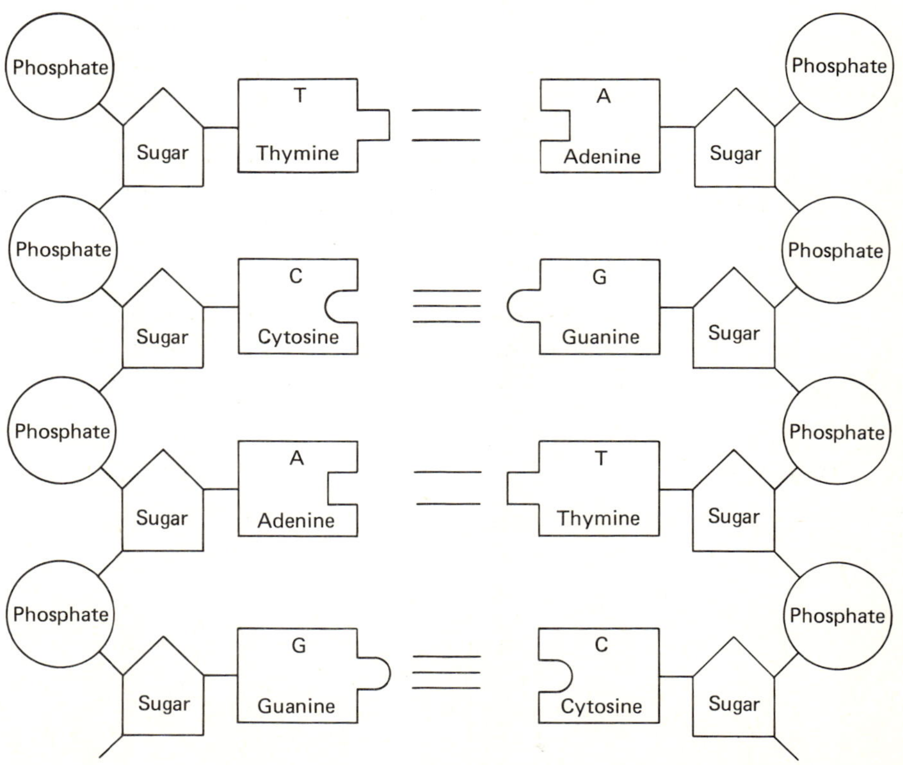

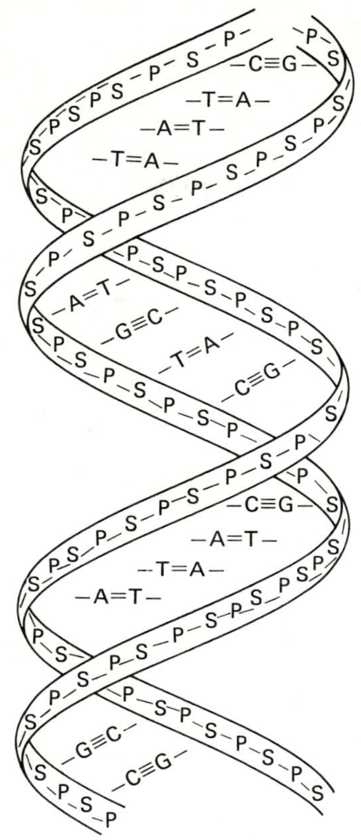

FIGURE 15-8
The helical arrangement of DNA in the Watson-Crick model. The S and P are the sugar and phosphate groups, respectively.

different from another type. The sequence of bases along the chain of the DNA molecule determines what proteins will be made, which in turn determines what characteristics a person will have.

EXAMPLE 15-5   Write the complementary base pairs for the following nitrogen bases of DNA.

   (a) adenine   (b) guanine
   (c) cytosine   (d) thymine

SOLUTION   (a) Adenine bonds with thymine.
                (b) Guanine bonds with cytosine.
                (c) Cytosine bonds with guanine.
                (d) Thymine bonds with adenine.

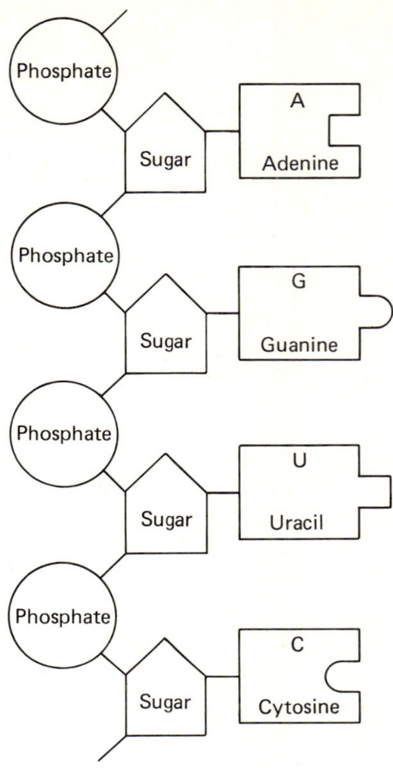

FIGURE 15-9
A long chain of RNA nucleotides makes up a molecule of RNA.

RNA

RNA molecules are single chains of nucleotides, unlike DNA molecules, which are double-coiled chains of nucleotides. RNA nucleotides are made up of a nitrogen-containing base, a five-carbon sugar which is ribose, and a phosphate group. When the phosphate group of one RNA nucleotide bonds to the sugar group on the next RNA nucleotide, a long chain of RNA nucleotides is formed. A long chain of RNA nucleotides makes up a molecule of RNA (Fig. 15-9). The nitrogen-containing bases of RNA are adenine, guanine, cytosine, and uracil. The complementary base pairs of RNA are adenine and uracil bonding with each other, and guanine and cytosine bonding with each other.

EXAMPLE 15-6   Write the complementary base pairs for the following nitrogen bases of RNA.

(a) adenine   (b) guanine
(c) cytosine   (d) uracil

SOLUTION

    (a) Adenine bonds with uracil.
    (b) Guanine bonds with cytosine.
    (c) Cytosine bonds with guanine.
    (d) Uracil bonds with adenine.

The Replication of DNA

When our cells are damaged and when we grow, we must produce new cells. The information needed to produce a new cell, which is exactly like the original cell, is contained in the DNA molecules that make up our chromosomes. The process of reproducing a new cell is called *mitosis*. In this process a cell divides forming two new cells. The new cells are called the *daughter cells* and the original cell is called the *parent cell*. The daughter cells will also undergo mitosis to form more identical cells. This is how damaged cells can be replaced and how growth can occur.

Each time a cell undergoes mitosis and reproduces itself, the DNA in the cell nucleus must also reproduce itself. The DNA in a cell directs the reproduction or replication of another identical DNA molecule. Let's see how this is done. (If you're not familiar with the parts of the cell, see Figure 18-2).

You might remember that a DNA molecule is a double-helical chain of DNA nucleotides which is held together by hydrogen bonds between nitrogen base pairs. When DNA replication occurs, an enzyme present in the cell causes the double-helical chain to uncoil, breaking the hydrogen bonds and separating the nitrogen-containing bases. When the helix is completely uncoiled, two separate DNA nucleotide chains remain. In the presence of the proper enzymes and the necessary nucleotides, two new DNA nucelotide chains which are complementary to each of the existing DNA nucleotide chains are formed. Each newly formed chain remains united with its complementary chain, and two new molecules of DNA which are exactly like the original DNA molecule are formed (Fig. 15-10). The process of DNA replication is completed.

The Synthesis of Proteins

Amino acids are the building blocks of proteins. In Chapter 13 we learned that specific sequences of amino acids make up the various proteins. We are now ready to learn exactly how proteins are synthesized by the body.

Amino acids enter the *cytoplasm* of a cell from the bloodstream. The cytoplasm is the main body of the cell. The amino acids must be connected in the correct sequences to synthesize the various proteins needed by the body. The proteins produced by the body are of two main types: genetic proteins and structural proteins. The *genetic proteins* carry the genetic information for that individual and the *structural proteins* maintain the structure of the various cells.

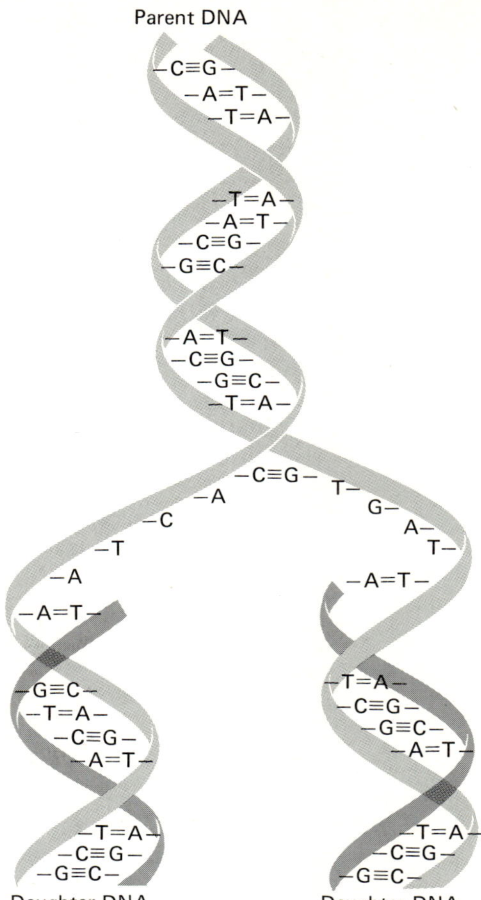

FIGURE 15-10

The replication of DNA. The parent DNA uncoils into two strands. New complementary strands are built on each of the parent strands, forming two new molecules of DNA, which are exactly like the original.

Let's stop for a minute and think about what wonders the cell can perform. The average cell is about 20 to 30 microns (0.02 to 0.03 mm) in diameter. This means that you would have to line up about 1000 cells to measure 1 inch. In the tiny nucleus of each cell are the 23 pairs of chromosomes, which are made partly of DNA. It is estimated that these 23 pairs of chromosomes carry the instructions to direct about 2 billion protein-making reactions.

The two nucleic acids, DNA and RNA, work together to synthesize proteins. The DNA operates only within the cell nucleus, so another mechanism is needed to transfer the protein-making information outside the cell nucleus. RNA is the substance that brings this information outside the cell nucleus to the cell's *ribosomes*. The *ribosomes* are the part of the cell where the actual protein synthesis occurs. The are two types of RNA, *messenger RNA* or *mRNA*, and *transfer RNA* or *tRNA*. Let's see how they operate.

A molecule of DNA uncoils to form single nucleotide chains, and one of the chains serves as a template so that the pattern of nitrogen-containing bases along the chain can be copied by the process of nitrogen base pairing. In this way a molecule of messenger RNA is formed alongside the single DNA chain. But remember, the nitrogen bases in RNA are adenine, guanine, cytosine, and *uracil*, whereas in DNA they are adenine, guanine, cytosine, and *thymine*. Therefore, the complementary base pairing between DNA and RNA is as follows:

1. Adenine on DNA pairs with uracil on RNA.

2. Guanine on DNA pairs with cytosine on RNA.

3. Cytosine on DNA pairs with guanine on RNA.

4. Thymine on DNA pairs with adenine on RNA (Fig. 15-11).

FIGURE 15-11
DNA acts as a template for the formation of *m*RNA, by the process of complementary base pairing.

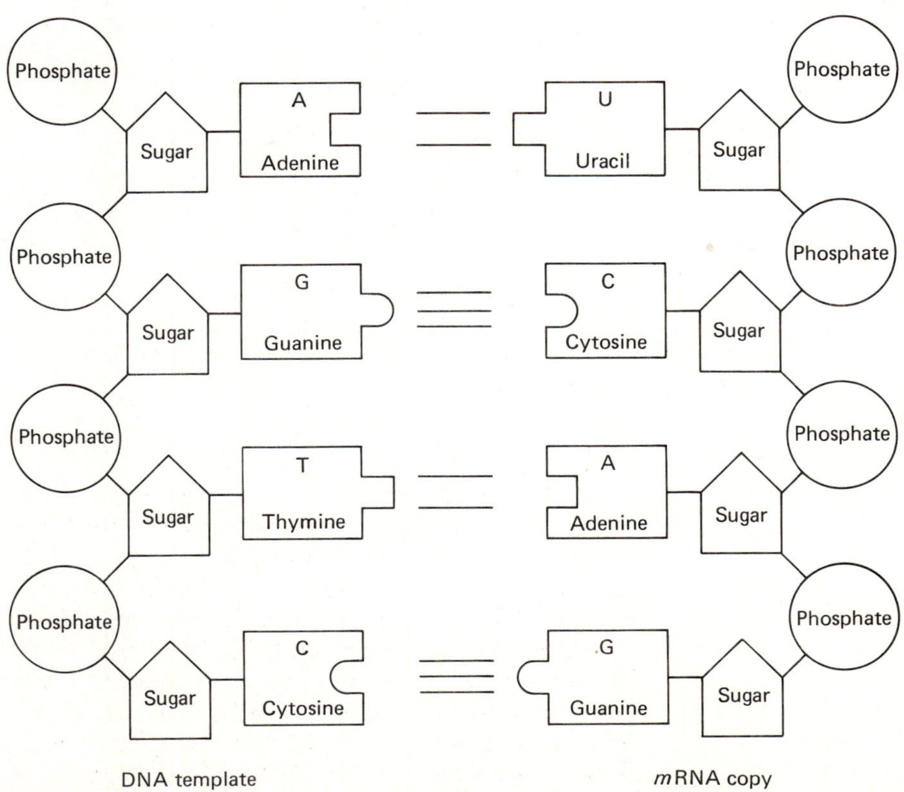

EXAMPLE 15-7   Write the complementary RNA base pairs for the following DNA bases.

(a) adenine   (b) guanine
(c) cytosine   (d) thymine

SOLUTION

DNA base	to	RNA base
adenine	to	uracil
guanine	to	cytosine
cytosine	to	guanine
thymine	to	adenine

After the *m*RNA chain forms in the nucleus it travels through the cytoplasm and ends up at the ribosomes where it becomes attached to them. Remember, the *m*RNA chain is a complementary copy of the DNA chain and is carrying the DNA's message to the place of protein synthesis, the ribosomes. There are many different proteins that must be synthesized by a cell, and an *m*RNA molecule is produced for each protein that must be synthesized.

While the *m*RNA is on its way to the ribosomes, amino acids are entering the cytoplasm of the cell from the bloodstream. There are 20 different amino acids found in the cytoplasm surrounding the ribosomes. Transfer RNA (*t*RNA) molecules are also found in the cytoplasm. It is their job to escort each amino acid to the ribosomes and to the *m*RNA for protein synthesis. Later in this chapter we will see how each *t*RNA molecule escorts a specific amino acid to the ribosomes. Once at the ribosomes, the *m*RNA molecule directs the formation of polypeptide chains from the combination of the proper sequence of amino acids. Let's see how this is accomplished.

The sequence of bases along the DNA nucleotide chain is really a code. Because it contains genetic information, it is called the *genetic code*. Every *three* nucleotides along the chain contains a message. A group of three nucleotides is called a *triplet*. The code is four letters, A, T, C, and G, which stand for adenine, thymine, cytosine, and guanine, the DNA bases. Every amino acid also has its own triplet code. To understand this, let's look at an experiment performed in 1961 to see how the code works.

During a protein synthesis reaction, a synthetic RNA molecule was substituted for *m*RNA. The synthetic RNA molecule contained only uracil nucleotides. Remember, each nucleotide contains a nitrogen base, so that these nucleotides contained only the base uracil, U. Therefore, the triplet code for this synthetic *m*RNA is UUU. The complementary base for uracil on RNA is adenine on DNA. So the DNA code that corresponds to UUU is AAA. At the end of the reaction the protein produced contained only the amino acid phenylalanine. This means that the code UUU for *m*RNA, which corresponds to AAA for DNA, represents the amino acid phenylalanine.

Another similar experiment took place in which a synthetic RNA molecule containing only adenine nucleotides was substituted for *m*RNA. The code for this group of nucleotides is AAA, which corresponds to the DNA nucleotide group, TTT. At the end of this reaction the amino acid lysine was produced. This means that the code AAA for *mRNA*, which corresponds to TTT for DNA, represents the amino acid lysine.

If these experiments sound a bit confusing to you, let's look at the first experiment again. Only, this time let's do it from the point of view of the DNA. Consider one strand of a DNA molecule composed of nucleotides that contain only adenine. We could represent the nucleotide sequence of such a molecule as

AAA   AAA   AAA   AAA   ...

where A represents the adenine in each nucleotide. You can see that the triplet code for such a molecule can only be AAA.

If an *mRNA* molecule was to form alongside this DNA strand, it would have to be complementary to it. In other words, its nucleotides would contain only uracil. We could represent the nucleotide sequence of such an *mRNA* molecule as

UUU   UUU   UUU   UUU   ...

where U represents the uracil in each nucleotide. You can see that the triplet code for this molecule can only be UUU.

The *mRNA* whose triplet code is UUU travels to the ribosomes, where it meets *tRNA* molecules. The *tRNA* molecules have their own triplet codes. The only *tRNA* molecules that can "react" with the *mRNA* are the ones that contain the complementary base. Therefore, because our *mRNA* contains only the triplet code UUU, it will accept only *tRNA* molecules that have the triplet code AAA. Such *tRNA* molecules carry the amino acid phenylalanine. Therefore, the protein produced would contain only phenylalanine.

EXAMPLE 15-8   Using your knowledge of complementary base pairing, write the proper *mRNA* code for the following DNA triplets. Use the abbreviations A for adenine, T for thymine, C for cytosine, G for guanine, and U for uracil.

(a) CAT   (b) GCA   (c) TAT

SOLUTION   (a) CAT on DNA corresponds to GUA on RNA.
(b) GCA on DNA corresponds to CGU on RNA.
(c) TAT on DNA corresponds to AUA on RNA

EXAMPLE 15-9   For the following nucleotide sequence on DNA, write the proper RNA code.

GCACATTAT

      Just use the complementary base pairs.

GCACATTAT on DNA   corresponds to   CGUGUAAUA on RNA

Translating the Genetic Code

Let's review what we've learned so far about the message carried by DNA and RNA. Inside the cell nucleus the double-helical DNA molecule uncoils and serves as a template so that *m*RNA can copy the information carried on one DNA nucleotide chain. The process of copying this information is called *transcription*. The *m*RNA now contains the genetic code carried on the DNA. The code is written in groups of three nucleotides, called triplets. Each triplet is called a *codon*. Each codon is a three-letter code word which has its own meaning and can be translated into its corresponding amino acid (Table 15-3).

In the cell, the *t*RNA molecules translate the messages held on *m*RNA. The *t*RNA molecules are made up of about 70 to 85 nucleotides, and they perform two functions:

1. They read the code carried by the *m*RNA molecules, translating it into the proper amino acid.

2. They pick up the amino acid and bring it to the ribosomes, where with the help of the *m*RNA, the amino acid is placed in its proper place.

Each *t*RNA molecule also contains groups of three nucleotide sequences, which are called *anticodons*. The nitrogen bases on the anticodons are complementary to the bases on the *m*RNA codons. The *t*RNA that carries lysine has the anti-codon UUC, which corresponds to the *m*RNA codon AAG.

EXAMPLE 15-10   Write the *t*RNA anticodons that correspond to these *m*RNA codons.

GCC   UAU   GCA

SOLUTION   The *t*RNA anticodons for GCC, UAU, and GCA are CGG, AUA, and CGU.

EXAMPLE 15-11   Using Table 15-3, write the amino acid sequence requested by the *m*RNA codons in Example 15-10.

SOLUTION   GCC is the code for alanine.
UAU is the code for tyrosine.
GCA is the code for alanine.

Therefore, the code GCC   UAU   GCA would have the amino acid sequence alanine-tyrosine-alanine.

TABLE 15-3   Translating the genetic code

*m*RNA CODON	TRANSLATION INTO AMINO ACID
AAA or AAG	Lysine
AAC or AAU	Asparagine
ACA, ACC, ACG, or ACU	Threonine
AGA, AGG, CGA, CGC, CGG, or CGU	Arginine
AGC, AGU, UCA, UCC, UCG, or UCU	Serine
AUA, AUC, or AUU	Isoleucine
AUG	Methionine (*Start*)
CAA or CAG	Glutamine
CAC or CAU	Histidine
CCA, CCC, CCG, or CCU	Proline
CUA, CUC, CUG, CUU, UUA, or UUG	Leucine
GAA or GAG	Glutamic acid
GAC or GAU	Aspartic acid
GCA, GCC, GCG, or GCU	Alanine
GGA, GGC, GGG, or GGU	Glycine
GUA, GUC, GUG, or GUU	Valine
UAC or UAU	Tyrosine
UGC or UGU	Cysteine
UGG	Tryptophan
UUC or UUU	Phenylalanine
UAA, UAG, or UGA	(*Stop*)

In the actual synthesis of proteins, the *m*RNA travels to the ribosome and unites with a *t*RNA that has the proper anticodon to match its codon. The *t*RNA molecule is held by the *m*RNA molecule at the ribosome. Remember that an amino acid is carried by the *t*RNA molecule corresponding to its coded message. Then the next *m*RNA unites with a *t*RNA with the proper amino acid and holds it. The amino acid held by the second *t*RNA is then bonded to the amino acid held by the first *t*RNA. As each *m*RNA attaches itself to the proper *t*RNA, the *t*RNA no longer holds its amino acid. The amino acid bonds to the amino acid next to it, thereby forming the proper sequence of amino acids, which form a polypeptide chain (Fig. 15-12).

When protein synthesis begins, the *m*RNA calls for a special amino acid called formylmethionine (fMet), whose codon is AUG. This is the starting

Part A

Complimentary strand	ATG	ATC	AGA	GTA	TTT	TAA	DNA
Informational strand	TAC	TAG	TCT	CAT	AAA	ATT	DNA
	AUG	AUC	AGA	GUA	UUU	UAA	*m*RNA

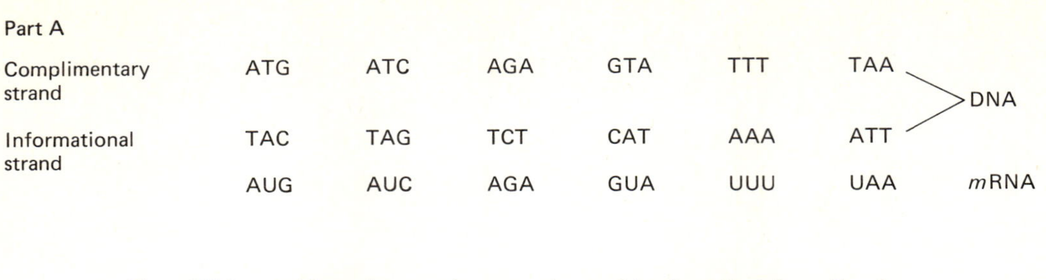

The *m*RNA strand forms by complementary base pairing from the informational strand of the DNA.

Part B

Starting signal	AUG	AUC	AGA	GUA	UUU	UAA	*m*RNA codons

Direction or reading the *m*RNA chain

The *m*RNA leaves the nucleus of the cell and travels to the ribosomes, where it is ready to direct the formation of a polypeptide chain with the help of the *t*RNA.

Part C

AUG	AUC	AGA	GUA	UUU	UAA	*m*RNA codons

The AUG codon on the *m*RNA initiates the protein formation by calling for the amino acid formylmethionine (fMet). The fMet is brought to the *m*RNA by the *t*RNA with the proper anticodon (UAC).

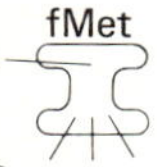

FIGURE 15-12
The formation of a protein.

The next amino acid is brought to the *m*RNA by its *t*RNA. The code AUC on the *m*RNA calls for the amino acid isoleucine. An enzyme will now cause the two amino acids, fMet and Ile to form a peptide bond.

Part E

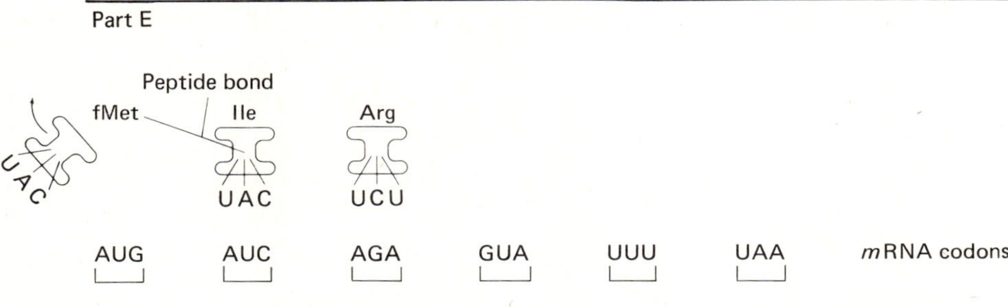

Once the fMet has formed a peptide bond with the Ile, its *t*RNA is released so that it may go back to the cell's cytoplasm to pick up another fMet amino acid. In the meantime the next amino acid called for by the *m*RNA is brought to it by the proper *t*RNA. It is the amino acid arginine.

Part F

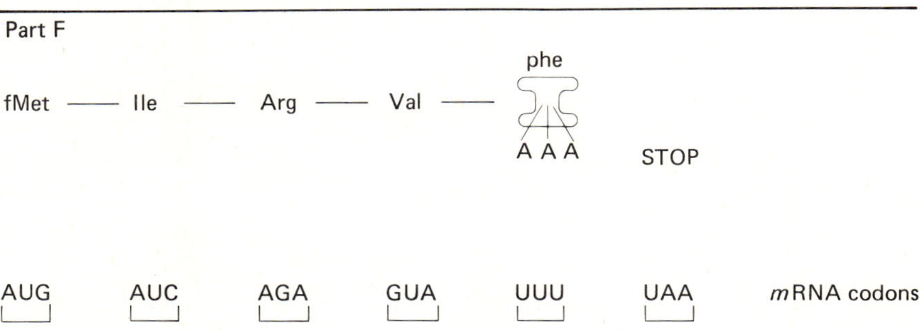

The process is repeated over and over again and the polypeptide chain is formed. When the *m*RNA codon reads UAA, the polypeptide formation stops and the protein is released.

signal for protein synthesis. As the process continues, the *mRNA* codons call for more amino acids in the proper sequence. Each amino acid bonds to the one next to it and the peptide chain grows. The last codon called for is called a *termination codon*. Its triplet code can be UGA, UAG, or UAA. When the termination codon is read, the synthesis of protein stops and the polypeptide chain is released from the ribosome.

EXAMPLE 15-12   For the following nucleotide sequence on DNA, write the proper *mRNA* code.

TAC   AGG   CAG   CGA   AGA   CGG   ATA   ATT

SOLUTION   The *mRNA* code for

TAC   AGG   CAG   CGA   AGA   CGG   ATA   ATT

is   AUG   UCC   GUC   GCU   UCU   GCC   UAU   UAA

EXAMPLE 15-13   For the *mRNA* code in Example 15-12, write the *tRNA* code. Then, using Table 15-3, write the proper sequence of amino acids.

SOLUTION   Let's rewrite the DNA code as given in Example 15-12. Underneath we'll write the *mRNA* code and under that we'll write the *tRNA* code. We'll then use Table 15-3 and the *mRNA* code to get the proper sequence of amino acids.

DNA: TAC   AGG   CAG   CGA   AGA   CGG   ATA   ATT

mRNA: AUG   UCC   GUC   GCU   UCU   GCC   UAU   UAA

tRNA: UAC   AGG   CAG   CGA   AGA   CGG   AUA   AUU

Amino acid sequence: fMet — Ser —— Val —— Ala —— Ser —— Ala —— Tyr   STOP

Regulating Protein Synthesis

The DNA of an organism contains directions for synthesizing very large numbers of proteins. But not all of these proteins are being synthesized at the same time. This is because there is a *repressor protein* bound to the DNA. This repressor protein is produced by a *regulatory gene*. The repressor protein prevents production of *mRNA* and therefore prevents protein synthesis. When a certain protein is needed, a process called *induction* begins. The repressor protein becomes inactive and protein synthesis begins. When no more of that protein is needed, the repressor protein again bonds to the DNA. In other words, the repressor protein becomes active again. Repression and induction is an on-and-off process. When the repressor protein bonds to the DNA, protein synthesis stops; and when the repressor protein is inactivated, protein synthesis begins. (The information known about gene regulation comes from studies done on *Escherichia coli* bacteria. Not much is known about gene regulation in higher animals.)

Mutations

When replication of DNA occurs, the new DNA that is produced must be exactly like the original DNA. This means that the sequence of bases along the nucleotide chains must not change. If it does change, then triplet codes will not be correct and the amino acid sequence called for during protein synthesis will be wrong. A change in the sequence of nucleotides in a DNA molecule is called a *mutation*.

Mutations can be caused by one set of complementary base pairs replacing another, or by adding an additional base pair or leaving out a base pair. Substituting one base pair for another will usually change just one amino acid for another during protein synthesis. But leaving out or adding a base pair means that the whole sequence of amino acids will be changed and the results are usually fatal, causing an entirely wrong protein to be synthesized.

Not all mutations occur by themselves. There are chemicals that can cause mutations. These chemicals are called *chemical mutagens*. Certain dyes and chemicals, such as nitrous acid and nitrogen mustard, are mutagens. Ultraviolet radiation, x rays, and gamma rays in high concentrations can also produce mutations.

Genetic Diseases

There are about 1500 genetic diseases which are caused by gene mutations. Because something happens to the sequence of bases along the DNA nucleotide chain, the wrong protein has been synthesized. These conditions are hereditary and the mistake in the DNA is carried from the parent to the child. Let's look at some of these genetic diseases.

Sickle cell anemia is a disease that occurs predominantly in blacks. In the United States, from 8 to 13% of the black population carries the sickle cell trait. This means that they have inherited the defective gene from one parent. People with the sickle cell trait experience mild discomfort and do not adapt well when the oxygen pressure is low. From 0.3 to 1.5% of the black population in the United States have sickle cell anemia. These people have inherited the gene for sickle cell anemia from both parents. Sickle cell anemia is caused by an error in one part of the DNA that controls the synthesis of hemoglobin. One of the bases in the nucleotide chain is incorrect and the wrong nitrogen base is included in the triplet code. The amino acid valine is substituted for glutamic acid in the peptide chain that is synthesized at the ribosomes. This causes the red blood cells to be sickle-shaped (Fig. 15-13).

Sickle-shaped red blood cells can't carry enough oxygen to the cells of the body because their surface area is not sufficient. Because their shape is distorted, the red blood cells can't pass through the small capillaries and arterioles of the body. Plugs of red blood cells block the vessels and blood clots occur. The sickled red blood cells are also more fragile than normal red blood cells and are often destroyed by the bouncing around they get while being circulated through the blood. This results in anemia. Although sickle cell anemia is a fatal disease, and few patients live past 40, people with sickled cells are protected from malaria. The microorganism that causes malaria cannot grow in a sickled cell, where it must spend part of its life after infecting its human host. So the person with sickled red blood cells is protected from malaria.

Phenylketonuria (PKU) is a genetic disease caused by the absence of an enzyme in the body which converts phenylalanine to tyrosine. The enzyme

FIGURE 15-13
Electron micrographs of normal red blood cells (a) and sickle cells (b). The magnification is about 5000. (Philips Electronic Instruments)

(a)

(b)

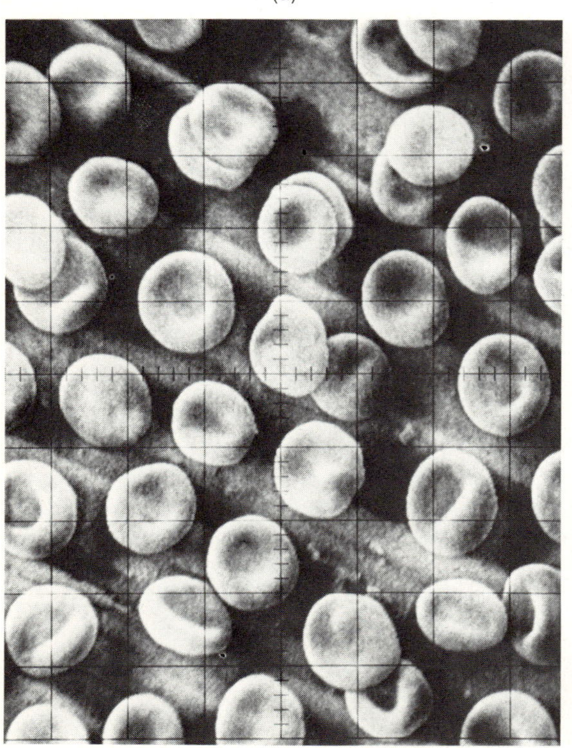

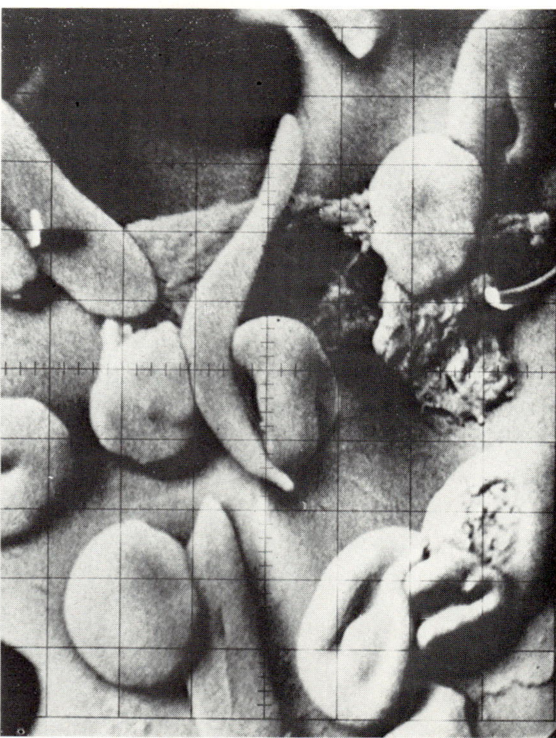

DISEASE	CLINICAL SYMPTONS	DEFECTIVE ENZYME OR PROTEIN
Albinism	White hair; pink iris.	Tyrosinase enzyme missing.
Galactosemia	In infants there is vomiting, growth failure, edema, and finally death as a result of wasting and malnutrition.	1-P uridyl transferase is not present, which is needed for the cellular utilization of galactose.
Glycogen storage disease (Type 1, also known as von Gierke's disease)	Liver enlargement, weight loss, followed by vomiting, hypoglycemia, convulsions, and coma.	Deficiency of glucose-6-phosphatase.
Hemophilia	Blood hemorrhages from trivial injuries.	Missing the antihemophilic (clotting) factor.
Hypervalinemia	Vomiting, mental retardation.	Valine transaminase is missing.
Pentosuria	Excess of pentose sugars in the body causes diabetic symptoms.	The enzyme xylulose dehydrogenase is missing.
Tay-Sachs disease	Mental retardation, blindness, muscular weakness, and death by the age of three.	The enzyme hexosaminidase, which breaks down glycolipids, is missing.

is phenylalanine hydroxylase, which is produced in the liver. A baby that is born with PKU may be unusually irritable, may have epileptic seizures, and may vomit. Phenylacetic acid appears in the urine and sweat, causing a specific odor. About two-thirds of the babies that have elevated phenylalanine levels in the blood show severe mental retardation. The disease is treated by lowering, but not eliminating, phenylalanine from the diet. Babies are fed Lofenalac, a formula fortified with vitamins, iron, and other minerals, instead of milk. Low-protein foods such as fruits, vegetables, and cereals are also included in the diet. With such treatment, mental retardation can be avoided and many of the other problems associated with this condition can be reversed.

Some of the other genetic diseases are summarized in Table 15-4.

SUMMARY

In this chapter we learned about nucleic acids. We began our discussion by looking at the two major types of nucleic acids, DNA and RNA. We saw that nucleic acids are composed of smaller units called nucleotides, and that each nucleotide is composed of a five-carbon sugar, a phosphate group, and a nitrogen-containing base. It is the nitrogen-containing bases that play a very important role in the functions of the nucleic acids.

We also looked at the structures of DNA and RNA and learned about the process of complementary nitrogen base pairing. We learned how DNA replicates itself and how DNA and RNA play a major role in the synthesis of proteins. Finally, we read about how mutations occur in the DNA code and how some of these mutations produce genetic diseases.

EXERCISES

1. Name the two types of nucleic acids.

2. Match the word on the left with its definition on the right.
   (1) amino acids   (a) The basic building blocks of nucleic
   (2) nucleotides   acids.
   (3) monosaccharides   (b) The basic building blocks of proteins.
                                      (c) The basic building blocks of
                                          polysaccharides.

3. Name the three subgroups that compose nucleotides.

4. Name the two five-carbon sugars and state the nucleic acids with which they are associated.

5. Place the words *single* and *double* in the appropriate places:
   The purines are ________ ring structures and the pyrimidines are ________ ring structures.

6. Place the words *purine* and *pyrimidine* in the appropriate places:
   Adenine and guanine are ________, and cytosine, thymine, and uracil are ________.

7. (a) Name the nitrogen-containing bases in DNA.
   (b) Name the nitrogen-containing bases in RNA.

8. The nucleic acid whose structure has a double helix is ________ (RNA or DNA).

9. Write the complementary base pairs for the following nitrogen bases of DNA.
   (a) thymine   (b) cytosine   (c) guanine   (d) adenine

10. Write the complementary base pairs for the following nitrogen bases of RNA.
    (a) uracil   (b) cytosine   (c) guanine   (d) adenine

11. Define the following terms.
    (a) mitosis   (b) parent cell   (c) daughter cell

12. Match the word on the left with its definition on the right.
    (1) cytoplasm   (a) This substance acts as a sort of amino acid
    (2) ribosomes   *escort.*
    (3) *m*RNA   (b) The main body of the cell (aside from the
    (4) *t*RNA   nucleus).
                         (c) The place in the cell where the actual protein
                             synthesis occurs.
                         (d) This substance directs the formation of the
                             polypeptide chain.

13. Write the complementary RNA base pairs for the following DNA bases.
     (a) adenine   (b) guanine   (c) cytosine   (d) thymine

14. Write the complementary DNA base pairs for the following RNA bases.
    (a) adenine   (b) guanine   (c) cytosine   (d) uracil

15. Write the proper *m*RNA code for the following DNA triplets.
    (a) TAG   (b) GAC   (c) TTG   (d) GAA

16. Write the proper *t*RNA anticodons that correspond to the following
    *m*RNA codons.

            AGG   GCA   UUC   UCG   AGC   CCC

17. Using Table 15-3, write the amino acid sequence requested by the
    *m*RNA codons in Exercise 16.

18. For the following nucleotide sequence on DNA:
    (a) Write the proper *m*RNA code.
    (b) Write the proper *t*RNA anticodons that correspond to the *m*RNA
        codons.
    (c) Write the amino acid sequence requested by the *m*RNA codons.

    DNA: TAC ATA CGG AGA CAG CGA TTT ATT

19. Match the word on the left with its definition on the right.
    (1) repressor protein   (a) A change in the sequence of nucleotides
    (2) regulatory gene   in the DNA molecule.
    (3) induction   (b) This process begins when a certain
    (4) mutation   protein is needed.
    (5) mutagen   (c) These substances are chemicals that
                                  can cause mutations.
                              (d) This substance prevents production
                                  of *m*RNA.
                              (e) This biological unit produces the
                                  repressor protein.

20. Match the genetic disease with its description or symptoms.

(1) sickle cell anemia   (a) Phenylacetic acid appears in the urine.

(2) phenylketonuria   (b) The enzyme necessary for the utilization of galactose is missing.

(3) galactosemia   (c) Causes mental retardation and death by age three as a result of accumulation of glycolipids.

(4) hemophilia

(5) Tay-Sachs

(d) This disease is caused by an error in the primary structure of hemoglobin.

(e) Blood hemorrhages from trivial injuries occur to individuals with this disease.

CHAPTER 16

Blood

The Flow of Life

You should be able to:

1. Name the two major types of body fluids: extracellular fluid and intracellular fluid.
2. State that water makes up about 60 to 70% of our total body weight.
3. State at least four major functions of the blood.
4. State that blood is composed of two parts, plasma and formed elements, and name each of the formed elements.
5. Calculate the specific gravity of blood given its mass and the mass of an equal volume of water.
6. Define the term hematocrit, and calculate the hematocrit of a sample of whole blood when given sufficient data.
7. Name the four important proteins found in blood plasma, and state the importance of each.
8. Define the terms hydrostatic pressure and osmotic pressure.
9. Define the terms, antigen, antibody, natural immunity, and artificial immunity.
10. Define the terms toxoid, antiserum, and vaccination.
11. State the three major steps in blood clot formation.
12. Explain the function of red blood cells in transporting oxygen through the body.
13. State where red blood cells are produced in the body.
14. Name the two types of white blood cells and state where they are produced.
15. State the four different blood types and the two Rh factors.
16. Tell which blood types can be mixed without agglutination occurring.
17. State the blood plasma antibodies for each type of blood.
18. Discuss the causes of the following disorders and diseases of the blood: anemia, polycythemia, leukopenia, leukocytosis, leukemia, and hemophilia.

Introduction

The fluids in our bodies are composed primarily of water, which makes up about 60 to 70% of our total body weight. Our body fluids are found in two locations: outside cells and inside cells. *Intracellular fluid* is found within our cells. It contains large amounts of potassium and phosphate ions as well as proteins, and provides the proper environment for the chemical reactions that take place in our cells. *Extracellular fluids* are found outside our cells. There are several extracellular fluids, such as *blood plasma*, which is the liquid portion of blood; *interstitial fluid*, which is found between the cells of the body; *lymph*, which is the fluid in the lymphatic ducts; *aqueous humor*, which is found in our eyes; and *cerebrospinal fluid*, which nourishes the brain and spinal cord (Fig. 16-1).

Although all these fluids play an important role in maintaining life, in this chapter we are going to study *blood*—the most important fluid in our bodies. Before we do, however, let's take a short look at water, a major component of blood and other body fluids.

FIGURE 16-1
The various types of body fluids.

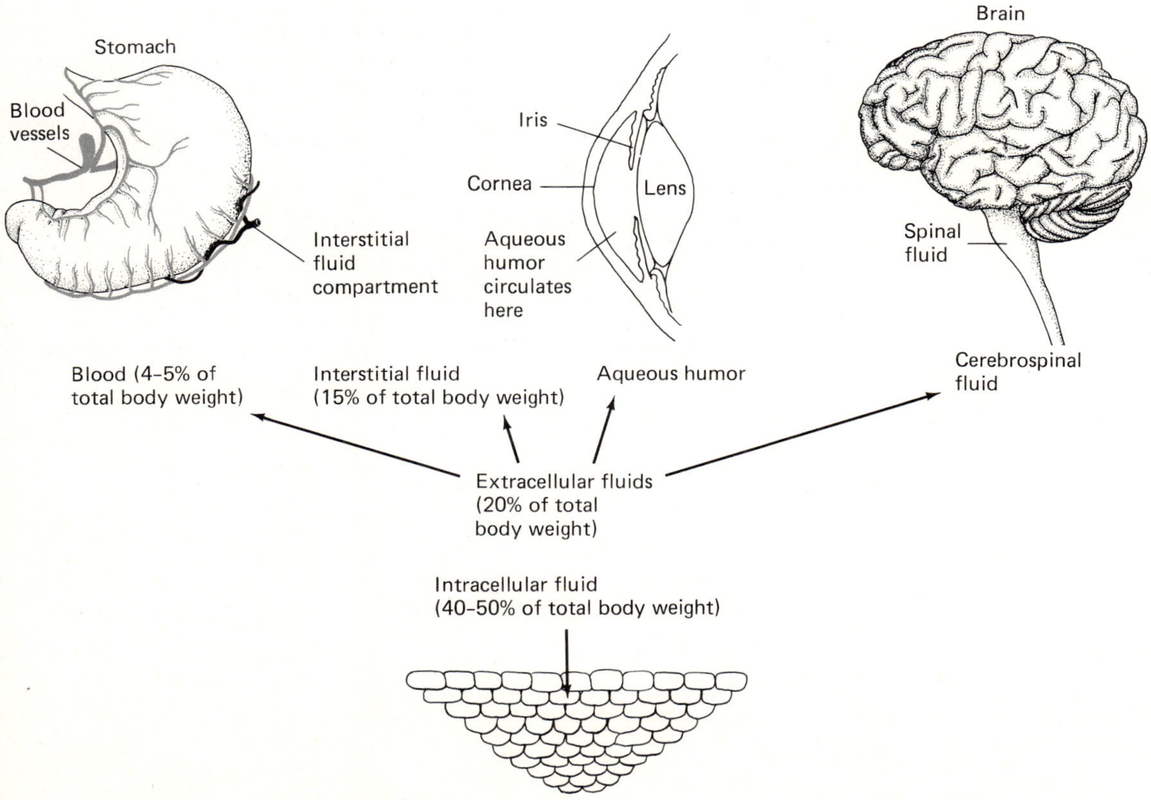

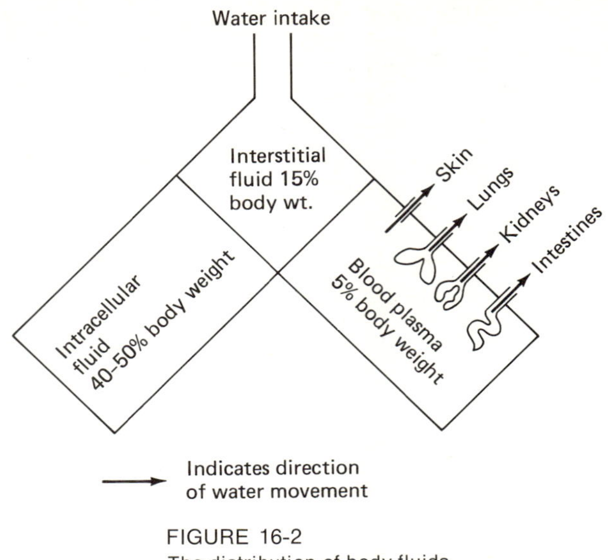

FIGURE 16-2
The distribution of body fluids.

Water: It's in Our Blood

Water is an important part of our diet. We obtain this water from the beverages we drink and the foods we eat. A small amount of water is also produced during metabolism. Where does this water go in our bodies? Most of the water is absorbed through our intestinal walls. It then goes into the interstitial fluid, which is the fluid found between our cells. Interstitial fluid makes up about 15% of our total body weight. From there the water passes into the blood plasma or into the cells. Blood plasma makes up about 5% of our total body weight. The fluid in our cells makes up about 40 to 50% of our total body weight. Water is finally excreted by the lungs, kidneys, intestines, and skin (Fig. 16-2 and Table 16-1).

TABLE 16-1   Average daily water balance for an adult

WATER INTAKE	AMOUNT		WATER OUTPUT	AMOUNT
Drinking water and beverages	1500 ml		Kidneys	1500 ml
			Lungs	350 ml
Water contained in foods	700 ml		Skin	450 ml
Metabolic water	300 ml		Intestines	200 ml
Total	2500 ml		Total	2500 ml

The Functions of Blood

Blood makes up about one-twelfth of the total body weight. A 120-pound person has about 10 pounds of blood. Let's look at the major functions of blood.

1. Blood carries the gases that we inhale away from our lungs and brings the gases that we must exhale to our lungs.

2. Blood transports foods as well as the end products of digestion throughout the body.

3. Blood transports waste products to the proper locations for excretion.

4. Blood distributes the various hormones, vitamins, and enzymes to the places where they are needed.

5. Blood protects us against many diseases.

6. Blood aids in the maintenance of the proper chemical and water balance in the body.

7. Blood helps regulate body temperature by moving (blood) either toward or away from the surface of the body.

8. Blood clots to prevent excessive loss of blood.

Now that we have an idea of what functions blood can carry out, let's look at the composition of blood.

The Composition of Blood

Blood is made of two parts: a fluid part, called *plasma*, and cells that are suspended in the plasma, called *formed elements*. The plasma and the formed elements together make up *whole blood*. The formed elements are the *red blood cells*, also known as *erythrocytes*; the *white blood cells*, also known as *leukocytes*; and the *blood platelets*, also known as *thrombocytes*. Plasma makes up 55 to 60% of the volume of whole blood, and the formed elements make up 40 to 45% of the whole blood volume (Table 16-2). An average adult has 4 to 5 liters of blood.

EXAMPLE 16-1   State the components of whole blood.

SOLUTION   The components of whole blood are plasma, red blood cells, white blood cells, and platelets. The red blood cells, white blood cells, and platelets are the formed elements.

TABLE 16-2   The composition of blood

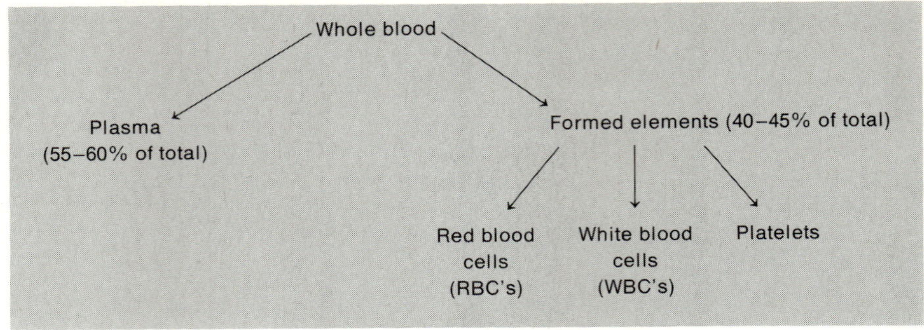

The Characteristics of Blood

The blood that circulates in our arteries is bright red. This is because it has a high oxygen content. The blood that circulates in our veins is darker red. This is because it has a low oxygen content. The pH of blood ranges from 7.35 to 7.45. The *specific gravity* of whole blood ranges from 1.035 to 1.075. Specific gravity refers to the weight of a certain volume of blood compared to the weight of the same volume of water; or

$$\text{Specific gravity (of blood)} = \frac{\text{mass of blood}}{\text{mass of equal volume of water}}$$

EXAMPLE 16-2   A 10.00-ml sample of blood has a mass of 10.50 g.
A 10.00-ml sample of water has a mass of 10.00 g.
What is the specific gravity of the blood?

SOLUTION   We'll use the formula that we've just discussed.

$$\text{Specific gravity (of blood)} = \frac{\text{mass of blood}}{\text{mass of equal volume of water}}$$

$$= \frac{10.50 \text{ g}}{10.00 \text{ g}} = 1.050$$

Blood is a chemical mixture. The compounds that make up blood are not combined chemically, so these compounds can be separated by physical means. Plasma can be separated from whole blood by centrifuging. The formed elements end up at the bottom of the test tube and the plasma remains above (Fig. 16-3).

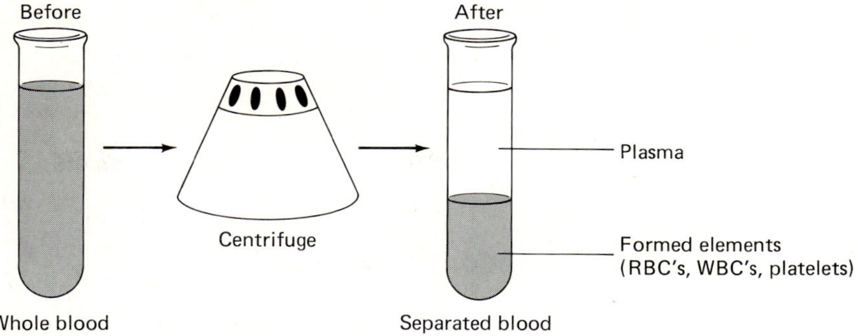

FIGURE 16-3
Centrifuging separates plasma and formed elements.

The percentage of formed elements in the blood is called the *hematocrit*. A sample of whole blood is centrifuged and the percentage of formed elements is calculated. Let's say that a 10.0-ml sample of whole blood is centrifuged and 4.50 ml of formed elements remains at the bottom of the tube. The percentage of formed elements is

$$\frac{4.50 \text{ ml}}{10.0 \text{ ml}} \times 100 = 45.0\%$$

The hematocrit of this sample is 45.0% (Fig. 16-4).

FIGURE 16-4
Obtaining the hematocrit of a sample of whole blood.

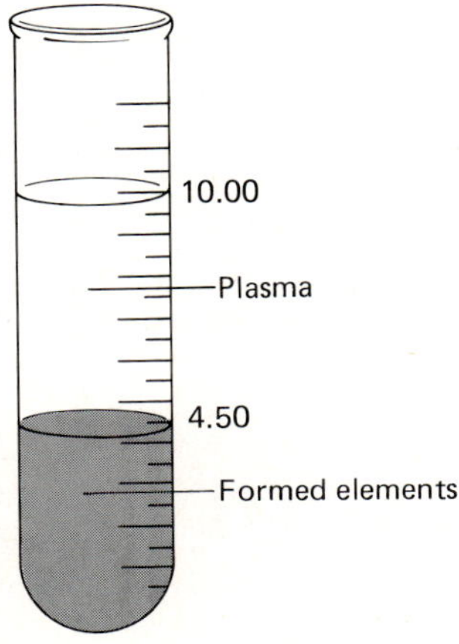

   Calculate the hematocrit for the following blood samples.

(a) 40 ml of whole blood containing 20 ml of formed elements
(b) 25 ml of whole blood containing 10 ml of formed elements

SOLUTION

(a) Percent formed elements $= \dfrac{20 \text{ ml formed elements}}{40 \text{ ml whole blood}} \times 100$

$$= 50\%$$

The hematocrit of this sample is 50%.

(b) Percent formed elements $= \dfrac{10 \text{ ml formed elements}}{25 \text{ ml whole blood}} \times 100$

$$= 40\%$$

The hematocrit of this sample is 40%.

The hematocrit is a measure of the amount of cells in the blood, especially red blood cells. When the hematocrit is too low, this indicates a low number of cells in the blood, which is characteristic of an anemic condition (Table 16-3). An anemic condition means that the oxygen and carbon dioxide carrying capacity of the blood is decreased. When the hematocrit is too high, other disorders such as polycythemia vera may be indicated. We'll discuss such disorders later in this chapter.

Now let's look at each of the components of blood in more detail.

Blood Plasma

Plasma is 90% water. Proteins, various salts, gases, hormones, enzymes, antibodies, nutritive materials, and waste products make up the remaining 10% (Table 16-4).

TABLE 16-3   Some hematocrit values

	MALE	FEMALE
Normal values	42–50%	38–45%
Abnormally high values (for example, in polycythemia vera	>55%	>55%
Abnormally low values (for example, in some forms of anemia)	<35%	<35%

TABLE 16-4   Some substances found in blood plasma
and their normal values

SUBSTANCE	MILLIGRAMS/100 ml
Carbon dioxide	(55–75% by volume)
Fibrinogen	200–400
Chloride (as NaCl)	580–630
Inorganic phosphorus	3–4.5
Total lipids	500–600
Triglycerides	40–150

There are four important proteins found in plasma. They are *fibrinogin*, *prothrombin*, *albumin*, and *globulin*. Except for one form of globulin, called *gamma globulin*, all of these proteins are produced by the liver. The gamma globulins are produced in the blood plasma.

EXAMPLE 16-4   Name the four proteins found in blood plasma.

SOLUTION   The four proteins found in blood plasma are fibrinogin, prothrombin, albumin, and globulin.

Albumin: A Plasma Protein

Albumin is the most abundant protein found in the plasma (59%). Albumin is a colloid with a large diameter. In fact, it is too large to pass through the walls of the blood vessels. Therefore, it remains in the blood plasma and creates a pressure on the blood vessels. This enables blood to maintain a high pulling pressure (osmotic pressure). Osmotic pressure "pulls" or "draws" water through a membrane from a region of low solute concentration to a region of high solute concentration. Because colloids, like albumin, are too large to leave the blood, the blood maintains a greater solute concentration than the surrounding interstitial fluid. Therefore, the blood maintains a higher osmotic pressure than the surrounding fluid. Let's see how this is important to the maintenance of the body.

The heart pumps blood through the blood vessels and blood is forced against the walls of the blood vessels. The pressure that results from a fluid being forced against a wall is called *hydrostatic pressure*. Hydrostatic pressure is a pushing pressure. It pushes fluid out of the blood vessels through the capillary walls. A capillary has two ends: an arterial end and a venous end

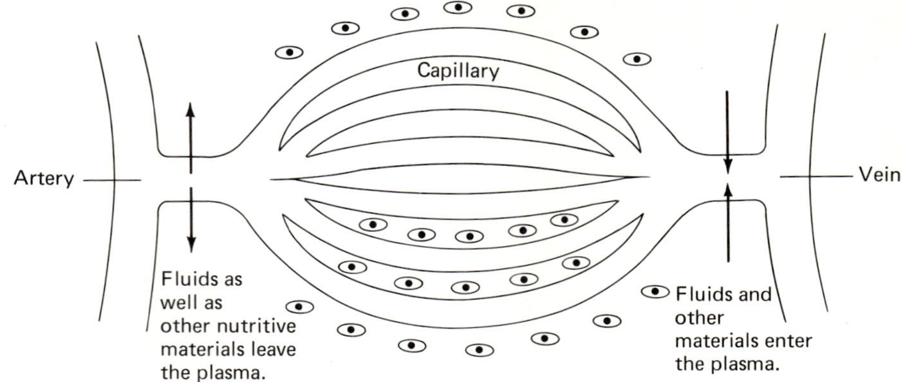

FIGURE 16-5

Movement of fluids in and out of the capillaries. At the arterial end of the capillary the hydrostatic pressure of the blood plasma is greater than the osmotic pressure and fluids and materials are pushed from the plasma to the interstitial fluid and then to the cells. At the venous end of the capillary the osmotic pressure of the blood plasma is greater than the hydrostatic pressure and fluids and other materials are pulled from the interstitial fluid into the blood plasma.

(Fig. 16-5). The arterial end comes from the arteries, which carry nutrients to the cells, and the venous end brings the waste products away from the cells. When blood enters the arterial end of the capillary its hydrostatic pressure is about 32 torr and its osmotic pressure (caused by colloids) is about 22 torr. Therefore, the "pushing force" is greater than the "pulling force" and fluids and materials are pushed out of the blood and into the interstitial fluid. At the venous end of a capillary the osmotic pressure is about 22 torr and the hydrostatic pressure is about 12 torr. Therefore, the pulling force is greater than the pushing force and fluids and waste materials are pulled into the blood from the interstitial fluid. A similar interaction of hydrostatic pressure and osmotic pressure moves nutrients into the cells from the interstitial fluid and moves wastes out of the cells into the interstitial fluid (Table 16-5).

TABLE 16-5   Pathway of nutrients and wastes

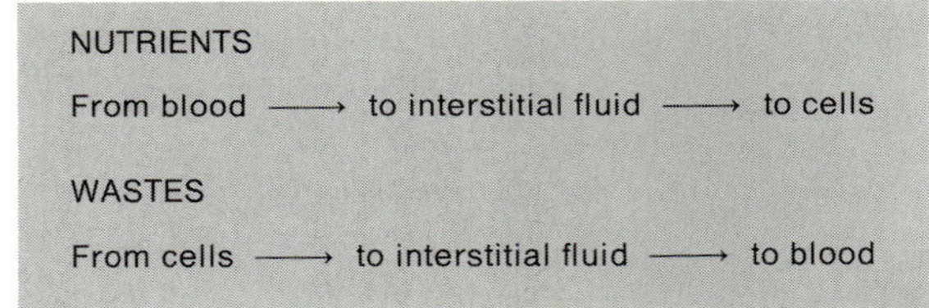

NUTRIENTS

From blood ⟶ to interstitial fluid ⟶ to cells

WASTES

From cells ⟶ to interstitial fluid ⟶ to blood

EXAMPLE 16-5   Complete each of the following statements with the words *hydrostatic pressure* or *osmotic pressure*.

    (a) The force that moves fluids out of the arterial end of the capillaries and into the interstitial fluid is known as __________.

    (b) The force that moves fluids from the interstitial fluid to the venous blood is known as __________.

SOLUTION   (a) This force is known as hydrostatic pressure.

    (b) This force is known as osmotic pressure.

Beside this very important function, albumin also transports molecules in the blood that are slightly soluble in water. These substances bind to the albumin, thereby increasing their solubility.

Globulin: A Plasma Protein

Globulin is found in the plasma in three forms: alpha, beta, and gamma. The gamma globulins are the most important of these proteins since they provide immunity against disease. It is for this reason that the gamma globulins are called *immunoglobulins*.

When certain foreign substances enter the body, specific proteins called *antibodies* are produced to counteract them. The gamma globulins act as antibodies against these foreign substances, which are known as *antigens*. Most antigens are proteins or substances that contain proteins. Microbes that cause disease, as well as pollen, snake venom, spores, and transplanted organs such as the heart and kidneys, are antigens. Not all antigens are poisonous to the body.

When antigens are present in the body they stimulate the production of very specific antibodies that react with the various antigens and neutralize them. Antigens stimulate the formation of antibodies with complementary shapes (Fig. 16-6). Thus there is a specific antibody for each type of antigen.

FIGURE 16-6
There is a specific antibody for each type of antigen.

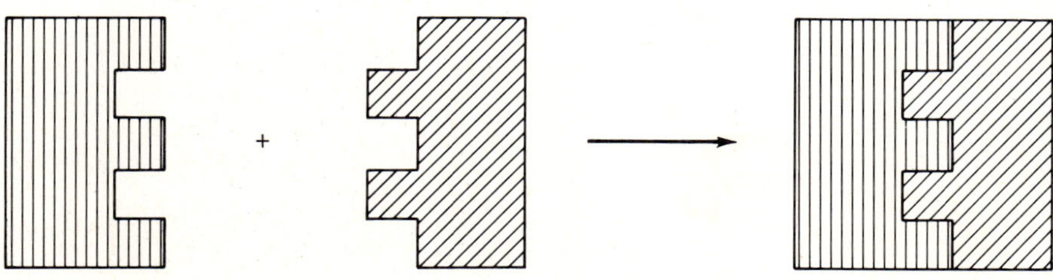

The gamma globulins are one type of antibody. Once formed, antibodies remain in the blood to protect it against the antigen that caused their formation. Antibodies provide immunity against diseases for varying lengths of time.

EXAMPLE 16-6   Define antigen and antibody.

SOLUTION   Antigens are foreign proteins that enter the body. Antibodies are proteins that are formed to counteract antigens. For example, the gamma globulins are antibodies.

There are two types of immunity against disease: *natural immunity* and *artificial immunity*. *Natural active immunity* is acquired by contracting a disease. The disease-causing organisms act as antigens which cause specific antibodies to be produced by the body. Although the patient must suffer with the disease, natural active immunity is the longest-lasting immunity. Newborns have a type of immunity called *natural passive immunity*. Antibodies from the mother's blood travel through the placenta into the child's blood. The child receives only the antibodies that the mother's blood contains. This type of immunity lasts only 4 to 6 months, during which time the child is immune to the same organisms as the mother.

There are two types of artificial immunity: artificial active immunity and artificial passive immunity. *Artificial active immunity* is acquired by injection of killed or weakened organisms in a process called *vaccination*. The body produces antibodies against the organism and immunity lasts from months to years. It often takes 7 to 10 days for the antibodies to be formed. Typhoid and pertussis (whooping cough) can be controlled by injecting a killed organism. Tetanus and diphtheria are controlled by injecting a *toxoid*. A toxoid is a toxin that has been treated with heat or chemicals so that it is no longer toxic, but still acts as an antigen for antibody production.

Artificial passive immunity is acquired by injection of an antiserum or antoxin. This gives immediate immunity to a person who has been exposed to or has contracted a disease such as infectious hepatitis, poliomyelitis, or measles. An *antiserum* is blood serum that contains specific antibodies. (Blood serum is plasma with the clotting factors removed.) An *antitoxin* is blood serum that contains antibodies for a specific toxin. The body doesn't have to build up antibodies because they are already present in the injection.

Fibrinogen and Prothrombin: Two More Plasma Proteins

The other proteins found in blood plasma are fibrinogen and prothrombin. These proteins are important in the clotting process. We'll look at how blood clots shortly, and we'll discuss in great detail how these two plasma proteins play such an important role in this process.

Nutritive Materials

The blood plasma carries nutritive materials to the cells of the body. These include amino acids from protein breakdown, monosaccharides from carbohydrate breakdown, and triglycerides.

The plasma carries small amounts of oxygen gas and large amounts of carbon dioxide in the form of bicarbonate (HCO_3^-) ions. Nitrogen gas is also carried in the plasma.

The plasma carries the waste products formed during chemical reactions to the kidneys, where they are removed by filtration.

Hormones secreted by the endocrine glands are carried to the locations where they are needed by the plasma. When they are no longer needed, the hormones are transported to the liver, where they are broken down. Enzymes are also brought to their specific reaction sites by the plasma, as is the case with the various salts that are needed by the body.

Blood Clotting

Blood clots form when the normal flow of blood must be stopped. Blood does not normally clot in the blood vessels because of the presence of a chemical called *heparin*, which is produced by the liver. When blood vessels are cut or damaged, a clot normally forms in 3 to 7 minutes. Here is how it happens.

In the blood are one type of formed elements called *platelets*. Platelets are small fragments of cells that have no definite shape, no nucleus, and no color. They contain a chemical that initiates the clotting of blood. When platelets come in contact with the cut edges of skin and blood vessels, they disintegrate and release substances that react to form *thromboplastin*. Thromboplastin is also released by the damaged tissue cells and the blood plasma. Thromboplastin is a protein which, together with calcium ions and another protein called *prothrombin*, generates an important enzyme. This important enzyme is called *thrombin*, and it catalyzes the formation of *fibrin* threads from a soluble plasma protein called *fibrinogen*. Fibrin is an insoluble protein which precipitates and forms long threads. These threads tangle up at the site of the injury and form the blood clot. The red blood cells also become tangled up in the clot and the bleeding stops (Table 16-6 and Fig. 16-7).

TABLE 16-6   Steps in blood clotting

STEP 1.   Thromboplastin released at site of injury.

STEP 2.   Prothrombin $\xrightarrow[\text{Ca}^{+2}, \text{ other factors}]{\text{thromboplastin}}$ thrombin

STEP 3.   Fibrinogen $\xrightarrow{\text{thrombin}}$ fibrin

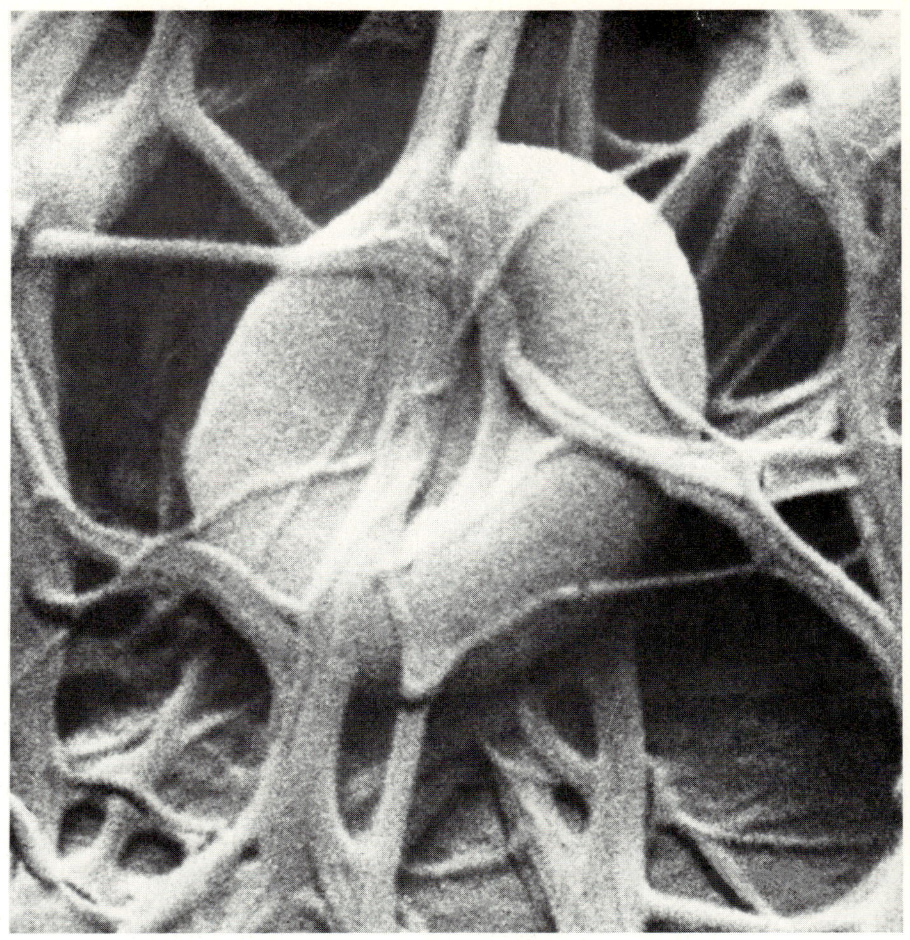

FIGURE 16-7

Scanning electron micrograph of an erythrocyte enmeshed in fibrin (about 20,500 ×). (Emil Bernstein and Eila Kairinen, Gillette Research Institute, Rockville, Maryland)

EXAMPLE 16-7   State the three major steps in blood clot formation.

SOLUTION

1. Thromboplastin is released at the site of the injury.
2. Prothrombin reacts to form thrombin in the presence of thromboplastin, calcium ions, and other factors.
3. Thrombin catalyzes the conversion of fibrinogen to fibrin.

Blood clotting is a very complicated process and there are several factors necessary during each step of the process. For example, the body must have a

The structure of dicumarol, an anticoagulant.

sufficient supply of vitamin K, since vitamin K is needed for the synthesis of prothrombin. There are also several factors which determine the clotting time. *Hemophilia* is a hereditary disease in which one of the factors necessary for the initiation of a blood clot is missing and the blood takes several hours to form a clot. A minor injury becomes a major one when bleeding doesn't stop.

There are several drugs that prevent blood from clotting after surgery or after a heart attack, when a blood clot could travel through the blood vessels and obstruct a major artery or vein. A clot formed in a blood vessel is called a *thrombus*. As we mentioned earlier, heparin is a substance found in the body that prevents clot formation. It interferes with the formation of thrombin from prothrombin. Dicumarol is a drug that acts like heparin to prevent blood from clotting (Fig. 16-8).

When blood is drawn, it automatically clots in the test tube unless chemicals, anticoagulants, are added to prevent clot formation. To prevent clotting in blood that will be used for transfusions a chemical called sodium citrate is added. This removes the calcium ions from the blood in the form of calcium citrate, thus eliminating the calcium ions from the clotting process.

Now that we've completed our discussion of the substances found in blood plasma, we are ready to look at the formed elements in the blood.

The Formed Elements: Red Blood Cells, White Blood Cells, and Platelets

Red Blood Cells

Red blood cells are involved in transporting oxygen to the cells of the body. They are also called *red blood corpuscles* and *erythrocytes*. The abbreviation for red blood cells is RBC's.

There are about 4.2 to 5.9 million RBC's in 1 cubic millimeter of blood. Imagine how small these cells are! There are about 25 trillion RBC's in the blood and millions of RBC's are destroyed every second, while millions are being replaced. Mature red blood cells have no nucleus, so they are incapable of reproducing themselves. They are produced in the *red bone marrow*. Bones contain a tissue called *bone marrow*. There are two types of bone marrow, yellow and red. Red bone marrow is found in almost every bone of a child, but when a person reaches adulthood, red bone marrow is found only in certain bones. It is found in certain parts of the bones of the upper arm and thigh, as well as in the vertebrae, the breastbone, the ribs, and the cranial bones.

The production of RBC's by the body is chemically regulated. When the body needs more oxygen or when the oxygen level is too low, a chemical is released into the bloodstream by the kidneys and is carried to the red bone marrow, where production of RBC's occurs. Vitamin B_{12} is important for the production of mature RBC's.

RBC's last from 90 to 120 days. These tiny RBC's undergo a great deal of mechanical movement as they travel through the blood and they wear out and are destroyed in the spleen, liver, and bone marrow.

Hemoglobin makes up about one-third of the weight of each red blood cell. In Chapter 13 we discussed the structure of this protein molecule. You might remember that it is made up of four subunits. Each subunit is made up of a protein called *globin*, which is attached to a prosthetic group called *heme*. Heme is an iron-containing group, and it is the iron part of each subunit which combines with the oxygen molecule to form *oxyhemoglobin*, the oxygenated form of hemoglobin. Since each hemoglobin molecule is made up of four subunits, each molecule of hemoglobin can carry four oxygen molecules. Each red blood cell has from 200 to 300 million hemoglobin molecules, so a single red blood cell can carry from 800 to 1200 million oxygen molecules.

The red blood cells have several important functions. Oxygen is transported around the body by combining with hemoglobin to form oxyhemoglobin. There is also a small amount of oxygen which dissolves in the plasma and is transported. Carbon dioxide is also transported around the body by combining with hemoglobin to form *carboxyhemoglobin*. In addition, carbon dioxide is transported away from the tissues in the plasma and in the red blood cells in the form of bicarbonate ($HCO_3{}^-$) ions.

EXAMPLE 16-8   How are oxygen and carbon dioxide transported by the RBC's?

SOLUTION   Oxygen combines with the hemoglobin in the RBC's and forms oxyhemoglobin. Carbon dioxide combines with the hemoglobin in the RBC's and forms carboxyhemoglobin. In addition, carbon dioxide is carried on the RBC's and in the plasma as bicarbonate ions.

White Blood Cells

White blood cells are involved in protecting the body against disease. They are also called *leukocytes* and are abbreviated WBC's.

White blood cells are larger than RBC's. They contain a nucleus but no hemoglobin, so they can't transport any gases. There are about 5000 to 10,000 WBC's per cubic millimeter of blood. When an infection occurs, the number of WBC's increases.

There are two types of WBC's: those with granules in the cytoplasm and those without granules in the cytoplasm. *Granular leukocytes*, or *granulocytes*, contain granules, or tiny grains, in the cytoplasm. They are produced in the

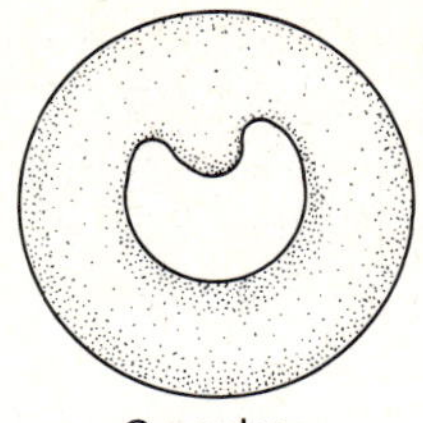 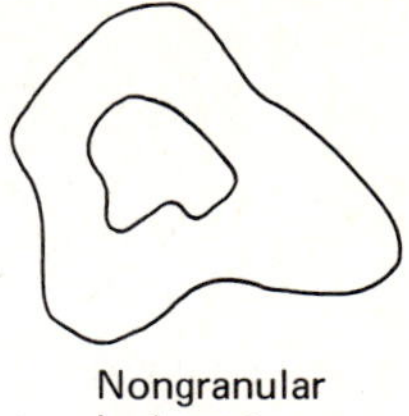

FIGURE 16-9
The structure of leukocytes.

red bone marrow. *Nongranular leukocytes* or *agranulocytes*, are produced in the spleen and lymph nodes (Fig. 16-9). Each type of leukocyte performs a different function in the blood

Leukocytes can move out of the capillaries and into the tissues, where they kill bacteria and surround damaged tissues. The mixture of dead bacteria and WBC's is called *pus*. Some WBC's act like antibodies which counteract antigens. They remain in the body to provide immunity against disease.

EXAMPLE 16-9   Name the two types of white blood cells and state where they are produced.

SOLUTION   The two types of white blood cells are granular leukocytes, which are produced in the red bone marrow, and nongranular leukocytes, which are produced in the spleen and lymph nodes.

Platelets

Platelets were discussed in connection with their role in blood clotting. They are cell fragments and have no specific shape, color, or nucleus. There are about 200,000 to 400,000 platelets per cubic millimeter of blood. They are produced in the red bone marrow and live from 5 to 9 days, at which time they are destroyed in the spleen (Fig. 16-10).

FIGURE 16-10
Blood platelets.

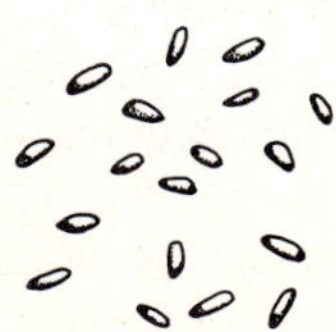

Platelets

Blood Transfusions and Blood Types

It is often necessary for a person who has lost a great deal of blood to receive a *blood transfusion* so that an adequate level of blood is maintained. A blood transfusion can consist of either whole blood or plasma being given to the *recipient*. The recipient is the person who receives the blood and the *donor* is the person who gives the blood. Whole blood is more desirable than plasma since it contains the formed elements. However, when receiving whole blood the donor and recipient must be cross-matched to be sure that the blood will

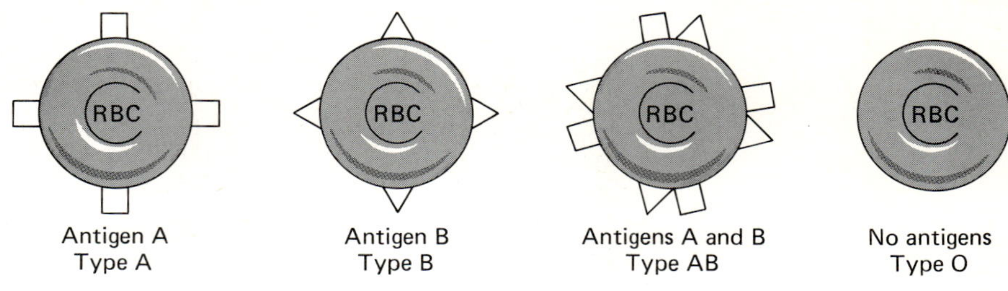

FIGURE 16-11
Antigens protrude from the red blood cells.

not agglutinate (clump) when mixed. Let's see why blood from a donor might cause a serious problem when given to a recipient.

Blood from a donor can clump when given to a recipient because of an antigen-antibody interaction. There are proteins attached to the membranes of the RBC's called antigens. The plasma contains antibodies that will counteract certain RBC antigens. The recipient's plasma must not contain antibodies that will react against the antigens in the donor's blood. Thus the donor's blood type and the recipient's blood type must be compatible.

There are four different blood types; A, B, AB, and O. The blood type tells you which antigen or antigens are present on the RBC's. Type A blood contains antigen A, type B blood contains antigen B, type AB blood contains antigens A and B, and type O blood contains no antigens. The antigens protrude from the RBC's (Fig. 16-11).

In the plasma are a group of gamma globulin antibodies and each blood type contains *antibodies which are opposite the antigens* present on the RBC's. Type A blood plasma has antibodies for type B blood, called anti-B. Type B blood plasma has antibodies for type A blood, called anti-A. Type O blood plasma has antibodies for types A and B blood, so it contains anti-A and anti-B. Type AB blood has no antibodies in the plasma (Table 16-7). During

TABLE 16-7   Blood antigen and antibodies

BLOOD TYPE	ANTIGEN(S) PRESENT	ANTIBODIES PRESENT	CAN RECEIVE BLOOD FROM:
A	A	Anti-B	A and O
B	B	Anti-A	B and O
AB[a]	A and B	None	A, B, AB, and O
O[b]	None	Anti-A and Anti-B	O

[a] Type AB is the universal receiver.
[b] Type O is the universal donor.

a transfusion the plasma antibodies will react against the antigens present on the RBC's if the antibody for a specific antigen is present. This interaction causes the formation of protruding structures on the RBC's which stop the RBC's from moving freely past each other. The RBC's then clump together. This condition can be fatal because agglutinated RBC's can't transport oxygen or carbon dioxide to the cells of the body (Fig. 16-12).

If we examine the antibodies and antigens present in the four blood types, we'll see that certain types can be mixed and others can't. A person with type A blood can receive only types A and O, because they both lack antigens that will interact with anti -B. A person with type B blood can receive only types B and O, because they both lack antigens that will interact with anti -A. A person with type AB blood can receive all types of blood, because type AB contains no antibodies and there can be no antigen-antibody interaction. Type AB is called the *universal recipient* for this reason. A person with type O blood can receive only type O, because it contains both anti -A and anti -B. However, because type O blood contains no antigens, it can serve as a donor for all types of blood and is often referred to as the *universal donor*.

There is one more factor to be considered in studying blood types. There is another protein that may be attached to the RBC membrane. It is called the *Rh antigen*. Eighty-five percent of the population have this antigen present on their RBC's and are termed Rh^+ (Rh positive). Fifteen percent of the population lack this antigen and are termed Rh^- (Rh negative). When blood is being cross-matched, the Rh factor must be considered. For example, if a donor was Rh^+ and a recipient was Rh^-, the recipient's blood would form antibodies to counteract the donor's blood. Therefore, a person who is, let's say, O^- may only receive O^- blood. (The symbol O^- means the blood is type O and Rh negative.)

If a mother is Rh^- and her unborn child is Rh^+ (the Rh^+ being inherited from the father), blood from the fetus can diffuse into the mother's blood and the mother will gradually build up antibodies against the Rh antigen. The number of antibodies will be low and will probably not affect the fetus. If,

FIGURE 16-12
Protruding structures form when like antigens and antibodies combine. This causes RBC's to clump together.

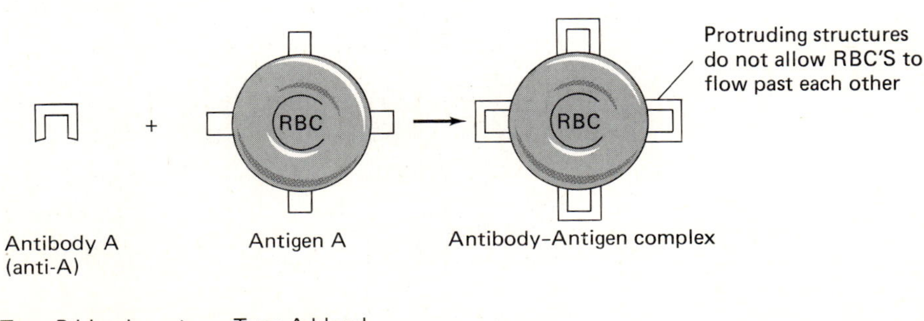

however, the mother becomes pregnant with another child who also inherits the Rh antigen from an Rh^+ father, a reaction between the Rh antibodies of the mother, which have now had enough time to reach significant levels, and the RBC's of the fetus can occur. The child's RBC's clump and rupture and the child is born dead. This condition is called *erythroblastosis fetalis*. Fortunately, a drug called *rhogam* has been developed which is given to the Rh^- mother of a firstborn Rh^+ child, shortly after the first child is born. This eliminates problems with subsequent pregnancies.

EXAMPLE 16-10   Name the four blood types and the two Rh factors.

SOLUTION   The four blood types are A, B, AB, and O. The two Rh factors are Rh^+ and Rh^-.

Disorders and Diseases of the Blood

Anemia

When the number of red blood cells or the hemoglobin level is too low, a condition called *anemia* results. The blood can't transport enough oxygen and carbon dioxide and the tissues don't receive sufficient oxygen. Removal of carbon dioxide from the tissues is also inadequate. As a result, the person can become weak and experience neurologic disorders as well as mental changes and stiffness in the extremities.

There are several types of disorders that cause anemic conditions. These include acute blood loss, iron deficiency, vitamin B_{12} deficiency, bone marrow failure, and hereditary disorders such as sickle cell anemia. A condition called *pernicious anemia* occurs when the red bone marrow can't produce enough RBC's. Vitamin B_{12} is necessary for the formation of RBC's, and when it is not present in the red bone marrow, the production of RBC's is affected. Some chemicals can affect the production of RBC's, as can x rays and atomic radiation. A decrease in RBC production due to such agents is called *aplastic anemia*.

Polycythemia

An increase in the number of RBC's circulating in the blood is called *polycythemia*. An elevated red blood cell count can be caused by a lack of oxygen in the tissues. This stimulates the production of RBC's, which causes an increase in the number of RBC's. There is a serious disease called *polycythemia vera*, which is a chronic, slowly progressive disease characterized by an increase in the total RBC mass. The thickness of the blood is increased and the rate of flow is decreased. This situation eventually leads to cerebral or coronary thrombosis (blood clots in the brain or heart).

Leukopenia

Leukopenia is a condition that results when the white blood cell number
is too low. There are less than 5000 WBC's per cubic millimeter of blood.
When the WBC count is too low, the person can't fight infections as easily
as a person with a normal WBC level.

Leukocytosis

Leukocytosis is a condition that results from an increased number of
white blood cells. An individual with this condition has more than 10,000
WBC's per cubic millimeter of blood. (Remember, the normal level is from
5000 to 10,000 WBC's per cubic millimeter of blood.) Mononucleosis, appen-
dicitis, and leukemia are conditions in which the WBC count is too high.

Leukemia

Leukemia is a disorder of the bone marrow. The cells of the bone marrow
become cancerous and produce excessive numbers of WBC's. This reduces
the number of RBC's that can circulate in the blood, because there is a limit
to the number of cells that can circulate in the blood. There are several types
of leukemias; some more serious than others. Early detection and treatment
are important in this condition.

Hemophilia

Hemophilias are bleeding disorders caused by inherited deficiencies or
abnormalities of coagulation factors in the blood. The disease occurs in males
and the abnormal gene is transmitted by the mother. The disease begins in
childhood and persists throughout life. Trivial injuries can cause severe,
life-threatening hemorrhages. The various types of hemophilia are caused by
a variety of factors. Some forms of hemophilia are more severe than others.

SUMMARY

In this chapter we learned about blood, one of the most important fluids
in our bodies. We began by looking at water, the major component of blood.
Next, we reviewed the functions of the blood. We learned about eight major
functions that the blood carries out. We then read about the composition of
the blood and learned that blood is made of two constituents: plasma and
formed elements. We also read about some of the characteristics of blood,
for example, its specific gravity and hematocrit.

In the second half of the chapter we looked at each of the components
of blood in greater detail. We began by looking at the blood plasma and the

four important proteins found in this plasma. We also looked at the nutritive materials found in blood plasma. We also discussed how the materials found in blood plasma help blood to clot when tissue cells are damaged. Next, we looked at the formed elements; red blood cells, white blood cells, and platelets, and we discussed the role of each of these substances.

We concluded the chapter with a discussion of blood types and blood transfusions, and looked at some disorders and diseases of the blood.

EXERCISES

1. Match the word on the left with its definition on the right.
   (1) intracellular fluid   (a) Fluid found between the cells of the
   (2) extracellular fluid   body.
   (3) blood plasma   (b) Fluid found outside our cells.
   (4) interstitial fluid   (c) Fluid that nourishes the brain and
   (5) lymph   spinal cord.
   (6) aqueous humor   (d) Fluid found in the lymphatic ducts.
   (7) cerebrospinal fluid (e) Fluid found within our cells.
                        (f) Fluid found in our eyes.
                        (g) The liquid portion of the blood.

2. A man who weighs exactly 150 pounds (68.2 kg) contains 90.0 pounds (40.9 kg) of water. What is the percentage of water by weight that composes this man?

3. State the eight major functions of blood.

4. Fill in the missing words.
   Blood is made up of two constituents: a fluid part, called _________, and cells that are suspended in the plasma, called _________.

5. Match the words on the left with the words on the right.
   (1) red blood cells   (a) leukocytes
   (2) white blood cells   (b) thrombocytes
   (3) blood platelets   (c) erythrocytes

6. Name the formed elements.

7. A 20.00-ml sample of blood has a mass of 22.00 g. A 20.00-ml sample of water has a mass of 20.00 g. What is the specific gravity of this blood?

8. Calculate the hematocrit for the following blood samples.
   (a) 75.0 ml of whole blood containing 25.0 ml of formed elements. Is this value high, low, or normal?
   (b) 50.0 ml of whole blood containing 30.0 ml of formed elements. Is this value high, low or normal?

9.   Name the four important proteins found in blood plasma.

10.   Complete each of the following statements with the terms *hydrostatic pressure* or *osmotic pressure*.
(a) The pushing force that moves fluids out of the arterial ends of the capillaries and into the interstitial fluid is known as __________.
(b) The pulling force that moves fluids from the interstitial fluid to the venous blood is known as __________.

11.   Match the protein on the left with its description on the right.
(1) albumin (a) This protein is involved in blood clotting.
(2) gamma globulin
(3) fibrinogen (b) This protein enables blood to maintain a high osmotic pressure.
(c) This protein provides immunity against certain diseases.

12.   Complete the following statements with the words *antigen* or *antibody*.
(a) A foreign substance that enters the body is known as an __________.
(b) The gamma globulins act as __________.

13.   Match the type of immunity on the left with its definition on the right.
(1) natural active immunity (a) Acquired by the injection of a killed or weakened organism.
(2) natural passive immunity (b) A child receives the antibodies that the mother's blood contains.
(3) artificial active immunity (c) Acquired by the injection of an antiserum or antitoxin.
(4) artificial passive immunity (d) Acquired by contracting the disease.

14.   Define the words toxoid, antiserum, and vaccination.

15.   List some of the nutritive materials carried by the blood plasma.

16.   State the three major steps in blood clot formation.

17.   How are oxygen and carbon dioxide transported by the RBC's?

18.   Name the two types of white blood cells. Where is each produced?

19.   Where are RBC's produced in the body? What vitamin is necessary for the production of mature RBC's?

20.   Match the formed element on the right with its major function on the left.
(1) RBC's (a) They protect the body against disease.
(2) WBC's (b) They are involved in blood clotting.
(3) platelets (c) They are involved in the transportation of oxygen.

21. For each of the blood types listed as recipient, write the name of the blood types that can act as donors.

RECIPIENT BLOOD TYPE	DONOR BLOOD TYPE
A	__________
B	__________
AB	__________
O	__________

22. For each of the blood types that follow, write the antigens and antibodies present.

BLOOD TYPE	ANTIGEN(S) PRESENT	ANTIBODIES PRESENT
A	__________	__________
B	__________	__________
AB	__________	__________
O	__________	__________

23. What is the condition know as erythroblastosis fetalis and how does it occur?

24. Match the symptom with the disease.

(1) anemia
(2) aplastic anemia
(3) polycythemia
(4) leukopenia
(5) leukocytosis
(6) leukemia
(7) hemophilia

(a) A condition that results when the WBC number is too low, for example, less than 5000 per cubic millimeter of blood.
(b) A decrease in RBC production due to x rays, chemicals, or atomic radiation.
(c) A bleeding disorder where blood does not clot normally.
(d) A condition that results from an increased number of WBC's.
(e) A condition that results when the number of RBC's is too low.
(f) The cells of the bone marrow become cancerous and produce excessive WBC's.
(g) A condition that results from an increase in the number of RBC's circulating in the blood.

CHAPTER 17

The Urinary System and Urine

The End Product

You should be able to:

1. Name the four parts of the urinary system and state the function or purpose of each.
2. Name the three major parts of the kidney and state the function or purpose of each.
3. State what is meant by kidney stones.
4. State what is meant by cystitis.
5. State the differences between the male and female urethra.
6. Define the terms nephron, Bowman's capsule, tubule, glomerulus, and glomerular filtrate.
7. State how urine is formed.
8. State what is meant by each of the following conditions: polyuria, oliguria, diuresis, and anuria.
9. State what the color, odor, clarity, specific gravity, and pH should be for normal urine.
10. Calculate the specific gravity of urine given sufficient data.
11. Calculate the pH of a sample of urine given the hydrogen ion concentration.
12. State the importance of each of the following substances in urine: urea, uric acid, creatinine, creatine, and hippuric acid.
13. State the condition from the symptoms for each of the following: diabetes mellitus, hematuria, nephritis, and pyuria.
14. State the purpose of each of the following urine tests: Benedict's test, Rothera's nitroprusside test, Albustix, and Hemastix Occultest.

Introduction

It is extremely important that the chemical composition of the body fluids be maintained at the proper level. The urinary system is the major system that is in charge of regulating which chemicals must be excreted and which chemicals must be retained by the body. As we've discussed in previous chapters, tremendous numbers of chemical reactions take place in our cells. Waste products are formed as a result of these reactions. These waste products are carried away from the cells by the blood. The lungs are one place where waste products leave the body. The urinary system is another means of exit for waste products. As well as providing a means of eliminating wastes, the urinary system also regulates the amount of water and salts that leave the body. This helps to maintain the proper chemical balance in our internal environment (Fig. 17-1).

Let's now take a look at the structure and function of the urinary system.

The Structure and Function of the Urinary System

The main parts of the urinary system are the *kidneys*, the *ureters*, the *urinary bladder*, and the *urethra* (Fig. 17-2).

EXAMPLE 17-1   Name the four parts of the urinary system.

FIGURE 17-1
The urinary system helps to maintain a proper chemical balance in our internal environment.

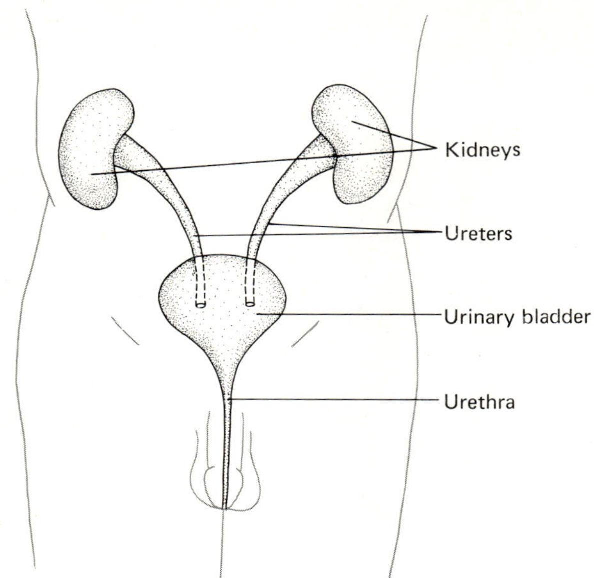

FIGURE 17-2
The parts of the urinary system.

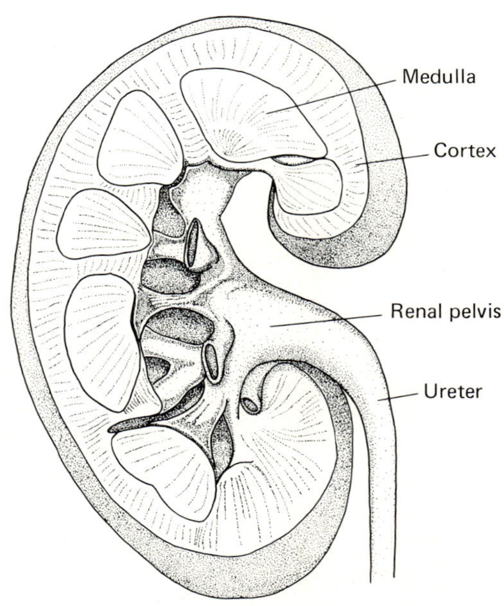

FIGURE 17-3
The kidney.

SOLUTION   The four parts of the urinary system are the kidneys, ureters, bladder and urethra.

The *kidneys* are found at the back of the body just above the waistline. There are two kidneys measuring about 11 cm long, 5 to 8 cm wide, and $2\frac{1}{2}$ cm thick (Fig. 17-3). Connective tissue surrounds the kidneys and connects them to the muscles which lie in that region.

The outer layer of the kidney is called the *cortex*. It is reddish brown in color. The inner layer of the kidney is called the *medulla*. It is deep red in color. The innermost region of the kidney is called the *renal pelvis*. The renal pelvis opens into the *ureter*, which is a tube that carries urine to the urinary bladder.

EXAMPLE 17-2   What are the three main parts of the kidney?

SOLUTION   The three main parts of the kidney are the cortex, the medulla, and the renal pelvis.

Each kidney has a tube called a *ureter*, which brings urine from the renal pelvis to the urinary bladder. Ureters are about 25 to 30 cm long and are made of smooth muscle. A mucous membrane lines the ureters, and this membrane also lines the kidneys and the bladder. Where the ureter enters the bladder, the mucous membrane folds to prevent urine from flowing back up the ureter. The ureter is made of smooth muscle which contracts and causes the urine to flow to the bladder. Sometimes, small deposits called *kidney stones* or *renal calculi* form in the renal pelvis. They can obstruct the kidney or they can travel to the ureters, where they may cause pain and bleeding. These stones can be composed of uric acid or calcium salts.

The *urinary bladder* is a hollow bag made of smooth muscle. It is an elastic structure which is lined with mucous membrane (Fig. 17-4). When the bladder is empty, the mucous membrane forms several folds. When the bladder is full, the folds expand to hold the urine. The bladder can hold about 1 liter of urine when fully expanded. When about 300 to 400 ml of urine collects in the bladder, the urge to urinate is felt.

When the mucous membrane that lines the bladder becomes inflamed, a condition known as *cystitis* results. This is often caused by an infection of the urethra which has traveled to the bladder. There is a burning sensation in the bladder and in the urethra during urination. It may also be difficult to urinate. Cystitis is often caused by staphylococci and can be treated with sulfa drugs.

The urethra is the tube that leads the urine from the bladder to outside the body. There is a difference in the male and female urethra.

The female urethra is from 2.5 to 5 cm in length. It is made of smooth muscle and is lined by a mucous membrane. It is located in front of the vagina and serves only to transport urine.

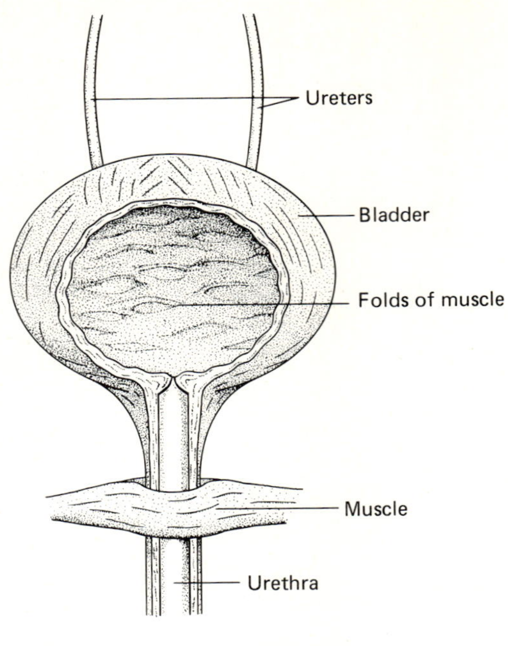

FIGURE 17-4
The urinary bladder.

The male urethra is about 20 cm long. It extends through the penis and opens to the outside of the body. It has two functions in the male. It transports urine and also carries semen, which is a mixture of sperm and other secretions.

EXAMPLE 17-3 State the functions of the ureters, urinary bladder, and urethra.

SOLUTION The ureters carry urine from the kidneys to the bladder.
The urinary bladder stores urine.
The urethra transports urine from the bladder to the outside of the body.

Now let's take a closer look at the kidneys to see how urine is formed.

The Formation of Urine

The *nephron* is the functional unit of the kidney. Each kidney contains about 2 million nephrons. A nephron looks like a small funnel which has a long curved tube attached to it (Fig. 17-5). The funnel-shaped structure is called *Bowman's capsule* and the long curved tube is called a *tubule*.

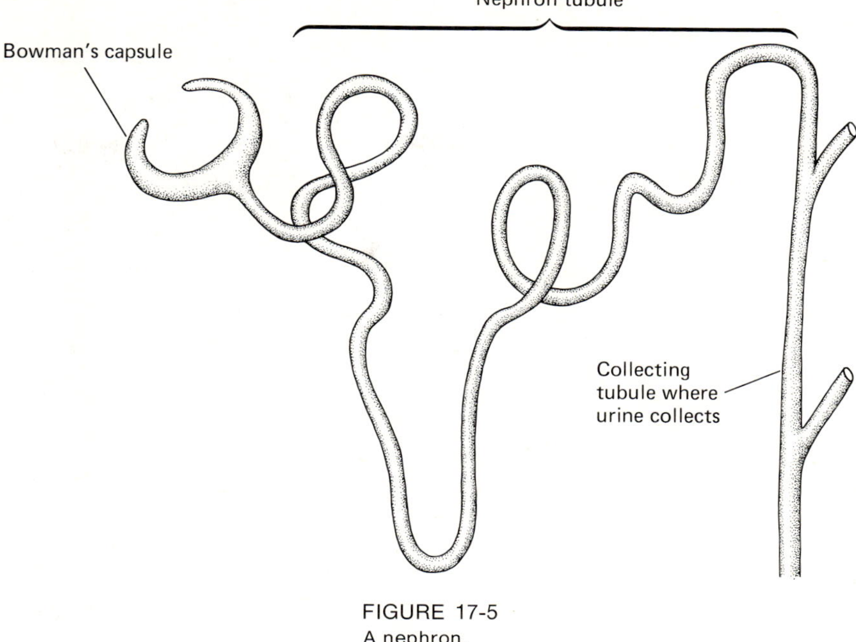

FIGURE 17-5
A nephron.

Bowman's capsule surrounds a group of capillaries called the *glomerulus*. The glomerulus receives blood which contains many fluids and dissolved substances. These fluids and dissolved substances are filtered out of the blood and enter Bowman's capsule. This fluid is called the *glomerular filtrate* and includes waste products as well as many important substances, such as water, potassium ions (K^{+1}), sodium ions (Na^{+1}), and glucose. These important substances must reenter the blood and the waste products must be excreted in the urine. Let's see how this takes place.

The glomerular filtrate leaves Bowman's capsule and enters the tubule, which is a very long, curved tube. Surrounding each section of the tubule are capillaries. Important substances that are needed by the body are reabsorbed by the blood through these capillaries. Any excess material or waste products remain in the tubule to form urine. Let's say that you eat a chocolate bar. The sugar in the chocolate bar appears as glucose in the blood after it has been digested. Glucose is one of the substances that is filtered out of the blood and into the glomerular filtrate. If the glucose level in the blood is too low, the required amount of glucose will be reabsorbed by the capillaries surrounding the tubules and will reenter the blood. Any excess glucose will not be reabsorbed and will remain in the tubules to be excreted in the urine. Traces of glucose will appear in the urine.

There are certain substances which are not part of the glomerular filtrate but must be excreted by the body to maintain the proper internal environment.

Such substances as hydrogen ions, ammonia, certain drugs, and dyes, as well as any excess potassium ions, are secreted into the nephron tubules from the capillaries surrounding these tubules. These unnecessary substances are excreted in the urine.

EXAMPLE 17-4   Trace the path of sodium ions in table salt from ingestion to excretion.

SOLUTION   Salt is ingested and ends up in the blood as sodium ions and chloride ions. The sodium ions are filtered out of the blood by the kidneys and become part of the glomerular filtrate. Most of the sodium ions are reabsorbed into the blood from the nephron tubules by the capillaries surrounding these tubules. Any excess sodium ions remain in the tubules and become part of the urine. The urine containing the excess sodium ions travels from the kidneys to the bladder by way of the ureters and exits through the urethra.

Properties of Urine

A laboratory examination of urine from a patient gives the health practitioner important information about the health of that patient. Let's look at some of the properties of urine which are examined in the laboratory.

Amount

The quantity of urine excreted by an adult in good health ranges from about 1000 to 1500 ml per day. A higher intake of fluids means a higher output of urine. In warm weather less urine is excreted because a great deal of water is lost through the skin in the sweat. In cold weather more urine is excreted because there is little water lost in the sweat.

There are several terms used to describe the elimination of abnormal amounts of urine. *Polyuria* is an increase in the amount of urine excreted. A disease called *diabetes insipidus* is characterized by polyuria. This disease is caused by the lack of a hormone called *antidiuretic hormone* (ADH), which is secreted and released by certain sections of the brain. This hormone travels to the nephron tubules and plays a part in the reabsorption of water into the blood from the nephron tubles in the kidneys. If ADH is not secreted, then excessive amounts of water are excreted.

Oliguria is a decrease in the amount of urine excreted. Dehydration may cause oliguria. *Retention* of water occurs when the proper amount of water has not been eliminated and is being retained by the body. An enlarged prostate can cause water retention. A *temporary* increase in the amount of urine excreted is called *diuresis*. This condition can occur when a diuretic drug is taken to eliminate water that has been retained. A very serious condition called *anuria* occurs when the kidneys are not functioning at all and no urine is formed.

EXAMPLE 17-5   Match the word on the left with the symptom on the right.

(1) polyuria   (a) A *temporary* increase in the amount of urine
(2) oliguria   excreted.
(3) diuresis   (b) A decrease in the amount of urine excreted.
(4) anuria   (c) An increase in the amount of urine excreted,
                 caused by a disease such as *diabetes insipidus.*
          (d) No urine is formed because the kidneys are not
                 functioning.

SOLUTION   The matching pairs are as follows:
      (a) 3   (b) 2   (c) 1   (d) 4

Color

Urine is normally yellow in color but ranges from very light yellow to dark brown. The more concentrated the urine, the darker the color. Diet also affects the color of urine, as do certain drugs. Blood or bile present in the urine also causes a darkened color.

Odor

Urine has a faint aromatic odor when freshly eliminated. It develops an odor characteristic of ammonia upon standing. The odor can be influenced by certain foods, drugs, or diseases.

Clarity

Urine is usually clear when freshly eliminated. However, upon standing, it becomes cloudy. If bacteria or mucus are present, the urine may be cloudy when freshly eliminated.

Specific Gravity

The normal range of specific gravity of urine is 1.008 to 1.030. The average is about 1.020. A reading higher than normal indicates too many dissolved solids in the urine. A reading lower than normal indicates that the urine is too watery. This is often seen in patients with diabetes insipidus.

EXAMPLE 17-6   A 10.00-ml sample of urine has a mass of 10.50 g.
A 10.00-ml sample of water has a mass of 10.00 g.
What is the specific gravity of the urine? Does this value fall into the normal range?

SOLUTION (*Hint*: If you have forgotten how to calculate specific gravity, see Chapter 16, p. 394.)

$$\text{Specific gravity (of urine)} = \frac{\text{mass of urine}}{\text{mass of equal volume of water}}$$

$$= \frac{10.50 \text{ g}}{10.00 \text{ g}} = 1.050$$

This value is above the normal range.

pH

The normal range of pH values of urine is 4.8 to 7.5. The pH decreases when a diet rich in proteins is eaten and increases when a diet rich in fruits and vegetables is eaten.

EXAMPLE 17-7 Calculate the pH of the following urine solutions. State whether the urine is acidic, basic, or neutral.

(a) $[H^+] = [10^{-5}]$
(b) $[H^+] = [10^{-8}]$
(c) $[H^+] = [10^{-7}]$

SOLUTION (*Hint*: If you have forgotten how to calculate pH, see Chapter 7, p. 138.)

(a) $pH = -\log[H^+] = -\log[10^{-5}] = 5$, which is acidic.

(b) $pH = -\log[H^+] = -\log[10^{-8}] = 8$, which is basic.

(c) $pH = -\log[H^+] = -\log[10^{-7}] = 7$, which is neutral.

The Normal and Abnormal Composition of Urine

Normal urine is 96% water and 4% solids. The solids include inorganic salts, nitrogenous organic compounds, and nonnitrogenous organic compounds. There is about 60 g of solids eliminated each day (Table 17-1). Let's take a look at some of the substances found in normal urine. We'll begin with some organic compounds.

FIGURE 17-6

The formula of urea, the major end product of protein metabolism.

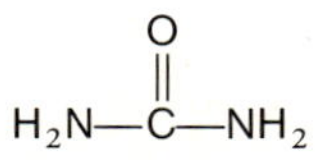

$$\underset{\text{Urea}}{H_2N-\overset{\overset{\displaystyle O}{\|}}{C}-NH_2}$$

Urea

Urea is the major end product of protein metabolism. About 25 to 30 g of urea is found in the urine each day (Fig. 17-6).

TABLE 17-1   The constituents of urine[a]

| Water, 96% | |
| Inorganic ions, nitrogenous organic compounds, non-nitrogenous organic compounds, 4% | |

INORGANIC IONS	AMOUNT EXCRETED PER DAY
CATIONS	
Na^{+1}	4 g
K^{+1}	1.5–2 g
Ca^{+2}	150 mg
Mg^{+2}	100–200 mg
NH_4^{+1}	700 mg
Fe^{+2}	
Fe^{+3}	
Cu^{+2}	Trace amounts
Zn^{+2}	
ANIONS	
Cl^{-1}	9–16 g
PO_4^{-3}	2 g
SO_4^{-2}	2.5 g
NO_3^{-1}	
SiO_2^{-1}	Trace amounts
F^{-1}	

NITROGENOUS ORGANIC COMPOUNDS	AMOUNT EXCRETED PER DAY
Urea	25–30 g
Uric acid	600 mg
Creatine	< 100 mg
Creatinine	15–25 mg/kg of body weight
Hippuric acid	40–80 mg
Amino acids	64–200 mg as N
Peptides	300–700 mg as N

TABLE 17-1 The constituents of urine (*cont.*)

NON-NITROGENOUS ORGANIC COMPOUNDS	AMOUNT EXCRETED PER DAY
Glucose	Trace amounts
Glucuronic acid	Trace amounts
Cholesterol	Trace amounts
Ketone bodies	$<$ 100 mg
Oxalates	Trace amounts
Citrate	0.2–1.2 g
Vitamins	Variable (depending on diet)
Sex hormones	Variable

[a] This is not a complete list. There are many substances found in the urine in trace amounts. We've tried to list the more important constituents of urine.

Uric Acid

Uric acid is produced from the metabolism of purines. Uric acid is slightly soluble in water and may settle out of urine that is allowed to stand. Certain types of kidney stones are composed of uric acid (Fig. 17-7). *Gout* is a disease characterized by deposits of uric acid and urates in the joints and tissues.

Creatinine and Creatine

Creatinine is formed from the breakdown of creatine. Creatine and creatinine are formed in muscle tissues. The amount of creatinine in the urine is a measure of the mass of your muscles (Fig. 17-8).

FIGURE 17-7
The formula of uric acid. Certain types of kidney stones are composed of this material.

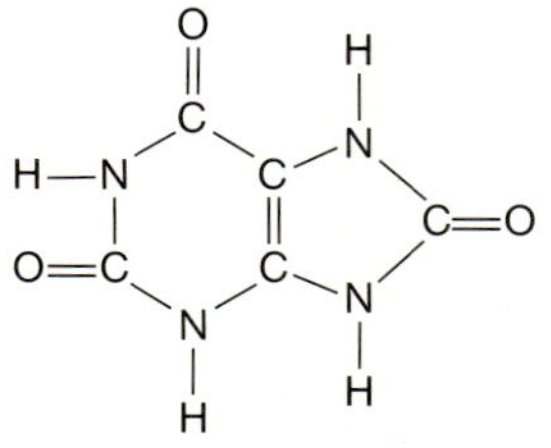

Uric acid

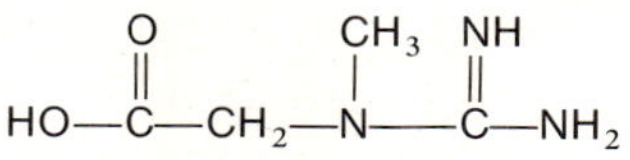

Creatine

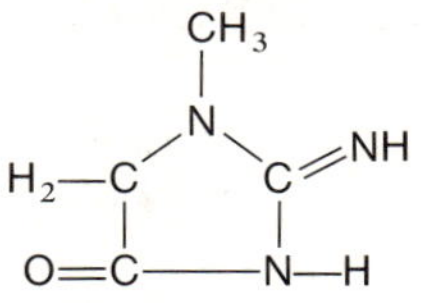

Creatinine

The formulas of creatine and creatinine. Creatine is produced from the amino acids arginine, methionine, and glycine. Creatinine is formed from the breakdown of creatine.

Creatine is produced from the amino acids arginine, methionine, and glycine. It is found in muscle, brain, and blood cells. When creatine is found in the urine, a condition called *creatinuria* results. This may happen to individuals who are starving. It may also happen to individuals who have prolonged fevers, wasting diseases, muscular dystrophy, or diabetes mellitus.

Hippuric Acid

Hippuric acid is another constituent of urine. It is the form in which benzoic acid is excreted. Benzoic acid enters the diet by ingestion of fruits and berries (Fig. 17-9).

Inorganic Compounds

There are several inorganic compounds which are excreted in the urine. Let's look at some of them.

Chloride ions (Cl^{-1}), mostly in the form of sodium chloride, are excreted in the urine. There is from 9 to 16 g of chloride ions excreted each day.

Sodium ions (Na^{+1}) are also found in the urine. About 4 g is excreted each day.

FIGURE 17-9
The formula of hippuric acid. This is the form in which benzoic acid is excreted by the body.

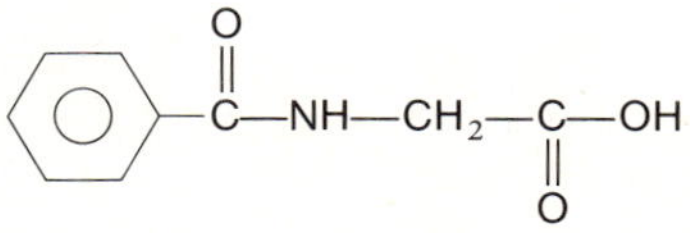

Hippuric acid

Phosphates are also present in the urine. The quantity of phosphate excreted depends upon how much is taken into the body. When foods high in phosphates are eaten, more phosphates are excreted.

Sulfates are produced by the breakdown of sulfur-containing proteins. Like phosphates, the concentration of sulfates excreted is also dependent upon how much is taken in through the diet.

Substances That Do Not Belong in the Urine

There are several substances which do not belong in urine. They include certain proteins, blood, pus, ketone bodies, and excessive amounts of glucose, to name a few.

Glucose is not normally found in the urine in significant quantities. *Glucosuria*, or glucose in the urine, is an indication of diabetes mellitus (see Chapter 11, p. 272).

Blood is not normally found in the urine except when there is disease or injury to the urinary system. Blood in the urine is called *hematuria*.

Pus in the urine is an indication of a possible infection in the urinary system. Pus in the urine is called *pyuria*.

Proteins are usually not present in the urine in more than trace amounts. The presence of proteins in the urine is called *proteinuria* and is usually an indication that the kidneys aren't functioning properly. Albumin and globulin are the proteins most often found in the urine when the kidneys aren't working properly. They pass through the glomerulus and become part of the glomerular filtrate. Proteins in the urine are characteristic of diseases of the kidney such as *nephritis* and *nephrosis*, where there is inflammation or degeneration of the kidneys.

Ketone bodies such as acetone, acetoacetic acid, and beta-hydroxybutyric acid appear in the urine when insufficient quantities of carbohydrates are metabolized by the body. They are present in the urine in diabetes mellitus, or when a diet with total lack of carbohydrates is eaten, or when a diet too high in fats is eaten.

EXAMPLE 17-8 Match the condition on the left with the symptom on the right.

(1) diabetes mellitus	(a)	Proteins are found in the urine.
(2) hematuria	(b)	Acetone is found in the urine.
(3) nephritis	(c)	Pus is found in the urine.
(4) pyuria	(d)	Blood is found in the urine.

SOLUTION

(a) Proteins are found in the urine. (3) nephritis
(b) Acetone is found in the urine. (1) diabetes mellitus
(c) Pus is found in the urine. (4) pyuria
(d) Blood is found in the urine. (2) hematuria

There are several tests which are routinely performed on urine specimens to determine how the body is functioning and to detect various disorders and diseases. Let's look at some of these tests.

Diabetics must routinely check their urine for excess glucose levels. Benedict's qualitative test can determine if sugar is present in the urine. Benedict's quantitative test can determine how much sugar is present in the urine. The reaction involves the oxidation of glucose to gluconic acid with a corresponding reduction of Cu^{+2} to Cu^{+1} (Fig. 17-10). The Cu^{+2} solution is blue, whereas the Cu^{+1} solution is yellow, orange, or red, depending upon how much Cu^{+1} (in the form of Cu_2O) is formed. A commercial product, Clinitest tablets, uses this reaction to test for glucose in urine.

The presence of ketone bodies in urine can be determined with Gerhardt's ferric chloride test and Rothera's nitroprusside test. The three ketone bodies, acetone, acetoacetic acid, and beta-hydroxybutyric acid, almost always occur together in urine, when a body malfunction occurs. The nitroprusside test is one of the most widely used tests for ketone bodies in urine. It is very sensitive to the presence of acetoacetic acid, but it will also detect the presence of acetone. A commercial product known as Acetest Reagent Tablets uses the nitroprusside reaction. A drop of urine is placed on the reagent tablet. A purple color indicates the presence of ketone bodies in the urine.

The presence of proteins in the urine can be detected with the use of various commercial products such as Albustix, Albutest, and Bumintest. Most of these tests are sensitive to the protein albumin in the urine.

Blood in the urine can be confirmed with the use of Hemastix Occultest. This commercial product is sensitive to hemoglobin in the urine.

Certain products are available which test for more than one substance in the urine (Fig. 17-11). Combistix test for protein, glucose, and pH. Hema-

FIGURE 17-10
The reaction involved in the Benedict's test for glucose. The original Benedict's solution is blue. However, if glucose is present in the urine, the Benedict's solution will turn to a yellow, orange, or red, depending upon how much glucose is present.

$$
\begin{array}{ccc}
\text{Glucose} & & \text{Gluconic } acid
\end{array}
$$

Glucose $+\ 2Cu^{+2} + 4OH^- \longrightarrow$ Gluconic acid $+\ Cu_2O + 2H_2O$

Benedict's solution (blue) → (yellow to red)

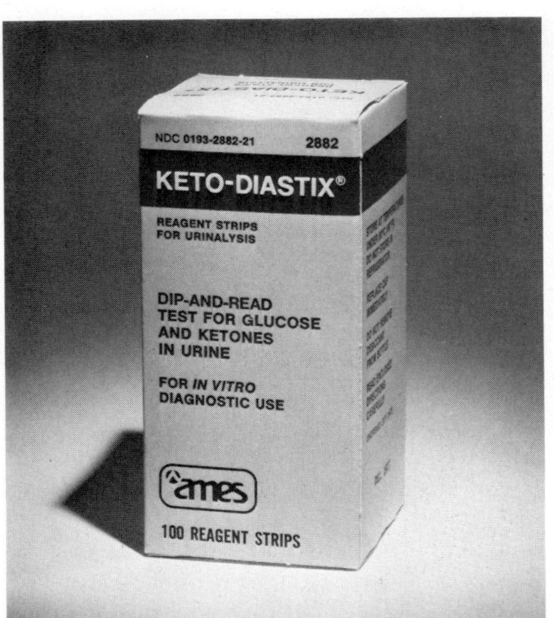
NDC 0193-2882-21   2882
KETO-DIASTIX®
REAGENT STRIPS
FOR URINALYSIS
DIP-AND-READ
TEST FOR GLUCOSE
AND KETONES
IN URINE
FOR IN VITRO
DIAGNOSTIC USE
ames
100 REAGENT STRIPS

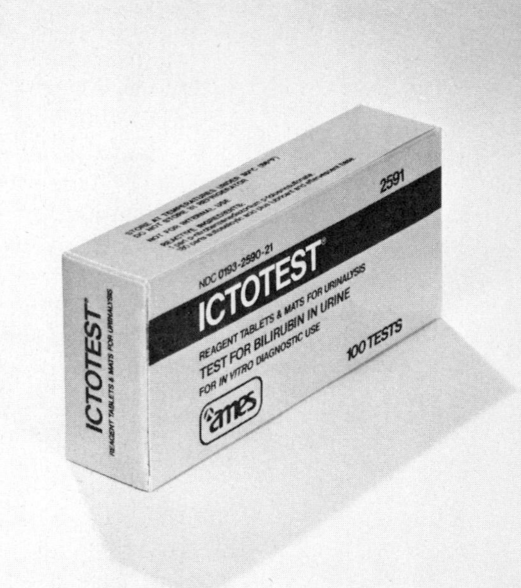
2591
NDC 0193-2590-21
ICTOTEST®
REAGENT TABLETS & MATS FOR URINALYSIS
TEST FOR BILIRUBIN IN URINE
FOR IN VITRO DIAGNOSTIC USE
100 TESTS
ICTOTEST
ames

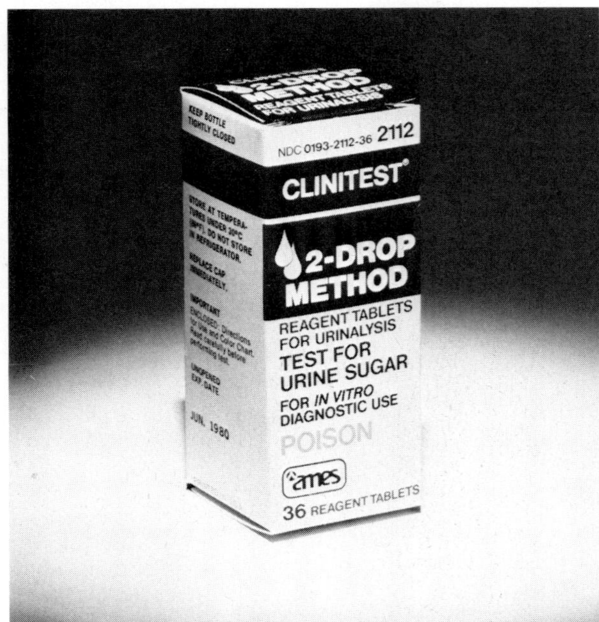
KEEP BOTTLE
TIGHTLY CLOSED
NDC 0193-2112-36   2112
CLINITEST®
2-DROP
METHOD
REAGENT TABLETS
FOR URINALYSIS
TEST FOR
URINE SUGAR
FOR IN VITRO
DIAGNOSTIC USE
POISON
JUN. 1980
ames
36 REAGENT TABLETS

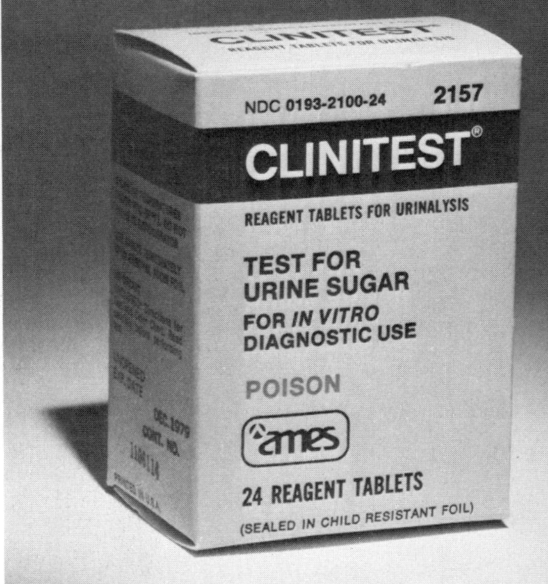
CLINITEST®
REAGENT TABLETS FOR URINALYSIS
NDC 0193-2100-24   2157
CLINITEST®
REAGENT TABLETS FOR URINALYSIS
TEST FOR
URINE SUGAR
FOR IN VITRO
DIAGNOSTIC USE
POISON
DEC. 1979
CONT. NO.
ames
24 REAGENT TABLETS
(SEALED IN CHILD RESISTANT FOIL)

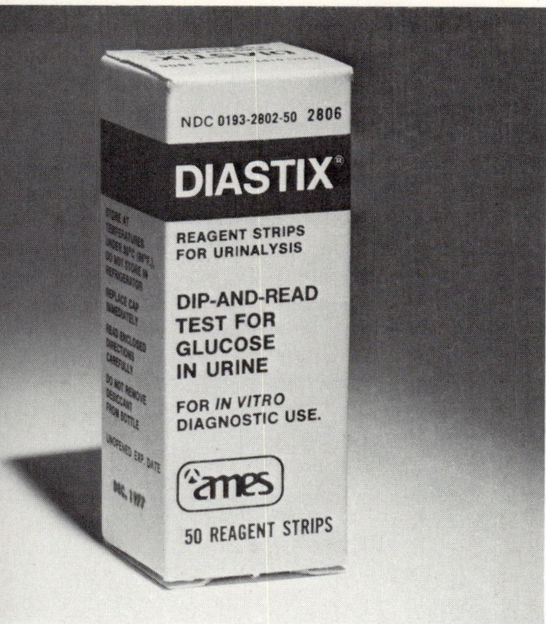

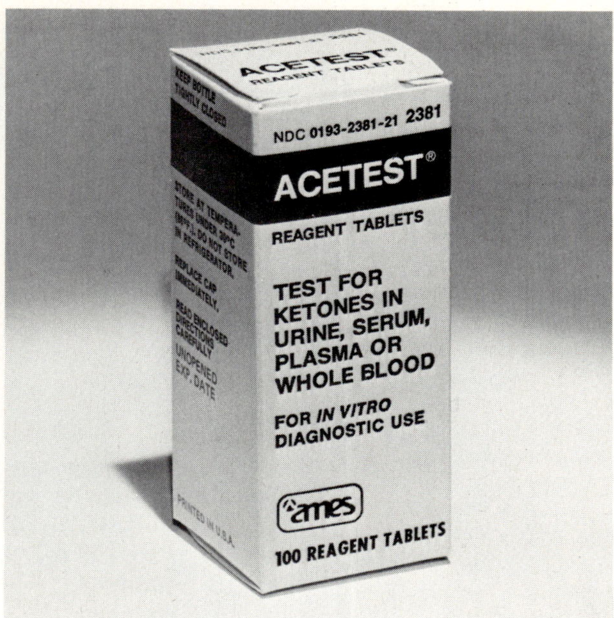

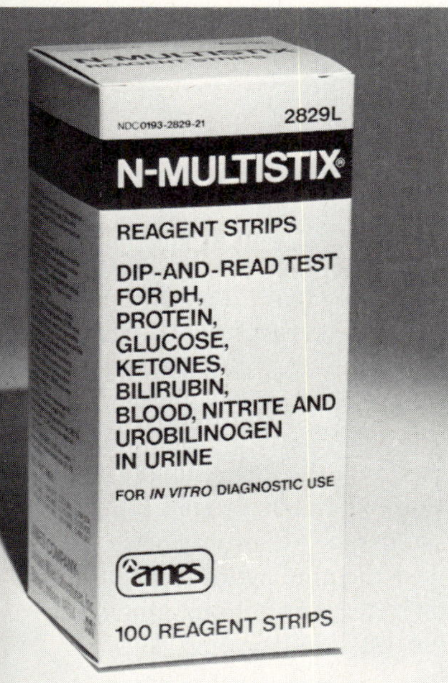

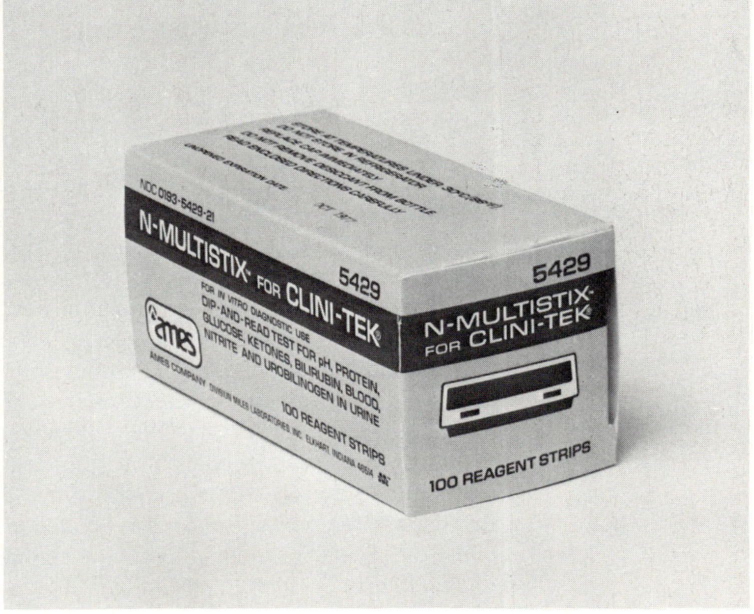

FIGURE 17-11
Some commercial products for testing urine. (Ames Company, Division of Miles Laboratories)

TABLE 17-2   Some commercial products used for testing urine

NAME	TEST
Acetest	Acetoacetic acid and acetone
Albustix, Albutest, and Bumintest	Protein
Azostix	Urea
Bili-Labstix	pH, protein, glucose, ketones, bilirubin, and blood
Clinitest tablets	Glucose
Clinistix, Tes-Tape	Glucose
Combistix	Protein, glucose, and pH
Hema-Combistix	Protein, glucose, pH, and blood
Hemastix Occultest	Blood
Labstix	Protein, glucose, pH, blood, and ketones

Combistix test for protein, glucose, pH, and blood. Labstix test for protein, glucose, pH, blood, and ketones. A summary of the commercial products used for testing urine can be found in Table 17-2.

SUMMARY

In this chapter we looked at urine and the entire urinary system. We began our study by looking at the various parts of the urinary system, including the kidneys, ureters, urinary bladder, and urethra. We read about the three main parts of the kidney, the cortex, the medulla, and the renal pelvis, and we traced the flow of urine from the kidney through the ureter to the bladder and finally out the urethra. We learned what could happen if kidney stones, also called renal calculi, form in the kidney.

In the second part of the chapter we learned about the formation of urine from the glomerular filtrate. We also looked at some of the properties of urine, including the amount, pH, color, odor, clarity, and specific gravity. We also learned the names of certain conditions that reflect the amount of urine eliminated from the body. Finally, we looked at some of the normal and abnormal constituents of urine and studied about some of the tests used for their detection.

EXERCISES

1. Name the four parts of the urinary system and state the purpose of each.

2. Match the part of the kidney with its description.
   (a) cortex   (a) The inner layer of the kidney.
   (2) medulla   (b) The outer layer of the kidney.
   (3) renal pelvis   (c) The innermost layer of the kidney which
                            opens into the ureter.

3. What are renal calculi? What specific compounds may form renal calculi?

4. Describe the symptoms associated with cystitis.

5. Match the organ with its function.
   (1) ureter   (a) Transports urine to the outside of the
   (2) urinary bladder   body.
   (3) urethra   (b) Carries urine from the kidneys.
                        (c) Stores the urine.

6. Describe the differences between the male and female urethra.

7. Match the term on the left with its definition on the right.
   (1) nephron   (a) A group of capillaries surrounded by
   (2) Bowman's capsule   Bowman's capsule.
   (3) tubule   (b) A long curved tube found in a nephron.
   (4) glomerulus   (c) The functional unit of the kidney.
   (5) glomerular filtrate (d) The fluid that enters Bowman's
                               capsule.
                           (e) The funnel-shaped structure that is
                               part of a nephron.

8. Trace the path of chloride ions in table salt from ingestion to excretion.

9. Complete the following sentences by filling in the terms *cold weather* or *warm weather*.
   (a) In __________ more urine is excreted.
   (b) In __________ less urine is excreted.

10. Match the word on the left with its symptom on the right.
    (1) oliguria   (a) No urine is formed because the kidneys are
    (2) anuria   not functioning.
    (3) polyuria   (b) A decrease in the amount of urine excreted.
    (4) diuresis   (c) An increase in the amount of urine excreted.
                      (d) A *temporary* increase in the amount of urine
                          excreted.

11. For each of the following statements, tell whether the urine is normal or abnormal.
    (a) The urine is very dark, with some dark red spots in it.
    (b) The urine has a faint aromatic odor.
    (c) The urine appears cloudy when freshly eliminated.
    (d) The specific gravity of the urine is 1.000.
    (e) The pH of the urine is 5.0.

12. A 100.0-ml sample of urine has a mass of 110.0 g. A 100.0-ml sample of water has a mass of 100.0 g. What is the specific gravity of the urine? Does this value fall into the normal range?

13. Calculate the pH of the following urine solutions. State whether the urine is acidic, basic, or neutral.
   (a) $[H^+] = [10^{-6}]$ (b) $[H^+] = [10^{-9}]$
   (c) $[H^+] = [10^{-7}]$ (d) $[H^+] = [10^{-8}]$

14. Tell how urine is formed.

15. Match the substance on the left with its description on the right.
   (1) urea   (a) Formed from the breakdown of creatine.
   (2) uric acid   (b) The form in which benzoic acid is excreted.
   (3) creatinine   (c) The major product of protein metabolism.
   (4) creatine   (d) This substance is produced from the amino
   (5) hippuric acid   acids arginine, methionine, and glycine.
                            (e) Formed from the metabolism of purines.

16. Name some of the inorganic ions found in urine.

17. Name some of the substances that do not belong in urine.

18. Define the following terms.
   (a) glucosuria   (b) hematuria
   (c) pyuria   (d) nephrosis

19. What disease may be expected if ketone bodies are found in the urine?

20. Match the urine test with its description.
   (1) Clinitest tablets   (a) Test for protein in urine.
   (2) Acetest   (b) Test for blood in urine.
   (3) Albustix   (c) Test for sugar in urine.
   (4) Hemastix   (d) Test for ketone bodies in urine.

CHAPTER 18

Metabolism

You Are What You Eat

You should be able to:

1. Define the processes of anabolism and catabolism.
2. Name the major parts of a cell and tell the function of each.
3. State the functions of ATP and ADP in metabolism.
4. Discuss the process of carbohydrate anabolism.
5. Discuss the process of aerobic catabolism and anaerobic catabolism (glycolysis).
6. Give the overall reaction for cellular respiration.
7. Give the overall reaction for anaerobic glycolysis.
8. Discuss the processes involved in the respiratory chain.
9. State the function of the cytochrome system.
10. Write the overall reaction for the respiratory chain.
11. State the major steps of the citric acid cycle (also known as the Krebs cycle).
12. Write the overall reaction for the citric acid cycle.
13. Discuss the energy efficiency of cellular respiration.
14. Discuss the process of lipid catabolism.
15. Discuss the process of lipid anabolism and be able to determine the total number of ATP molecules produced from the beta oxidation of a lipid.
16. Discuss the major features of protein anabolism.
17. Discuss the major features of protein catabolism, including deamination and decarboxylation.

Introduction

We have spent several chapters looking at how the lipids, carbohydrates, and proteins that we ingest are broken down in the body. As soon as these foods are digested to form their basic chemical molecules, they are transported to cells throughout the body. Once inside the cells, these molecules undergo many chemical reactions. The chemical reactions that these molecules undergo within the cells are called *metabolism*.

What happens to these basic chemical molecules once they enter the cells? There are two paths which they may follow: anabolism or catabolism. The process of *anabolism* occurs when organic molecules such as sugars, amino acids, fatty acids, and glycerol, which have been absorbed by the cell, serve as the building blocks for the synthesis of more complex substances such as polysaccharides, lipids, nucleic acids, and proteins. In other words, anabolism is the building up of body tissues in the presence of specific enzymes and sufficient energy.

Many of the organic molecules that are absorbed by cells do not take part in the process of building up the body tissues. Instead, they are broken down chemically into even smaller molecules. Molecules such as sugars, amino acids, fatty acids, and glycerol are broken down into simpler molecules. Sometimes the breakdown continues until molecules such as water, carbon dioxide, and ammonia remain. This process is known as *catabolism*. The molecules formed as a result of the chemical breakdown contain much less energy than the original molecules. So *catabolism* is the breaking down of food substances accompanied by a release of energy from the molecules that are broken down (Fig. 18-1).

Let's now take a brief look at some of the parts of the cell which are involved in metabolism.

Specialized Parts of a Cell

We can think of a cell as being like a small factory (Fig. 18-2). Raw materials are taken in and utilized so that various products can be produced. There are special areas within the factory that perform various specialized functions. Waste products are produced and must be eliminated. A cell is like a factory in all these ways.

A cell is enclosed by a membrane called the *cell membrane* or *plasma membrane*. It is made of a double layer of proteins and lipids, sort of like a sandwich (Fig. 18-3). There are several small openings or pores in the membrane. The cell membrane regulates what compounds can enter and leave the cell.

The specialized cell structures are suspended in the *cytoplasm*, which is a watery fluid surrounding the nucleus. The nucleus is the control center of the cell that determines which reactions the cell undergoes. The nucleus also contains the genetic information for the organism.

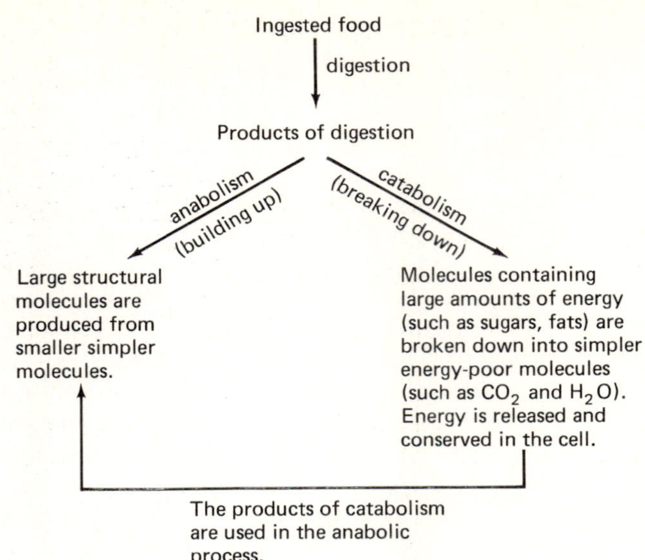

FIGURE 18-1
How food is used.

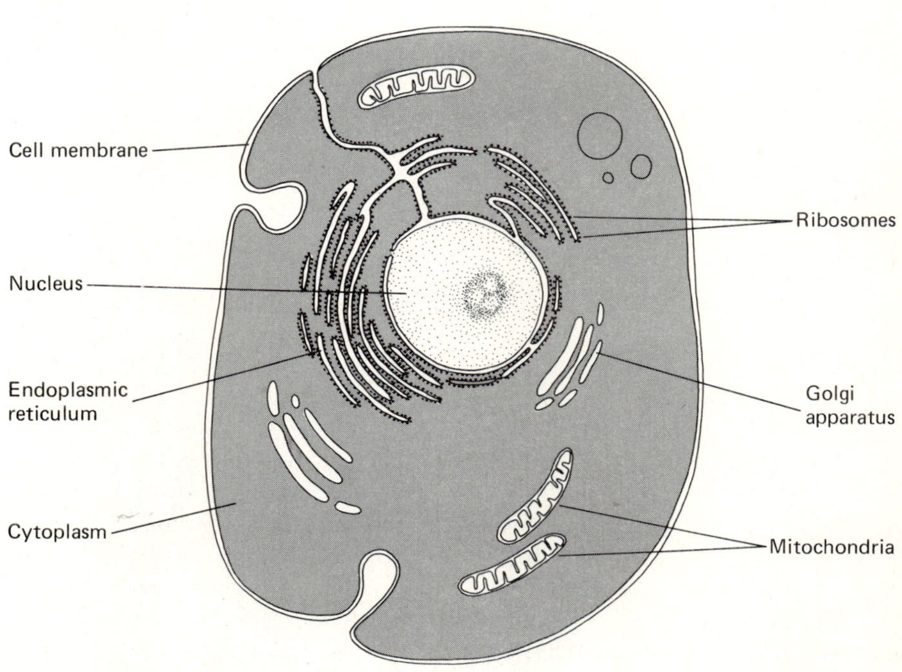

FIGURE 18-2
A cell.

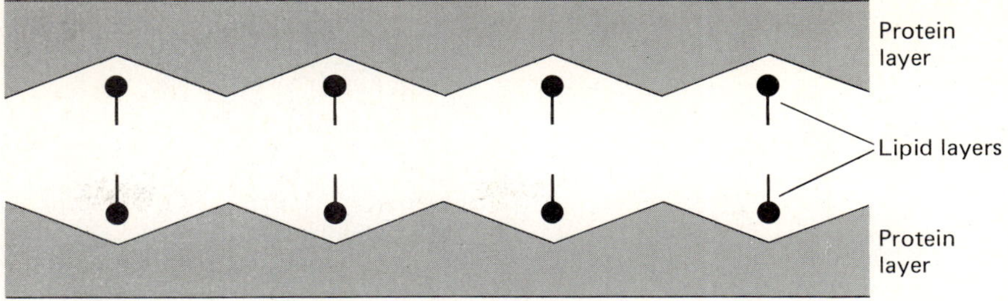

FIGURE 18-3
A cell membrane.

Embedded in the cytoplasm is the *endoplasmic reticulum* (E.R., for short). This is a branched interconnected system of membranes which extends from the nucleus of the cell to the cell membrane. Attached to the E.R. are *ribosomes.* *Ribosomes* are the site of protein synthesis in the cell. The E.R. transports the proteins that are produced in the ribosomes to the various locations inside and outside the cell (Fig. 18-4).

The *golgi complex* is also found in the cytoplasm. It is a series of flattened membranes whose purpose is to store proteins, lipids, and carbohydrates and to secrete them when necessary (Fig. 18-5).

The *mitochondria* are the "power plants" of the cell. These structures are found in the cytoplasm. Their main function is to produce energy that can be

FIGURE 18-4
Endoplasmic reticulum with attached ribosomes.

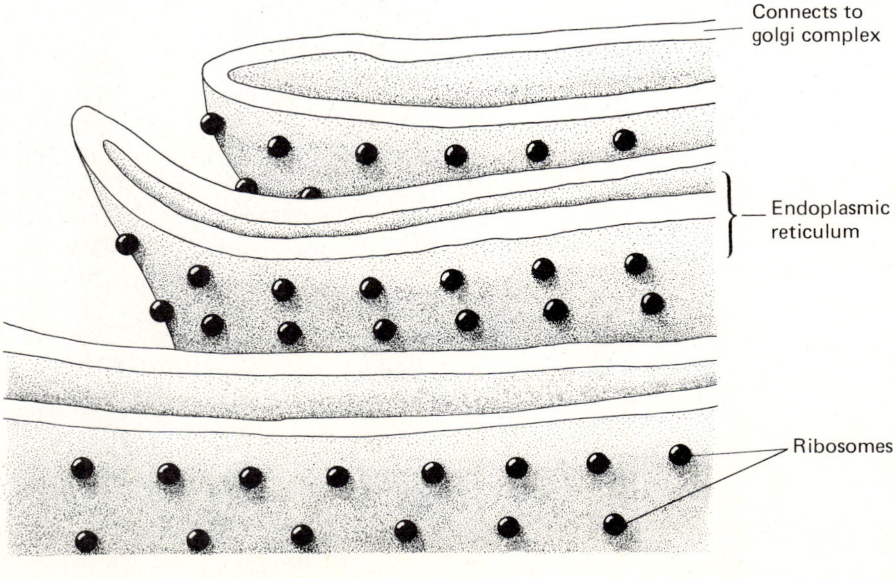

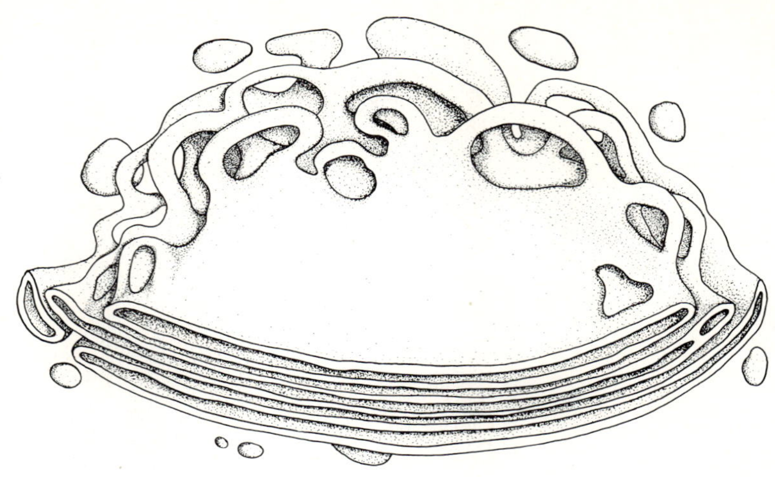

FIGURE 18-5
The golgi complex.

used by the cell to carry out various cell processes. Mitochondria have a double membrane: an outer membrane and an inner membrane. Attached to the inner membrane are enzymes which catalyze the oxidation-reduction reaction that produces energy for the cell. The mitochondria convert the potential energy contained within various digested foods into a form of energy that can be used by the cell to carry out its functions (Fig. 18-6). Let's see how this works.

In Chapter 11 we studied the following reaction:

$$C_6H_{12}O_6 + 6O_2 \xrightarrow{enzymes} 6CO_2 + 6H_2O + energy$$

glucose

In this overall reaction a monosaccharide molecule reacts with oxygen in the presence of specific enzymes to produce carbon dioxide, water, and energy. This is an oxidation-reduction reaction that produces energy in the form of adenosine triphosphate (ATP). ATP is a compound that can store large amounts of energy in its bonds. A reaction that is very similar to the one studied earlier occurs in the mitochondria. In the cytoplasm, glucose, a monosaccharide, is broken into two pyruvic acid molecules (Fig. 18-7). The pyruvic acid molecules move into the mitochondria and the following reaction occurs:

$$2 \text{ pyruvic acid molecules} + 6O_2 \xrightarrow{enzymes} 6CO_2 + 6H_2O + energy \text{ (as ATP)}$$

The ATP that is formed by this reaction leaves the mitochondria and travels throughout the cell to areas where energy is needed. If a muscle cell must contract or a nerve cell must transmit impulses, ATP provides the energy to carry out these functions. Cells that require large amounts of energy in order to function contain large numbers of mitochondria. In this way more ATP can be produced to power those cells.

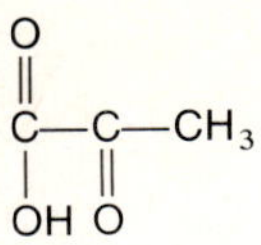

Pyruvic acid

FIGURE 18-7
A pyruvic acid
molecule.

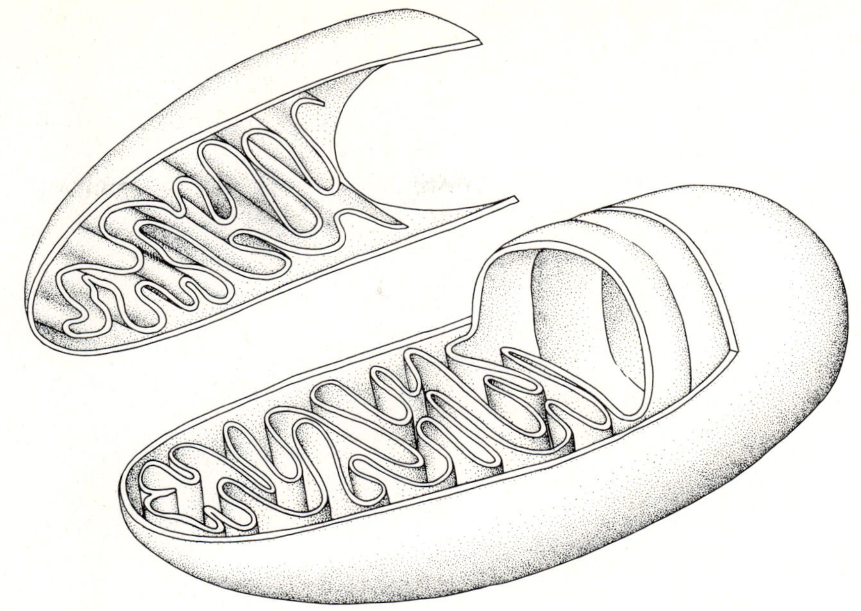

FIGURE 18-6
Mitochondria.

EXAMPLE 18-1   Match the cell part with its function.

(1) cell membrane
(2) ribosomes
(3) golgi complex
(4) mitochondria
(5) cytoplasm
(6) endoplasmic
    reticulum

(a) Fluid in which cell structures are suspended.
(b) Regulates what compounds enter and leave the cell.
(c) Transports proteins produced by the ribosomes.
(d) The site of protein synthesis in the cell.
(e) The places where proteins, carbohydrates, and lipids are stored.
(f) The structure that produces energy for the cell.

SOLUTION   The matching pairs are as follows:

(a) 5   (b) 1   (c) 6   (d) 2   (e) 3   (f) 4

Energy, ATP, and ADP

We took a brief look at the nature of ATP in the last section. Let's now take a closer look at ATP, ADP (adenosine diphosphate), and the energy stored in the cell.

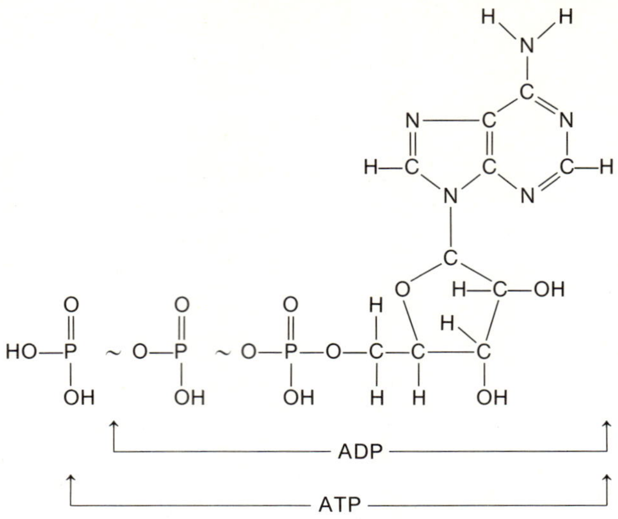

ADP + phosphate $\longrightarrow$ ATP

$\sim$ indicates a high energy bond

FIGURE 18-8
The structure of ADP and ATP.

When a molecule such as glucose or pyruvic acid is oxidized to produce a less complex molecule, energy is given off. This energy is conserved by the cell with the production of a molecule of ATP. ATP consists of the purine adenine, ribose (a five-carbon sugar), and three phosphate groups (Fig. 18-8). The second and third phosphate groups are attached by high-energy bonds, which are indicated by wavy lines ($\sim$). These bonds are able to store large amounts of energy. When a cell needs energy, the ATP molecule undergoes hydrolysis; a high-energy bond breaks which releases energy and a molecule of ADP. The equation for this reaction is

$$\text{ATP} \xrightarrow{\text{hydrolysis}} \text{ADP} + \text{phosphate group} + 8000 \text{ cal}$$

(The phosphate group is often abbreviated P_i.)

When a mole of ATP is hydrolyzed, about 8000 cal of energy are released to the cell. When a mole of ATP is formed, about 8000 cal of energy are stored in the cell.

Now that we have a basic knowledge of some of the reactions and processes that occur in the cell, we can look at the metabolism of carbohydrates, lipids, and proteins in more detail.

EXAMPLE 18-2 We know that when ATP undergoes hydrolysis a molecule of ADP is produced, along with a phosphate group, and energy is released. Write the reverse reaction for the formation of ATP from ADP and a phosphate group.

$$ADP + P_i + 8000 \text{ cal} \longrightarrow ATP$$

Carbohydrate Metabolism

Carbohydrate metabolism is very important to the body. Carbohydrates are broken down to produce energy as a result of catabolic reactions (in other words, the process of catabolism). Excess amounts of carbohydrates are converted to glycogen and fat by anabolic reactions (the process of anabolism) and are stored in the body. We will now study these two processes.

Anabolism of Carbohydrates

Carbohydrate anabolism occurs primarily in the liver. It is here that the process of *glycogenesis* takes place. *Glycogenesis* is the combination of monosaccharides to form glycogen. Glycogen is stored sugar. Insulin is important in this process, because it helps to bring glucose into the liver cells so that glycogen can be synthesized. Insulin also helps bring glucose into the skeletal muscles for the synthesis of glycogen. When energy is needed for a muscle to contract, glycogen breaks down and energy is released. In addition, it is possible for simple carbohydrate molecules to be converted to lipids and stored in fat cells in the body.

In summary, carbohydrate anabolism occurs in the liver and in muscle tissue, where glucose is converted to glycogen for storage. Simple carbohydrates can also be converted to lipids and stored until needed (Fig. 18-9).

Carbohydrate Catabolism

Carbohydrates are the main source of energy for the body. By undergoing catabolism, carbohydrates break down and release energy which can be stored as ATP and used when necessary by the cells.

Catabolism is accomplished by two processes: anaerobic catabolism and aerobic catabolism. *Anaerobic catabolism* does not require oxygen and produces relatively small amounts of energy. *Aerobic catabolism* requires oxygen and produces large amounts of energy. Both processes are involved in the breakdown of carbohydrates in the cells. Let's look at their roles in the body as we study cellular respiration.

Cellular Respiration

Cellular respiration is the oxidation of organic compounds that occurs within cells. It is a very complicated process that takes place in the mitochondria and requires the presence of oxygen and several enzymes and coenzymes. The overall reaction for cellular respiration is

$$C_6H_{12}O_6 + 6O_2 \longrightarrow 6CO_2 + 6H_2O + \text{energy (as ATP)}$$

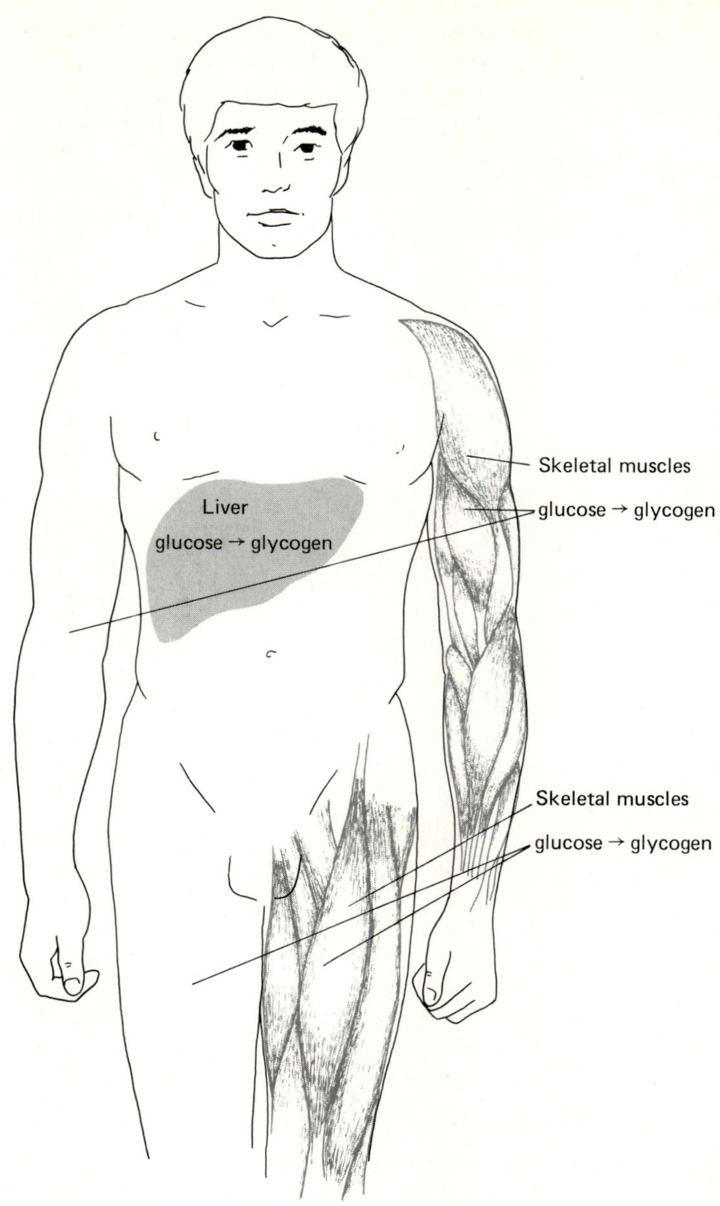

FIGURE 18-9
Glycogenesis.

In cells, the most common type of oxidation occurs when hydrogens are removed from a substance. Since oxidation involves the loss of electrons, the removal of hydrogen atoms (H $\longrightarrow$ H$^+$ + e^-) accomplishes this process. The coenzymes NAD and FAD are the primary oxidizing agents in cells. (NAD stands for nicotinamide adenine dinucleotide and FAD stands for flavin adenine dinucleotide. The structures of NAD and FAD are shown in Figure 18-10.) These two coenzymes can remove hydrogen atoms from various com-

FIGURE 18-10
The structures of NAD and FAD.

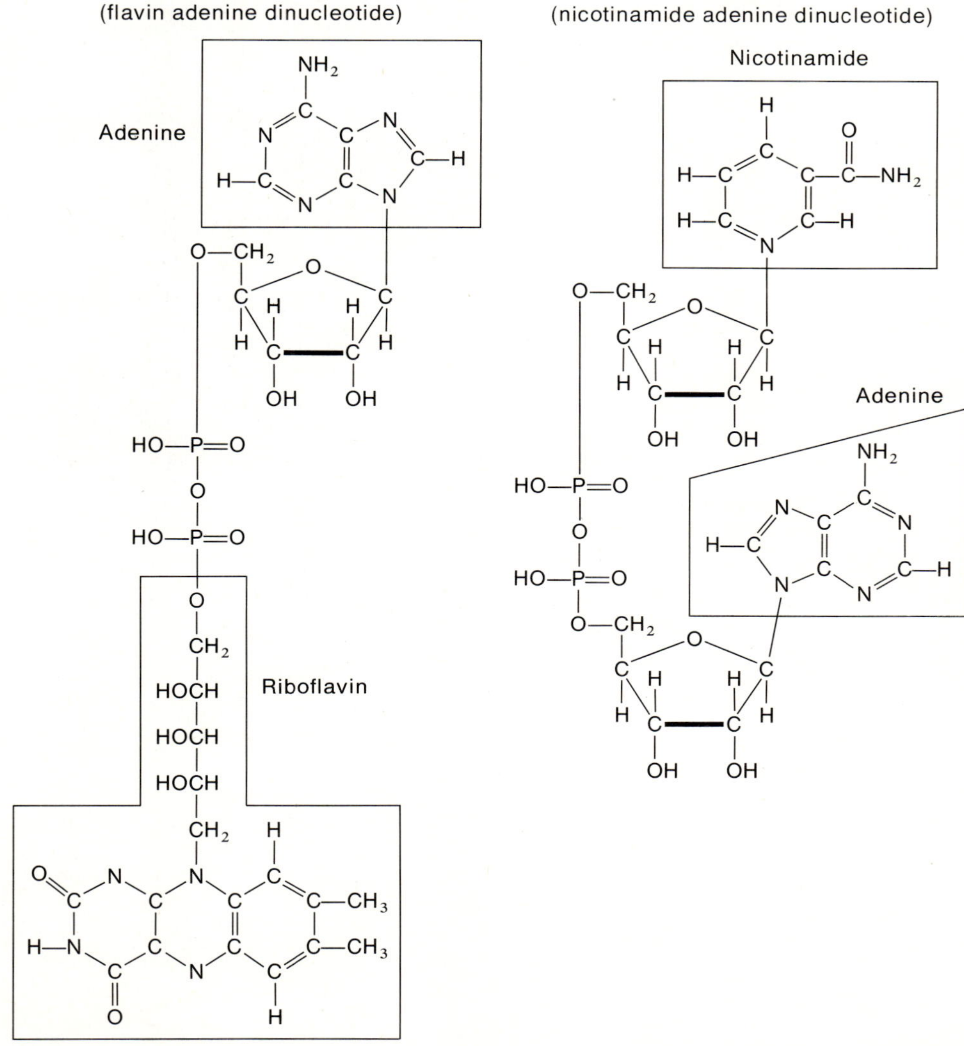

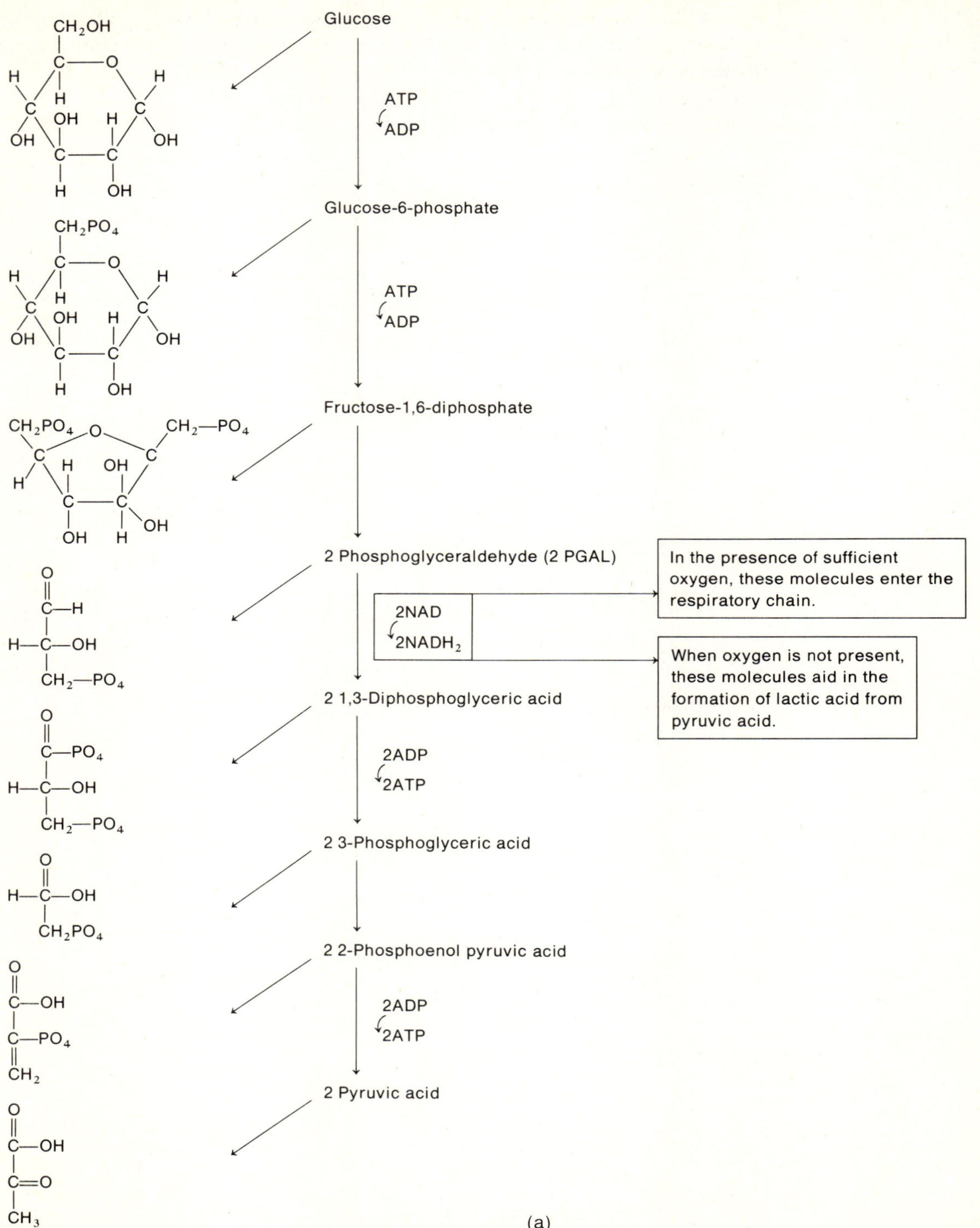

CH2OH
Glucose
ATP
ADP
Glucose-6-phosphate
CH2PO4
ATP
ADP
Fructose-1,6-diphosphate
CH2PO4
CH2—PO4
OH
2 Phosphoglyceraldehyde (2 PGAL)
In the presence of sufficient oxygen, these molecules enter the respiratory chain.
2NAD
2NADH2
When oxygen is not present, these molecules aid in the formation of lactic acid from pyruvic acid.
2 1,3-Diphosphoglyceric acid
2ADP
2ATP
2 3-Phosphoglyceric acid
2 2-Phosphoenol pyruvic acid
2ADP
2ATP
2 Pyruvic acid
(a)

pounds. In doing so, they are reduced to $NADH_2$ and $FADH_2$, respectively. Keeping this information in mind, let's continue to try and understand how cellular respiration occurs.

Glycolysis: Anaerobic Catabolism

We begin our story with a glucose molecule, a simple carbohydrate, which will ultimately be broken down to form CO_2, H_2O, and energy (Fig. 18-11). The first step in the breakdown process is the activation of a glucose molecule so that it can be respired. The glucose molecule must be joined to a phosphate group in a process called *phosphorylation*, which occurs in the presence of a specific enzyme. The phosphate group is donated to the glucose molecule by a molecule of ATP, which breaks a bond and transfers its third phosphate group and the energy that it contains to the glucose.

$$ATP \longrightarrow ADP + P_i$$

The simple glucose molecule now becomes glucose-6-phosphate. This molecule contains more energy than glucose. Another phosphate group from a second molecule of ATP is added to the glucose-6-phosphate in the presence of a specific enzyme, and fructose-1,6-diphosphate forms. This molecule then splits into two halves, forming two molecules of phosphoglyceraldehyde (PGAL, for short). Oxidation of two molecules of PGAL removes two pairs of hydrogens, which are accepted by two molecules of NAD, forming two molecules of $NADH_2$. Another reaction produces two molecules of phosphoglyceric acid (PGA, for short) and two molecules of ATP. Two more chemical transformations occur, resulting in the formation of two more molecules of ATP and two molecules of pyruvic acid.

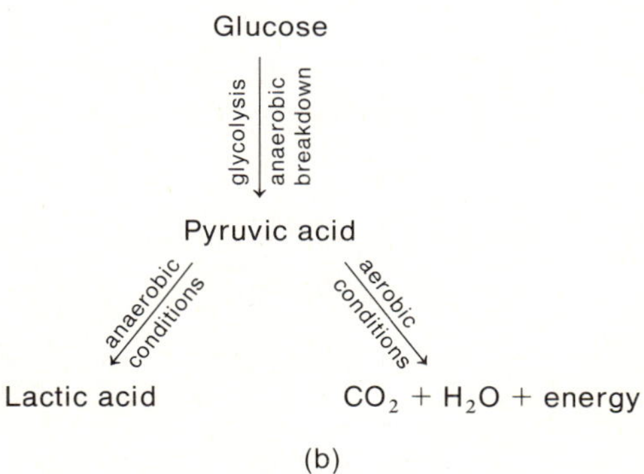

FIGURE 18-11

(a) Glycolysis. The first steps in cellular respiration involve the breakdown of glucose without the presence of oxygen, in a process called glycolysis. During glycolysis, glucose is broken down to pyruvic acid. (b) When sufficient oxygen is *not* present, pyruvic acid is broken down to form lactic acid. When sufficient oxygen is present, pyruvic acid is broken down to form CO_2, H_2O, and large amounts of energy (as ATP).

Let's look at the energy changes in this process. We start out by "plugging" two molecules of ATP into the system to activate it and get it going. Four molecules of ATP are produced. The conversion of glucose to pyruvic acid is an anaerobic process. It requires no oxygen and is often called *anaerobic glycolysis*. In this process there is a net gain of two molecules of ATP for every molecule of glucose used. This takes place in the cytoplasm. Under anaerobic conditions, the two $NADH_2$ molecules formed during glycolysis are reoxidized. This occurs when pyruvic acid is reduced to lactic acid in the reaction:

$$\underset{\substack{\text{Pyruvic}\\\text{acid}}}{2\,\overset{\displaystyle COOH}{\underset{\displaystyle CH_3}{C}{=}O}} + 2NADH_2 \longrightarrow \underset{\substack{\text{Lactic}\\\text{acid}}}{2H{-}\overset{\displaystyle COOH}{\underset{\displaystyle CH_3}{C}}{-}OH} + 2NAD$$

The production of lactic acid is the end product of anaerobic glucose metabolism (glycolysis).

When we are physically active, the circulatory system sometimes can't supply enough oxygen to our muscle cells to satisfy our demand for energy. In this case we get the energy we need from anaerobic glycolysis. The glucose we need to start the process comes from glycogen, which is stored in our muscle cells. The end product of this reaction is lactic acid and energy. The accumulation of high concentrations of lactic acid in muscle tissue is not good because it lowers the pH of the cell and the extracellular fluid as well. Large amounts of lactic acid cause the muscle to become incapable of further action. At this point we must stop and rest and let our cells recover.

EXAMPLE 18-3   How many moles of ATP are produced for every mole of glucose broken down during anaerobic glycolysis?

SOLUTION   Two moles of ATP are produced per mole of glucose utilized.

EXAMPLE 18-4   Write the overall reaction for anaerobic metabolism (glycolysis).

SOLUTION   Glucose $\longrightarrow$ 2 lactic acid molecules + 2ATP

Aerobic Catabolism

In the presence of sufficient quantities of oxygen in the cell, aerobic catabolism occurs. In aerobic catabolism, glucose is completely broken down,

$$2H^+ + 2e^- + \tfrac{1}{2}O_2 \longrightarrow H_2O + \text{energy (ATP)}$$

FIGURE 18-12
The overall reaction for the respiratory chain.

to form CO_2, H_2O, and large amounts of energy. (Remember, during anaerobic catabolism, glucose is broken down to form lactic acid.) In aerobic catabolism the two $NADH_2$ molecules formed initially (during glycolysis) enter what is called the respiratory chain, where they ultimately form H_2O and energy. In aerobic catabolism the pyruvic acid molecules formed are completely broken down, to form CO_2, H_2O, and large amounts of energy by means of the *citric acid cycle*. We'll have more to say about this cycle shortly, but right now we want to take a closer look at the respiratory chain.

The Respiratory Chain

The *respiratory chain* is a mechanism by which electrons are transferred from one molecule to another. The final goal of the respiratory chain is the formation of ATP and the production of water from the combination of electrons, oxygen, and hydrogen (Fig. 18-12). Because water has lower energy than oxygen and hydrogen, energy is released. It is stored in the cell by the formation of ATP. This process takes place in the mitochondria, and it is an energy-producing process. In the respiratory chain we find two components: *hydrogen carriers* and *electron carriers*. The primary hydrogen carriers in the respiratory chain are the coenzymes NAD and FAD. These substances are oxidizing agents and can oxidize certain organic molecules by removing two hydrogens from them (Fig. 18-13). The two hydrogens join with NAD and FAD, forming $NADH_2$ and $FADH_2$. You will remember from our discussion of oxidation–reduction reactions that when one substance is oxidized, another substance is reduced. An oxidizing agent is reduced by the substance that it oxidizes. So when NAD and FAD oxidize various molecules, they are reduced to $NADH_2$ and $FADH_2$, respectively.

When a substrate is oxidized by NAD, two hydrogens (two electrons and their hydrogen nuclei) are removed, forming $NADH_2$. These hydrogens are then transferred down the chain from $NADH_2$ to the other hydrogen acceptor, FAD. FAD then becomes $FADH_2$ and $NADH_2$ is reoxidized to NAD. The NAD can now remove hydrogens from another substrate (Fig. 18-14).

As if this isn't complicated enough, let's see what happens next. The hydrogens that are part of $FADH_2$ are now separated into two electrons and two protons ($2H \longrightarrow 2H^+ + 2e^-$) by a substance called coenzyme Q. Coenzyme Q accepts the hydrogens from $FADH_2$. The $FADH_2$ is then reoxidized to FAD, which is then ready to accept the next hydrogens that come along. The protons ($2H^+$) are separated from the electrons ($2e^-$) and the electrons enter the *cytochrome system*.

FIGURE 18-13
NAD and FAD remove hydrogens from organic molecules and oxidize them.

$$2H \longrightarrow 2e^- + 2H^+$$

$$\text{Substrate} \cdot H_2 \xrightarrow{\text{NAD}} \text{substrate} + 2H$$

A substrate containing hydrogen is oxidized
by NAD and loses the hydrogens.

$$\text{NAD} \xrightarrow{2H} \text{NADH}_2$$

NAD is reduced to $NADH_2$

$$\text{FAD} \xrightarrow{2H} \text{FADH}_2$$

FAD then oxidizes $NADH_2$ and
becomes $FADH_2$.

$$\text{NADH}_2 \longrightarrow \text{NAD}$$

$NADH_2$ is reoxidized to NAD
and is available again.

FIGURE 18-14
The action of NAD and FAD.

The Cytochrome System

The *cytochrome system* consists of a series of electron acceptors. The cytochromes are enzymes that contain iron and copper. Free electrons are passed separately down the cytochrome chain. In this process two metals, iron and copper, are alternately reduced and reoxidized from Fe^{+3} to Fe^{+2} back to Fe^{+3} and from Cu^{+2} to Cu^{+1} back to Cu^{+2}. The cytochromes have different names. From coenzyme Q, electrons are passed to cytochrome b to cytochrome c, to cytochrome a, and finally to cytochrome a_3. In the final step of the chain electrons, protons and oxygen combine to form a water molecule (Fig. 18-15).

Why such a complicated process to produce a mole of water? With each successive step in the respiratory chain energy is liberated. The energy is used to form 3 moles of ATP when NAD is the initial hydrogen acceptor. It is also possible for FAD to be the initial hydrogen acceptor. In this case 2 moles of ATP are formed. You remember that a mole of ATP stores about 8000 calories of energy. So tremendous amounts of energy can be provided for use by the cell as a result of the respiratory chain.

EXAMPLE 18-5   What are the components of the respiratory chain?

SOLUTION   The components of the respiratory chain are hydrogen acceptors and electron acceptors. The coenzymes NAD and FAD are hydrogen acceptors. The cytochromes are the electron acceptors.

EXAMPLE 18-6   What is the goal of the respiratory chain and how is this goal accomplished?

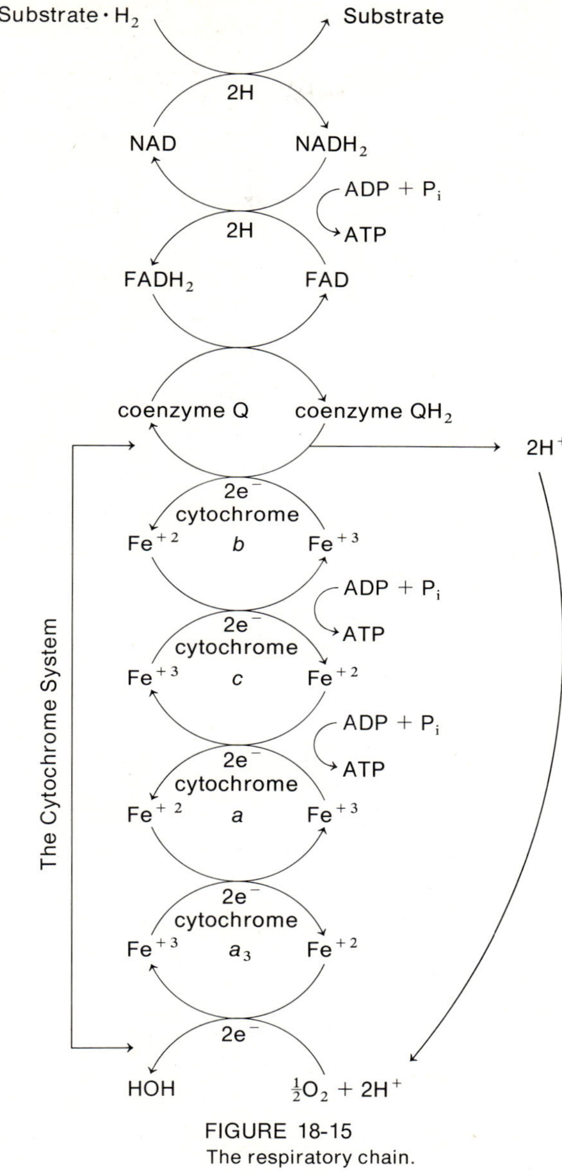

FIGURE 18-15
The respiratory chain.

SOLUTION The goal of the respiratory chain is the production of 2 or 3 moles of ATP. This goal is accomplished by the step-by-step breakdown of certain organic molecules and the release of energy from these molecules. In the end, electrons, hydrogens, and oxygens combine to form water as the final product.

EXAMPLE 18-7 Write the overall reaction for the respiratory chain. How much energy is produced?

SOLUTION The overall reaction is

$$2H^+ + 2e^- + \tfrac{1}{2}O_2 \longrightarrow H_2O + \text{energy (as ATP)}$$

Three moles of ATP are produced per mole of NAD, when $NADH_2$ is the initial hydrogen acceptor. Therefore, 24,000 calories (3 moles × 8000 cal/mole ATP) are produced.

Two moles of ATP are produced per mole of FAD when $FADH_2$ is the initial hydrogen acceptor. Therefore, 16,000 calories (2 moles × 8000 cal/mole ATP) are produced.

The Citric Acid Cycle

The *citric acid cycle* (often known as the *Krebs cycle*) completes the process of oxidizing glucose so that the maximum amount of energy can be removed and stored. You remember that in glycolysis, glucose is broken down into two molecules of pyruvic acid. The citric acid cycle continues to break down pyruvic acid until only CO_2 and H_2O are left. Just as in glycolysis and in the respiratory chain, the citric acid cycle occurs in a stepwise fashion. Let's go through the steps of the citric acid cycle (Fig. 18-16).

First, pyruvic acid is decarboxylated and oxidized. This means that it loses a molecule of CO_2 as well as two hydrogens. Acetic acid is formed. The two hydrogens enter the respiratory chain and combine with NAD, forming $NADH_2$ and three molecules of ATP. The acetic acid is now activated by combining with a coenzyme called coenzyme *A*. A compound called acetyl coenzyme A (acetyl-CoA) is formed and enters the citric acid cycle.

In the citric acid cycle two types of reactions occur: decarboxylation (loss of CO_2) and dehydrogenation (loss of 2H). The hydrogens are passed through the respiratory chain and ATP is produced.

The next step in the process is the combination of acetyl-CoA with oxaloacetic acid to produce citric acid. Citric acid is found in citrus fruits. The citric acid molecule is decarboxylated and dehydrogenated, forming alpha-ketoglutaric acid. The hydrogens produced enter the respiratory chain and produce ATP.

Alpha-ketoglutaric acid is decarboxylated and dehydrogenated to form succinic acid. Again the hydrogens enter the respiratory chain to produce ATP. Succinic acid is dehydrogenated to produce fumaric acid. The hydrogens enter the respiratory chain. Fumaric acid is hydrated to form malic acid. Malic acid is dehydrogenated to form oxaloacetic acid. The hydrogens enter the respiratory chain. This is one whole turn of the citric acid cycle. Four times during the cycle hydrogens have entered the respiratory chain to combine with NAD and form $NADH_2$. Each molecule of $NADH_2$ produces three molecules of ATP, so 12 molecules of ATP are produced by each molecule of $NADH_2$. Because not just one, but two molecules of pyruvic acid are broken down, 24 molecules of ATP are produced by the citric acid cycle.

EXAMPLE 18-8 Write the overall reaction for the citric acid cycle, including the energy produced.

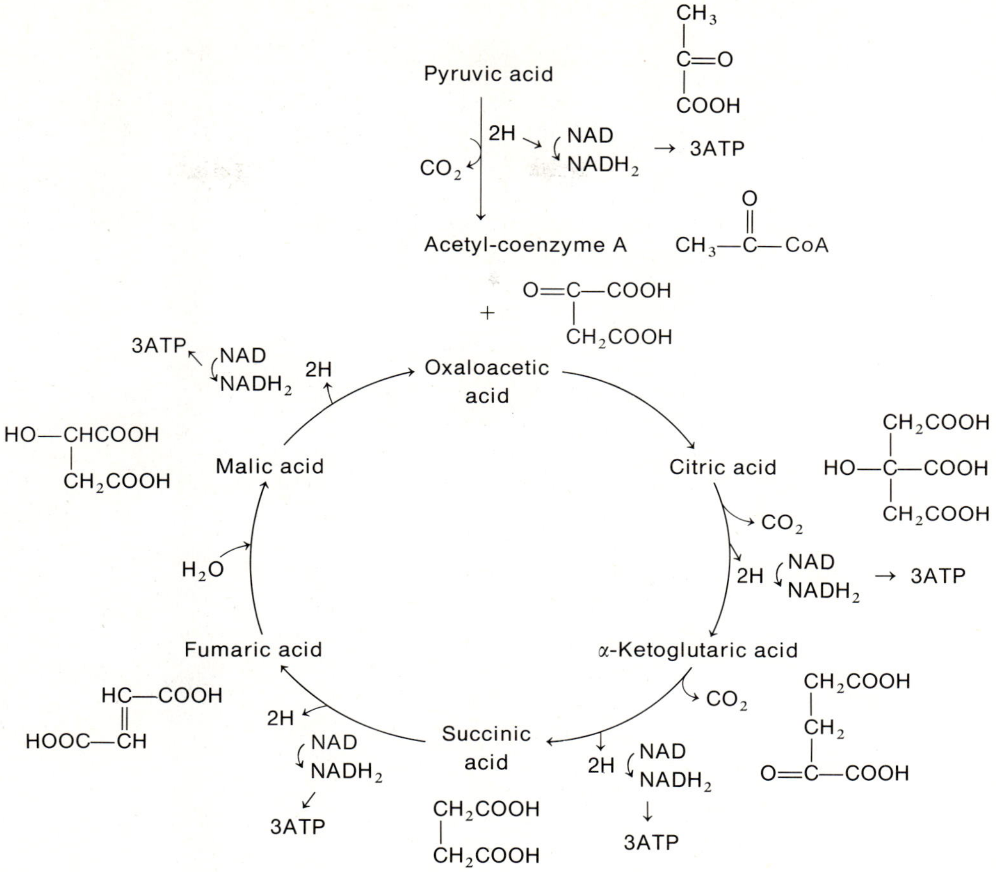

FIGURE 18-16
The citric acid cycle.

SOLUTION

$$2CH_3COOH + 4O_2 \longrightarrow 4CO_2 + 4H_2O + 24ATP$$

Acetic acid

Conservation of Energy:
A Summary of Anaerobic and Aerobic Catabolism

We're now going to do some bookkeeping to show the number of ATP's produced during anaerobic and aerobic catabolism (Table 18-1).

Anaerobic catabolism (glycolysis) produces a net yield of two ATP's. This is a small amount of energy which is used to sustain muscle activity during vigorous physical exertion. This occurs when the circulatory system can't supply sufficient oxygen to the muscle tissue.

Aerobic catabolism produces a total of 38 ATP's. Two ATP's are formed during glycolysis. The two $NADH_2$'s produced during glycolysis enter the

TABLE 18-1   ATP bookkeeping

	ATP PRODUCED	ATP USED
GLYCOLYSIS		
1. Glucose $\xrightarrow{ATP \quad ADP}$ glucose-6-phosphate		1
2. Glucose-6-phosphate $\xrightarrow{ATP \quad ADP}$ fructose-1,6-diphosphate		1
3. Two molecules of phosphoglyceraldehyde $\xrightarrow{2NADH_2 \quad 2NAD}$ 2 molecules of 1,3-diphosphoglyceric acid	6	
4. Two molecules of 1,3-diphosphoglyceric acid $\xrightarrow{2ADP \quad 2ATP}$ 2 molecules of 3-phosphoglyceric acid	2	
5. Two molecules of phosphoenolpyruvic acid $\xrightarrow{2ADP \quad 2ATP}$ pyruvic acid	2	
DECARBOXYLATION OF PYRUVIC ACID		
6. Two molecules of pyruvic acid $\xrightarrow{2NADH_2 \quad 2NAD}$ Acetyl-CoA	6	
CITRIC ACID CYCLE		
7. Two molecules of acetic acid + $2O_2 \longrightarrow 2CO_2 + 2H_2O$	24	
Total	38 ATP's produced	

respiratory chain and produce six ATP's. The decarboxylation of pyruvic acid to acetyl-CoA produces two $NADH_2$'s, which enter the respiratory chain and produce six more ATP's. The citric acid cycle produces 24 ATP's.

Energy Efficiency of Cellular Respiration

When fully oxidized, a mole of glucose yields 686,000 calories of energy. The step-by-step oxidation of glucose in cellular respiration allows much of this energy to be stored in chemical molecules. Let's calculate just how much energy is conserved.

The respiration of 1 mole of glucose produce 38 moles of ATP. When 1 mole of ATP is formed from ADP, about 8000 calories of energy are stored. For 38 ATP's this equals 304,000 calories stored (38 ATP × 8000 cal/ATP = 304,000 cal). This means that the efficiency of conversion of the chemical energy of glucose into the chemical energy of ATP is 44%

$$\% \text{ efficiency} = \frac{304,000 \text{ cal}}{686,000 \text{ cal}} \times 100 = 44\%$$

This is an impressive efficiency!

Metabolism of Lipids

The metabolism of lipids takes place in the liver. Let's look at the anabolic and catabolic reactions.

Lipid Catabolism

When lipids undergo hydrolysis, fatty acids and glycerol result. The fatty acids undergo a process known as *beta oxidation*. In this process the beta carbon is oxidized, which causes a two-carbon segment to break off of the fatty acid molecule. (The beta-carbon atom is the one that is one removed from the carboxyl group.) The two-carbon segment combines with coenzyme A (CoA), forming acetyl-CoA. The acetyl-CoA enters the citric acid cycle and ATP is produced. This process continues until all the two-carbon segments of the fatty acid are broken off, transformed to acetyl-CoA, and oxidized in the citric acid cycle. A lipid molecule can produce tremendous amounts of ATP because fatty acid chains are so long and there are three fatty acids per lipid molecule (Fig. 18-17).

Lipid Anabolism

Lipid synthesis occurs in the liver and adipose tissue. The acetyl-CoA molecules that form when a fatty acid breaks down may be used for lipid synthesis. Substances such as fatty acids, cholesterol, steroids, and triglycerides may be formed.

EXAMPLE 18-9 Calculate the energy formed when a 16-carbon fatty acid breaks down.

SOLUTION A 16-carbon fatty acid will form eight acetyl-CoA's by way of beta oxidation.

$$\underset{1}{C-C}+\underset{2}{C-C}+\underset{3}{C-C}+\underset{4}{C-C}+\underset{5}{C-C}+\underset{6}{C-C}+\underset{7}{C-C}+C-C$$

The beta oxidation must occur seven times. This produces seven $FADH_2$ and seven $NADH_2$ molecules. The eight acetyl-CoA's will go through the citric acid cycle eight times.

Tally sheet

	MOLECULES PRODUCED	ATP'S PRODUCED*
Beta oxidation	7 $FADH_2$	14
	7 $NADH_2$	21
Citric acid cycle	8 acetyl-CoA	96
		131 ATP's

7 $FADH_2$ molecules $\times$ 2 ATP/$FADH_2$ = 14 ATP's
7 $NADH_2$ molecules $\times$ 3 ATP/$NADH_2$ = 21 ATP's
8 acetyl-CoA molecules $\times$ 12 ATP/acetyl-CoA = 96 ATP's

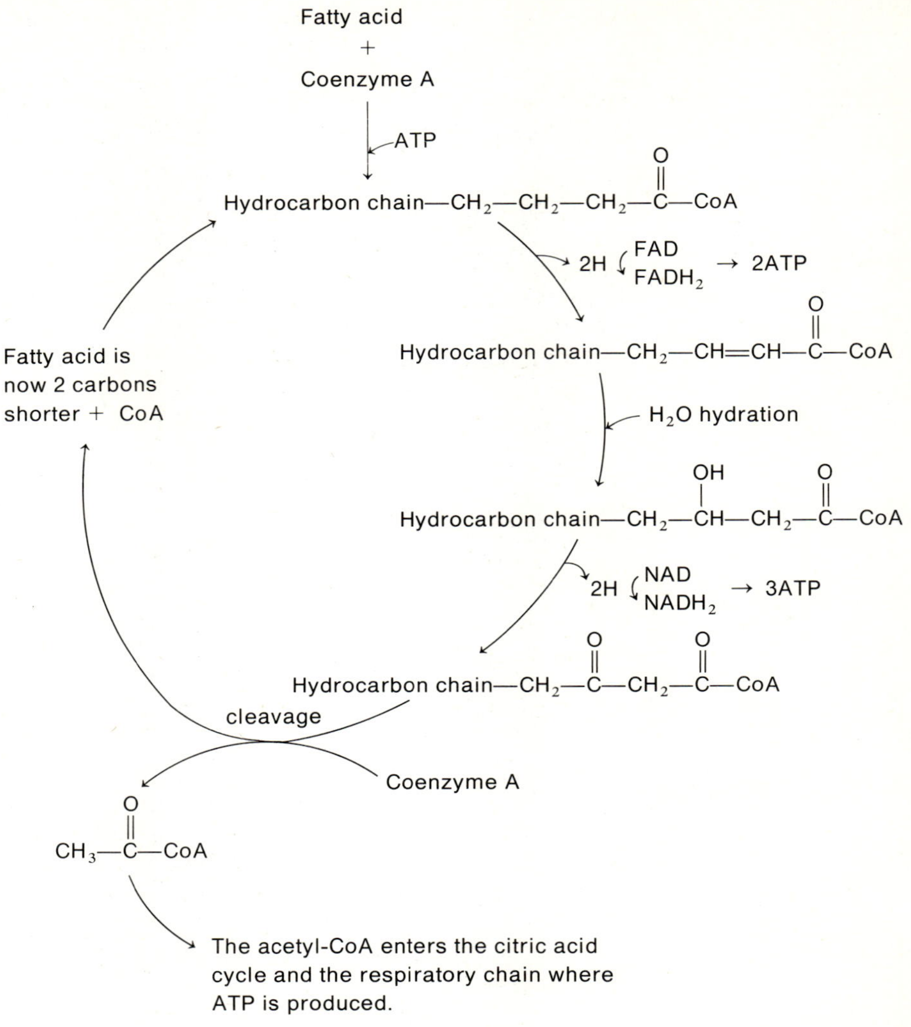

FIGURE 18-17
The fatty acid cycle.

There are 131 ATP's produced. (However, one ATP is needed for the initial activation of the system, so there is a net gain of 130 ATP's.)

Protein Metabolism

Amino acids build the structural units of the body as well as enzymes and hormones. We'll first study the more important process of protein anabolism. Then we'll look at protein catabolism.

TABLE 18-2   Various components of the citric acid cycle can be produced from the deamination of amino acids

CITRIC ACID CYCLE COMPONENT	AMINO ACIDS THAT PRODUCE THIS COMPONENT BY DEAMINATION
Pyruvic acid	Alanine, cysteine, glycine, serine, threonine
Acetic acid	Leucine, lysine, tryptophan
α-Ketoglutaric acid	Arginine, glutamic acid, glutamine, histidine, proline
Succinic acid	Isoleucine, methionine, valine
Fumaric acid	Phenylalanine, tyrosine
Oxaloacetic acid	Aspartic acid

Protein Anabolism

Proteins are synthesized from amino acids. The amino acids are the end products of digestion. They enter the cytoplasm and are combined in the proper sequence to produce proteins. (See Chapter 15, p. 373.)

Catabolism of Amino Acids

Amino acids are the end products of protein digestion. In the liver amino acids are broken down. *Deamination* is the primary catabolic reaction that amino acids undergo. Deamination involves the removal of the amino group ($-NH_2$) from an amino acid, and the following reaction occurs:

$$NH_2 + H \longrightarrow NH_3$$

$$2NH_3 + CO_2 \xrightarrow{\text{enzymes}} NH_2-\overset{\overset{\textstyle O}{\|}}{C}-NH_2$$

$$\text{Urea}$$

Amino acids can also be decarboxylated (a CO_2 group is removed) to form alpha-keto acids. These acids can enter the citric acid cycle and undergo further breakdown and production of energy (Table 18-2).

SUMMARY

In this chapter we surveyed the topic of metabolism. We looked at the processes of anabolism and catabolism.

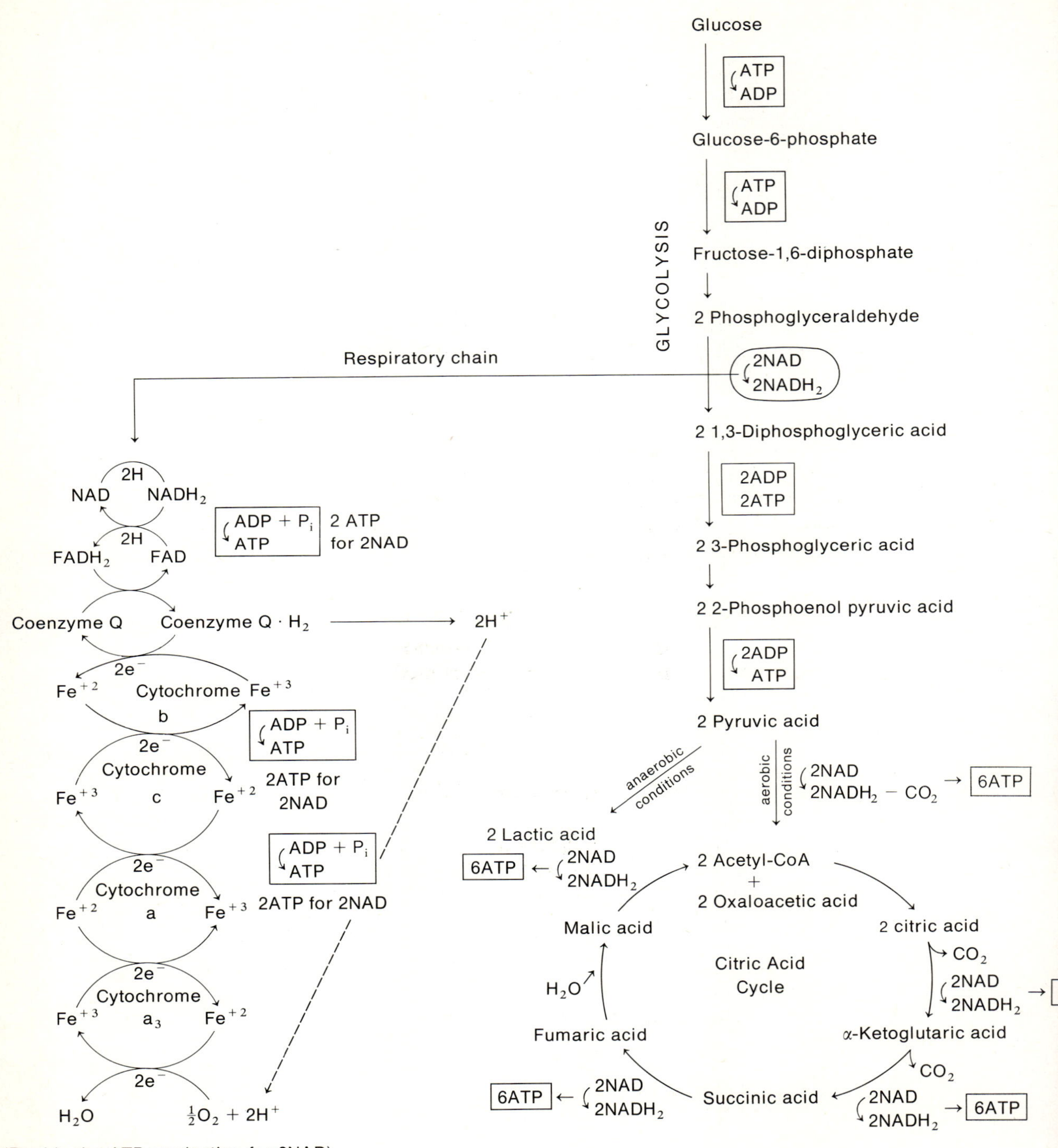

FIGURE 18-18

All the steps involved in cellular respiration. (Adapted from Kimball, *Biology*, 2nd ed., © 1968, Addison-Wesley, Reading, Mass., p. 196, Fig. 11-6. Reprinted with permission.)

We began our study by learning about the various parts of the cell. Next, we learned how ATP and ADP are used by the cell to produce and store energy. We then studied the metabolism of carbohydrates. We learned that carbohydrate anabolism occurs primarily in the liver, where the process of glycogenesis takes place. We also learned about carbohydrate catabolism and the two processes by which it can be accomplished: aerobic catabolism and anaerobic catabolism. We saw that these two processes are tied in with cellular respiration.

Next, we studied about the respiratory chain whose final goal is the production of ATP. We also learned how the cytochrome system helps with this ATP production. We also took a close look at the citric acid cycle and saw how large amounts of ATP can be produced from pyruvic acid (Fig. 18-18). Finally, we looked briefly at the metabolism of lipids and proteins.

EXERCISES

1. Define the terms anabolism and catabolism.

2. Match the cell part with its function.
   (1) endoplasmic reticulum
   (2) mitochondria
   (3) ribosomes
   (4) cell membrane
   (5) golgi complex
   (6) cytoplasm

   (a) Fluid in which the cell structures are suspended.
   (b) Regulates what compounds enter and leave the cell.
   (c) Transports proteins produced by the ribosomes.
   (d) The site of protein synthesis in the cell.
   (e) The place where proteins, carbohydrates, and lipids are stored.
   (f) The structure that produces energy for the cell.

3. What is the function of ATP and ADP in metabolism?

4. Complete the following equation:

$$ATP \xrightarrow{\text{hydrolysis}}$$

5. Name the major place in the body of carbohydrate anabolism. What is the name of the process that takes place there to store sugar?

6. Write the overall reaction for cellular respiration.

7. Complete the following statements.
   (a) When NAD is reduced, it becomes __________.
   (b) When FAD is reduced, it becomes __________.

8. What is the first step of glycolysis? What molecule is formed?

9. Complete the following statement: The conversion of glucose to pyruvic acid is an __________ (aerobic/anaerobic) process.

10. Name the final product of anaerobic catabolism, and state how many moles of ATP are produced for every mole of glucose broken down.

11. What major process occurs in the respiratory chain?

12. Name the major components of the respiratory chain.

13. Complete the following reaction, which represents the overall reaction of the respiratory chain.

$$2H^+ + 2e^- + \tfrac{1}{2}O_2 \longrightarrow$$

14. Name the two types of reactions that occur in the citric acid cycle.

15. Write the overall reaction of the citric acid cycle, including the energy produced.

16. Explain why lipids can produce tremendous amounts of energy.

17. What happens in beta oxidation?

18. Calculate the energy formed when an 18-carbon fatty acid breaks down by beta oxidation and goes through the citric acid cycle.

19. Complete the following reactions, which refer to amino acid catabolism.

$$NH_2 + H \longrightarrow \ ?$$

$$? \ + \ ? \ \xrightarrow{\text{enzymes}} \ NH_2\overset{\overset{\displaystyle O}{\|}}{-}C-NH_2$$

20. In what part of the body are amino acids broken down?

Basic Mathematics
The Numbers Game

Some Things You Should Know After Reading This Supplement

You should be able to:

1. Add numbers algebraically.
2. Subtract numbers algebraically.
3. Multiply exponential numbers.
4. Divide exponential numbers.
5. Write numbers in scientific notation.
6. Solve problems using the factor-unit method.
7. Solve an algebraic equation for an unknown.
8. Solve a word equation for an unknown letter.

Introduction

You say that you're rusty in mathematics and that you need some review. Well, we've got just what you need; a brief review of the math that you'll need for this course. Just peruse the material and try the sample problems. See if it doesn't refresh your memory.

Adding and Subtracting Algebraically

In the study of chemistry you will come into contact with positive and negative numbers. You may have to add and subtract these numbers, so remember the following rules.

Addition Rule

    (a) *When the signs are alike* When the signs are the same, add the numbers and keep the same sign. For example,

$$+5 + 2 = +7$$
$$-5 - 2 = -7$$

    (b) *When the signs are different* When the signs are not the same, subtract the numbers and keep the sign of the larger number. For example.

$$-5 + 2 = -3$$
$$+5 - 2 = +3$$

Subtraction Rule

    Change the sign of the *number being subtracted* and follow the rules for addition. (*Note:* Plus signs before positive numbers are usually left out, but minus signs before negative numbers are always written.) For example.

$$3 - (-2) \text{ becomes } \quad 3 + 2 = 5$$
$$3 - (+2) \text{ becomes } \quad 3 - 2 = 1$$
$$-3 - (+2) \text{ becomes } -3 - 2 = -5$$
$$-3 - (-2) \text{ becomes } -3 + 2 = -1$$

EXAMPLE A-1   Add the following numbers algebraically.

    (a) $25 - 60$ (b) $-25 - 60$
    (c) $-25 + 60$ (d) $25 + 60$

SOLUTION   Follow the rules for algebraic addition.

    (a) $25 - 60 = -35$ (b) $-25 - 60 = -85$
    (c) $-25 + 60 = 35$ (d) $25 + 60 = 85$

EXAMPLE A-2   Subtract the following numbers algebraically.

      (a) $30 - (+50)$ (b) $30 - (-50)$
      (c) $-30 - (+50)$ (d) $-30 - (-50)$

SOLUTION   Follow the rules for algebraic subtraction.

      (a) $30 - (+50)$ becomes $30 - 50 = -20$
      (b) $30 - (-50)$ becomes $30 + 50 = 80$
      (c) $-30 - (+50)$ becomes $-30 - 50 = -80$
      (d) $-30 - (-50)$ becomes $-30 + 50 = 20$

Exponents: Those Numbers Above

Exponents are numbers written to the right of another number (called the *base number*) and above it. They tell you to perform an operation on the base number. For example, $(3)^2$ means to multiply 3 by itself: 3×3. Three is the base number and 2 is the exponent.

$$(3)^2 \longleftarrow \text{Exponent}$$
$$\uparrow \text{Base number}$$

Sometimes exponents are seen as negative numbers. For example, $(3)^{-2}$ actually means $\dfrac{1}{(3)^2}$, or if the number is $\dfrac{1}{(3)^{-2}}$, this actually means $\dfrac{(3)^2}{1}$.

Let's solve some problems using exponents.

EXAMPLE A-3   Perform the indicated operation.

      (a) $(4)^2$ (b) $(5)^3$

      (c) $(5)^1$ (d) $(2)^{-4}$

      (e) $\dfrac{1}{(2)^5}$ (f) $\dfrac{1}{(3)^{-4}}$

SOLUTION   (a) $(4)^2 = 4 \times 4 = 16$

      (b) $(5)^3 = 5 \times 5 \times 5 = 125$

      (c) $(5)^1 = 5$

      (d) $(2)^{-4} = \dfrac{1}{(2)^4} = \dfrac{1}{2 \times 2 \times 2 \times 2} = \dfrac{1}{16}$

(e) $\dfrac{1}{(2)^5} = \dfrac{1}{2 \times 2 \times 2 \times 2 \times 2} = \dfrac{1}{32}$

(f) $\dfrac{1}{(3)^{-4}} = \dfrac{(3)^4}{1} = 3 \times 3 \times 3 \times 3 = 81$

There will be times when we have to multiply exponential numbers that have the same base. The rule for doing this is to *keep the base number and add the exponents algebraically.* Let's see how this is done.

EXAMPLE A-4   Perform the indicated operation.

(a) $(3)^3(3)^1$ (b) $(3)^3(3)^{-1}$ (c) $(3)^{-3}(3)^1$
(d) $(3)^{-3}(3)^{-1}$ (e) $(2)^4(2)^1$ (f) $(2)^4(2)^{-1}$

SOLUTION   We will keep the same base and add the exponents algebraically.

(a) $(3)^3(3)^1 = (3)^4$ (b) $(3)^3(3)^{-1} = (3)^2$

(c) $(3)^{-3}(3)^1 = (3)^{-2}$ (d) $(3)^{-3}(3)^{-1} = (3)^{-4}$

(e) $(2)^4(2)^1 = (2)^5$ (f) $(2)^4(2)^{-1} = (2)^3$

There will be times when we have to divide exponential numbers that have the same base. The rule for doing this is to *keep the base number and subtract the exponents algebraically.* Let's see how this is done.

EXAMPLE A-5   Perform the indicated operation.

(a) $\dfrac{(3)^7}{(3)^2}$ (b) $\dfrac{(3)^7}{(3)^{-2}}$

(c) $\dfrac{(3)^{-7}}{(3)^2}$ (d) $\dfrac{(3)^{-7}}{(3)^{-2}}$

SOLUTION   We will keep the same base and subtract the exponents algebraically.

(a) $\dfrac{(3)^7}{(3)^2} = (3)^{7-2} = (3)^5$

(b) $\dfrac{(3)^7}{(3)^{-2}} = (3)^{7-(-2)} = (3)^{7+2} = (3)^9$

(c) $\dfrac{(3)^{-7}}{(3)^2} = (3)^{-7-2} = (3)^{-9}$

(d) $\dfrac{(3)^{-7}}{(3)^{-2}} = (3)^{-7-(-2)} = (3)^{-7+2} = (3)^{-5}$

There will be times when we have to multiply or divide more than two exponential numbers. Let's try some of these examples.

EXAMPLE A-6   Perform the indicated operation.

(a) $(2)^3(2)^7(2)^4$ (b) $(2)^3(2)^7(2)^{-4}$

(c) $\dfrac{(3)^2(3)^5}{(3)^3}$ (d) $\dfrac{(a)^8(a)^{-3}}{(a)^2(a)^7}$

SOLUTION

(a) $(2)^3(2)^7(2)^4 = (2)^{14}$

(b) $(2)^3(2)^7(2)^{-4} = (2)^6$

(c) $\dfrac{(3)^2(3)^5}{(3)^3} = \dfrac{(3)^7}{(3)^3} = (3)^{7-3} = (3)^4$

(d) $\dfrac{(a)^8(a)^{-3}}{(a)^2(a)^7} = \dfrac{(a)^5}{(a)^9} = (a)^{5-9} = (a)^{-4}$

Writing Numbers as Powers of 10: Scientific Notation

In the physical sciences we often deal with numbers that are very large and other numbers that are very small: for example, 100,000,000,000 (a hundred billion) or 0.0000008 (eight ten-millionths). The number 100,000,000,000 can be written as 1×10^{11}, and 0.0000008 as 8×10^{-7}. In the first example we move the decimal point 11 places to the left.

$$100,000,000,000. = 1 \times 10^{11}$$

This corresponds to multiplying by a positive power of 10. In the second example we moved the decimal point seven places to the right.

$$0.0000008 = 8 \times 10^{-7}$$

This corresponds to multiplying by a negative power of 10. The exponent indicates the number of places that we moved the decimal point. The exponent is written as a positive number if the decimal point is moved to the left and a negative number if the decimal point is moved to the right. When you write numbers in scientific notation it is only necessary to write the nonzero numbers, since the zeros are incorporated into the exponent. (Of course, a zero appearing between two nonzero numbers must be written. For example, the number 805,000 is written as 8.05×10^5.) Also, the usual rule for writing numbers in scientific notation is to move the decimal point so that only one number remains to the left of the decimal point. Review the following examples and see if you get the idea.

EXAMPLE A-7   Write the following numbers in scientific notation.

(a) 350,000,000,000   (b) 807,000,000
(c) 4,205,000,000   (d) 0.00000765
(e) 0.00204   (f) 0.001008

SOLUTION

(a) $350,000,000,000. = 3.5 \times 10^{11}$

(b) $807,000,000. = 8.07 \times 10^8$

(c) $4,205,000,000. = 4.205 \times 10^9$

(d) $0.00000765 = 7.65 \times 10^{-6}$

(e) $0.00204 = 2.04 \times 10^{-3}$

(f) $0.001008 = 1.008 \times 10^{-3}$

The Factor-Unit Method, or How to Work with Units

When you work with numbers in the physical sciences they are almost always accompanied by units. You should learn how to work with these units as well as with their numbers. For example, let's say that we wanted to change 4 feet into inches. You could probably do this in your head, but we're going to learn how to use a process called the factor-unit method. This method involves setting up the problem in the following manner.

What I want to know = (quantity given)(factor unit)

The idea is that when you multiply the "quantity given" by the "factor unit," the proper units cancel to give the desired quantity. Let's see how this works.

EXAMPLE A-8   Change 4 feet into inches by the factor-unit method.

   In this problem the factor unit is 12 inches = 1 foot, which can be written in two ways:

$$\frac{12 \text{ inches}}{1 \text{ foot}} \quad \text{or} \quad \frac{1 \text{ foot}}{12 \text{ inches}}$$

Our job is to choose the proper factor unit so that our answer comes out in feet. We proceed as follows:

$$\text{What I want to know} = (\text{quantity given})(\text{factor unit})$$

$$? \text{ inches} = (4 \text{ feet})\left(\frac{12 \text{ inches}}{1 \text{ foot}}\right)$$

$$= 48 \text{ inches}$$

Notice how the term "feet" cancels to give inches. If we had chosen the other factor unit, the terms wouldn't have canceled. Let's do it so you can see for yourself.

$$? \text{ inches} = (4 \text{ feet})\left(\frac{1 \text{ foot}}{12 \text{ inches}}\right)$$

$$= 0.33 \frac{\text{feet}^2}{\text{inches}}$$

The units don't cancel and we don't get the answer *inches*.

In doing problems by the factor-unit method, always write the factor unit so that the term you want is in the numerator, and the term you want to cancel is in the denominator of the factor unit. See what we do in the next example.

EXAMPLE A-9   Change 60 inches into feet by the factor-unit method.

SOLUTION   Again, the factor unit is 12 inches = 1 foot.

$$\text{What I want to know} = (\text{quantity given})(\text{factor unit})$$

$$? \text{ feet} = (60 \text{ inches})\left(\frac{1 \text{ foot}}{12 \text{ inches}}\right)$$

$$= 5 \text{ feet}$$

EXAMPLE A-10   Solve each of the following conversions by the factor-unit method.

(a) 27 feet = ? yards

(b) 0.5 yard = ? feet

$$\text{(c)} \quad \frac{60 \text{ miles}}{1 \text{ hour}} = \frac{? \text{ feet}}{1 \text{ second}}$$

SOLUTION

(a) The factor unit we must know is 3 feet = 1 yard.

What I want to know = (quantity given)(factor unit)

$$? \text{ yards} = (27 \text{ feet})\left(\frac{1 \text{ yard}}{3 \text{ feet}}\right)$$

$$? \text{ yards} = 9 \text{ yards}$$

(b) Again, the factor unit is 3 feet = 1 yard.

What I want to know = (quantity given)(factor unit)

$$? \text{ feet} = (0.5 \text{ yard})\left(\frac{3 \text{ feet}}{1 \text{ yard}}\right)$$

$$? \text{ feet} = 1.5 \text{ feet}$$

(c) In this problem we need two factor units: one to change miles to feet, and one to change hours to seconds. These factor units are

$$1 \text{ mile} = 5280 \text{ feet}$$

$$1 \text{ hour} = 3600 \text{ seconds}$$

We set the conversion up as follows:

What I want to know = (quantity given)(factor units)

$$\frac{? \text{ feet}}{1 \text{ second}} = \left(\frac{60 \text{ miles}}{1 \text{ hour}}\right)\left(\frac{5280 \text{ feet}}{1 \text{ mile}}\right)\left(\frac{1 \text{ hour}}{3600 \text{ seconds}}\right)$$

$$= \frac{88 \text{ feet}}{1 \text{ second}}$$

Solving for the Unknown in an Algebraic Equation

In your study of chemistry you will have to know how to solve simple algebraic equations. The following examples will help you to refresh your memory.

EXAMPLE A-11 Solve each of the following equations for the unknown quantity.

(a) $4a = 8$ (b) $6x + 5 = 53$

(c) $8y - 3 = 69$ (d) $\dfrac{5b}{3} = \dfrac{10}{4}$

SOLUTION
(a) $4a = 8$ Isolate the unknown by dividing each side of the equation by 4.

$$\frac{\cancel{4}a}{\cancel{4}} = \frac{8}{4}$$

$$a = 2$$

(b) $6x + 5 = 53$ Remove the 5 from the left-hand side of the equation by subtracting five from both sides of the equation.

$$6x + 5 - 5 = 53 - 5$$

$$6x = 48$$

Now divide each side of the equation by 6.

$$\frac{\cancel{6}x}{\cancel{6}} = \frac{48}{6}$$

$$x = 8$$

(c) $8y - 3 = 69$ Remove the 3 from the left-hand side of the equation by adding three to both sides of the equation.

$$8y - 3 + 3 = 69 + 3$$

$$8y = 72$$

Now divide both sides of the equation by 8.

$$\frac{\cancel{8}y}{\cancel{8}} = \frac{72}{8}$$

$$y = 9$$

(d) $\dfrac{5b}{3} = \dfrac{10}{4}$ Remove the 3 from the left-hand side of the equation by multiplying both sides of the equation by 3.

$$(\cancel{3})\left(\frac{5b}{3}\right) = \left(\frac{10}{4}\right)(3)$$

$$5b = \frac{30}{4}$$

Remove the 5 from the left-hand side of the equation by dividing both sides of the equation by 5.

$$\frac{5b}{5} = \frac{30}{(4)(5)}$$

$$b = \frac{30}{20} = 1.5$$

Many of the mathematical equations we use in chemistry are word equations, such as

$$\text{Density} = \frac{\text{mass}}{\text{volume}}$$

which is usually abreviated as

$$D = \frac{m}{V}$$

As written, the equation is set up so that we can solve for the density of a substance if we are given its mass and volume. However, in some problems we are given the density and volume of a substance and asked to solve for its mass. In other problems we may be given the density and mass of a substance and asked to solve for its volume. What do we do then?

EXAMPLE A-12   Given $D = \dfrac{m}{V}$ solve this equation for m.

SOLUTION $D = \dfrac{m}{V}$ To isolate the m, multiply both sides of the equation by V.

$$(D)(V) = \left(\frac{m}{\cancel{V}}\right)(\cancel{V})$$

$$(D)(V) = m$$

EXAMPLE A-13   Given $D = \dfrac{m}{V}$ solve this equation for V.

SOLUTION   The V is in the denominator, so we must get it into the numerator. We can do this by multiplying each side of the equation by V. We've already done this in Example A-12 and obtained

$$(D)(V) = m$$

To isolate the V, we must divide each side of the equation by D.

$$\frac{(\cancel{D})(V)}{(\cancel{D})} = \frac{m}{D}$$

$$V = \frac{m}{D}$$

SUMMARY

In these few pages we've reviewed the basic mathematics that you will need in your health science chemistry course. Our purpose here was to refresh your memory and give you a place to look up some basic math rules in case you got stuck. If you want to be sure that you've mastered the topics in this appendix, try the exercises that follow.

EXERCISES

1. Add the following numbers algebraically.
   (a) $18 + 35$ (b) $-64 - 13$ (c) $-20 + 59$
   (d) $30 - 19$ (e) $19 - 30$

2. Subtract the following numbers algebraically.
   (a) $18 - (+35)$ (b) $-64 - (-13)$ (c) $-20 - (+59)$
   (d) $30 - (-19)$ (e) $19 - (-30)$

3. Perform the indicated operation.
   (a) $(3)^2$ (b) $(4)^3$ (c) $(6)^1$

   (d) $(3)^{-4}$ (e) $\dfrac{1}{(2)^4}$ (f) $\dfrac{1}{(2)^{-4}}$

4. Perform the indicated operation.
   (a) $(2)^3(2)^1$ (b) $(2)^3(2)^{-1}$ (c) $(2)^{-3}(2)^1$
   (d) $(2)^{-3}(2)^{-1}$

5. Perform the indicated operation.

   (a) $\dfrac{(2)^5}{(2)^8}$ (b) $\dfrac{(2)^5}{(2)^{-8}}$ (c) $\dfrac{(2)^{-5}}{(2)^8}$

   (d) $\dfrac{(2)^{-5}}{(2)^{-8}}$

6. Perform the indicated operation.

   (a) $\dfrac{(3)^5(3)^7}{(3)^2}$ (b) $(4)^8(4)^{-5}$

   (c) $\dfrac{(2)^{14}(2)^{-10}}{(2)^3(2)^4}$ (d) $\dfrac{(y)^{-5}(y)^3}{(y)^{-2}(y)^5}$

7. Write the following numbers in scientific notation.
   (a) $789,000,000,000,000$ (b) 0.0000001003
   (c) $3,050,000$ (d) 0.0021

8. Perform the indicated operation and leave the answer in scientific notation.

(a) $(3 \times 10^5)(2 \times 10^3)$ (b) $(3 \times 10^5)(2 \times 10^{-3})$

(c) $(3 \times 10^{-5})(2 \times 10^3)$ (d) $(3 \times 10^{-5})(2 \times 10^{-3})$

(e) $\dfrac{8 \times 10^9}{2 \times 10^5}$ (f) $\dfrac{8 \times 10^9}{2 \times 10^{-5}}$

(g) $\dfrac{8 \times 10^{-9}}{2 \times 10^5}$ (h) $\dfrac{8 \times 10^{-9}}{2 \times 10^{-5}}$

9. Solve each of the following problems by the factor-unit method.
   (a) 32 pints $= ?$ quarts (*Hint:* 2 pints $= 1$ quart.)
   (b) 21,120 feet $= ?$ miles (*Hint:* 1 mile $= 5280$ feet.)
   (c) 25 yards $= ?$ feet $= ?$ inches (*Hint:* 1 yard $= 3$ feet, 1 foot $= 12$ inches.)
   (d) 500 cm $= ?$ m (*Hint:* This is a metric conversion. We are going to discuss the metric system in Appendix B. However, see if you can solve this problem with the following hint: 1 m $= 100$ cm.)

10. Solve each of the following equations for the unknown quantity.
    (a) $3x = 21$ (b) $5a + 15 = 45$

    (c) $2c + 3 = 24 - 5c$ (d) $\dfrac{8m}{5} = \dfrac{160}{2}$

11. Given the formula $PV = nRT$, solve for each of the following in terms of the other letters.
    (a) $P =$
    (b) $V =$
    (c) $n =$
    (d) $R =$
    (e) $T =$

APPENDIX B

The Metric System
and Measurement

The Backbone
of the Sciences

Some Things You Should Know After Reading This Supplement

You should be able to:

1. Convert between the units of mm, cm, dm, m, and km.
2. Convert between the units of mg, cg, dg, g, and kg.
3. Convert between the units of ml, cl, dl, liters, and kl.
4. Convert from a metric unit to a corresponding English unit, and vice versa.
5. Use the density formula, $D = \dfrac{m}{V}$, to calculate the density, mass, or volume of a substance when given the other two values.
6. Convert Celsius temperatures to Fahrenheit, and vice versa.
7. Define exothermic and endothermic reactions.
8. Define enthalpy and calorie.
9. Calculate the calories absorbed by a sample of water when given its mass and temperature change.
10. Calculate the ΔH of a substance when given the experimental data.
11. Convert the ΔH of a substance from kcal/mole to kcal/g, and vice versa.

Introduction

The *metric system* was developed by the French in the late 1700's. The metric system is currently used in almost all parts of the world and in just about every branch of science. It is a very easy system to use since it is based on units of 10, much like our American money system. This means, for example, that conversion between one metric unit of length and another is simply a move of the decimal point (in other words, a multiplication or division of a power of 10). This is much simpler than the English system of measurement, where if you want to convert from one unit of length to another you must *memorize* all the "odd-ball" conversion factors (for example, 12 inches = 1 foot, 3 feet = 1 yard, 5280 feet = 1 mile). As we review the metric system over the next few pages you will see just how easy it is.

The Metric System: A Short Course

The metric system—like the English system—has units of length, mass, and volume. These are the *meter*, *gram*, and *liter*, respectively. For a mass that is smaller than a gram, we simply use a prefix (Tables B-1 and B-2). A milligram (mg) is equal to one one-thousandth of a gram (1/1000 or 0.001 g). A kilogram (kg) is a thousand times as large as a gram and is therefore equal to 1000 g. The following examples show how easy it is to convert from one metric unit to another.

EXAMPLE B-1   Change 8 grams to milligrams.

TABLE B-1   Prefixes and abbreviations used in the metric system

nano-	= 0.000000001	nanometer	= nm
micro-	= 0.000001	micrometer	= μm (Greek lowercase mu)
milli-	= 0.001	millimeter	= mm
centi-	= 0.01	milliliter	= ml
deci-	= 0.1	milligram	= mg
deca-	= 10	centimeter	= cm
kilo-	= 1000	centigram	= cg
		decimeter	= dm
		decigram	= dg
		kilometer	= km
		kilogram	= kg

TABLE B-2   The metric system

LENGTH

1 millimeter = 0.001 meter = 1/1000 meter	1 meter = 1000 millimeters
1 centimeter = 0.01 meter = 1/100 meter	1 meter = 100 centimeters
1 decimeter = 0.1 meter = 1/10 meter	1 meter = 10 decimeters
1 kilometer = 1000 meters	1 meter = 0.001 kilometer

MASS

1 microgram = 0.000001 gram	1 gram = 1,000,000 micrograms
1 milligram = 0.001 gram	1 gram = 1000 milligrams
1 centigram = 0.01 gram	1 gram = 100 centigrams
1 decigram = 0.1 gram	1 gram = 10 decigrams
1 kilogram = 1000 grams	1 gram = 0.001 kilogram

VOLUME

1 milliliter = 0.001 liter	1 liter = 1000 milliliters
1 milliliter = 1 cubic centimeter (cubic centimeter is abbreviated cc or cm^3)	1 liter = 1000 cubic centimeters
1 centiliter = 0.01 liter	1 liter = 100 centiliters
1 deciliter = 0.1 liter	1 liter = 10 deciliters
1 kiloliter = 1000 liters	1 liter = 0.001 kiloliter

SOLUTION   We will use the factor-unit method that we discussed in Appendix A. The factor unit that we *must* know is

$$1 \text{ gram} = 1000 \text{ milligrams} \quad \text{(Table B-2)}$$

$$? \text{ milligrams} = (8 \cancel{\text{ grams}}) \left(\frac{1000 \text{ milligrams}}{1 \cancel{\text{ gram}}} \right)$$

$$= 8000 \text{ milligrams}$$

EXAMPLE B-2   Change 3500 centimeters to meters.

SOLUTION   We'll use the same method as before, only this time we'll use abbreviations for the units (in other words, m for meters and cm for centimeters). The factor unit we *must* know is

$$1 \text{ m} = 100 \text{ cm} \quad \text{(Table B-2)}$$

$$? \text{ m} = (3500 \text{ cm}) \left(\frac{1 \text{ m}}{100 \text{ cm}} \right)$$

$$= 35 \text{ m}$$

As you can see, this isn't really too difficult. Let's try some additional examples.

EXAMPLE B-3   Change 35,000 ml into
(a) liters   (b) centiliters   (c) deciliters   (d) kiloliters

SOLUTION   First, we'll change the 35,000 ml into liters. Then we'll change the liters into cl, dl, and kl. The reason that we'll do it in this order is because we have the necessary conversion factors in Table B-2.

(a) The factor unit we *must* know is

$$1 \text{ liter} = 1,000 \text{ ml}$$

$$? \text{ liters} = (35,000 \text{ ml}) \left(\frac{1 \text{ liter}}{1000 \text{ ml}} \right)$$

$$= 35 \text{ liters}$$

(b) The factor unit we *must* know is

$$1 \text{ liter} = 100 \text{ cl}$$

$$? \text{ cl} = (35 \text{ liters}) \left(\frac{100 \text{ cl}}{1 \text{ liter}} \right)$$

$$= 3500 \text{ cl}$$

(c) The factor unit we *must* know is

$$1 \text{ liter} = 10 \text{ dl}$$

$$? \text{ dl} = (35 \text{ liters}) \left(\frac{10 \text{ dl}}{1 \text{ liter}} \right)$$

$$= 350 \text{ dl}$$

(d) The factor unit we *must* know is

$$1 \text{ kl} = 1000 \text{ liters}$$

$$? \text{ kl} = (35 \text{ liters}) \left(\frac{1 \text{ kl}}{1000 \text{ liters}} \right)$$

$$= 0.035 \text{ kl}$$

EXAMPLE B-4   Change 255 mg to dg.

SOLUTION   First, we'll change the 255 mg to grams. Then we'll change the grams into dg. (Of course, you may go directly from mg to dg if you know that there are 100 mg in 1 dg.)

$$? \text{ g} = (255 \text{ mg}) \left(\frac{1 \text{ g}}{1000 \text{ mg}} \right)$$

$$= 0.255 \text{ g}$$

$$? \text{ dg} = (0.255 \text{ g}) \left(\frac{10 \text{ dg}}{1 \text{ g}} \right)$$

$$= 2.55 \text{ dg}$$

We can now see that conversion between units of the metric system is easy. But what if you want to convert metric into English units, or vice versa? Table B-3 should help you in this task. Let's use this table and the factor-unit method to do some English-metric and metric-English conversions.

EXAMPLE B-5   Convert 12 inches into centimeters.

SOLUTION   Table B-3 tells us that 1 inch = 2.54 cm.

$$? \text{ cm} = (12 \text{ inches}) \left(\frac{2.54 \text{ cm}}{1 \text{ inch}} \right)$$

$$= 30.5 \text{ cm}$$

EXAMPLE B-6   Convert 2270 grams into pounds.

TABLE B-3   Conversion factors for English-metric and metric-English conversions.

LENGTH	MASS	VOLUME
1 inch = 2.54	1 ounce = 28.35 g	1 liter = 1.06 quarts
1 foot = 0.30 m	1 pound = 454 g	

Note: The conversions in this table can be used to go from English to metric or metric to English simply by writing the conversions as factor units, for example, $\dfrac{1 \text{ inch}}{2.54 \text{ cm}}$ or $\dfrac{2.54 \text{ cm}}{1 \text{ inch}}$.

SOLUTION   Table B-3 tells us that 1 pound = 454 grams.

$$? \text{ pounds} = (2270 \text{ g}) \left(\frac{1 \text{ pound}}{454 \text{ g}} \right)$$

$$= 5 \text{ pounds}$$

Although it is always handy to be able to convert between systems, it is more important for you to be able to associate metric measurements with objects in the world around you. You've got to begin to *think metric*. Only in this way will the metric system mean something to you. Here are some things to think about to help you begin to think metric.

A meter stick is slightly longer than a yard stick (about $3\frac{3}{8}$ inches longer).
A liter is just a little more than a quart. (Think of a quart milk container; a liter of milk would be almost the same size.)
A standard-size paper clip weighs about 1 gram.

Density: An Interesting Property of Matter

Which is heavier: glass, iron, or wood from an oak tree? Naturally, it depends on the size of each piece. However, what if all three were the same size? In other words, what if we got hold of cubes of glass, iron, and oak wood, each with a volume of 1 cm³? (Remember, a cube with a volume of 1 cm³ would be 1 cm long, 1 cm wide, and 1 cm high; see Fig. B-1.) Let's say that we weigh each cube to determine which is the heaviest. We find that 1 cm³ of iron weighs 7.9 g, 1 cm³ of glass weighs 2.4 g, and 1 cm³ of oak wood weighs 0.6 g. We conclude that, for a *particular volume*, the iron has the greatest mass. We have now developed the idea of *density*.

Density is the mass per unit volume of a substance. Mathematically, we define density as

$$D = \frac{m}{V}$$

FIGURE B-1
Three cubes of equal volume.

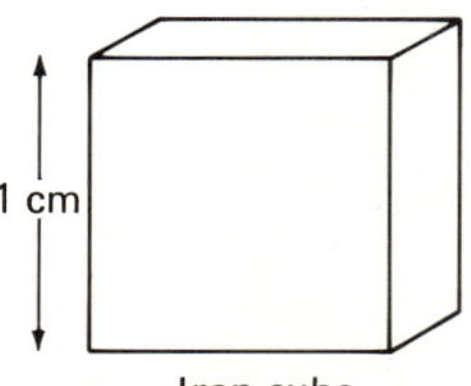

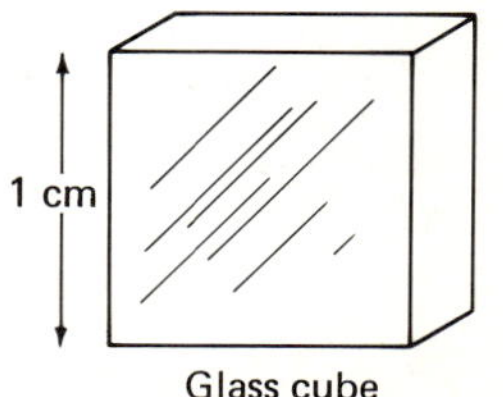

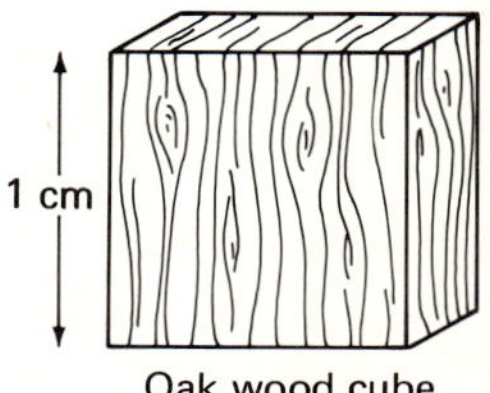

where D is density, m is mass, and V is volume. If we find the mass of our substance in grams and the volume in cubic centimeters, then the units of density become

$$\frac{\text{grams}}{\text{cubic centimeter}} \quad \text{or} \quad \frac{\text{g}}{\text{cm}^3}$$

Let's see how we would calculate the density of the object in the following example.

EXAMPLE B-7   A block of the element barium is 5 cm long, 3 cm wide, and 1 cm high, and it weighs 52.5 g. What is the density of the element barium?

SOLUTION   First, calculate the volume of the block.

$$\text{Volume} = \text{length} \times \text{width} \times \text{height}$$

$$V = 5 \text{ cm} \times 3 \text{ cm} \times 1 \text{ cm} = 15 \text{ cm}^3$$

We now know that $V = 15 \text{ cm}^3$ and $m = 52.5$ g, so all we have to do is solve the density formula for D.

$$D = \frac{m}{V} = \frac{52.5 \text{ g}}{15 \text{ cm}^3} = 3.5 \frac{\text{g}}{\text{cm}^3}$$

(which also can be thought of as $\dfrac{3.5 \text{ g}}{1 \text{ cm}^3}$, which reads that there is 3.5 g for every cubic centimeter of barium.)

EXAMPLE B-8   Suppose you are told that 400 g of alcohol occupies a volume of 500 ml. What is the density of the alcohol?

SOLUTION   In this problem we are given the mass and volume of the substance whose density we are asked to find. But note that the volume of the alcohol is given in milliliters (ml). Remember that in the metric system 1 *milliliter is equal to 1 cubic centimeter*. That is, 1 ml = 1 cm^3 (look back at Table B-2). So it's okay to interchange the terms ml and cm^3. Now let's solve this problem.

$$D = \frac{m}{V}$$

$$= \frac{400 \text{ g}}{500 \text{ cm}^3}$$

$$= 0.8 \frac{\text{g}}{\text{cm}^3}$$

Sometimes you may know the volume and density of an object and you may be asked to calculate its mass. Let's see how we use the density formula to do this.

EXAMPLE B-9   A piece of aluminum in the shape of a perfect cube measures 10 cm on each side. The density of aluminum is 2.7 g/cm^3. What is the mass of this block of aluminum?

SOLUTION   First calculate the volume of the block.

$$\text{Volume} = \text{length} \times \text{width} \times \text{height}$$

$$V = (10 \text{ cm})(10 \text{ cm})(10 \text{ cm}) = 1000 \text{ cm}^3$$

We now know that the volume is 1000 cm^3 and the density is 2.7 g/cm^3. We may now solve the density formula for m, plug in the numbers, and obtain the mass of the aluminum.

$$D = \frac{m}{V}; \qquad \text{therefore, } (D)(V) = m$$

$$m = (D)(V)$$

$$= \left(2.7 \, \frac{\text{g}}{\text{cm}^3}\right)(1000 \text{ cm}^3)$$

$$= 2700 \text{ g}$$

Sometimes you may be given the mass and density of an object and be asked to solve for the volume. Let's see how we do this.

EXAMPLE B-10   The density of blood serum is about 1.014 g/cm^3. What would be the volume of 5070 g of blood serum (which is about the mass of blood in an adult)?

SOLUTION   Let's solve the density formula for V.

$$D = \frac{m}{V}; \qquad \text{therefore, } (D)(V) = m \text{ and finally } V = \frac{m}{D}$$

$$V = \frac{m}{D}$$

$$= \frac{5070 \text{ g}}{1.014 \text{ g/cm}^3}$$

$$= 5000 \text{ cm}^3 \quad (\text{or 5 liters})$$

Temperature: The Intensity of Hotness or Coldness

When we light an oven and the gas starts to burn, we know that heat energy is being generated. How can we find out how intense this heat is? Well, naturally, when we want to know how hot something is, we measure its temperature. What we find out by measuring the temperature is the *intensity* of the object's hotness or coldness. Temperature tells us nothing about the *quantity* of heat that has entered into a substance or a place; we will learn about that later in this supplement.

In 1724 a German scientist by the name of Gabriel Daniel Fahrenheit devised a way of measuring the heat intensity of various substances. He used a sealed tube containing mercury (a thermometer). When he placed the thermometer in a substance, the mercury in the thermometer would move up or down the tube. To give some meaning to the height of the mercury column, Fahrenheit set up a temperature scale. On his scale, the temperature of a mixture of ice and salt water is zero degrees. The temperature of a mixture of ice and water is 32 degrees (the freezing point of water), and the temperature of a water–steam mixture (at normal atmospheric pressure) is 212 degrees (which is called the normal boiling point of water). Notice that on the Fahrenheit scale there are 180 equal divisions between the freezing point and boiling point of water. Each division corresponds to a degree Fahrenheit. But the Fahrenheit scale, though very useful, was soon replaced by a new temperature scale devised by a Swedish astronomer named Anders Celsius.

Celsius devised his Celsius or centigrade temperature scale in 1742. Scientists liked it better than Fahrenheit's, since the Celsius scale uses zero degrees as the freezing point of water and 100 degrees as the boiling point of water. There are 100 equal divisions between these two points; each division is called 1 degree Celsius (or centigrade) (Fig. B-2).

Scientists throughout the world use the Celsius thermometer to measure temperatures, mainly because of the convenience of the 100-degree scale. It fits right in with the metric system of measurement, as you can see. Just travel outside the United States and ask any health practitioner, "What's normal body temperature?" Don't be surprised when he tells you that it's 37! That's 37 degrees Celsius (37°C), of course.

Converting Celsius and Fahrenheit Degrees

As we were saying, on the Celsius scale there are 100 divisions between the boiling point and the freezing point of water. Therefore, 100 Celsius degrees are equal to 180 Fahrenheit degrees, or 1°C = 1.8°F. The formula you may use to convert from one system to another is

$$9°C = 5°F - 160$$

[Some instructors prefer to use the following two formulas for temperature conversions.

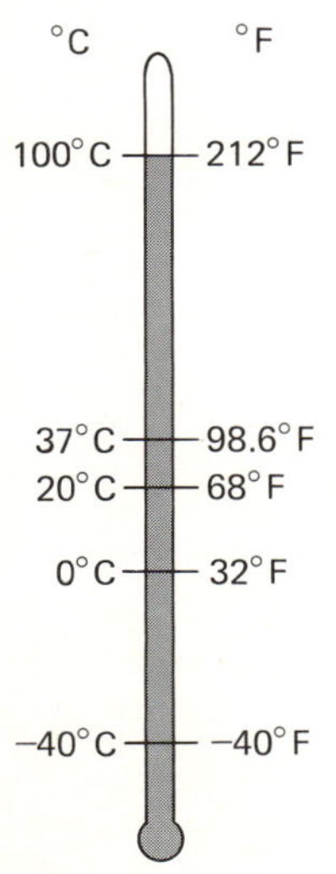

FIGURE B-2
The Fahrenheit and Celsius temperature scales.

$$°F = (1.8 \times °C) + 32$$

and
$$°C = \frac{°F - 32}{1.8}$$

You may use these two formulas if you wish, but we'll use our formula in the following examples.] That looks pretty odd, but let's use this formula to convert from Fahrenheit to Celsius and Celsius to Fahrenheit degrees.

EXAMPLE B-11   Convert 68°F (average room temperature) into °C.

SOLUTION   Substitute 68°F into the conversion formula and solve for degrees Celsius.

$$9°C = 5°F - 160$$
$$= (5)(68°F) - 160$$
$$°C = \frac{(5)(68) - 160}{9}$$
$$= \frac{340 - 160}{9}$$
$$= \frac{180}{9}$$
$$= 20$$

Therefore, 68°F = 20°C.

EXAMPLE B-12   Convert 37°C into °F.

SOLUTION   Substitute 37°C into the conversion formula and solve for degrees Fahrenheit.

$$9°C = 5°F - 160$$
$$(9)(37°C) = 5°F - 160$$
$$(9)(37) + 160 = 5°F$$
$$\frac{(9)(37) + 160}{5} = °F$$
$$\frac{333 + 160}{5} = °F$$
$$\frac{493}{5} = °F$$
$$98.6 = °F$$

Therefore, 37°C = 98.6°F (normal body temperature).

Heat and Chemical Reactions

Every substance—foods, minerals, and even people—has a specific amount of energy associated with it: a sort of chemical potential energy. When a substance reacts to form a new substance, it can either lose energy to the surroundings or gain energy from the surroundings. This is because, in chemical reactions, old bonds between atoms are broken and new bonds are formed. Breaking a bond usually requires additional energy, while forming a bond usually releases energy. Depending on the number of bonds broken and the number of bonds formed, as well as the strength of the bonds, energy is either released to the surroundings or absorbed from them.

Reactions that release energy to the surroundings are called *exothermic reactions.* For example, the explosion of a stick of dynamite is a chemical reaction that releases a tremendous amount of heat energy to the environment. Reactions that absorb energy from the surroundings are called *endothermic reactions.* For example, plants live and grow by a process called photosynthesis. In this process the plants use carbon dioxide, water, and *energy from the sun* (sunlight) to make glucose and oxygen.

We can usually calculate the energy changes that result from reactions by measuring the *heat* released or absorbed during the reaction. This is because each chemical substance has a certain heat content associated with it. The heat content is called *enthalpy,* and has the symbol H. One mole of a chemical substance has a definite heat content (enthalpy). (If you've forgotten what a mole is, see Chapter 2.) Although we can't measure the heat content of specific substances, we can measure the *change* in heat content during chemical reactions. The change in heat content during a chemical reaction is called the *heat of reaction.* We measure this heat in units of *calories. One calorie is the amount of heat needed to raise the temperature of one gram of water 1°C.* Large heat changes are measured in *kilocalories* (1000 cal = 1 kcal).

EXAMPLE B-13   You have a liter of water (1000 g) that you heat from 20°C to 80°C. How many calories of heat energy did the water absorb?

SOLUTION   Based on our definition, every gram of water that we heat 1°C gains a calorie of heat energy. Therefore, we can derive an equation that says:

(for water)   calories = (grams of water)(change in °C) or in abbreviated form, cal = $(m)(\Delta t)$. The symbol Δ means "change in."

$$? \text{ cal} = (m)(\Delta t)$$

$$= (1000 \text{ g})(80°C - 20°C)$$

$$= (1000 \text{ g})(60°C)$$

$$= 60{,}000 \text{ cal} \quad \text{or} \quad 60 \text{ kcal}$$

As you can see, it's easy to measure the calories of heat gained by a sample of water. But how can we measure the change in heat content of other substances when they undergo chemical reactions? We can, by using a device called a *calorimeter* (Fig. B-3). The calorimeter is frequently used to measure changes in the heat content of substances (for example, foods) as they react with oxygen. The way such a determination is carried out is as follows:

We want to measure the *heat of reaction* of glucose, as it reacts with oxygen. A sample of glucose is obtained, for example 1.8 g. It is placed inside the special heat-transferring container that fits into the calorimeter. The container is positioned in the calorimeter and is attached to an ignition switch which will be used to cause a spark that will allow the sugar and oxygen to react. The calorimeter is filled with a known amount of water, for example 1000 g. The temperature of the water is measured and is found to be 25°C. The ignition switch is pressed and a great amount of heat is liberated by the reaction between glucose and oxygen.

$$C_6H_{12}O_6 + 6O_2 \longrightarrow 6CO_2 + 6H_2O + energy$$

The heat energy passes through the special heat-transferring container to the water, whose temperature increases to 31.7°C. With this information we can now calculate the calories of heat liberated by the 1.8-g sample of glucose. We proceed as follows:

$$? \text{ cal (absorbed by water)} = (m_{water})(\Delta t_{water})$$
$$= (1000 \text{ g})(31.7°C - 25°C)$$
$$= (1000 \text{ g})(6.7°C)$$
$$= 6700 \text{ cal} \quad \text{or} \quad 6.7 \text{ kcal}$$

FIGURE B-3
A calorimeter.

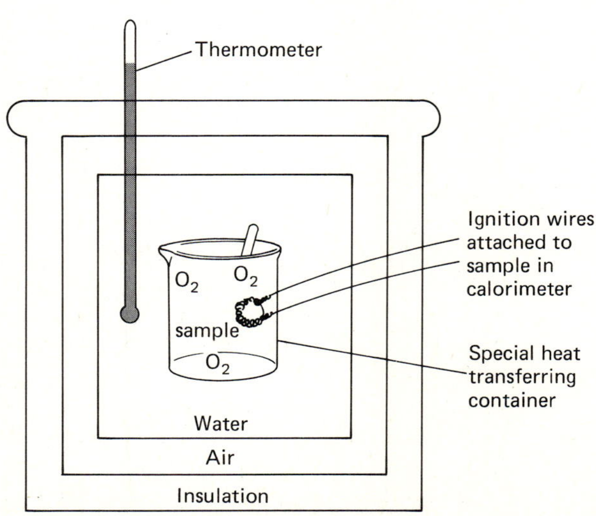

Therefore, the 1.8-g sample of glucose liberated 6.7 kcal of heat energy to the water. Therefore, the 1.8-g sample of glucose lost 6.7 kcal of energy. We usually like to express *heats of reaction* in kcal per mole. The molecular weight of glucose is 180. So we can obtain the *heat of reaction* in kcal/mole as follows:

$$? \frac{\text{kcal}}{\text{mole}} = \left(\frac{6.7 \text{ kcal}}{1.8 \text{ g}}\right)\left(\frac{180 \text{ g}}{1 \text{ mole}}\right)$$

$$= 670 \frac{\text{kcal}}{\text{mole}}$$

The symbol for heat content (enthalpy) is H. We represent the *change* in heat content by the symbol ΔH. For exothermic reactions the ΔH value is always preceded by a *minus* sign. For endothermic reactions the ΔH has a *positive* value. In our example of glucose we express the heat of reaction as follows:

$$\Delta H = -670 \text{ kcal/mole}$$

indicating it was an exothermic reaction.

The nutritional value of many food substances are expressed in kcal/mole. However, sometimes they are also expressed in kcal per gram (kcal/g).

EXAMPLE B-14   The nutritional value of glucose is 670 kcal/mole. Express this in kcal/g.

SOLUTION   We will use the factor-unit method. We must also know the molecular weight of glucose, which is 180. In other words, 1 mole of glucose = 180 g

$$? \frac{\text{kcal}}{\text{g}} = \left(\frac{670 \text{ kcal}}{1 \text{ mole}}\right)\left(\frac{1 \text{ mole}}{180 \text{ g}}\right) = 3.7 \frac{\text{kcal}}{\text{g}}$$

SUMMARY

The purpose of this supplement was to give you some review of measurement in the health sciences. We began with a discussion of the metric system and looked at the basic components of this system. We then looked at the concept of density and learned how you would go about measuring the density of an object in the laboratory. Next, we discussed the Fahrenheit and Celsius temperature scales and practiced switching from one scale to the other. Finally, we discussed the concept of heat and learned how to measure heats of reactions in calories. See if you've mastered the major points of this supplement by trying the exercises that follow.

1. Convert 855 mm into
   (a) cm   (b) dm   (c) m   (d) km

2. Convert 4.38 liters into
   (a) ml   (b) cl   (c) dl   (d) kl

3. Find the area of a rug that measures 5 m by 2 m. Report the answer in (a) m^2 and (b) cm^2.

4. Find the volume of a cube that measures 5 cm on each side. Report the answer in (a) cm^3 and (b) mm^3.

5. Express 6 ft in
   (a) m   (b) cm   (c) mm

6. Express 2 quarts in (a) ml (b) liters

7. Express 2 liters in
   (a) quarts   (b) fluid ounces

8. A block of gold in the shape of a cube measures 20 cm on each side. The density of gold is 19.3 g/cm^3. Determine the mass of this block of gold in (a) grams and (b) pounds.

9. A 71-g sample of diethyl ether has a volume of 100 ml. What is the density of this compound?

10. The density of isopropyl alcohol is 0.79 g/ml. What is the volume of 158 g of this compound?

11. Convert the following temperatures to °C.
    (a) 50°F   (b) 185°F   (c) −40°F

12. Convert the following temperatures to °F.
    (a) 230°C   (b) 490°C   (c) −148°C

13. A 500-g sample of water is heated from 20°C to 50°C. How many calories of heat did the water absorb?

14. A 1-g sample of sucrose raises the temperature of 500 g of water from 25°C to 33.2°C. Determine the enthalpy change for sucrose in kcal/mole. (*Hint:* The molecular weight of sucrose is 342.)

15. The caloric value of vanillin is 914 kcal/mole. Determine its caloric value in kcal/g. (*Hint:* The molecular weight of vanillin is 152.)

16. The caloric value of ethyl alcohol is 7.12 kcal/g. Determine its caloric value in kcal/mole. (*Hint:* The molecular weight of ethyl alcohol is 46.)

Important Tables

TABLE 1   Prefixes and abbreviations used in the metric system

nano-	= 0.000000001	nanometer	= nm
micro-	= 0.000001	micrometer	= μm (Greek lowercase mu)
milli-	= 0.001	millimeter	= mm
centi-	= 0.01	milliliter	= ml
deci-	= 0.1	milligram	= mg
deca-	= 10	centimeter	= cm
kilo-	= 1000	centigram	= cg
		decimeter	= dm
		decigram	= dg
		kilometer	= km
		kilogram	= kg

TABLE 2   The metric system

LENGTH

1 millimeter = 0.001 meter = 1/1000 meter	1 meter = 1000 millimeters
1 centimeter = 0.01 meter = 1/100 meter	1 meter = 100 centimeters
1 decimeter = 0.1 meter = 1/10 meter	1 meter = 10 decimeters
1 kilometer = 1000 meters	1 meter = 0.001 kilometer

MASS

1 microgram = 0.000001 gram	1 gram = 1,000,000 micrograms
1 milligram = 0.001 gram	1 gram = 1000 milligrams
1 centigram = 0.01 gram	1 gram = 100 centigrams
1 decigram = 0.1 gram	1 gram = 10 decigrams
1 kilogram = 1000 grams	1 gram = 0.001 kilogram

VOLUME

1 milliliter = 0.001 liter	1 liter = 1000 milliliters
1 milliliter = 1 cubic centimeter (cubic centimeter is abbreviated cc or cm^3)	1 liter = 1000 cubic centimeters
1 centiliter = 0.01 liter	1 liter = 100 centiliters
1 deciliter = 0.1 liter	1 liter = 10 deciliters
1 kiloliter = 1000 liters	1 liter = 0.001 kiloliter

TABLE 3   Conversion factors for English-metric
and metric-English conversions

LENGTH	MASS	VOLUME
1 inch = 2.54	1 ounce = 28.35 g	1 liter = 1.06 quarts
1 foot = 0.30 m	1 pound = 454 g	

Note: The conversions in this table can be used to go from English to metric or metric to English simply by writing the conversions as factor units, for example, $\dfrac{1 \text{ inch}}{2.54 \text{ cm}}$ or $\dfrac{2.54 \text{ cm}}{1 \text{ inch}}$.

TABLE 4 Naturally occurring isotopes of the first 15 elements

NAME	SYMBOL	ATOMIC NUMBER	MASS NUMBER	PERCENTAGE NATURAL ABUNDANCE
Hydrogen-1	$^{1}_{1}H$	1	1	99.985
Hydrogen-2	$^{2}_{1}H$	1	2	0.015
Hydrogen-3	$^{3}_{1}H$	1	3	Negligible
Helium-3	$^{3}_{2}He$	2	3	0.00013
Helium-4	$^{4}_{2}He$	2	4	99.99987
Lithium-6	$^{6}_{3}Li$	3	6	7.42
Lithium-7	$^{7}_{3}Li$	3	7	92.58
Beryllium-9	$^{9}_{4}Be$	4	9	100
Boron-10	$^{10}_{5}B$	5	10	19.6
Boron-11	$^{11}_{5}B$	5	11	80.4
Carbon-12	$^{12}_{6}C$	6	12	98.89
Carbon-13	$^{13}_{6}C$	6	13	1.11
Nitrogen-14	$^{14}_{7}N$	7	14	99.63
Nitrogen-15	$^{15}_{7}N$	7	15	0.37
Oxygen-16	$^{16}_{8}O$	8	16	99.759
Oxygen-17	$^{17}_{8}O$	8	17	0.037
Oxygen-18	$^{18}_{8}O$	8	18	0.204
Fluorine-19	$^{19}_{9}F$	9	19	100
Neon-20	$^{20}_{10}Ne$	10	20	90.92
Neon-21	$^{21}_{10}Ne$	10	21	0.257
Neon-22	$^{22}_{10}Ne$	10	22	8.82
Sodium-23	$^{23}_{11}Na$	11	23	100
Magnesium-24	$^{24}_{12}Mg$	12	24	78.70
Magnesium-25	$^{25}_{12}Mg$	12	25	10.13
Magnesium-26	$^{26}_{12}Mg$	12	26	11.17
Aluminium-27	$^{27}_{13}Al$	13	27	100
Silicon-28	$^{28}_{14}Si$	14	28	92.21
Silicon-29	$^{29}_{14}Si$	14	29	4.70
Silicon-30	$^{30}_{14}Si$	14	30	3.09
Phosphorus-31	$^{31}_{15}P$	15	31	100

Source: Alan Sherman, Sharon Sherman, and Leonard Russikoff, *Basic Concepts of Chemistry*, Houghton Mifflin Company, Boston, 1976. By permission of the publisher.

ELEMENT	ATOMIC NUMBER	ELECTRON CONFIGURATION						
		K	L	M	N	O	P	Q
H	1	1						
He	2	2						
Li	3	2	1					
Be	4	2	2					
B	5	2	3					
C	6	2	4					
N	7	2	5					
O	8	2	6					
F	9	2	7					
Ne	10	2	8					
Na	11	2	8	1				
Mg	12	2	8	2				
Al	13	2	8	3				
Si	14	2	8	4				
P	15	2	8	5				
S	16	2	8	6				
Cl	17	2	8	7				
Ar	18	2	8	8				
K	19	2	8	8	1			
Ca	20	2	8	8	2			
Sc	21	2	8	9	2			
Ti	22	2	8	10	2			
V	23	2	8	11	2			
Cr	24	2	8	13	1			
Mn	25	2	8	13	2			
Fe	26	2	8	14	2			
Co	27	2	8	15	2			
Ni	28	2	8	16	2			
Cu	29	2	8	18	1			
Zn	30	2	8	18	2			

ELEMENT	ATOMIC NUMBER	ELECTRON CONFIGURATION						
		K	*L*	*M*	*N*	*O*	*P*	*Q*
Ga	31	2	8	18	3			
Ge	32	2	8	18	4			
As	33	2	8	18	5			
Se	34	2	8	18	6			
Br	35	2	8	18	7			
Kr	36	2	8	18	8			
Rb	37	2	8	18	8	1		
Sr	38	2	8	18	8	2		
Y	39	2	8	18	9	2		
Zr	40	2	8	18	10	2		
Nb	41	2	8	18	12	1		
Mo	42	2	8	18	13	1		
Tc	43	2	8	18	14	1		
Ru	44	2	8	18	15	1		
Rh	45	2	8	18	16	1		
Pd	46	2	8	18	18			
Ag	47	2	8	18	18	1		
Cd	48	2	8	18	18	2		
In	49	2	8	18	18	3		
Sn	50	2	8	18	18	4		
Sb	51	2	8	18	18	5		
Te	52	2	8	18	18	6		
I	53	2	8	18	18	7		
Xe	54	2	8	18	18	8		
Cs	55	2	8	18	18	8	1	
Ba	56	2	8	18	18	8	2	
La	57	2	8	18	18	9	2	
Ce	58	2	8	18	20	8	2	
Pr	59	2	8	18	21	8	2	
Nd	60	2	8	18	22	8	2	
Pm	61	2	8	18	23	8	2	

ELEMENT	ATOMIC NUMBER	ELECTRON CONFIGURATION						
		K	*L*	*M*	*N*	*O*	*P*	*Q*
Sm	62	2	8	18	24	8	2	
Eu	63	2	8	18	25	8	2	
Gd	64	2	8	18	25	9	2	
Tb	65	2	8	18	27	8	2	
Dy	66	2	8	18	28	8	2	
Ho	67	2	8	18	29	8	2	
Er	68	2	8	18	30	8	2	
Tm	69	2	8	18	31	8	2	
Yb	70	2	8	18	32	8	2	
Lu	71	2	8	18	32	9	2	
Hf	72	2	8	18	32	10	2	
Ta	73	2	8	18	32	11	2	
W	74	2	8	18	32	12	2	
Re	75	2	8	18	32	13	2	
Os	76	2	8	18	32	14	2	
Ir	77	2	8	18	32	15	2	
Pt	78	2	8	18	32	17	1	
Au	79	2	8	18	32	18	1	
Hg	80	2	8	18	32	18	2	
Tl	81	2	8	18	32	18	3	
Pb	82	2	8	18	32	18	4	
Bi	83	2	8	18	32	18	5	
Po	84	2	8	18	32	18	6	
At	85	2	8	18	32	18	7	
Rn	86	2	8	18	32	18	8	
Fr	87	2	8	18	32	18	8	1
Ra	88	2	8	18	32	18	8	2
Ac	89	2	8	18	32	18	9	2
Th	90	2	8	18	32	18	10	2
Pa	91	2	8	18	32	20	9	2
U	92	2	8	18	32	21	9	2

TABLE 5 Electron configurations of the elements (*cont.*)

ELEMENT	ATOMIC NUMBER	ELECTRON CONFIGURATION						
		K	L	M	N	O	P	Q
Np	93	2	8	18	32	22	9	2
Pu	94	2	8	18	32	24	8	2
Am	95	2	8	18	32	25	8	2
Cm	96	2	8	18	32	25	9	2
Bk	97	2	8	18	32	26	9	2
Cf	98	2	8	18	32	28	8	2
Es	99	2	8	18	32	29	8	2
Fm	100	2	8	18	32	30	8	2
Md	101	2	8	18	32	31	8	2
No	102	2	8	18	32	32	8	2
Lr	103	2	8	18	32	32	9	2
Rf	104	2	8	18	32	32	10	2
Ha	105	2	8	18	32	32	11	2

TABLE 6 Table of ions frequently used in chemistry

+1		+2		+3	
Hydrogen	H^{+1}	Calcium	Ca^{+2}	Iron(III)	Fe^{+3}
Lithium	Li^{+1}	Magnesium	Mg^{+2}	Aluminium	Al^{+3}
Sodium	Na^{+1}	Barium	Ba^{+2}		
Potassium	K^{+1}	Zinc	Zn^{+2}		
Mercury(I)	Hg^{+1}	Mercury(II)	Hg^{+2}		
Copper(I)	Cu^{+1}	Tin(II)	Sn^{+2}		
Ammonium	$(NH_4)^{+1}$	Iron(II)	Fe^{+2}		
Silver	Ag^{+1}	Lead(II)	Pb^{+2}		
		Copper(II)	Cu^{+2}		

−1		−2		−3	
Fluoride	F^{-1}	Oxide	O^{-2}	Nitride	N^{-3}
Chloride	Cl^{-1}	Sulfide	S^{-2}	Phosphate	$(PO_4)^{-3}$
Hydroxide	$(OH)^{-1}$	Sulfite	$(SO_3)^{-2}$	Arsenate	$(AsO_4)^{-3}$
Nitrite	$(NO_2)^{-1}$	Sulfate	$(SO_4)^{-2}$		
Nitrate	$(NO_3)^{-1}$	Carbonate	$(CO_3)^{-2}$		
Acetate	$(C_2H_3O_2)^{-1}$	Chromate	$(CrO_4)^{-2}$		

TABLE 7   The activity series

METALS	NONMETALS
Lithium	Fluorine
Potassium	Chlorine
Calcium	Bromine
Sodium	Iodine
Magnesium	
Aluminum	
Zinc	
Chromium	
Iron	
Nickel	
Tin	
Lead	
Hydrogen[a]	
Copper	
Mercury	
Silver	
Platinum	
Gold	

[a] Hydrogen is in italic type because the activities of the other elements were calculated relative to hydrogen.

	Acetate	Arsenate	Bromide	Carbonate	Chlorate	Chloride	Chromate	Hydroxide	Iodide	Nitrate	Oxide	Phosphate	Sulfate	Sulfide
Aluminum	W	a	W	—	W	W	—	A	W	W	a	A	W	d
Ammonium	W	W	W	W	W	W	W	W	W	W..	—	W..	W..	W
Barium	W	w	W	w	W	W	A	W	W	W	W	A	a	d
Cadmium	W	A	W	A	W	W	A	A	W	W	A	A	W	A
Calcium	W	w	W	w	W	W	W	W	W	W	W	w	w..	w
Chromium	W	—	W[b]	W	—	I	—	A	W	W	a	w	W[c]	d
Cobalt	W	A	W	A	W	W	A	A	W..	W	A	A	W	A
Copper(II)	W	A	W	—	W	W	—	A	a	W	A	A	W	A
Hydrogen	W	W	W	—	W	W	—	—	W	W	W	W	W	W
Iron(II)	W	A	W	w	W	W	—	A	W	W	A	A	W	A
Iron(III)	W	A	W	—	W	W	A	A	W	W	A	w	w	d
Lead(II)	W	A	W	A	W	W	A	w	w	W	w	A	w	A
Magnesium	W	A	W	w	W	W	W	A	W	W	A	w	W	d
Mercury(I)	w	A	A	A	W	a	w	—	A	W	A	A	w	I
Mercury(II)	W	w	W	—	W	W	w	A	w	W	w	A	d	I
Nickel	W	A	W	w	W	W	A	w	W	W	A	A	W	A
Potassium	W	W	W	W	W	W	W	W	W	W	W	W	W	W
Silver	w	A	a	A	W	a	w	—	I	W	w	A	w	A
Sodium	W	W	W	W	W	W	W	W	W	W	d	W	W	W
Strontium	W	w	W	w	W	W	w	W	W	W	W	A	w	W
Tin(II)	d	—	W	—	W	W	A	A	W	d	A	A	W	A
Tin(IV)	W	—	W	—	—	W	W	w	d	—	A	—	W	A
Zinc	W	A	W	w	W	W	w	A	W	W	w	A	W	A

[a] W = soluble in water; A = insoluble in water, but soluble in acids; w = only slightly soluble in water, but soluble in acids; a = insoluble in water, and only slightly soluble in acids; I = insoluble in both water and acids; d = decomposes in water.
[b] $CrBr_3$.
[c] $Cr_2(SO_4)_3$.
Source: Reprinted in part from *The Handbook of Chemistry and Physics*, Chemical Rubber Company, Cleveland, Ohio., 1977–78.

TEMPERATURE (°C)	PRESSURE (torr)	TEMPERATURE (°C)	PRESSURE (torr)
0	4.58	32	35.66
5	6.54	33	37.73
10	9.21	34	39.90
15	12.79	35	42.18
16	13.63	36	44.56
17	14.53	37	47.07
18	15.48	38	49.69
19	16.48	39	52.44
20	17.54	40	55.32
21	18.65	45	71.88
22	19.83	50	92.51
23	21.07	55	118.04
24	22.38	60	149.38
25	23.76	65	187.54
26	25.21	70	233.7
27	26.74	75	289.1
28	28.35	80	355.1
29	30.04	85	433.6
30	31.82	90	525.8
31	33.70	95	633.9
		100	760.0

Glossary

The number in parentheses that follows each entry indicates the chapter in the text in which the subject is introduced.

Absolute zero (6)   The coldest possible temperature that matter can reach, which is $0°K$ or $-273°C$.

Acetals (10)   A compound of the type

$$R-\underset{\underset{O-R''}{|}}{\overset{\overset{H}{|}}{C}}-O-R',$$

in which two ether groups are attached to the same carbon atom.

Acid (7)   A substance that releases hydrogen ions in solution.

Activation energy (14)   The energy necessary to get the initial substances in a chemical reaction to the point where they may begin to react.

Activator (14)   The term used to describe a cofactor that is a metal ion.

Active site (14)   The place where the substrate attaches to the enzyme during the chemical reaction.

Alcohols (10)   A class of organic compounds with the general formula R—OH.

Aldehydes (10)   A class of organic compounds with the general formula
$$R-\underset{\underset{\text{O}}{\|}}{C}-H.$$

Alkanes (9)   A group of organic hydrocarbons containing only single carbon-carbon bonds.

Alkenes (9)   A group of organic hydrocarbons each of which contains a carbon-carbon double bond in its structure.

Alkynes (9)   A group of organic hydrocarbons each of which contains a carbon-carbon triple bond in its structure.

Alpha-helix (13)   A coiled protein chain in which there are 3.6 amino acids per turn of the helix.

Alpha rays (8)   Helium atoms with their electrons removed.

Amides (10)   A class of organic compounds with the general formula
$$R-\underset{\underset{\text{O}}{\|}}{C}-NH_2.$$

Amines (10)   A class of organic compounds with the general formula R—NH$_2$.

Amino acids (13)   Difunctional organic compounds that contain a carboxyl group at one end of the molecule and an amino group at the other end of the molecule.

Anabolism (18)   The process in which organic molecules such as sugars, amino acids, fatty acids, and glycerol, which have been absorbed by the cells in the body, serve as the building blocks for the synthesis of more complex organic substances such as polysaccharides, lipids, nucleic acids, and proteins.

Anemia (16)   A condition in which the number of red blood cells or the hemoglobin level is too low.

Antibodies (16)   Specific types of proteins produced by the body to counteract certain foreign substances that enter the body.

Anticodon (15)   The groups of three nucleotide sequences which are contained by each *t*RNA molecule.

Antigens (16)   Any substance which when introduced into the body stimulates the production of antibodies.

Apoenzyme (14)   The protein part of the enzyme.

Aromatic hydrocarbons (9)   A series of cyclic hydrocarbons that contain alternating single and double bonds.

Atom (2)   The smallest part of an element that can enter into chemical combinations.

Atomic mass (3)   The mass of an element in relation to the mass of an atom of carbon-12, which has a mass of 12 amu's.

Atomic mass unit (amu) (3)   The unit used to describe the mass of an atom. (1 amu $= 1.66053 \times 10^{-24}$g)

Blood sugar (11)   Glucose (dextrose) found in the bloodstream.

Boiling point (6)   The temperature at which the equilibrium vapor pressure of a liquid equals the atmospheric pressure.

Bowman's capsule (17)   The funnel-shaped part of the nephron.

Buffer solution (7)   A solution prepared by mixing a weak acid and its salt or a weak base and its salt.

Calorie (Appendix B)   A measure of heat energy. The amount of heat needed to raise the temperature of 1 gram of water 1 degree Celsius.

Carbohydrates (12)   Organic compounds which are generally composed of the elements C, H, and O in the ratio $1:2:1$.

Carboxylic acids (10)   A class of organic compounds with the general formula

$$R-\underset{\underset{O}{\|}}{C}-OH.$$

Catabolism (18)   The breaking down of food substances into smaller molecules, usually accompanied by a release of energy from the molecules that are broken down.

Catalysts (14)   Substances that affect the rate at which chemical reactions occur.

Cell membrane (18)   The membrane that encloses the cell. It is made up of a double layer of proteins and lipids.

Celsius temperature scale (Appendix B)   A temperature scale on which the freezing point of water is $0°C$ and the normal boiling point of water is $100°C$.

Chain reaction (8)   A self-sustaining nuclear reaction.

Chemical formula (2)   The combination of symbols of particular elements that form a chemical compound.

Cholesterol (12)   A sterol which has been linked to the development of atherosclerosis and high blood pressure.

Chromosomes (15)   Substances in the cells of the body that hold the genetic information that we inherit from our parents.

Citric acid cycle (18)   A series of biochemical reactions that take place in the body to produce large amounts of energy.

Codon (15)   The *m*RNA genetic code, which is written as groups of three nucleotides.

Coenzyme (14)   A cofactor which is a complex organic molecule.

Cofactor (14)   The nonprotein part of a conjugated enzyme.

Colloidal dispersion (7)   A type of solution in which the solute particles don't dissolve to the point that they do in a true solution.

Combination reactions (5)   Reactions in which two or more substances combine to form a more complex substance.

Competitive inhibition (of enzymes) (15)   The type of inhibition where the substrate and inhibitor compete for the same active site on the enzyme.

Conjugated enzyme (14)   An enzyme composed of a protein part and a nonprotein part.

Covalent bond (4)   A bond formed by the sharing of electrons between two atoms.

Crenation (7)   Cell shrinkage due to water moving out of the cell.

Curie (8)   A measure of the activity of a radiation source. (1 Ci $= 3.7 \times 10^{10}$ disintegrations per second.)

Cytochrome system (18)   An enzyme system which uses Cu^{+2} ions and Fe^{+3} ions and is part of the respiratory chain.

Cytoplasm (15)   The watery fluid that surrounds the nucleus of the cell.

Deamination (18)   The removal of an NH_2 group from an amino acid.

Decomposition reactions (5)   Reactions that involve the breakdown of a complex substance into simpler substances.

Dehydrating agent (12)   A substance that removes water from a molecule.

Dehydration synthesis (11)   The reaction by which a disaccharide is formed from two monosaccharides by means of the loss of a water molecule.

Denaturation (of proteins) (13)   An alteration in the secondary, tertiary, or quaternary structure of the protein.

Density (Appendix B)   The mass per unit volume of a substance.

Deoxyribonucleic acid (DNA) (15)   The genetic material found in the nucleus of each cell, which contains all the information needed for the development of that individual.

Deuterium (3)   An isotope of hydrogen which has a mass of 2 amu's.

Diabetes insipidus (17)   A disease in which abnormally high amounts of urine are excreted each day.

Diabetes mellitus (11)   A disease in which there is a higher-than-normal blood sugar level (hyperglycemic condition), usually caused by insufficient insulin being produced by the pancreas.

Dialysis (7)   A means by which particles of a solution can be separated from the particles of a colloid through the use of a membrane.

Diatomic elements (4)   Elements that are found naturally as molecules with two atoms apiece, for example, H_2, O_2, and N_2.

Diffusion (7)   The movement of a substance in solution from an area of high concentration to an area of low concentration.

Disaccharide (11)   A carbohydrate containing two sugar groups in the molecule.

Dosimeter (8)   A device used to measure the amount of radiation an individual receives.

Double bond (4)   A covalent bond in which each atom donates two electrons to form the bond.

Double helix (15)   A secondary protein structure found, for example, in DNA.

Double replacement reaction (5)   Reactions in which two compounds exchange ions with each other.

Electrolytic solution (7)   A solution that conducts electric current.

Electron (3)   A particle with a relative negative charge of one unit and a mass of 0.0005486 amu.

Electron configuration (3)   The positions of the electrons in the various energy levels of an atom.

Electronegativity (4)   The attraction that an element has for shared electrons in a molecule.

Elements (2)   The basic building blocks of matter.

Emulsifying agent (12)   Substances used to hold fatty or oily substances in suspension.

Endoplasmic reticulum (18)   A branched interconnected system of membranes which extends from the nucleus of the cell to the cell membrane.

Endothermic reaction (Appendix B)   A reaction that absorbs energy from the surroundings.

Energy (2)   The ability to do work, Energy appears in many forms, for example, heat, chemical, electrical, mechanical, and radiant energy.

Enthalpy (Appendix B)   The heat content of a chemical substance, represented by the symbol H.

Enzymes (14)   A group of proteins that regulate biological reactions in our bodies.

Equilibrium vapor pressure (6)   The pressure exerted by a vapor when it is in equilibrium with its liquid at any temperature.

Erg (8)   A unit of work in the metric system.

Esters (10)   A class of organic compounds with the general formula

$$R—\overset{\displaystyle O}{\underset{\displaystyle \|}{C}}—O—R'.$$

Ethers (10)   A class of organic compounds with the general formula R—O—R'.

Exothermic reaction (Appendix B)   A reaction that releases energy to the surroundings.

Extracellular fluid (16)   Fluid found outside the cell.

FAD (flavin adenine dinucleotide) (18)   One of the primary oxidizing agents in cells.

Fahrenheit temperature scale (Appendix B)   A temperature scale on which the freezing point of water is $32°F$ and the normal boiling point of water is $212°F$.

Fats (12)   A type of lipid composed of fatty acids and glycerol.

Fat soluble vitamins (12)   Vitamins A, E, K, and D. They are stored and carried by fats in the body.

Fatty acids (12)   Organic compounds composed of a long chain of carbon atoms connected to a carboxyl group.

Fermentation (11)   The process in which simple sugars are treated with yeast in the presence of bacteria and turned into ethanol and carbon dioxide.

Fibrous proteins (13)   Proteins that have a protective or structural function in the body. They are water-insoluble.

Fission (nuclear) (8)   The splitting of an atom into two or more different atoms when bombarded with neutrons.

Formed elements (in blood) (16)   The red blood cells, white blood cells, and platelets.

Freezing point (6)   The temperature at which the solid and liquid phase of a substance exist together.

Functional group (10)   The reactive part of an organic molecule.

Gamma globulins (16)   Blood plasma proteins which provide immunity against disease.

Gamma rays (8)   A high-energy form of electromagnetic radiation given off by many radioactive substances.

Gaucher's disease (12)   A hereditary disorder in which glucose is substituted for galactose i glycolipid molecules.

Genes (15)   Substances found on the chromosomes which carry the information for each specific characteristic that a person has.

Genetic code (15)   The sequence of bases along the DNA nucleotide chain.

Genetic diseases (15)   Hereditary diseases caused by an incorrect sequence of bases along the DNA nucleotide chain.

Globular proteins (13)   Those proteins which are water-soluble and carry out the chemical processes of the body.

Glycerides (12)   Fats composed of long chains of fatty acids bonded to glycerol.

Glycolysis (18)   A process that involves the breakdown of glucose by anaerobic catabolism.

Glycosuria (11)   A condition in which sugar appears in the urine.

Golgi complex (18)   A part of the cell. It is a series of flattened membranes whose purpose is to store proteins, lipids, and carbohydrates and secrete them when necessary.

Gram (Appendix B)   The metric unit of mass.

Half-life (8)   The time required for half the atoms originally present in a radioactive sample to decay.

Heat of reaction (Appendix B)   The change in heat content of a substance during a chemical reaction.

Hemiacetal (10)   An organic compound having an ether group and alcohol group off the same carbon atom. The general formula is

$$
\begin{array}{c}
H \\
| \\
R-C-O-R'. \\
| \\
O-H
\end{array}
$$

Hemoglobin (13)   The oxygen-carrying protein of the blood.

Hemolysis (7)   The expansion of a cell due to water flowing into it.

Hemophilias (16)   Bleeding disorders due to inherited deficiencies or abnormalities of coagulation factors in the blood.

Hepatitis (14)   A disease which can cause liver damage.

Heterocyclic compounds (11)   A group of cyclic organic compounds composed of carbon plus an additional element in the ring.

Heterogeneous (2)   Having different parts with different properties.

Holoenzyme (14)   The whole enzyme; apoenzyme plus cofactor.

Homogeneous (2)   Having similar consistency throughout.

Homologous series (9)   A series of compounds of the same chemical type which differ only by fixed increments of the constituent elements.

Hydrocarbons (9)   Organic compounds composed only of carbon and hydrogen.

Hydrostatic pressure (16)   A pushing force, for example, when blood is forced against the walls of the blood vessels.

Hyperglycemia (11)   A condition in which there is a higher-than-normal blood sugar level.

Hypertension (12)   High blood pressure.

Hypertonic solution (7)   A solution whose osmotic pressure is higher than that of the cells.

Hypoglycemia (11)   A condition in which there is a low blood sugar level.

Hypotonic solution (7)   A solution whose osmotic pressure is lower than that of the cells.

Immunoglobin (16)   Proteins in the body that protect against disease.

Insulin (11)   The hormone that controls blood sugar level in the body.

Insulin shock (11)   The overproduction or overinjection of insulin, which can lead to very low blood sugar levels, causing convulsions or coma.

Interstitial fluid (16)   The fluid found between the cells of the body.

Intracellular fluid (16)   The fluid found within the cells of the body.

Iodine number (12)   The number of grams of iodine absorbed by 100 g of fat or oil.

Ionic bond (4)   A bound formed by the transfer of electrons from one atom to another.

Ionization (7)   The breakup into ions of some substances when they dissolve in solution.

Ions (4)   Atoms that have gained or lost electrons and therefore have a positive or negative charge.

Isoenzyme (14)   Isomeric forms of a particular enzyme.

Isomers (9)   Compounds with the same molecular formula but different structural formulas.

Isotonic solution (7)   A solution whose osmotic pressure is the same as that of the cells.

Isotopes (3)   Atoms that have the same number of protons and electrons but different numbers of neutrons.

Kelvin temperature scale (6)   Degrees Celsius + 273.

Ketones (10)   A class of organic compounds with the general formula

$$R\!-\!\underset{\underset{\textstyle O}{\|}}{C}\!-\!R'.$$

Lecithins (12)   Phospholipids composed of glycerol, fatty acids, phosphoric acid, and choline.

Leukemia (16)   Cancer of the bone marrow in which excessive white blood cells are produced.

Lewis dot structure (4)   A notation that shows the symbol of the element and the number of outer electrons in an atom of the element.

Lipids (12)   Biochemical substances found in plant and animal tissues. They include fats and other compounds that resemble fats in physical properties.

Liter (Appendix B)   The basic unit of volume in the metric system.

Mass number (3)   The relative weight of an isotope of an element equal to the sum of its protons and neutrons.

Matter (2)   Anything that occupies space and has mass.

Melting point (6)   The temperature at which the solid and liquid phases of a substance exist together.

Messenger-RNA (mRNA) (15)   The substance that directs the formation of proteins in the cells.

Metabolic acidosis (7)   An abnormally high concentration of hydrogen ions due to the ingestion of an acidic substance, kidney failure, or diabetes mellitus.

Metabolic alkalosis (7)   An abnormally high hydroxide ion concentration due to some internal disorder or ingestion of an alkaline substance.

Metabolism (11)   The process by which the food we eat is transformed in the body, by a series of complex chemical reactions, to provide material for cellular function and energy to power the body.

Metals (2)   A class of elements that conduct electricity and heat, have luster, and take on a positive oxidation number when they bond.

Metalloids (2)   A class of elements that have some of the properties of metals and some of the properties of nonmetals.

Meter (Appendix B)   The basic unit of length in the metric system.

Mitochondria (18)   The structures that are found in the cytoplasm of cells. Their function is to produce energy that can be used by the cell to carry out the various cell processes.

Mitosis (15)   The process where one cell divides and becomes two.

Mixture (2)   A material that consists of two or more substances, with each substance retaining its own characteristic properties.

Molarity (7)   A concentration unit for solutions: moles of solute per liter of solution.

Mole (2) 6.023×10^{23} items.

Molecule (2)   The smallest part of a compound that can enter into chemical combinations.

Monosaccharide (11)   A carbohydrate containing one sugar group.

Mutation (15)   A change in the sequence of nucleotides in a DNA molecule.

NAD (nicotinamide adenine dinucleotide) (18)   One of the primary oxidizing agents in cells.

Niemann-Pick disease (12)   An inherited disorder in which sphingomyelins build up in the brain, spleen, and liver, resulting in mental retardation and early death.

Nephron (17)   The functional unit of the kidney.

Neutralization reactions (5)   Reactions in which an acid and base react to form a salt and water.

Neutron (3)   A particle with zero charge and a mass of 1.008665 amu.

Nitrogen bases (of DNA and RNA) (15)   Cyclic compounds containing both nitrogen and carbon in the ring. There are five nitrogen-containing bases found in DNA and RNA: adenine, guanine, cytosine, thymine, and uracil.

Noncompetitive inhibition (14)   A substance that can deactivate an enzyme by binding to it in a place other than its active site.

Nonmetals (2)   A class of elements that are poor conductors of heat and electricity, and usually take on a negative oxidation number when they bond with metals.

Nuclear fission (8)   The splitting of an atom into two or more different atoms when it is bombarded by neutrons.

Nuclear fusion (8)   The combination of two atoms to form a new, heavier atom. The process releases huge amounts of energy.

Nuclear radiation (8)   Alpha, beta, and gamma rays emitted from the nuclei of certain atoms.

Nucleic acids (15)   The major components of chromosomes.

Nucleotides (15)   The subunits of nucleic acids.

Octet rule (4)   When there are eight electrons in the outermost energy level of an atom, that atom tends to be nonreactive.

Organic compounds (9) Compounds that contain carbon, along with such elements as hydrogen, oxygen, nitrogen, sulfur, and the group VIIA elements.

Osmosis (7) The passage of water through a semipermeable membrane.

Osmotic pressure (7) The amount of pressure that must be applied to prevent the flow of water through a membrane.

Oxidation (5) The loss of electrons by a substance undergoing a chemical reaction.

Oxidation number (4) The combining capacity of an element.

Oxidizing agent (5) A substance that causes something else to be oxidized.

Pellagra (14) A condition resulting from insufficient niacin in the diet. The individual feels weak and suffers from inflammation of the skin, mouth, and tongue.

Peptide bond (13) The bond that holds amino acids together in a protein molecule.

pH (7) The negative logarithm of the hydrogen ion concentration, in moles per liter.

pH scale (7) The scale that expresses the acid–base strength of a solution.

Phospholipids (12) A class of lipids that contain the element phosphorus.

Phosphorylation (18) The process that occurs when a glucose molecule is joined to a phosphate group to become glucose-6-phosphate.

Photosynthesis (11) A chemical reaction involving carbon dioxide (from air) and water (from soil) in the presence of sunlight, to produce a simple sugar (a carbohydrate) and oxygen.

Plasma (blood) (16) The fluid part of the blood.

Platelets (16) One of the formed elements in the blood. They contain a chemical that initiates the clotting process.

Polyatomic ions (4) A charged group of covalently bonded atoms.

Polypeptides (13) A molecule consisting of three or more amino acids bonded together.

Polysaccharides (11) Carbohydrates containing many sugar groups per molecule.

Polyunsaturated fats (12) Those fats that contain a large number of fatty acids having double bonds.

Prostaglandins (12) Cyclic fatty acids containing 20 carbon atoms that form a five-membered ring in the center of the molecule. They have important physiological properties in the body.

Prosthetic group (14)   A cofactor that is tightly bound to the protein part of a conjugated enzyme.

Protein (13)   Important biological molecules composed of amino acids linked together.

Proton (3)   A particle with a relative positive charge of 1 unit and a mass of 1.0072766 amu.

Purines (15)   The double-ring nitrogen-containing bases adenine and guanine.

Pyrimidines (15)   The single-ring nitrogen-containing bases cytosine, thymine, and uracil.

R group (9)   Any number of carbon atoms attached to the reactive part of an organic molecule.

Rad (radiation absorbed dose) (8)   The amount of radiation absorbed by a living tissue, regardless of the type of radiation.

Radioactive decay (8)   The energy released by the nucleus of an atom as it undergoes nuclear decay. It usually involves the release of alpha, beta, or gamma radiation.

Radioisotope (8)   An isotope of an element that emits nuclear radiation.

Reactants (5)   The starting materials in a chemical reaction.

Red blood cell (16)   One of the formed elements in blood responsible for carrying oxygen throughout the body.

Reducing agent (5)   A substance that causes something else to be reduced.

Reducing sugar (11)   Small carbohydrate molecules that have a free aldehyde group in solution that can act as reducing agents.

Reduction (5)   The gain of electrons by a substance undergoing a chemical reaction.

Rem (röentgen equivalent man) (8)   A weighted unit of radiation.

Renal pelvis (17)   The innermost region of the kidney.

Representative elements (3)   The A-group elements in the periodic table.

Rh antigen (16)   A protein attached to the RBC membrane.

Ribonucleic acid (RNA) (15)   One of the two major nucleic acids.

Ribosome (15)   The place in the cell where protein synthesis occurs.

Rickets (12)   A condition caused by a lack of vitamin D, which causes improper development of bones and teeth.

Röentgen (8)   A measure of the ionizing ability of x rays and gamma rays.

Saccharide group (11)   A simple sugar molecule, for example $C_6H_{12}O_6$.

Saline solution (7)   A sodium chloride in water solution.

Saturated hydrocarbon (9)   When all the carbon-carbon bonds in the hydrocarbon are single bonds.

Saturated solution (7)   A solution in which no more solute can dissolve.

Scintillation counter (8)   A device that counts and determines the energy of a radioactive source.

Scurvy (14)   A deficiency disease caused by a lack of vitamin C. The symptoms of the disease are loose teeth and bleeding gums.

Semipermeable membrane (7)   A substance like cellophane or cell walls that allows the passage of certain substances through it.

Sickle cell anemia (15)   A hereditary disease in which the RBC's are sickle-shaped.

Simple enzymes (14)   Enzymes composed entirely of proteins.

Solid (6)   One of the states of matter. Solids have definite shape and definite volume.

Solute (7)   The substance that is being dissolved in a solution.

Solution (2)   A homogeneous mixture of two or more substances.

Solvent (7)   The substance that is doing the dissolving in a solution.

Specific gravity (16)   The mass of a substance divided by the mass of an equal volume of water.

Sphingomyelins (12)   A type of phospholipid found in the brain and nervous tissue.

Steroids (12)   Physiologically active molecules having the following basic design:

Sterols (12)   Steroid alcohols, for example, cholesterol.

Substrate (14)   The substance upon which an enzyme works.

Suspension (7)   A mixture of a liquid and solid in which the solid settles to the bottom of the liquid if the mixture is not constantly shaken.

Tay-Sachs disease (12)   A hereditary disorder in which infants can't break down glycolipids because they lack a specific enzyme.

Temperature (Appendix B)   A measure of the intensity of heat.

Terpenes (12)   Polymers of isoprene, for example, vitamins A, E, and K.

Torr (6)   A unit of pressure. One torr is the amount of pressure necessary to raise the level of mercury in a barometer 1 mm.

Toxin (16)   Any of the various organic poisons produced in living or dead organisms, or any specific poisonous substance generated by pathogenic microorganisms which acts as the causative agent in various diseases, such as tetanus or diphtheria.

Toxoid (16)   A toxin that has been treated with heat or chemicals so that it is no longer toxic but still acts as an antigen for antibody production.

Transcription (15)   The process where mRNA copies the information on the DNA strand.

Transfer-RNA (tRNA) (15)   The RNA that escorts amino acids from the cytoplasm to the ribosomes for protein synthesis.

Transition elements (3)   The B-group elements in the periodic table.

Triplet (code) (15)   The sequence of three nucleotides along the DNA chain.

Tyndall effect (7)   The cloudiness seen in a colloidal dispersion when a beam of light passes through it.

Unit cell (6)   A repetitive unit in a crystal lattice.

Universal donor (16)   A person with type O blood.

Universal recipient (16)   A person with type AB blood.

Ureters (17)   The tube in the kidney which brings urine from the renal pelvis to the urinary bladder.

Urethra (17)   The tube that leads urine from the bladder to outside the body.

Valence (4)   The combining capacity of an element when it enters into chemical combination.

White blood cells (16)   One of the formed elements in the blood. They are responsible for fighting infections.

X rays (8)   Electromagnetic radiation of extremely short wavelength.

Zwitterion (13)   The dipolar form of an amino acid:

$$\overset{+}{H_3N}-\overset{\overset{\displaystyle R}{|}}{\underset{\underset{\displaystyle H}{|}}{C}}-\overset{\overset{\displaystyle}{}}{\underset{\underset{\displaystyle O}{\|}}{C}}-O^-$$

Zymogen (14)   The inactive form of an enzyme.

Answers
to Selected Exercises

Chapter 2

2. (a) mixture   (b) mixture   (c) element   (d) element   (e) mixture
   (f) compound   (g) mixture   (h) compound   (i) element   (j) mixture
3. (a) atom   (b) molecule
5. (a) The atomic weight of bromine would be 2.
   (b) The atomic weight of neon would be $\frac{1}{2}$.
6. 412,600 years
7. (a) 160   (b) 54   (c) 98   (d) 174   (e) 149   (f) 132
8. (a) 46 g   (b) 0.46 g   (c) 23 g   (d) 8.0 g
9. (a) 2.0 moles   (b) 0.20 mole   (c) $1\bar{0}$ moles   (d) 15 moles
10. 0.0100 mole
11. 0.050 mole

Chapter 3.

1. protons, electrons and neutrons
3. (a) K, L, M, N, O, P, and Q (b) 72 electrons
5. (a) 92 p, 92 e, 143 n   (b) 48 p, 48 e, 71 n
   (c) 38 p, 38 e, 52 n   (d) 66 p, 66 e, 94 n
6. (a) K has two, L has eight and M has two electrons.
   (b) K has two and L has six electrons.
   (c) K has two, L has eight, and M has eight electrons.

7. (a) Ca has two electrons in its outermost energy level.
   (b) Cs has one electron in its outermost energy level.
   (c) I has seven electrons in its outermost energy level.
   (d) Kr has eight electrons in its outermost energy level.
   (e) Sn has four electrons in its outermost energy level.
9. Element 111 would be similar to Cu, Ag, and Au.
10. (a) Ne   (b) K
11. (a) $^{24}_{12}Mg$ (b) $^{81}_{35}Br$ (c) $^{346}_{97}Bk$ (d) $^{9}_{4}Be$

Chapter 4

2. (a) $:\ddot{\underset{..}{C}}l:^{-1}$ (b) $:\ddot{\underset{..}{S}}:^{-2}$ (c) $:\ddot{\underset{..}{P}}:^{-3}$

4. (a) $:\ddot{\underset{..}{B}}r:\ddot{\underset{..}{B}}r:$ (b) $H:\ddot{\underset{..}{B}}r:$ (c) $H:\underset{H}{\overset{}{C}}::\underset{H}{\overset{}{C}}:H$

   (d) $H:\underset{H}{\overset{..}{P}}:H$ (e) $:\ddot{\underset{..}{C}}l:C::C:\ddot{\underset{..}{C}}l:$

6. octet, two
7. (a) double   (b) triple   (c) single
8. two, one
9. (a) covalent   (b) ionic   (c) ionic   (d) ionic   (e) covalent
10. H_2Te, H_2Se, H_2S, and H_2O is most polar.
12. (a) Na_3AsO_4 (b) K_2SO_3 (c) $BaCl_2$ (d) $Fe(NO_3)_2$
    (e) $CuSO_4$ (f) Al_2S_3 (g) Rb_2O (h) $Co(NO_2)_2$
    (i) CCl_4 (j) SO_2 (k) $HC_2H_3O_2$ (l) PCl_5
13. (a) silver bromide   (b) copper(II) sulfide   (c) iron(II) nitrate
    (d) iron(III) nitrate   (e) calcium hydroxide   (f) aluminum sulfite
    (g) zinc phosphide   (h) copper(II) chromate   (i) mercury(II) phosphate
    (j) cobalt(II) chloride   (k) osmium(IV) oxide   (l) copper(I) nitride
14. (a) $+2$ (b) $+8$ (c) $+5$ (d) $+6$
15. (a) polar, polar   (b) polar, polar   (c) nonpolar, nonpolar   (d) polar, polar

Chapter 5

1. (a) $2K + 2H_2O \longrightarrow 2KOH + H_2$
   (b) $Mg + Cu(NO_3)_2 \longrightarrow Mg(NO_3)_2 + Cu$
   (c) $3H_2SO_4 + 2Fe(OH)_3 \longrightarrow Fe_2(SO_4)_3 + 6H_2O$
   (d) $2Na + Br_2 \longrightarrow 2NaBr$
   (e) $2Al + 3Hg(C_2H_3O_2)_2 \longrightarrow 2Al(C_2H_3O_2)_3 + 3Hg$
2. (a) single replacement   (b) single replacement   (c) double replacement
   (d) combination   (e) single replacement
3. (a) $2Mg + O_2 \longrightarrow 2MgO$ (c) $SO_3 + H_2O \longrightarrow H_2SO_4$
4. (b) $2HgO \longrightarrow 2Hg + O_2$ (d) $2NaCl \longrightarrow 2Na + Cl_2$
5. (b) no reaction   (d) $3Mg + 2H_3PO_4 \longrightarrow Mg(PO_4)_2 + 3H_2$
6. (a) $2HBr + Mg(OH)_2 \longrightarrow MgBr_2 + 2H_2O$ (c) no reaction
7. (a) Magnesium is oxidized and hydrogen is reduced.
   (b) Sodium is oxidized and hydrogen is reduced.

(c) Mercury is oxidized and oxygen is reduced.

(d) Aluminum is oxidized and iron is reduced.

9. (a) single replacement   (b) single replacement
    (c) combination   (d) single replacement

10. (a) $2K + 2H_2O \longrightarrow 2KOH + H_2$

    (c) $2LiI \longrightarrow 2Li + I_2$

    (e) $3CuCl_2 + 2Na_3PO_4 \longrightarrow 6NaCl + Cu_3(PO_4)_2$

    (g) $2H_2SO_3 + O_2 \longrightarrow 2H_2SO_4$

Chapter 6

4. molecular solid
5. (a) increase   (b) decrease
6. The melting point of a substance is not affected very much by changes in atmospheric pressure.
10. decrease
11. increases
12. Henry's law
13. Boyle's law
14. (a) 0.500 atm   (b) 0.200 atm   (c) 2.50 atm   (d) 6.00 atm
15. (a) 1140 torr   (b) 76 torr   (c) 570 torr   (d) 2660 torr
16. (a) $298°K$ (b) $310°K$ (c) $248°K$ (d) $0°K$
17. (a) $100°C$ (b) $27°C$ (c) $-273°C$ (d) $-173°C$
18. $V_f = 57,528$ liters (or about 57,500 liters to three significant figures)
19. $V_f = 217$ ml
20. $V_f = 144$ ml
21. partial pressure of $CO_2 = 38$ torr; partial pressure of $O_2 = 722$ torr

Chapter 7

2. (a) suspension   (b) true solution
3. (a) 12.5%   (b) 30%   (c) 20%
4. 75 g of KCl
5. (a) $2.00\ M$ (b) $0.400\ M$ (c) $1.0\ M$
6. 88.8 g (or about 89 g to two significant figures)
7. 2.0 liters
8. $2\overline{0}0$ mg
12. $H_3O^+ + Cl^-$
13. (a) $HCl + NaOH \longrightarrow NaCl + H_2O$
    (c) $2HC_2H_3O_2 + Ca(OH)_2 \longrightarrow Ca(C_2H_3O_2)_2 + 2H_2O$
17. (a) $10^{-8}\ M$ (b) $10^{-5}\ M$ (c) $10^{-13}\ M$
18. (a) 6   (b) 9   (c) 1
19. (a) 0   (b) 2   (c) 12   (d) 10
20. (a) left   (b) right   (c) right
22. acidosis
23. alkalosis
25. (a) isotonic   (b) hypertonic   (c) hypotonic   (d) hypertonic
26. (a) no effect   (b) crenation   (c) hemolysis   (d) crenation

Chapter 8

2. (a) $^4_2$He   (b) $_{-1}^{\ 0}e$
5. (a) $_{-1}^{\ 0}e + {}^{14}_{7}$N   (b) $_{-1}^{\ 0}e + {}^{24}_{12}$Mg   (c) $^4_2$He $+ {}^{234}_{90}$Th
   (d) $^1_0 n + {}^{86}_{36}$Kr   (e) $^4_2$He $+ {}^{208}_{82}$Pb
6. 12.5 mg
7. 112 years old
11. (a) stage 1   (b) stage 2   (c) stage 1
12. bone marrow and reproductive organs
15. 1.1×10^{13} dps
16. 400 rads
17. 0.0004 rad

Chapter 9

3. (a) The corners of the tetrahedrons touch.
    (b) There is a common edge between the two tetrahedrons.
    (c) There is a common side between the two tetrahedrons.
8. (a) *n*-heptane   (b) 2-methylhexane
   (c) 3-methylhexane   (d) 2,3-dimethylpentane
   (e) 2,4-dimethylpentane   (f) 2,2-dimethylpentane
   (g) 3,3-dimethylpentane   (h) 2,2,3-trimethylbutane
   (i) 2-methyl-2-ethylbutane
10. (a) 2,4-dimethylnonane   (b) 2,2-dimethylpropane
   (c) 4,4-dimethyl-3-propylhexane
11. (a) 2-pentene   (b) 2-pentene   (c) 1-pentene   (d) 2,5-dimethyl-3-heptene
12. (a) 2-ethyl-3-hexyne   (b) 4-ethyl-2-hexyne   (c) 5,6-dimethyl-4-ethyl-2-nonyne
13. (a) cycloheptane   (b) cyclooctyne   (c) 1,4,6-trimethylcyclooctane
   (d) cyclobutene   (e) 1,4-dimethylbenzene (*p*-xylene)

14. (a) 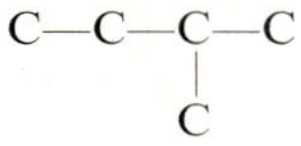 C—C—C—C   (c)
          | C
          C |
                           C—C—C
                              |
                              C

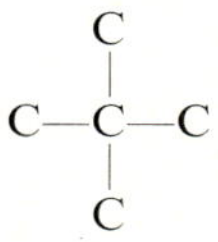

   (e) C—C=C—C—C—C   (g) 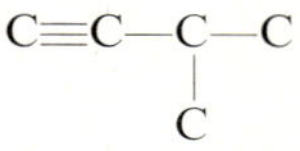 C≡C—C—C
                                          |
                                          C

15. (a) △   (c) 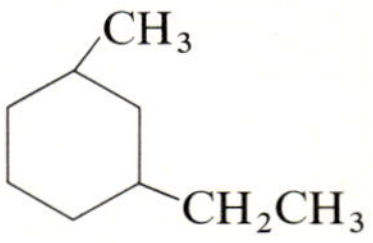 (cyclohexane ring with CH_3 and CH_2CH_3 substituents)

   (e) 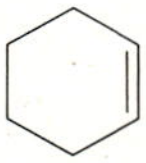 (cyclohexene)   (g) 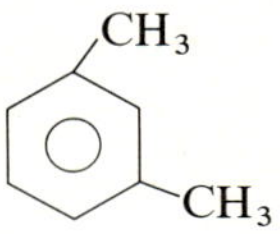 (dimethylbenzene ring with two CH_3 substituents)

Chapter 10

2. (a) 1-butanol or *n*-butyl alcohol   (b) 3-hexanol   (c) cycloheptanol
   (d) 2-methyl-1-propanol   (e) 1-ethyl-1-cyclohexanol

3. (a) primary   (b) secondary   (c) secondary   (d) primary   (e) tertiary

4. (a) dimethyl ether or methoxymethane
   (c) methyl phenyl ether or methoxybenzene

5. (a) $2CH_3CH_2\text{—}OH \xrightarrow{H_2SO_4} CH_3CH_2\text{—}O\text{—}CH_2CH_3 + H_2O$

7. (a) hexanal or hexaldehyde (or caproaldehyde)   (c) 3-methylpentanal

8. (a) $CH_3\text{—}\underset{\underset{O}{\parallel}}{C}H + H_2 \xrightarrow{Pt} CH_3CH_2OH$ (reduction)

9. (a) 3-pentanone or diethyl ketone   (c) 3-octanone

10. (a) $CH_3CH_2\text{—}\underset{\underset{O}{\parallel}}{C}\text{—}CH_3 \xrightarrow{Pt,\ H_2} CH_3CH_2\text{—}\underset{\underset{OH}{|}}{C}H\text{—}CH_3$

11. (a) hexanoic acid   (c) 3-methylpentanoic acid

12. (a) $CH_3CH_2\text{—}\underset{\underset{O}{\parallel}}{C}\text{—}OH + HO\text{—}CH_2CH_3 \xrightarrow{H_2SO_4}$

$CH_3CH_2\text{—}\underset{\underset{O}{\parallel}}{C}\text{—}O\text{—}CH_2CH_3 + H_2O$

13. (a) methylpentanoate   (c) phenylpropanoate or phenyl propionate

14. (a) 3-aminohexane   (c) 1-aminocyclohexane or cyclohexylamine

15. (a) $CH_3\text{—}\underset{\underset{O}{\parallel}}{C}\text{—}OH + NH_3 \longrightarrow CH_3\text{—}\underset{\underset{O}{\parallel}}{C}\text{—}NH_2 + H_2O$

16. (a) *N*-methyl-*N*-ethylpentanamide
    (b) *N*-phenylpropanamide or *N*-phenylpropionamide

17. (a) primary   (b) tertiary   (c) primary   (d) secondary

18. (a) $CH_3CH_2\text{—}\underset{\underset{OH}{|}}{C}H\text{—}CH_2CH_3$ (b) 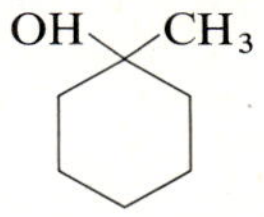 (cyclohexane with OH and CH₃)

(e) $CH_3CH_2CH_2\text{—}\underset{\underset{CH_3}{|}}{C}H\text{—}\underset{\underset{O}{\parallel}}{C}\text{—}H$ (f) 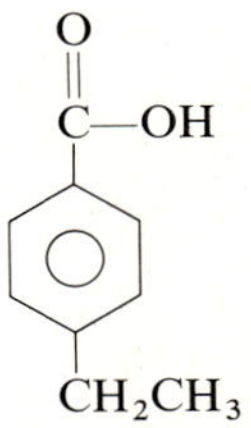 (benzene ring with C—OH and CH₂CH₃)

(i) $CH_3\text{—}\underset{\underset{O}{\parallel}}{C}\text{—}O\text{—}CH_2CH_2CH_3$ (j) $CH_3\text{—}\underset{\underset{CH_3}{|}}{N}\text{—}CH_3$

Chapter 11

2. 4 kcal/g
4. structures that support the plant

7. (a) monosaccharides contain one sugar group per molecule
   (b) disaccharides contain two sugar groups per molecule
   (c) polysaccharides contain many sugar groups per molecule
8. (a) two simple sugars plus water   (b) many simple sugars plus water
9. (a) monosaccharides form disaccharides   (b) disaccharides form polysaccharides
14. glucose, galactose, and fructose
17. (a) Using Fischer formulas: in the alpha form of glucose, the OH group on
       carbon 1 is written on the right-hand side. In the beta form, the OH group
       on carbon 1 is written on the left-hand side.
   (b) Using Haworth formulas: in the alpha form of glucose, the OH group on
       carbon 1 is written below the plane of the ring. In the beta form, the OH
       group on carbon 1 is written above the plane of the ring.
19. 70 to 100 mg of glucose per 100 ml of blood
23. (a) Lactose is formed from glucose and galactose.
   (b) Maltose is formed from two glucose molecules.
   (c) Sucrose is formed from fructose and glucose.
26. (a) aldopentose   (b) ketotriose   (c) ketopentose   (d) aldotriose
27. (a) beta form   (b) alpha form
28. (a) alpha form   (b) beta form

Chapter 12

3. (a) fat   (b) glycolipid   (c) wax   (d) lecithin   (e) sphingomyelin
4. (a) $C_nH_{2n+1}COOH$ (b) $C_nH_{2n-1}COOH$ (c) $C_nH_{2n-3}COOH$
   (d) $C_nH_{2n-5}COOH$ (e) $C_nH_{2n-7}COOH$
5. unsaturated
6. (a) unsaturated, liquid, arachidonic
   (b) saturated, solid, saturated fatty acid series
   (c) unsaturated, liquid, oleic
7. linoleic, linolenic, arachidonic
10. (a) diglyceride   (b) triglyceride   (c) monoglyceride
12. The matching pairs are as follows:
   (a) 3   (b) 2   (c) 5   (d) 1   (e) 4
13. (a) steroid   (b) terpene   (c) terpene   (d) steroid
14. unsaturated
16. The matching pairs are as follows:
   (a) 4   (b) 1   (c) 2   (d) 3
18. saturated

Chapter 13

3. (a) H   (b) CH_3
4. lysine, tryptophan, phenylalanine, methionine, threonine, leucine, isoleucine, and
   valine

8. (a) alanine

$$^{+}H_3N\!-\!\underset{\underset{\displaystyle H}{|}}{\overset{\overset{\displaystyle CH_3}{|}}{C}}\!-\!\underset{\underset{\displaystyle O}{\|}}{C}\!-\!O^{-}$$

10. Tyr-Cys-Gly-Ser

11. Met-Thr-Asn   Asn-Thr-Met   Thr-Met-Asn   Asn-Met-Thr
     Thr-Asn-Met   Met-Asn-Thr
13. (a) primary   (b) tertiary   (c) secondary   (d) quaternary
14. (a) breaks its tertiary structure
     (b) destroys protein's secondary and tertiary structure
     (c) causes protein to coagulate

Chapter 14

2. (a) conjugated   (b) conjugated   (c) simple
3. (a) sucrase   (b) maltase
4. The matching pairs are as follows:
     (a) 4   (b) 1   (c) 3   (d) 5   (e) 2
6. The matching pairs are as follows:
     (a) 7   (b) 4   (c) 1   (d) 3   (e) 5   (f) 2   (g) 6
7. A catalyst lowers the activation energy of reaction.
8. (a) decrease   (b) increase   (c) decrease   (d) decrease
9. forward and reverse
12. The matching pairs are as follows:
     (a) 2   (b) 3   (c) 1
16. water-soluble and fat-soluble
17. (a) hypovitaminosis   (b) avitaminosis   (c) hypervitaminosis
18. coenzymes
19. The matching pairs are as follows:
     (a) 3   (b) 1   (c) 5   (d) 2   (e) 4

Chapter 15

1. DNA and RNA
2. The matching pairs are as follows:
     (a) 2   (b) 1   (c) 3
4. ribose found in RNA and deoxyribose found in DNA
5. double, single
6. purines, pyrimidines
9. (a) thymine-adenine   (b) cytosine-guanine
     (c) guanine-cytosine   (d) adenine-thymine
10. (a) uracil-adenine   (b) cytosine-guanine
     (c) guanine-cytosine   (d) adenine-uracil
12. The matching pairs are as follows:
     (a) 4   (b) 1   (c) 2   (d) 3
13. (a) adenine-uracil   (b) guanine-cytosine
     (c) cytosine-guanine   (d) thymine-adenine
14. (a) adenine-thymine   (b) guanine-cytosine
     (c) cytosine-guanine   (d) uracil-adenine
15. (a) AUC   (b) CUG   (c) AAC   (d) CUU
16. UCC CGU AAG AGC UCG GGG
17. Arg-Ala-Phe-Ser-Ser-Pro
18. (c) fMet-Tyr-Ala-Ser-Val-Ala-Lys-STOP
19. The matching pairs are as follows:
     (a) 4   (b) 3   (c) 5   (d) 1   (e) 2

20. The matching pairs are as follows:
(a) 2   (b) 3   (c) 5   (d) 1   (e) 4

Chapter 16

1. The matching pairs are as follows:
(a) 4   (b) 2   (c) 7   (d) 5   (e) 1   (f) 6   (g) 3
2. 60.0%
4. plasma, formed elements
5. The matching pairs are as follows:
(a) 2   (b) 3   (c) 1
7. Specific gravity = 1.10.
8. (a) 33.3%   (b) 60.0%
9. fibrinogin, prothrombin, albumin, and globulin
10. (a) hydrostatic pressure   (b) osmotic pressure
11. The matching pairs are as follows:
(a) 3   (b) 1   (c) 2
12. (a) antigen   (b) antibodies
13. The matching pairs are as follows:
(a) 3   (b) 2   (c) 4   (d) 1
18. granular leukocytes produced in the red bone marrow and nongranular leukocytes produced in the spleen and lymph nodes
19. red bone marrow, vitamin B_{12}
20. The matching pairs are as follows:
(a) 2   (b) 3   (c) 1
24. The matching pairs are as follows:
(a) 4   (b) 2   (c) 7   (d) 5   (e) 1   (f) 6   (g) 3

Chapter 17

1. kidneys, ureters, urinary bladder, urethra
2. The matching pairs are as follows:
(a) 2   (b) 1   (c) 3
5. The matching pairs are as follows:
(a) 3   (b) 1   (c) 2
7. The matching pairs are as follows:
(a) 4   (b) 3   (c) 1   (d) 5   (e) 2
9. (a) cold weather   (b) warm weather
10. The matching pairs are as follows:
(a) 2   (b) 1   (c) 3   (d) 4
11. (a) abnormal   (b) normal   (c) abnormal
(d) abnormal   (e) normal
12. Specific gravity = 1.10, above normal range.
13. (a) pH = 6   (b) pH = 9   (c) pH = 7   (d) pH = 8
15. The matching pairs are as follows:
(a) 3   (b) 5   (c) 1   (d) 4   (e) 2
19. diabetes mellitus
20. The matching pairs are as follows:
(a) 3   (b) 4   (c) 1   (d) 2

Chapter 18

2. The matching pairs are as follows:
   (a) 6   (b) 4   (c) 1   (d) 3   (e) 5   (f) 2
4. $ADP + P_i + 8000$ cal
5. liver, glycogenesis
6. $C_6H_{12}O_6 + 6O_2 \longrightarrow 6CO_2 + 6H_2O +$ energy (as ATP)
7. (a) $NADH_2$ (b) $FADH_2$
8. phosphorylation, glucose-6-phosphate
9. anaerobic
10. lactic acid, 2 moles
12. Hydrogen acceptors NAD and FAD. Cytochromes are electron acceptors.
13. $H_2O +$ energy (as ATP)
14. decarboxylation and dehydrogenation
15. $2CH_3COOH + 4O_2 \longrightarrow 4CO_2 + 4H_2O + 24ATP$
18. There are 148 ATP's formed.

19. $NH_2 + H \longrightarrow NH_3 \qquad 2NH_3 + CO_2 \longrightarrow NH_2\overset{\displaystyle O}{\overset{\|}{-C-}}NH_2$
20. the liver

Appendix A

1. (a) 53   (b) -77 (c) 39   (d) 11   (e) -11
2. (a) -17 (b) -51 (c) -79 (d) 49   (e) 49
3. (a) 9   (b) 64   (c) 6   (d) $\frac{1}{81}$ (e) $\frac{1}{16}$ or 0.063
   (f) 16
4. (a) $(2)^4$ (b) $(2)^2$ (c) $(2)^{-2}$ (d) $(2)^{-4}$
5. (a) $(2)^{-3}$ (b) $(2)^{13}$ (c) $(2)^{-13}$ (d) $(2)^3$
6. (a) $(3)^{10}$ (b) $(4)^3$ (c) $(2)^{-3}$ (d) y^{-5}
7. (a) 7.89×10^{14} (b) 1.003×10^{-7} (c) 3.05×10^6 (d) 2.1×10^{-3}
8. (a) 6×10^8 (b) 6×10^2 (c) 6×10^{-2} (d) 6×10^{-8}
   (e) 4×10^4 (f) 4×10^{14} (g) 4×10^{-14} (h) 4×10^{-4}
9. (a) 16 quarts   (b) 4 miles   (c) 75 feet, 900 inches   (d) 5 m
10. (a) $x = 7$ (b) $a = 6$ (c) $c = 3$ (d) $m = 50$

11. (a) $P = \dfrac{nRT}{V}$ (b) $V = \dfrac{nRT}{P}$ (c) $n = \dfrac{PV}{RT}$

   (d) $R = \dfrac{PV}{nT}$ (e) $T = \dfrac{PV}{nR}$

Appendix B

1. (a) 85.5 cm   (b) 8.55 dm   (c) 0.855 m   (d) 0.000855 km
2. (a) 4380 ml   (b) 438 cl   (c) 43.8 dl   (d) 0.00438 kl
3. 10 m^2 (b) 1×10^5 cm^2
4. (a) 125 cm^3 (b) 1.25×10^5 mm^3
5. (a) 1.8 m   (b) 180 cm   (c) 1800 mm
6. (a) 1890 ml   (b) 1.89 liters
7. (a) 2.12 quarts   (b) 67.8 fluid ounces

8. (a) 1.54×10^5 g   (b) 340 pounds
9. 0.71 g/ml
10. 200 ml
11. (a) 10°C   (b) 85°C   (c) −40°C
12. (a) 446°F   (b) 914°F   (c) −234°F
13. 15 kcal
14. $\Delta H = -1402$ kcal/mole
15. 6.01 kcal/g
16. 327.5 kcal/mole

Index

oxygen, 48, 107–108, 109–110
  hyperbaric therapy, 110–112
  in lungs, 107–110
oxyhemoglobin, 404

PABA (*see* para-aminobenzoic acid)
para-aminobenzoic acid (PABA),
    250, 351
para-aminosalicylic acid, 351
partial charge, 63
partial pressure, Dalton's law of,
    106–108
partial pressure of respiratory gases,
    table, 108
particles, subatomic, 27–31
passive immunity, 400
Pauling, Linus, 53
pectin, 274
pellagra, 355–356
n-pentane, 200
pentosans, 278
pentose, 264
pepsin, 338, 344, 346
peptide bond, 252, 326–329
peptide linkage, 321 (*see* also peptide
    bond)
percent concentration by weight,
    122–123
periodicity, 38–40
periodic table, 38–41
PGAL, 445
pH, 138–140
  of urine, 422
  values of some common substances,
    table, 139
phenacetin, 253
phenol, 229
phenyl group, 231
phenylacetic acid, 385
phenylketonuria, 384–385
philosophers, 4
  Greek, 4
phosphate group (in nucleotides),
    364–365, 367
1-phosphofructoaldose, 274
phosphoglycerides, 304
phospholipids, 304–308
  definition, 304
  lecithins, 304–306
  phosphoglycerides, 304
phosphoric acid, 129
phosphorylation, 445
photosynthesis, 262
physiological saline solution, 123
PKU, 384–385
plasma, blood, 392, 396–401
  albumin, 397–399
  fibrinogen, 401–403
  globulin, 399–400
  membrane, 435
  prothrombin, 397, 401–403
platelets, 393, 401–403
pleated sheet conformation, 333
poison nerve gas, 352
polar covalent bond, 56–57
polyatomic ions, 57
  table of, 58
polycythemia, 408

polycythemia vera, 396, 408
polyhydroxyaldehydes, 263
polyhydroxyketones, 263
polypeptides, 328
polysaccharides, 262–263, 278–281
  cellulose, 262, 279–280
  dextrin, 280–281
  glycogen, 261, 272
  heparin, 261, 281
  hydrolysis products, 264
  starch, 261, 262, 278–279
polyunsaturated fats and oils, 299
polyuria, 420
powers of ten, 464–465
pressure, osmotic, 149
pressure, water vapor, 497
pressure of gases, 100
primary alcohols, 225
primary amines, 247
primary protein structure, 331–332
products of reactions, 71
  definition, 71
progesterone, 239, 309
propanal, 236
propane, 198
n-propanol, 225, 238
propanone, 238, 239
n-propyl alcohol, 225
properties of solids, liquids, and
    gases, table, 91
prostaglandins, 296–297
prosthetic group, 345
proteases, 347
proteins, 320–340
  amino acids, 321–329
  anabolism, 455
  classification, 330–331
    by function, table, 330
  complete protein, 323–324
  definition, 320
  denaturation, 337–339
  functions, 331
  incomplete proteins, 323–324
  primary structure of, 331–332
  quaternary structure, 335–337
    table, 337
  secondary structure, 332–334
  structure, 331
  tertiary structure, 335
  tests, 339
proteinuria, 426
prothrombin, 397, 401–403
protium, 32
protons, 27–31
ptomaines, 250
purines, 366–367
pus, 426
pyranose, 269
pyrimidines, 366–367
pyruvic acid, 446, 447, 450
pyuria, 426

quaternary protein structure,
    335–337

R group, 197
rad, 175–176

radiation (*see* nuclear radiation)
radioactive decay, 162–163
radioactivity, nuclear (*see* nuclear
    radiation)
radioisotopes, 178
  used in medical therapy, table,
    179–186
radiopharmaceutical therapy, 178
rancidity of fats, 301
RDA (*see* recommended dietary
    allowances)
reactants, definition, 71
reaction, chain, 166–167
reactions, chemical (*see* chemical
    reactions)
recommended dietary allowances,
    table, 357–358
red blood cells, 393, 403–404
red bone marrow, 403–404
reducing agent, 84–85
reducing sugars, 275
reduction, 83
regulatory gene, 382–383
rem, 175, 176
renal calculi, 417
renal pelvis, 417
renal threshold, 272
renaturation, 337
rennin, 346
replication of DNA, 373
representative elements, 40
repressor protein, 382–383
respiration, 49, 107–108, 109–110,
    145–146
respiratory acidosis, 145
respiratory alkalosis, 145
respiratory chain, 447–450
Rh antigen, 407–408
rhodopsin, 312
rhogam, 408
ribonucleic acid (RNA), 367–368,
    374–375, 376, 378–382
  definition, 367
  messenger RNA, 374–375, 376,
    378–382
  structure of, 372–373
  transfer RNA, 375, 377–380
ribose, 267, 365–366
ribosomes, 364, 374–375, 379
rickets, 311
Ringer's lactated solution, 125
RNA (*see* ribonucleic acid)
  *m*RNA (*see* messenger ribonucleic
    acid)
  *t*RNA (*see* transfer ribonucleic
    acid)
Rocky Mountain spotted fever, 250
roentgen, 175
Roentgen, Wilhelm, Konrad, 160
Rothera test (nitroprusside), 427
Rutherford, Lord, 161, 165

saccharic acid, 282
saccharide group, 263 (*see also*
    mono-, di-, and
    polysaccharides)
saline solution, 123